SUPERCONDUCTIVITY IN d- AND f-BAND METALS

SUPERCONDUCTIVITY IN d- AND f-BAND METALS

Second Rochester Conference

Edited by D. H. Douglass

University of Rochester
Rochester, New York

Plenum Press · New York and London

Library of Congress Cataloging in Publication Data

Rochester Conference on Superconductivity in d- and f-Band Metals, 2d, 1976.
  Superconductivity in d- and f-band metals.

  Proceedings of the 2d of a series of meetings; proceedings of the 1st are entered
under title: Superconductivity in d- and f-band metals.
  Includes index.
  1. Superconductivity—Congresses. 2. Superconductors—Congresses. I. Douglass,
David H., 1932-   II. Title.
QC612.S8R58 1976   537.6'23   76-46953
ISBN 0-306-30994-7

Proceedings of the Second Rochester Conference on Superconductivity in
d- and f-Band Metals held in Rochester, New York, April 30 - May 1, 1976

© 1976 Plenum Press, New York
A Division of Plenum Publishing Corporation
227 West 17th Street, New York, N.Y. 10011

All rights reserved

No part of this book may be reproduced, stored in a retrieval system, or transmitted,
in any form or by any means, electronic, mechanical, photocopying, microfilming,
recording, or otherwise, without written permission from the Publisher

Printed in the United States of America

Foreword

The occurrence of superconductivity among the d- and f-band metals remains one of the unsolved problems of physics. The first Rochester conference on this subject in October 1971 brought together approximately 100 experimentalists and theorists, and that conference was considered successful; the published proceedings well-represented the current research at that time and has served as a "handbook" to many.

In the four and one half years since the first conference, impressive progress has been made in many areas (although Berndt Matthias would be one of the first to point out that raising the maximum transition temperature by a significant amount was not one of them). For a variety of reasons, I decided that it was time for a Second Rochester Conference on Superconductivity in d- and f-Band Metals and it was held on April 30 and May 1, 1976. It would appear that this conference was even more successful judging from the quality of the talks and various comments made to me. I believe that this was due to the fact that the subject matter is exciting and that the timing was particularly appropriate for several areas of research that were discussed. However, I cannot rule out other factors such as the one advanced humorously by J. Budnick---that these conferences are successful because the meeting room had two podiums; one in the front for the speaker and one in the back next to the slide projectionist for those who preferred to have their dialogues with the speaker and others from that position. The papers presented at this conference comprise this volume.

I wish to thank the Energy Research and Development Administration for their financial support of this conference. Also, I wish to acknowledge the competent assistance of A. Ghosh, C. Manheimer, and D. Miller in the preparation of the manuscripts.

D.H. Douglass

Contents

POSSIBILITY OF EXCITONIC SUPERCONDUCTIVITY 1
   J. Bardeen

SOME COMMENTS ON THE EXCITONIC MECHANISM
OF SUPERCONDUCTIVITY 7
   M.L. Cohen and S. G. Louie

MAXIMUM T_c - OPTIMISTIC EVIDENCE 15
   R.C. Dynes and P.B. Allen

A THEORY OF THE ELECTRON-PHONON INTERACTION
AND THE SUPERCONDUCTING TRANSITION TEMPERATURE,
T_c, IN STRONGLY SCATTERING SYSTEMS 29
   B.L. Gyorffy

UNEXPECTED SUPERCONDUCTIVITY IN rf. SPUTTERED
Nb/Ge SAMPLES 59
   A.K. Ghosh and D.H. Douglass

GAP ANISOTROPY AND T_c ENHANCEMENT: GENERAL
THEORY, AND CALCULATIONS FOR Nb, USING FERMI
SURFACE HARMONICS 73
   W.H. Butler and P.B. Allen

ANISOTROPY PHENOMENA IN CUBIC d-BAND
SUPERCONDUCTORS 121
   H. Teichler

THE EFFECT OF HIGH PRESSURE ON SUPERCONDUCTING
TERNARY MOLYBDENUM CHALCOGENIDES 137
   R.N. Shelton

HIGH CRITICAL FIELD SUPERCONDUCTORS 161
   S. Foner

ON THE UPPER CRITICAL FIELD OF THE TERNARY
MOLYBDENUM CHALCOGENIDES 175
  Ø. Fischer, M. Decroux, R. Chevrel, and
  M. Sergent

PHONON SPECTRA OF A-15 COMPOUNDS AND TERNARY
MOLYBDENUM CHALCOGENIDES 189
  B.P. Schweiss, B. Renker, E. Schneider, and
  W. Reichardt

INELASTIC NEUTRON SCATTERING STUDIES OF THE
PHONON SPECTRA OF CHEVREL PHASE SUPERCONDUCTORS . . . 209
  S.D. Bader, S.K. Sinha, and R.N. Shelton

PHONON ANOMALIES IN TRANSITION METALS, ALLOYS
AND COMPOUNDS 223
  H.G. Smith, N. Wakabayashi, and M. Mostoller

THEORY OF THE LATTICE DYNAMICS OF STRONG
COUPLING SYSTEMS IN THE RIGID MUFFIN TIN
APPROXIMATION 251
  W.E. Pickett and B.L. Gyorffy

PHONON ANOMALIES IN d-BAND METALS AND THEIR
RELATIONSHIP TO SUPERCONDUCTIVITY 269
  S.K. Sinha and B.N. Harmon

ELECTRON AND PHONON PROPERTIES OF A-15 COMPOUNDS
AND CHEVREL PHASES 297
  F.Y. Fradin, G.S. Knapp, S.D. Bader,
  G. Cinader, and C. W. Kimball

DIRECT CORRELATION OF OBSERVED PHONON ANOMALIES
AND MAXIMA IN THE GENERALIZED SUSCEPTIBILITIES
OF TRANSITION METAL CARBIDES 313
  M. Gupta and A.J. Freeman

CALCULATIONS OF THE SUPERCONDUCTING PROPERTIES
OF COMPOUNDS 339
  B.M. Klein, D.A. Papaconstantopoulos, and
  L.L. Boyer

PRESSURE DEPENDENCE OF T_c FOR CARBIDES OF THORIUM,
YTTRIUM AND SCANDIUM 361
  A.L. Giorgi, H.H. Hill, E.G. Szklarz, and
  R.W. White

SUPERCONDUCTIVITY IN NIOBIUM: IMPLICATIONS FOR
STRONG COUPLING SUPERCONDUCTIVITY THEORY 367
    J. Bostock, K.H. Lo, W.N. Cheung, V. Diadiuk,
    and M.L.A. MacVicar

TUNNELING STUDY OF NIOBIUM USING ALUMINUM-
ALUMINUM OXIDE-NIOBIUM JUNCTIONS 381
    B. Robinson, T.H. Geballe, and J.M. Rowell

CALCULATION OF THE ELECTRON-PHONON SPECTRAL
FUNCTION OF NIOBIUM 391
    B.N. Harmon and S.K. Sinha

EXTENDED DEFECTS IN A-15 SUPERCONDUCTORS 413
    J.C. Phillips

IMPURITY STABILIZATION OF Nb_3Ge 421
    J.R. Gavaler

SUPERCONDUCTIVE TUNNELING INTO NIOBIUM-TIN
THIN FILMS - ABSTRACT 429
    D.F. Moore, J.M. Rowell, and M.R. Beasley

THE STRUCTURE OF SPUTTERED SUPERCONDUCTING Nb_3Ge
FILMS REVEALED BY X-RAY DIFFRACTION, TRANSMISSION
ELECTRON DIFFRACTION, AND BY TUNNELING 431
    P.H. Schmidt, E.G. Spencer, D.C. Joy, and
    J.M. Rowell

HIGH TEMPERATURE RESISTIVE PHASE TRANSITION IN
A-15 HIGH TEMPERATURE SUPERCONDUCTORS 453
    C.W. Chu, C.Y. Huang, P.H. Schmidt, and
    K. Sugawara

STRUCTURAL STUDIES OF ORDER, DISORDER, AND
STOICHIOMETRY IN SOME HIGH T_c Nb-BASE A-15
SUPERCONDUCTORS 461
    D.E. Cox, S. Moehlecke, A.R. Sweedler,
    L.R. Newkirk, and F.A. Valencia

DEFECT PRODUCTION AND STOICHIOMETRY IN A-15
SUPERCONDUCTORS 489
    J.M. Poate, R.C. Dynes, L.R. Testardi, and
    R.H. Hammond

EMPIRICAL RELATIONS IN TRANSITION METAL
SUPERCONDUCTIVITY. 507
    C.M. Varma and R.C. Dynes

STUDIES OF THE TRANSITION TEMPERATURE AND NORMAL
STATE RESISTIVITY OF Nb_3Ge AND Nb FILMS 535
   H. Lutz, H. Weismann, M. Gurvitch, A. Goland,
   O.F. Kammerer, and M. Strongin

SATURATION OF THE HIGH TEMPERATURE NORMAL STATE
ELECTRICAL RESISTIVITY OF SUPERCONDUCTORS 545
   Z. Fisk and G.W. Webb

EVIDENCE FOR SELECTIVE ELECTRON-PHONON SCATTERING
IN THE A-15 SUPERCONDUCTORS Nb_3Sn, Nb_3Sb, and V_3Si . . 551
   S.J. Williamson and M. Milewits

INTERBAND SCATTERING CONTRIBUTIONS TO THE
RESISTIVITY OF A-15 METALS 567
   S.D. Bader and F.Y. Fradin

SUPERCONDUCTIVITY IN METALLIC HYDROGEN 583
   R.P. Gupta and S.K. Sinha

ELECTRON-PHONON INTERACTION AND SUPERCONDUCTIVITY
IN METALLIC HYDROGEN 593
   A.C. Switendick

DIRECT MEASUREMENT OF $\alpha^2 F$ IN NORMAL METALS
USING POINT-CONTACTS: NOBLE METALS 607
   A.G.M. Jansen, F.M. Mueller, and P. Wyder

CALORIMETRIC OBSERVATION OF A PHASE TRANSITION
IN THE SUPERCONDUCTING STATE IN $Gd_{1.2}Mo_6Se_8$ 625
   R.W. McCallum, D.C. Johnston, R.N. Shelton,
   and M.B. Maple

SOME SURPRISES IN SUPERCONDUCTIVITY 635
   B.T. Matthias

INDEX 643

POSSIBILITY OF EXCITONIC SUPERCONDUCTIVITY

John Bardeen

University of Illinois

Urbana - Champaign

I. INTRODUCTION

The purpose of this talk is to review the present status of the theory of the possibility of getting an effective attraction between electrons by means of virtual electronic excitation in such a way as to enhance the phonon contribution and thus the superconducting transition temperature, T_c. This is an old idea that has been discussed for many years by Little, Ginzburg and others, but as yet there is no experimental evidence that it actually occurs. Two or three years ago, Allender, Bray and I (ABB)[1] proposed a study of a single interface between a very thin metallic film on a semiconductor substrate. The purpose was to try to observe the effect under rather ideal conditions, not to get a high T_c. If an effect were observed, one could possibly make use of the excitonic mechanism in composite materials which have very thin alternating layers of metal and semiconductor. In suitable metals, there may be an intrinsic low-lying electronic excitation that could be used.

At a metal-semiconductor interface, it is necessary to have close coupling between the metal and substrate so that the wave functions of electrons near the Fermi surface of the metal can penetrate into the gap region of the semiconductor. This really implies that the metal is chemically bonded to the substrate.

A model of the ideal interface, similar to one suggested by Heine some years ago [2], is shown in Figure 1. Wave functions of the electrons of the metal near the Fermi surface decay exponentially in the gap region of the semiconductor. The average penetration depth, D, of the density of states, obtained by averaging over

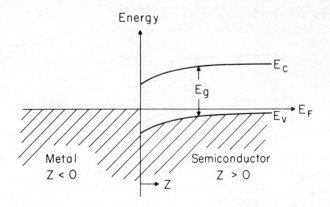

Figure 1: Metal-semiconductor interface. E_c and E_V are averages of the bottom of the conduction band and top of the valence band, respectively, while E_F marks the Fermi level, and E_g is the average band gap.

transverse wave vectors and for energies throughout the gap region, is for this simple model [1]:

$$D = \frac{\pi}{3} \frac{E_F}{E_g} \frac{2}{k_F} , \qquad (1)$$

where E_F is the Fermi energy of the metal, E_g is the energy gap of the semiconductor and k_F the wave vector at the Fermi surface. For silicon, $E_F \sim 12$ ev, $k_F \sim 2$ Å and $E_g \sim 4$ ev, which gives $D \sim 3$ Å. This is close to the value obtained recently by Louie and Cohen [3] in a detailed calculation of a jellium model of a metal in contact with a silicon surface. The average gap in germanium estimated from the peak in the imaginary part of the dielectric function, $\varepsilon_2(\omega)$, is about 3 ev, giving $D \sim 4$ Å. An earlier estimate for PbTe is about 5 Å. Thus only the first one or two atomic layers near the semiconductor surface take part.

Experiments have been carried out by Miller, Strongin, Kammerer and Streetman [4] on thin metallic films on substrates of single crystals of PbTe and of Te with negative results. The films were monitored by LEED and by Auger spectroscopy. A number of the metals, including Pb, grow epitaxially on PbTe, so that the desired close coupling was achieved. However, a proximity effect, giving a decrease in T_c with decreasing film thickness was observed rather than an enhancement expected from the exciton effect. It is believed that the reason for this is that the minimum of the conduction band in PbTe is pulled down below the Fermi level of the metal, such that the metallic wave functions extend into the semiconductor for depths of the order of 100 Å. This implies that there is sufficient scattering at the interface or in the metallic film to spoil the conservation of transverse wave vector. The low energy states in the conduction band (or the holes near the top of the valence band) occupy only a small fraction of k-space but apparently are coupled to all or nearly all of the states near the Fermi level of the metal.

The conditions for observing the exciton effect are thus very stringent. One wants a semiconductor with a small average gap. As illustrated in Figure 2, one envisages virtual excitation of electron-hole pairs with a wave vector difference Q. One may start from a state with any wave vector $\vec{Q}_V$ in the valence band and excite the electron to a state $\vec{Q}_c = \vec{Q}_V - \vec{Q}$ in the conduction band. Through such virtual excitation, an electron pair $(\vec{k}\uparrow, -\vec{k}\downarrow)$ in the metal coupled to the semiconductor, is scattered to the pair $[(\vec{k} + \vec{Q})\uparrow, -(\vec{k} + \vec{Q})\downarrow]$. Generally $\vec{Q}$ lies outside of the first Brillouin zone and the scattering is an Umklapp process with $\vec{Q} = \vec{q} + \vec{K}$, where $\vec{K}$ is a reciprocal lattice vector and $\vec{q}$ is in the first zone.

Rough estimates of the magnitude of the exciton effect were made by using a free electron model for the metal and a model for semiconductor in which gaps are introduced by weak psuedopotentials to form the valence and conduction bands. The Coulomb vertex, $4\pi e^2/Q^2$, is screened by the density of valence electrons and reduced (approximately) by the Fermi-Thomas dielectric function $\varepsilon_{FT}(Q)$:

$$\varepsilon_{FT}(Q) = (Q^2 + q_s^2)/Q^2 \qquad (3)$$

where

$$q_s^2 = 6\pi e^2 n/E_F . \qquad (4)$$

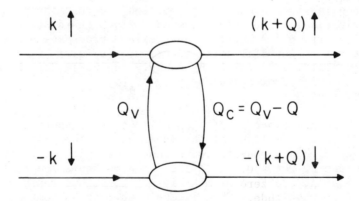

Figure 2: Diagram for interaction by virtual excitation of electron-hole pairs in a semiconductor. The reduced wave vector is q = Q + K, where K is a reciprocal lattice vector. The vertices represent screened Coulomb interaction.

We showed [1] by sum rule arguments that an approximate expression for the effective interaction is the screened Coulomb repulsion plus an attractive contribution from the process illustrated in Figure 2.

$$V_c(Q) = \frac{4\pi e^2}{Q^2 \varepsilon_{FT}(Q)} \left[1 - \frac{\alpha \Omega_{ep}^2}{\varepsilon_{FT} \omega_g^2} \right] \quad (5)$$

Here Ω_{ep} is the electron plasma frequency, ω_g an average gap frequency, and α is a numerical factor somewhat less than unity that gives the fraction of the oscillator strength for transitions across the gap. It is questionable how well a calculation based on bulk properties applies to a thin surface region of a few Angstroms, but it should be qualitatively correct.

This expression is very similar to one derived by M.L. Cohen and P.W. Anderson [5] for the phonon mechanism. In their formulation Ω is the ion plasma frequency and ω is an average phonon frequency. They show that because of Umklapp processes and local field effects, the second term can be larger than the first, giving a net attractive interaction. Allender [6] has shown that similar arguments can be used for the exciton mechanism, also leading to the form (5). In this case α represents the fraction of the total valence electron density in the bond positions. In the Cohen-Anderson treatment, one is interested in the local field acting on an ion. In Allender's, the local field is that acting on electrons in the valence bonds.

Exciting an electron across the gap corresponds to moving it from a bonding to a non-bonding position, so that the physical effect is polarization of the valence bonds. Since the distance involved is small, large Q are involved and Umklapp processes play the predominant role. The dielectric function, $\varepsilon_{KK'}(q,\omega)$, is a tensor in reciprocal lattice vectors. Presumably one could derive the effective interaction from $\Sigma_{K'}(4\pi e^2/Q^2)(\varepsilon^{-1})_{KK'}$, but this would be a difficult task.

One can not expect that λ_{exc} will be large under any circumstances, since it will be of the order of μ. When renormalized to the range of phonon energies, μ becomes μ^*, which is typically only of the order of 0.1 - 0.15. The exciton mechanism hopefully could reduce $\mu^* - \lambda^*_{exc}$ to zero or make it negative, but would not change the order of magnitude. Thus it would have to be combined with the phonon effect to get a high T_c. Further, if superconductivity arises from narrow d-bands, one could not expect to have the desired penetration of the metallic wave functions into the semiconductor. However, in view of the importance of finding superconductors which could be cooled with liquid hydrogen, all possibilities should be explored.

REFERENCES

1. D.W. Allender, J.M. Bray and John Bardeen, Phys. Rev. B7, 1020 (1973); Phys. Rev. B8, 4433 (1973).
2. It is type (3) in a recent classification of metal-semiconductor barriers of J.M. Andrews and J.C. Phillips, Phys. Rev. Lett. 35, 56 (1975).
3. Steven G. Louie and Marvin L. Cohen, Phys. Rev. 13B, 2461 (1976).
4. D.L. Miller, Myron Strongin, O.F. Kammerer and B.G. Streetman, Phys. Rev. B (to be published).
5. M.L. Cohen and P.W. Anderson, Superconductivity in d- and f-Band Metals, David H. Douglass, Ed. A.I.P. Conf. Proceedings No. 4, American Institute of Physics, New York, 1972.
6. D.W. Allender,"Model for an Exciton Mechanism of Superconductivity in Planar Geometry", Doctoral Thesis, University of Illinois at Urbana-Champaign, 1975.

QUESTIONS AND COMMENTS

C. Varma: A few years ago there were some very optimistic estimates of λ for excitons on the basis of a Thomas-Fermi calculation for screening at the interface. Would you please explain to us the difference between that and what you said in your talk.

J. Bardeen: We probably got a little enthusiastic in making plots for λ going up to values a lot larger than 0.5, but if you read the print we said that 0.5 was an optimistic estimate. The difficulty is that the Coulomb interaction is relatively small; μ is probably between 1/3 and certainly not more than 1/2 and that is the order of magnitude of the effect we are working with. You can not expect to get much larger values of λ_{exc}.

D. Gray: What sort of systems do you think are most likely to give both this excitonic effect and the usual electron-phonon effect?

J. Bardeen: One would like a system such as germanium in which the gap region is populated with electrons to about half the gap width. One way of approaching that would be with a composite structure, but it is possible we could have similar matrix elements for interband transitions within the substance itself. People have tried putting in highly polarizable molecules in the metals in a hope to increase the T_c but that did not work. It is a tricky business. You have to have everything just right to see the effect.

SOME COMMENTS ON THE EXCITONIC MECHANISM OF SUPERCONDUCTIVITY*

Marvin L. Cohen and Steven G. Louie[†]

Department of Physics, University of California

and

Materials and Molecular Research Division
Lawrence Berkeley Laboratory
Berkeley, California 94720

ABSTRACT

The Inkson and Anderson (IA) objection to the calculation of Allender, Bray and Bardeen (ABB) is investigated. The IA model dielectric functions are compared with the Lindhard and band structure dielectric functions; the consequences of the use of these dielectric functions is discussed. It is also demonstrated that a dielectric function calculated for Ge gives a repulsive superconductivity kernel for all frequencies. The relation of this result to the calculation of ABB and the role of local-fields or umklapp processes is discussed.

I. INTRODUCTION

Allender, Bray and Bardeen [1] (ABB) have explored the possibility of using electronic polarizability to induce Cooper pair formation and superconductivity in a system consisting of a thin metal layer on a semiconductor surface, i.e., a Schottky barrier. The process considered involves the tunneling of metal electrons at the Fermi surface into the semiconductor gap where they interact by exchanging "virtual excitons" [2].

Shortly after ABB introduced their model, Inkson and Anderson [3] (IA) used a dielectric function approach to estimate the pairing interaction, and reported that the attractive interaction between electron pairs was stronger in the metal side of the Schottky barrier than in the semiconductor side. In reply ABB [4] questioned the detailed structure of the IA semiconductor dielectric function and its appropriateness with respect to the ABB model.

This paper deals mainly with the IA objection to ABB and discusses the pairing interaction in general. It is shown that the IA model for the metallic dielectric function does yield a more attractive pairing interaction than their model semiconductor dielectric function; however, the pairing interactions differ from those calculated here. It is also demonstrated that a semiconductor dielectric function based on a pseudopotential band calculation does not yield an attractive interaction. It is proposed that if an attractive interaction is possible via the exchange of excitons, umklapp processes or local fields are necessary.

II. CALCULATION

Following ABB, the N(0)V parameter of BCS can be written as

$$N(0)V \simeq \lambda_{ex} - \mu \qquad (1)$$

where λ_{ex} is the attractive electron-electron coupling constant arising from exciton exchange and μ is the repulsive Coulomb-parameter.

In analogy with the phonon induced [5] effective interaction ABB arrive at the following expression

$$-N(0)V \simeq \mu\left[1 - \frac{\beta\omega_p^2}{\varepsilon(\vec{q})\omega_g^2}\right] = \mu - \lambda_{ex}, \qquad (2)$$

where ω_p is the electron plasma frequency in the semiconductor, ω_g is the average semiconductor gap, $\varepsilon(\vec{q})$ is a wavevector dependent dielectric constant for metal of equal electron density and β is a numerical factor which accounts for the decay of the metallic electron wave functions into the semiconductor and the fraction of time the metal electrons spend in the semiconductor. ABB introduce a screening factor, a, and the exciton coupling constant becomes

$$\lambda_{ex} = ab\mu\, \omega_p^2/\omega_g^2 . \qquad (3)$$

In favorable cases, ABB estimate $\lambda_{ex} \sim 0.2 - 0.5$. These values would give substantial increases in the superconducting transition temperatures of metal films.

IA approach the pairing interaction from a different point of view. They argue that the total interaction, attractive exciton and repulsive coulomb, can be treated using the wavevector and frequency dependent dielectric function, $\varepsilon_s(\vec{q},\omega)$, appropriate to the semiconductor. They express the total interaction as

$$V_t(\vec{q},\omega) = \frac{4\pi e^2}{q^2 \varepsilon_s(\vec{q},\omega)} . \qquad (4)$$

The IA form for ε_s is

$$\varepsilon_s^{IA} = 1 + \frac{A}{1 + AB} \qquad (5)$$

where $A = \varepsilon_0 - 1$, $B = q^2/k^2 - \omega^2/\omega_p^2$, ε_0 is the static electronic dielectric constant, and k^{-1} and ω_p are the screening length and plasmon energy of an equivalent electron density metal.

If $\varepsilon_0 \to \infty$, then it is expected that $\varepsilon_s^{IA} \to \varepsilon_m^{IA} = 1 + B^{-1}$, a dielectric function appropriate for a metallic system. Therefore using the above expressions,

$$\frac{1}{\varepsilon_s^{IA}} \approx \frac{1}{\varepsilon_m^{IA}} + \frac{1}{\varepsilon_0} \qquad (6)$$

for $\varepsilon_0 \gg 1$. Eqs. (4) and (6) show that the total interaction in the semiconductor is equal to the total interaction in a metal plus an added repulsive term. IA therefore conclude that the semiconductor is less favorable than the metal for superconductivity.

To obtain the $N(0)V$ parameter or the frequency-dependent kernel of the BCS equation, $K(\delta)$ [6], it is necessary to do a Fermi surface average of the wavevector dependent interaction, V_t,

$$K(\delta) \sim \int_{|\vec{k}-\vec{k}'|}^{|\vec{k}+\vec{k}'|} q \, V_t(q,\delta) \, dq \qquad (7)$$

where $\delta = \hbar\omega/E_F$, and $\vec{k}$ and $\vec{k}'$ are the initial and scattered electron wavevector, i.e., $\vec{k}' = \vec{k} + \vec{q}$. It is $K(\delta)$ which must have attractive regions for the pairing interaction to be positive. It is not sufficient to have negative regions of the wavevector dependent interaction, V_t.

To calculate $K(\delta)$ [7], we assume a metal-semiconductor interface with electron densities appropriate to Aℓ and Si, i.e. $r_s \sim 2$ and $\varepsilon_0 \sim 10$. We first evaluate $K(\delta)$ for the IA model dielectric functions ε_s^{IA} and ε_m^{IA}. In Fig. 1(a) the kernels appropriate to ε_s and ε_m are displayed. The IA metal kernel is more favorable for superconductivity since it is less repulsive at low frequencies and the attractive region is larger than the attractive region obtained using the IA semiconductor dielectric function. This is in accord with the IA calculations.

However, a more relevant question is how good are the IA approximations to begin with. The ε_m^{IA} is constructed to approximate the the frequency and wavevector dependent dielectric function for a metal. ε_m^{IA} coincides with the RPA or Lindhard [8] dielectric function for $\vec{q} = 0$ and for $\omega = 0$, $\vec{q} \ll k_F$. A better approximation for the metal kernel would be to use the Lindhard dielectric function in V_t. The results (Fig. 1(b)) show that $K(\delta)$ is <u>repulsive</u> for all δ. Thus the attractive region found using ε_m^{IA} is <u>a result of their</u> model and is not found in the more realistic Lindhard model.

It is also possible to do a more realistic calculation of $K(\delta)$ for a semiconductor dielectric function. We have computed $K(\delta)$ using the numerical values for the $\varepsilon(\vec{q},\omega)$ of Ge. The calculation [9] for the Ge $\varepsilon(\vec{q},\omega)$ was based on a pseudopotential calculation of the energy band structure and wavefunctions, and is therefore expected to be more realistic than ε_s^{IA}. The results which appear in Fig. 1(c) indicate that $K(\delta)$ is repulsive for all frequencies.

III. DISCUSSIONS AND CONCLUSIONS

So we have explored the IA objection to the ABB model based on ε_s^{IA} and ε_m^{IA} and we have shown that the kernel of the BCS equation $K(\delta)$, (and therefore the BCS parameter $N(0)V$), is repulsive for all frequencies if the total interaction used is based on a realistic semiconductor dielectric function. What does this imply about ABB?

In their reply to IA, ABB emphasized that the reason that IA did not obtain a favorable result was that the pole of ε_s^{IA} did not

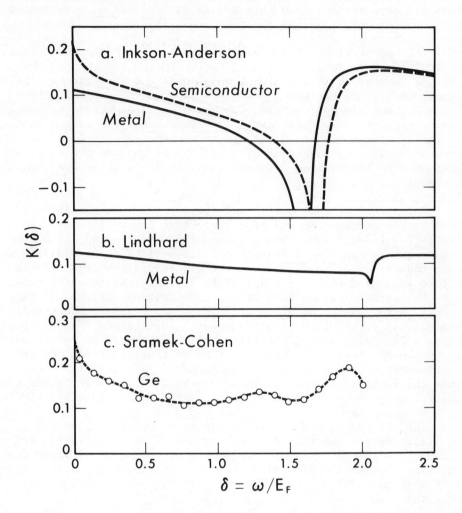

Figure 1: (a) The frequency dependent kernel $K(\delta)$ ($N(0)V$ parameter) for a screened Coulomb interaction using the semiconductor and metallic dielectric function model of Inkson and Anderson. Parameters appropriate for Aℓ and Si were used.
(b) The frequency dependent kernel $K(\delta)$ ($N(0)V$ parameter) for a screened Coulomb interaction using the Lindhard dielectric function (parameters are appropriate for Aℓ).
(c) The frequency dependent kernel $K(\delta)$ ($N(0)V$ parameter) for a screened Coulomb interaction using a dielectric function calculated from a pseudopotential band structure for Ge.

have the proper $\vec{q}$ dependence. However, as we have shown, the problems are more serious than this and in fact ε_s^{IA} is more favorable for superconductivity than the more realistic Ge calculation.

The essential point is that the peak in the dielectric function will give a zero in K(δ). A qualitative reason for this is that the peak in the dielectric function signals a transverse excitation (electron-hole or excitonic resonance) and in this approximation the electrons are not coupling to this mode. The strongest coupling comes near the zero of the dielectric function i.e. plasmon exchange. A similar effect occurs in the electron-phonon interaction. Electrons interact only with longitudinal modes unless umklapp processes are invoked. In the exciton case the coupling could arise via local-field effects.

One method of including local-field effects in the dielectric function is through the use of a dielectric tensor [10,11], $[1/\varepsilon]_{\vec{G}\vec{G}'}$. IA addressed themselves to this problem and calculate the frequency dependence of some off-diagonal terms in the dielectric tensor. They conclude that the dielectric tensor still has a pole at ω_g and that the coupling to the excitons is still zero, i.e. the kernel will be zero at ω_g. Two problems arise: (1) The formalism for using the dielectric tensor to evaluate the kernel and pairing interaction is not adequately discussed. For the phonon case, a generalized susceptibility will have a pole at the transverse phonon frequencies, yet it is known that electrons couple to transverse phonon modes (via umklapps). (2) It is not clear that the approximate calculations for the IA dielectric tensor are sufficiently accurate to rule out attractive pairing interactions.

The ABB approach circumvents (1) by computing λ_{ex} using eq. (3). We presume that ABB have assumed that local fields are included in this expression. This coupling constant is large for small ω_g, but small ω_g usually implies small local fields in covalent systems. ABB suggest PbTe which is partially ionic to overcome this problem.

It would be useful in estimating the coupling to use a local-field semiconductor dielectric function which was computed for a realistic semiconductor for ω and q. To our knowledge the only semiconductor local field dielectric function [12] in the literature is given as a function of ω, but for $\vec{q} = 0$. To show why this approach or some other is desirable to obtain reliable values for λ_{ex} we can make estimates assuming the ABB form to be correct for the interaction including local fields.

If we take the Phillips-Van Vechten [13] model for the dielectric function and identify ω_g with the Phillips average gap, then

EXCITONIC MECHANISM OF SUPERCONDUCTIVITY

$$\varepsilon_0 = 1 + \frac{\omega_p^2}{\omega_g^2} \qquad (8)$$

and assuming eq. (3) to be valid, we obtain

$$\lambda_{ex} = ab\mu(\varepsilon_0 - 1) . \qquad (9)$$

ABB estimate a ~ 1/5 to 1/3, b ~ 0.2, μ ~ 1/3 to 1/2 and using ε_0 ~ 5 to 30, then λ_{ex} ~ .05 to 1.0. Leaving out phonons this would give estimates of the transition temperature from zero (repulsive total N(0)V) to extrodinarily large values. Estimates for b by ABB are consistent with recent self-consistent Schottky barrier calculations[14] for the penetration of metallic electrons. The parameters a and μ can be evaluated more carefully, but it would still be more reassuring to use a total local-field dielectric function and/or some other method to estimate N(0)V for the exciton interaction.

In conclusion, our calculation of the total semiconductor kernel yields a repulsive interaction. This together with the IA arguments would suggest that the ABB results should be reconsidered; however, we feel that the umklapp contribution should be included explicitly before a firm conclusion is reached.

Part of this work was done under the auspices of the U.S. Energy Research and Development Administration.

REFERENCES

* Supported in part by the National Science Foundation Grant DMR72-03206-A02.
† Supported by a National Science Foundation Fellowship.
1. J. Bardeen, this conference; and D. Allender, J. Bray and J. Bardeen, Phys. Rev. B7, 1020 (1973).
2. Excitonic states are not essential to the argument; for most of the models considered, the excitations are not excitons but unbound electron-hole pairs. We will not distinguish between these in this paper.
3. J.C. Inkson and P.W. Anderson, Phys. Rev. B8, 4429 (1973).
4. D. Allender, J. Bray and J. Bardeen, Phys. Rev. B8, 4433 (1973).
5. M.L. Cohen and P.W. Anderson, in Superconductivity in d- and f-Band Metals, edited by D.H. Douglass, (AIP), New York, 1972, p. 17.
6. e.g. M.L. Cohen, Phys. Rev. 134, A511 (1964).

7. In all cases, it is the real part of the kernel which is being computed.
8. J. Lindhard, Kgl. Danske Videnskab. Selskab, Mat.-Fys. Medd. $\underline{28}$, 8 (1954).
9. S.J. Sramek and M.L. Cohen, Phys. Rev $\underline{B6}$, 3800 (1972).
10. S.L. Adler, Phys. Rev. $\underline{126}$, 413 (1962).
11. N. Wiser, Phys. Rev. 129, 62 (1963).
12. S.G. Louie, J.R. Chelikowsky and M.L. Cohen, Phys. Rev. $\underline{28}$, 8 (1954).
13. J.C. Phillips, Rev. Mod. Phys. $\underline{42}$, 317 (1970).
14. S.G. Louie and M.L. Cohen, Phys. Rev. $\underline{B13}$, 2461 (1976).

QUESTIONS AND COMMENTS

F. Mueller: The obvious question---do you plan to do the calculation to show whether or not there is an attractive interaction coming from the d electrons coupled to the transverse phonons?

M. Cohen: We hope to. We calculated $\varepsilon_{gg'}(\omega)$ for q going to zero because we were looking at optical processes; we then started to put in the q-dependence, but at that point we got involved with surface physics (as Professor Bardeen mentioned) so we have been currently involved with the electronic properties of surfaces and we never finished calculating $\varepsilon_{gg'}(q)$. But we can do it.

MAXIMUM T_c - OPTIMISTIC EVIDENCE

R.C. Dynes

Bell Laboratories

Murray Hill, New Jersey   07974

and

P.B. Allen*

Department of Physics, State University of New York

Stony Brook, New York   11790

ABSTRACT

A reanalysis of Eliashberg theory in the strong coupling limit, and the relationship between the superconducting transition temperature T_c of a material and its normal state properties is presented. At weak and intermediate coupling ($\lambda \lesssim 1.5$) we reproduce McMillan's earlier results but at stronger coupling we find the functional form differs from that contained in the McMillan equation. We find in the asymptotic limit of very large λ, $T_c = 0.15 \, (\lambda \langle \omega^2 \rangle)^{1/2}$. The $\lambda = 2$ limit predicted by McMillan disappears in the correct theory because it is a result of the functional dependence of T_c on λ and $\langle \omega \rangle$ used by McMillan which is valid only if $\lambda < 1.5$. Finally, we solve the Eliashberg equations for a wide variety of phonon spectra and find that the ratio $T_c/\omega_{\log}$ as a function of λ shows surprisingly little dependence on the choice of shape for $\alpha^2(\omega) F(\omega)$ for values of $\lambda \lesssim 1.5$.

---

*Work supported in part by NSF grant #DMR-7307578 A01.

INTRODUCTION

In 1968 McMillan [1] presented an extensive study of the relationship between the critical temperature of a superconductor T_c, the phonon density of states modulated by the electron-phonon coupling function $\alpha^2 F(\omega)$ and the effective Coulomb repulsion μ^*. In this work, it was shown that the critical temperature could be written

$$T_c = \frac{<\omega>}{1.20} \exp\left[-\frac{1.04(1+\lambda)}{\lambda-\mu^*-.62\lambda\mu^*}\right] \qquad (1)$$

where λ is a dimensionless measure of the electron-phonon coupling strength given by

$$\lambda = 2 \int_0^\infty d\omega \, \frac{\alpha^2 F(\omega)}{\omega} \qquad (2)$$

μ^* is the effective Coulomb repulsion reduced from its instantaneous value μ by the logarithm of the ratio of electron-to-phonon frequencies, i.e.

$$\frac{1}{\mu^*} = \frac{1}{\mu} + \ln\left(\frac{\omega_e}{\omega_{ph}}\right) \qquad (3)$$

and $<\omega>$ is given by the first moment ($<\omega^1>$) of

$$<\omega^n> = \frac{2}{\lambda} \int_0^\infty d\omega \, \alpha^2 F(\omega) \omega^{n-1} \qquad (4)$$

From superconducting tunneling experiments [2], the spectral function $\alpha^2 F(\omega)$, μ^*, and T_c can be extracted and the validity of McMillan's equation (Eq. 1) tested. In Fig. 1 we show the experimental situation as obtained from tunneling measurements and its comparison with the McMillan equation. The crosses are data obtained from the Tℓ-Pb-Bi system [3] and the closed circles are all other crystalline materials where tunneling data is available. It is this comparison which was the motivation for the present work. Until the work on the Pb based alloys was completed the maximum

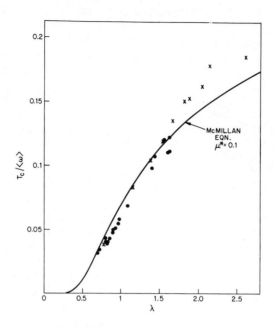

Figure 1: $T_c/\langle\omega\rangle$ versus λ. The solid curve is that calculated from McMillan's equation (Eq. 1) for $\mu^* = .1$. The crosses are data obtained from tunneling into the Tℓ-Pb-Bi alloy system [3]. The closed circles are determined from tunneling measurements into other metals and alloys [7].

λ's observed for crystalline metals were for Pb [2] and Hg [4] (1.55 and 1.6 respectively). From Fig. 1 it is clear that the T_c of these materials is satisfactorily described by Eq. 1. With the completion of the measurements on the Pb-based alloys where the maximum λ was increased to 2.6, it was evident that the agreement was less than satisfactory and a reanalysis of Eliashberg theory out to stronger coupling was in order. It should be emphasized that the maximum value of λ that McMillan considered in his analysis was $\lambda = 1.5$. Others have previously observed that experiment indicates an almost linear relationship between $T_c/\langle\omega\rangle$ and λ; empirical T_c equations [5,6] based on this observation have been proposed. The present work is also motivated by those empirical observations. The details of the calculation will not be presented here, only the results. For mathematical details, the reader is referred to an earlier paper [7].

CALCULATIONS AND RESULTS

There are two equivalent versions of the Eliashberg gap equations [8]: a matrix equation for $\Delta(i\omega_n)$ on the Matsubara imaginary frequency points $\omega_n = (2n+1)\pi T$ or the more physically meaningful integral equation for $\Delta(\omega)$ on the real frequency domain. McMillan used the latter form, but for computational ease in the strong coupling regime, we have solved the equations using the Matsubara matrix representation. This form of the gap equation has been previously solved by Owen and Scalapino [9] and Bergmann and Rainer [10], and we chose the form presented by the latter authors. As a first test of our calculation we solve for T_c given the Nb spectrum used by McMillan [1]. Within approximately 1% computational error we reproduce his results for $\lambda < 1.5$. Next we generate solutions for T_c vs λ utilizing three widely varying shapes for $\alpha^2 F(\omega)$. The empirical observation from Fig. 1 suggests that $T_c/\langle\omega\rangle$ does not depend in a critical way on the shape of the spectra function; i.e. $T_c/\langle\omega\rangle$ vs λ shows a general monotonic trend for a wide range of materials with differently shaped $\alpha^2 F(\omega)$. We solve for T_c using the three functions illustrated in Fig. 2 which

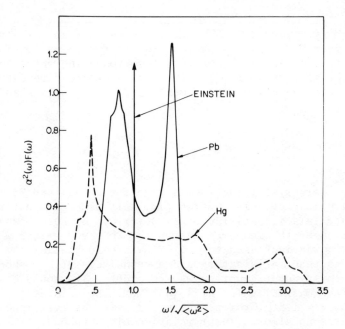

Figure 2: Electron-phonon spectral functions $\alpha^2 F(\omega)$ as measured by tunneling in Refs. 2 and 4. The frequency ω has been scaled to the rms frequency $\sqrt{\langle\omega^2\rangle}$ in order to illustrate the relative deviations from the Einstein spectrum (also shown).

represent, we feel, the extremes one is likely to face, from the dispersionless Einstein spectrum, to the rather typical shape given for Pb [2], to the case of wide dispersion in Hg [4]. The frequencies in Fig. 2 are normalized to $\langle\omega^2\rangle^{\frac{1}{2}}$ where $\langle\omega^2\rangle$ is given by the second moment of equation 4.

The results of our calculation for the various shapes illustrated are given in Fig. 3 where we show the data shown in Fig. 1 compared with the McMillan equation and our exact solutions. What is evident is that the experimental data points are bracketed by our curves representing the extreme shapes of $\alpha^2 F(\omega)$, and the McMillan tendency for underestimating T_c at larger values of λ is not reflected in the exact solutions. The McMillan equation predicts that in the limit of large λ, T_c is independent of λ and and goes as $T_c = 0.28 \langle\omega\rangle$ (for $\mu^* = 0.1$). For this reason the McMillan equation solution is showing a tendency to saturate at large λ in Figs. 1 and 3. On the other hand, our solutions show (and it can also be rigorously shown [7]) that in the case limit when solved properly $T_c = .15(\lambda\langle\omega^2\rangle)^{\frac{1}{2}}$ (for $\mu^* = 0.1$) independent of the shape chosen for $\alpha^2 F(\omega)$. This disagreement occurs because McMillan solved the Eliashberg equations only in the range $\lambda = 0 \to 1.5$ where the functional form fitted (Eq. 1) was wholly appropriate. For larger λ, deviations from this begin until finally at the asymptotic limit it is of a completely different form ($\sim \lambda\langle\omega^2\rangle)^{\frac{1}{2}}$. Thus the disagreement at large λ between, experiment and exact solutions on the one hand, and McMillan's equation on the other hand, can be interpreted as the onset of the crossover toward the asymptotic regime.

We can reduce the shape dependence of T_c as observed in Fig. 3 by plotting T_c/ω_{log} vs λ where

$$\omega_{log} = \lim_{n \to 0} \langle\omega^n\rangle^{\frac{1}{n}} = \exp \frac{2}{\lambda} \int_0^\infty \frac{d\omega}{\omega} \alpha^2 F(\omega) \ln\omega. \quad (5)$$

To a good approximation [7], $\langle\omega^n\rangle^{\frac{1}{n}}$ is a linear function of n and so in terms of more familiar quantities we can write

$$\omega_{log} \simeq 2\langle\omega\rangle - \langle\omega^2\rangle^{\frac{1}{2}} \quad (6)$$

In Fig. 4 we show a plot of T_c/ω_{log} vs λ. With this scaling, the shape dependence has been virtually eliminated for $\lambda < 2$.

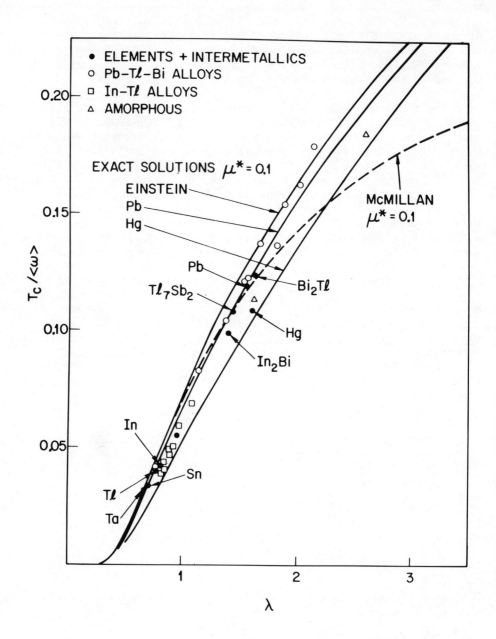

Figure 3: $T_c/\langle\omega\rangle$ plotted versus λ. The solid curves are calculated for the various shapes in Fig. 2. The dashed curve is McMillan's equation. The data points are the same as Fig. 1.

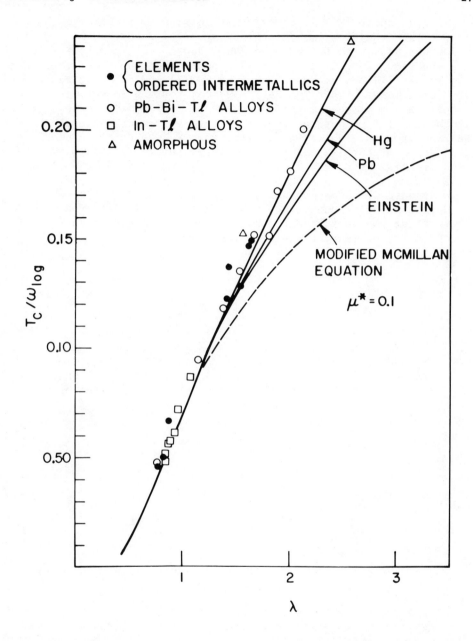

Figure 4: T_c/ω_{log} plotted versus λ. The solid curves and the data points are the same as in Fig. 3. The dashed curve is McMillan's equation with prefactor $\omega_{log}/1.20$. Note that the scaling T_c/ω_{log} has produced a universal curve describing all the experiments and calculations for $\lambda < 1.5$.

The dashed curve is McMillan's equation with a prefactor $\omega_{log}/1.2$. We see that the experimental points are well described by our calculations and that for $\lambda > 2$ the shape dependence is rather mild. The equations have also been solved for a spectrum similar to that for Nb_3Sn [11] and amorphous $Pb_{.45}Bi_{.55}$ [7] and the resultant curves are very close to that for Hg. Hence, for a wide variety of shapes of $\alpha^2 F(\omega)$ we expect T_c/ω_{log} to be very weakly shape dependent. It should be reemphasized that for large values of λ the appropriate scaling frequency shifts gradually from ω_{log} to $\langle \omega^2 \rangle^{\frac{1}{2}}$. The shape dependence of T_c/ω_{log} observed in Fig. 4 for $\lambda > 2$ is a manifestation of this effect.

Because of the success and simplicity of McMillan's equation, we do not propose abandoning it but rather providing correction factors for the shape and the strong coupling modifications. Accordingly, we rewrite equation 1 as

$$T_c = f_1 f_2 \frac{\omega_{log}}{1.2} \exp\left(-\frac{1.04(1+\lambda)}{\lambda - \mu^* - .62\lambda\mu^*}\right) \qquad (7)$$

where f_1 and f_2 are unity for small λ thus reducing equation 7 to McMillan's equation. The strong coupling correction f_1 ensures that $T_c \rightarrow \lambda^{\frac{1}{2}}$ at large λ and we have chosen a scaling factor

$$f_1 = \left\{1 + [\lambda/(2.46 + 9.35\mu^*)]^{\frac{3}{2}}\right\}^{\frac{1}{3}}$$

The "shape correction" f_2 on the other hand ensures that $T_c \propto \langle \omega^2 \rangle^{\frac{1}{2}}$ and is given by

$$f_2 = 1 + \frac{(\langle \omega^2 \rangle^{\frac{1}{2}}/\omega_{log} - 1)\lambda^2}{\lambda^2 + (1.82 + 11.5\mu^*)^2 (\langle \omega^2 \rangle/\omega_{log}^2)}$$

These two corrections ensure the asymptotic behavior $T_c \sim (\lambda\langle \omega^2 \rangle)^{\frac{1}{2}}$ at large λ. The numerical coefficients have been least-squares fit to numerous numerical calculations of T_c, which are represented to $\sim 5\%$ rms accuracy.

DISCUSSION

McMillan showed [1] rigorously that λ could be written in the form

$$\lambda = \frac{N(0)\langle I^2 \rangle}{M\langle \omega^2 \rangle} = \frac{\eta}{M\langle \omega^2 \rangle} \qquad (8)$$

where $\langle I^2 \rangle$ is the average over the Fermi surface of the electron phonon matrix element, $N(0)$ is the density of states at the Fermi surface and M is the atomic mass. He also noted from observations in the column V and VI bcc transition elements, that η was approximately a constant and so $M\lambda\langle\omega^2\rangle$ was invariant in this system. Using this invariance, it follows from equation 1 that for increasing λ, the prefactor $\langle\omega\rangle$ finally dominates and T_c vs λ shows a broad peak at $\lambda = 2.6$. We find that this peak does not occur and that the maximum T_c occurs at $\lambda \to \infty$. This is most clearly illustrated in Fig. 5 where we show plots of $T_c/(\lambda\langle\omega^2\rangle)^{\frac{1}{2}}$ vs λ for McMillan's equation and our solutions. The vertical axis is proportional to T_c <u>if we assume $\lambda\langle\omega^2\rangle$ is fixed</u>. This $\lambda = 2$ limit is another manifestation of the underestimate of McMillan's equation in Fig. 1 and is again based on a functional extrapolation which fails for $\lambda > 1.5$. In fact the McMillan equation underestimates the maximum value of $T_c/(\lambda\langle\omega^2\rangle)^{\frac{1}{2}}$ by about 40%. There is clearly reason to believe that increasing λ above 2 will result in higher values of T_c/ω_{log}.

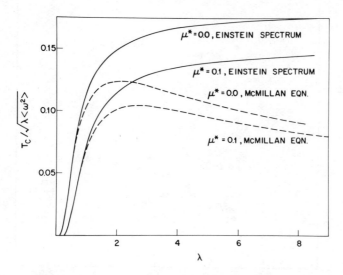

Figure 5: Variation of T_c with λ keeping $\lambda\langle\omega^2\rangle$ fixed. Exact solutions with an Einstein spectrum (solid curves) are compared with the results from McMillan's Eq. 1 (dashed curves) for $\mu^* = 0$ and $\mu^* = \underline{0.1.}$ The value of T_c is normalized to the constant $\sqrt{\lambda\langle\omega^2\rangle}$. The maxima at $\lambda = 2$ and $\lambda = 2.6$ found from McMillan's equation with $\mu^* = 0$ and $\mu^* = 0.1$ are spurious. The true maxima are at $\lambda = \infty$.

In all of the previous discussion we were concerned with the predictions of Eliashberg theory [8] and how it related to observations. It was implicitly assumed that all of the variables could be arbitrarily adjusted and the dependence of T_c on these parameters determined. For example, it was shown that $T_c = .15 \times (\lambda \langle \omega^2 \rangle)^{\frac{1}{2}}$ for values of $\lambda > 10$. The problem of increasing λ in real systems is a problem of stability as it is now generally understood that associated with very strong coupling is some form of lattice and/or electronic instability. This point can be best illustrated if we look at the values of η extracted from tunneling measurements. It was observed by McMillan that within a class of materials this electronic parameter was approximately constant, but now there is evidence to suggest the contrary. In Fig. 6 the closed circles are values of η determined from tunneling measurements and the

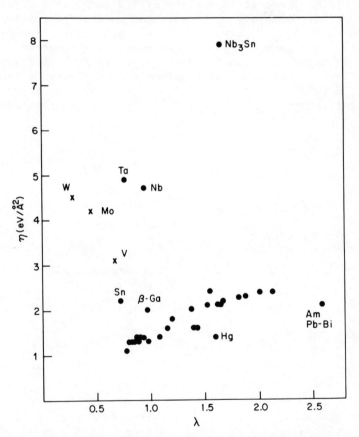

Figure 6: η vs λ for various s_p and d materials. The closed circles are values obtained from tunneling measurements while the crosses are empirical values derived from estimates of λ and $\langle \omega^2 \rangle$. Note the values for η are larger in transition metals than in simple metals.

crosses are empirical values obtained from estimates of λ (from Equation 7) and $\langle\omega^2\rangle$ (from measured phonon spectra). The following points are notable. Firstly, the s-p metals show rather low values of η with an increasing tendency with increasing λ. Also with increasing η there is a stronger trend toward covalent instability. The maximum value observed in the simple metals is 2.4 eV/A^2. Attempts to push η to values greater have resulted in instability and phase changes. In the transition metals on the other hand, larger values of η occur because the d wave functions are not as extended as the s-p functions and have a strong interaction with the core. Also because of the strong attraction to the core, covalent instabilities are less likely and so larger values of η can be achieved before instabilities.

At this stage one can only empirically observe that the increase in λ in Nb$_3$Sn over the transition elements is due to an increase in η. Coupled with a not so dramatic change in ω_{log}, the result is the high T$_c$ for this A15 compound. The reason for this enhancement in the A15 compounds is unclear at this time but undoubtedly is associated with the unique structure and the close packing of the Nb on the cube face chains.

CONCLUSIONS

From a numerical solution of the Eliashberg equations it is concluded that the McMillan equation underestimates the transition temperature T$_c$ in the strong coupling limit $\lambda > 1.5$. It is also concluded that the "$\lambda = 2$ limit" does not exist within Eliashberg theory and there are good reasons for continued efforts to increase λ above 2. Available tunneling data support these conclusions and continue to provide evidence for the accuracy of Eliashberg theory for all superconductors studied so far.

NOTE ADDED AFTER THE MEETING

Just before this paper was submitted in manuscript, an interesting article by Karakozov et al. [12] reached us in translation. These authors have done extensive numerical solutions of the Eliashberg equations, primarily with the aim of studying the dependence of T$_c$ on the shape of $\alpha^2 F$. Their conclusions are consistent with ours. In particular they have analytic arguments and numerical support for the use of ω_{log} in the prefactor of McMillan's equation. The breakdown of McMillan's equation when λ is very large is visible in their figures 2 and 3, although the physics is obscured by the fact that the large λ limit was studied only for peculiar shapes of $\alpha^2 F$; thus they failed to recognize the generality of this breakdown. Finally they present interesting analytic arguments which illuminate the reasons for the breakdown of the two-square-well solution by McMillan for fitting purposes.

REFERENCES

1. W.L. McMillan, Phys. Rev. 167, 331 (1968).
2. W.L. McMillan and J.M. Rowell, in Superconductivity, edited by R.D. Parks (Marcel Dekker, New York, 1969).
3. R.C. Dynes and J.M. Rowell, Phys. Rev. B11, 1884 (1975).
4. W.N. Hubin and D.M. Ginsberg, Phys. Rev. 188, 716 (1969).
5. C.R. Leavens and J.P. Carbotte, J. Low Temp. Phys. 14, 195 (1974). See Fig. 4.
6. J.M. Rowell, private communication.
7. P.B. Allen and R.C. Dynes, Phys. Rev. 12, 905 (1975).
8. G.M. Eliashberg, JETP 11, 696; 12, 1000 (1961).
9. C.S. Owen and D.J. Scalapino, Physica 55, 691 (1971).
10. G. Bergmann and D. Rainer, Z. Phys. 263, 59 (1972).
11. L.Y.L. Shen, Phys. Rev. Lett. 29, 1082 (1972).
12. A.E. Karakozov, E.G. Maksimov, and S.A. Mashkov, Zh. Eksp. Teor. Fiz. 68, 1937 (1975). [Soviet Physics - JETP 41, 971 (1976).]

QUESTIONS AND COMMENTS

S. Nam: I noticed an interesting point---the calculation was done using the Einstein spectrum, which is very close to realistic experimental data. Presumably, McMillan's calculations were based on a realistic spectrum but the Einstein spectrum gives you a better fit. Do you have any physical intuitions? Or, why when one starts from a realistic spectrum is the agreement worse?

R. Dynes: You are asking why if you start from the non-realistic spectrum you get a better fit than a realistic spectrum? I am not sure that is true. The solution for the Einstein spectrum is here [see figure 4].

S. Nam: It is certainly better than McMillan's.

R. Dynes: Oh, it is certainly better than McMillan's because the functional form assumed by McMillan, $\omega e^{-(1+\lambda)/(\lambda-\mu^*)}$, is an extrapolation above $\lambda = 1$ and does not accurately describe T_c. McMillan uses a realistic spectrum and it works fine for small λ, but at large λ, T_c follows a different functional form.

S. Nam: My point is: doesn't this illustrate that the Einstein spectrum which does not have any information about the phonon structure gives you very nice numbers. In other words, the spectrum doesn't affect the transition temperature much.

M. Cohen: If I may interrupt, the Einstein spectrum with the correct solution of the Eliashberg equation is better than the McMillan solution of the Eliashberg equation with the Nb-like spectrum. The difference between the two solutions is the important thing, not so much the spectrum. The McMillan equation does not represent a good solution of the Eliashberg equations in the large λ limit.

D. Papaconstantopoulos: The title of your talk has presented some optimistic evidence. Do you have any feeling of where to look for high η in order to increase T_c?

R. Dynes: Where to look for what?

D. Papaconstantopoulos: Higher η. Presumably from your conclusion, if we can find materials which have large η's then you can increase λ and T_c.

R. Dynes: I think that is a fair conclusion to be taken. One can certainly look at that viewgraph [see figure 6] and conclude that immediately. We have reliable data on Nb_3Sn and that certainly is the maximum T_c on that viewgraph. You know as well as I do how to increase η. I do not have a better answer---but everyone in this room has his own belief as to how to increase T_c. I think I am not going to change anybody else's mind.

M. Cohen: Is the λ for the amorphous lead-bismuth alloys the largest that you have ever seen?

R. Dynes: It is the largest λ that I have measured by tunneling. There are some other tunneling measurements on amorphous materials with λ's in this range.

M. Cohen: But the world's record for λ is around 2.6?

R. Dynes: Yes, somewhere in the range 2.6 to 3.0.

D. Ginsberg: It has been reported that μ^* for niobium is negative. Would that come into this kind of picture and give you an even higher transition temperature or is this something you do not believe?

R. Dynes: Well, it certainly will come into the picture. All of these solutions that I show are for μ^* equal to 0.1.

D. Ginsberg: Well, you had one graph where you had μ^* equal to zero.

R. Dynes: Yes, this viewgraph [see figure 5]. As you can see, $\mu^* = 0$ does increase the critical temperature. I think we are going to hear a lot more in the next couple of days on positive and negative μ^*'s and let's get into it when the data is presented.

A THEORY OF THE ELECTRON-PHONON INTERACTION AND THE SUPERCONDUCTING TRANSITION TEMPERATURE, T_c, IN STRONGLY SCATTERING SYSTEMS

B.L. Gyorffy

H.H. Wills Physics Laboratory, University of Bristol

Bristol BS8 1TL, United Kingdom

I. INTRODUCTION

In this paper I will have in mind three somewhat different reasons for studying the superconducting transition temperature T_c.

(a) One might be interested in T_c as a physical property of metals, like the resistivity or the elastic constants, etc. One would, of course, keep an eye on the possibility of discovering better superconductors. From this point of view a useful theory is semi-quantitative and predicts trends.

(b) Recognizing that T_c and the superconducting tunneling current are the two most sensitive probes of the electron-phonon interaction one might study these in order to learn about the electron-phonon interaction as such. Though not as prevalent as (a) this approach is attractive because of all the physical consequences of the electron-phonon interaction like the resistivity and ultrasonic attenuation T_c and the tunneling current are the two most accessible to accurate theoretical treatment. Obviously, to make progress in this direction one must do very accurate calculations on relatively simple systems like pure metals.

(c) Superconductivity is one of the best understood phenomena in solid state physics. We certainly understand it better than metallic magnetism. And yet, it still seems possible to learn more about superconductivity itself by studying the normal to superconductor transition. Here, the interest is in search for mechanisms other than the electron-phonon interaction as the cause of superconductivity and pairing in other than the usual relative s-state. I will have something to say only in connection with the latter.

The theory I will present is first principles in the sense that it aims to avoid adjustable parameters. It was developed explicitly for d-band metals. Nevertheless, at least in principle, it is appicable to s-p and f-band metals as well. However, no calculations for such systems have been performed as yet. Therefore I shall not refer to this aspect of the problem.

This paper is mainly about calculating the mass enhancement factor λ but, at times, it will be necessary to refer to other elements in the theory of T_c. In the hope that it will add clarity to my remarks I will outline a way of viewing that theory.

II. THE SUPERCONDUCTING INSTABILITY AND THE RIGID-MUFFIN-TIN APPROXIMATION

If one is not interested in the properties of the superconducting state and satisfied with a theory which gives T_c only, the most efficient way of proceeding is to ask: at what temperature, if any, will the normal state of the electron gas in a metal become unstable to formation of Cooper pairs? [1,2] To answer this question one must consider two electrons in the normal state, near the Fermi energy, scattering from each other at finite temperatures. Technically this means that we have to calculate the scattering vertex $\Gamma(\underline{k},\underline{k}';T)$, [3] which is roughly speaking the scattering amplitude that takes a pair of electrons from the state $|\underline{k},-\underline{k}\rangle$ to the state $|\underline{k}',-\underline{k}'\rangle$. Note that the center of mass of the two electrons is stationary and for the moment I neglect to specify the spins. The effective potential $v^{eff}(\underline{k},\underline{k}';\omega)$ which gives rise to $\Gamma(\underline{k},\underline{k}';\omega)$ is the sum of the Coulomb repulsion, properly screened by the other electrons, and the partially attractive, retarded potential due to exchange of phonons. If the net interaction $v^{eff}(\underline{k},\underline{k}';\omega)$ is attractive for low frequencies there will be a temperature T_c where $\Gamma(\underline{k},\underline{k}';T)$ diverges. As usual, the divergence of the scattering amplitude can be attributed to the appearance of a bound state. In this case two electrons with total energy $2\varepsilon_F$ will form a bound state with an energy $E_b < 2\varepsilon_F$. These bound states are the Cooper pairs and their formation leads to the instability of the whole electron system with the result that the superconducting state is formed. Thus T_c is the superconducting transition temperature. As far as finding T_c is concerned this approach can be made completely equivalent to solving the strong coupling Eliashberg equations [4]. See also Appel and Kohn [5].

Neglecting for the moment the residual screened Coulomb repulsion a diagrammatic representation of the integral equation which determines $\Gamma(\underline{k},\underline{k}';\omega)$ is shown in Fig. 1a. According to the Migdal theorem [6], to the order $\sqrt{m/M}$, where m is the electronic mass and M is the nuclear mass, the irreducible scattering vertex with one phonon line is sufficient to describe the particle-particle scatter-

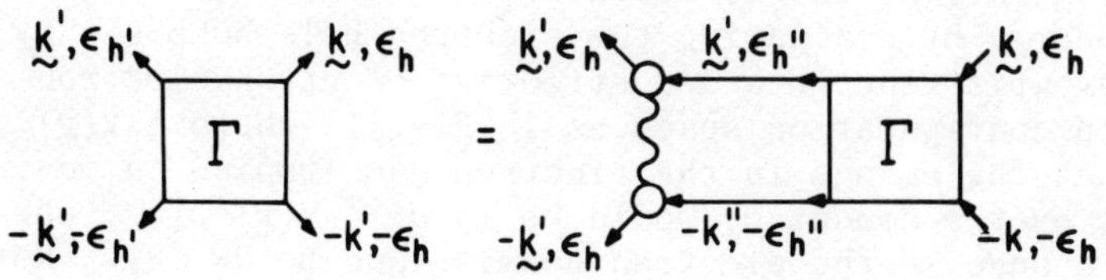

ELECTRON-PHONON INTERACTION

The integral equation which determines the finite temperature particle-particle vertex function $\Gamma(\underline{k}',\underline{k};\epsilon_h-\epsilon_{h'})$

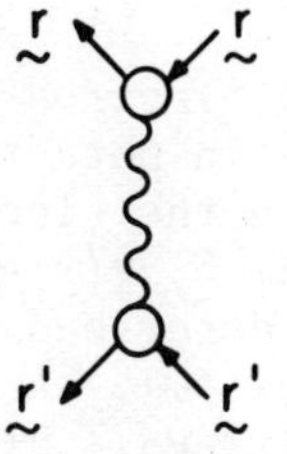

$$= V^{eff}(\underline{r},\underline{r}';\omega) = \underline{\delta}_l V(\underline{r}) \underline{\underline{D}}(l,l';\omega) \underline{\delta}_{l'} V(\underline{r}')$$

The one phonon exchange effective pairing potential.

Figure 1

ing. Hence, one of the central issues of the theory is to know what to take for the electron-phonon vertex γ. In pseudo-potential theory where one works to all orders in the electron-electron interaction but only to second order in the electron-ion interaction the question has been settled rigorously [7]. For transition metals the problem is more difficult since the electron-ion potential also has to be taken into account to all orders. For the particle-hole interaction this problem has been treated properly by Sham [8] in connection with lattice dynamics. There, it turns up when, in the harmonic approximation, one is calculating the electron induced phonon self-energy by evaluating the electron-hole bubble. By analogy we may write the retarded effective electron-electron interaction in configuration space as in Fig. 1b where $\delta_\ell V(r)$ is defined so that the change in the electron Hamiltonian on moving the ion at R_ℓ by the amount U_ℓ would be $\int dr U_\ell \cdot \delta_\ell V(r) \delta\rho(r)$, where $\delta\rho(r)$ is the change in the electron density due to U_ℓ [8]. Note that $\delta_\ell V(r)$ does not depend on time. This is the consequence of the Born-Oppenheimer theorem.

The rigid muffin-tin approximation consists of relacing $\delta_\ell V(r)$ by the gradient of the muffin-tin potential of band theory at the site R_ℓ namely by $\nabla v(r-R_\ell)$. Consequently

$$v^{eff}(r,r';\omega) = \sum_{\ell\ell'} \nabla v(r-R_\ell) \, D(\ell,\ell';\omega) \, \nabla v(r'-R_{\ell'}) \qquad (1)$$

This result may be interpreted as follows: A Bloch electron at r' moves the ion at $R_{\ell'}$. As the ion moves the muffin-tin potential moves with it rigidly and hence its interaction with the electron is through the local change in the crystal potential $U_\ell \nabla v(r'-R_{\ell'})$. The response of the ions to small perturbations is described by the fully renormalized phonon Green's function $D(\ell,\ell';\omega)$. A displacement at $R_{\ell'}$ causes a displacement $D(\ell,\ell';\omega)\nabla v(r'-R_{\ell'})$ at R_ℓ. This results in a change of the crystal potential at R_ℓ, $\nabla v(r-R_\ell)D(\ell,\ell';\omega)\nabla v(r'-R_{\ell'})$. This change of potential is then seen by a second electron at r. The sum of all such processes is the effective electron-electron interaction potential. There is no place here to argue the case at length. Let it suffice to say that in principle $v(r-R_\ell)$ is the self-consistent potential in the local density approximation to the many electron problems [8,9] but, in practice, one uses whatever potential is appropriate for determining the Bloch states. This is consistent with the pseudo-potential result that the Coulomb screened and vertex corrected electron phonon interaction contains the same dressed pseudopotential which determines the Fermi Surface [7].

Of course the great attraction of the rigid muffin-tin approximation is its relative simplicity. Being the simplest non-trivial first principles thing to do it is reasonable to pursue it until it breaks down. One might also draw courage from the fact that it has been very useful in calculations of the resistivity for transition metals [10]. In the present context it was first used by Gaspari and Gyorffy [11]. The foregoing discussion was included merely to point out where we have to return if we find it failing.

In terms of the Fermi Surface Harmonics introduced recently by Allen [12] the integral equation depicted in Fig. 1a with the effective electron-electron interaction given in Eq. (1) may be written as

$$\Gamma_{J,J'}(\varepsilon_n) = v^{eff}_{J,J'}(\varepsilon_n) + \sum_{n'} \sum_{J'',J'''} [v^{eff}_{J,J''}(\varepsilon_n - \varepsilon_{n'})$$

$$\cdot F_{J'',J'''}(\varepsilon_{n'})\Gamma_{J''',J'}(\varepsilon_{n'} - \varepsilon_n)] \quad (2)$$

where J is a three component index used to enumerate the Fermi Surface Harmonics, $\Gamma_{J,J'}$, $v^{eff}_{J,J'}$, and $F_{J,J'}$ are the "matrix elements" of the scattering amplitude, the effective potential and the product of two fully renormalized Green's function $G(\underline{k},\varepsilon_n) G(-\underline{k},-\varepsilon_n)$ with respect to the Fermi Surface Harmonics. In fact they are integrals over the Fermi surface of the corresponding Bloch state matrix elements weighted with polynomials of the Fermi velocity components such as $\left(\frac{\partial}{\partial k_x} \varepsilon_{\underline{k}}\right)^{J_x} \left(\frac{\partial}{\partial k_y} \varepsilon_{\underline{k}}\right)^{J_y} \left(\frac{\partial}{\partial k_z} \varepsilon_{\underline{k}}\right)^{J_z}$ and $\varepsilon_n = \frac{\pi}{\beta}(2n+1)$

is the usual Matsubara frequency [3] for fermions with β the inverse temperature and n an integer.

Using present day computational techniques Eq. (2) could be and should be solved numerically for pure cubic metals. However, I will now make a second approximation by assuming that $v^{eff}_{J,J'}$ is diagonal. Then as far as the problem of finding the temperature at which the s-component $\Gamma_{oo}(T)$ diverges is concerned Eq. (2) is entirely equivalent to the linearized strong coupling Eliashberg equation [4,12]. The relevant component of v^{eff} is

$$v^{eff}_{oo}(\varepsilon_m) = \frac{\sum_{\nu,nn'} \int \frac{d^2k}{v_F} \int \frac{d^2k'}{v'_F} \frac{|\langle\psi_{\underline{k},n}|\nabla v|\psi_{\underline{k}',n'}\rangle|^2}{2MN\omega_{\underline{k}-\underline{k}';\nu}} \frac{2\omega_{\underline{k}-\underline{k}';\nu}}{\varepsilon_m^2 + \omega^2_{\underline{k}-\underline{k}';\nu}}}{\int (d^2k/v_F)} \quad (3)$$

where $|\psi_{k,n}\rangle$'s are the Bloch states with wave vector k and band index n and $\omega_{k-k';\nu}$ is a fully renormalized and therefore experimentally observable phonon frequency with wave vector $k-k'$ and mode index ν, the integrals $\int (d^2k/v_F)$ refer to integrals over the Fermi Surface with $v_F = |\nabla \epsilon_k|$ at each k point, M is the mass of the ions and N is their number per unit volume. Note that M,N and $\omega_{k-k';\nu}$ occur as a result of the lattice Fourier Transform of the phonon Green's Function $D(\ell,\ell';\omega)$ in Eq. (1) and the Fermi Surface integrals are the consequence of having expended $v^{eff}(k,k';\epsilon_n) = \langle k-k|v^{eff}|k'-k'\rangle$ in Fermi Surface Harmonics [12].

Since v_{oo}^{eff} is not expected to diverge, in looking for T_c we can neglect the first term on the right-hand side of Eq. (2). The same can be done with the off-diagonal elements of $\Gamma_{o,J}$ which couple to Γ_{oo}. Then, for $\Gamma_{oo}(\epsilon_n)$, in the diagonal v^{eff} approximation, Eq. (2) can be simply transformed [4] into the equation used by Bergman and Rainer [14] and Allen and Dynes [15] for their investigation of T_c in the strong coupling limit. A difference arises from my having neglected, for the sake of simpler presentation, the residual direct Coulomb repulsion between the electrons. In what follows I assume that this interaction has been put back into the problem. Although the Allen and Dynes [15] equation gives more accurate solutions than the more familiar McMillan formula [16], which is a solution of the real frequency version of the Elashberg equation, for the rest of the paper I shall use the latter. My purpose in setting up Eq. (2) was to demonstrate that what is left out of the McMillan formula can be put back in by reasonably well known means albeit at the expense of some fairly heavy numerical work.

Thus from now on I take it that $\Gamma_{oo}(T)$ diverges at the McMillan value

$$T_c = \frac{\theta_D}{1.45} e^{-1.04 \frac{1+\lambda}{\lambda - \mu^*(1+.62\lambda)}} \tag{4}$$

where μ^* is the Coulomb interaction parameter and λ is the mass enhancement factor. It is conventional to write this as

$$\lambda = 2 \int_0^\infty \frac{d\omega}{\omega} \alpha^2(\omega) F(\omega) \tag{5}$$

with $\alpha^2(\omega) F(\omega) \equiv$

$$\sum_{n,n';\nu} \int \frac{d^2k}{v_F} \int \frac{d^2k'}{v_{F'}} \frac{|\psi_{k,n}|\nabla v|\psi_{k',n'}|^2}{(2\pi)^3 2MN\omega_{k-k';\nu}} \delta(\omega - \omega_{k-k';\nu}) \bigg/ \int \frac{d^2k}{v_F} \tag{6}$$

where $F(\omega)$ is the phonon density of states and $\alpha^2(\omega)$ is a frequency dependent coupling constant defined by the above relation.

Note that since $\frac{1}{(2\pi)^3}\int\frac{d^2k}{v_F}$ is the density of states at the Fermi energy, $n(\varepsilon_F)$,

$$\lambda = n(\varepsilon_F) v_{oo}^{eff}(o). \qquad (7)$$

That is to say λ is a direct measure of the zero frequency component of that pairing potential. That Eqs. (4) and (7) together should give a formula for T_c which is practically identical to the one in the original BCS theory [2] should be a tribute to those authors. However, in the present discussion its simplicity does not hide untold numbers of omissions. In fact there were only three since the use of the McMillan formula was only a matter of convenience. Of these, neglecting the off diagonal elements of v^{eff} (this is equivalent to assuming an isotropic gap) and describing the residual Coulomb repulsion by a μ^* only are probably not serious. Consequently, so far the rigid muffin-tin approximation is the only major simplification.

III. THEORY OF THE MASS ENHANCEMENT FACTOR

λ as a Product of an Electronic and a Phonon Factors

For simple metals λ has been evaluated frequently from Eqs. (5) and (6) using perturbation theory [17]. However, for transition metals one cannot use perturbation theory and it is still rather difficult to carry out the Fermi Surface integrals in Eq. (6) k-point by k-point [18]. To simplify matters it is customary to note that the first frequency moment of $\alpha^2(\omega)F(\omega)$ does not depend on the phonon coordinates and therefore define

$$\eta \equiv 2\int_0^\infty d\omega\, \omega\, \alpha^2(\omega) F(\omega) \qquad (8)$$

which turns out to be a purely electronic property and is given by

$$\eta = \frac{1}{(2\pi)^3}\sum_{n,n'}\int\frac{d^2k}{v_F}\int\frac{d^2k'}{v_{F'}}|\langle\psi_{k,n}|\nabla v|\psi_{k',n'}\rangle|^2 \bigg/ \int\frac{d^2k}{v_F}. \qquad (9)$$

One can then write

$$\lambda = (1/M\langle\omega^2\rangle)\,\eta \tag{10}$$

and hope that $(1/M\langle\omega^2\rangle)$ defined as

$$\frac{1}{M\langle\omega^2\rangle} \equiv \int_0^\infty \frac{d\omega}{\omega}\alpha^2(\omega)F(\omega) \Big/ \int_0^\infty d\omega\,\omega\alpha^2(\omega)F(\omega) \tag{11}$$

does not depend too much on the electronic variables. There is empirical [16] as well as calculational evidence [10] that such hope is freqently justified and there are various methods in the literature for estimating $(1/M\langle\omega^2\rangle)$ from the knowledge of the phonon frequencies only [15,16,19]. One of the more popular methods is to assume that α^2 does not depend on ω and therefore write [16]

$$\frac{1}{M\langle\omega^2\rangle} \simeq \int_0^\infty \frac{d\omega}{\omega} F(\omega) \Big/ \int_0^\infty d\omega\,\omega F(\omega)\,. \tag{12}$$

In this paper I will also make use of other prescriptions but will make no attempt at a critical assessment of the various methods. They are all merely different approximations to Eq. (1).

Perhaps it is worthwhile to point out that in a tunneling experiment where V vs I characteristics is inverted to yield $\alpha^2(\omega)F(\omega)$, η is measured directly. Moreover, as I shall argue, η is the quantity which can be calculated most accurately. Hence, when one is interested mainly in the electron phonon interaction as such confrontation between theory and experiments can most efficiently take place at the level of η not of λ or T_c.

Theory of η in the Rigid Muffin-Tin Approximation

Whatever method was used to construct the crystal potential, for easy visualization, it is permissible to regard it as arising from overlapping atomic potentials centered at the lattice sites. After these potentials have been overlapped they are spherically averaged about each site to obtain a set of spherically symmetric muffin-tin wells separated by a constant potential which is set equal to the spatial average of the potential in the interstitial region. The radius of the wells r_{MT} is chosen to minimize the interstitial volume. This construction is illustrated in Fig. 2a. In what follows we measure all energies from the interstitial potential usually referred to as the muffin-tin zero V_{MTZ}. This is usually below the eigen energies of those bound states of the atomic potential from which the conduction band arises. Thus we are

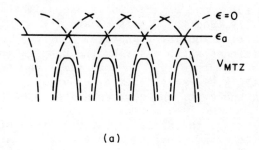

(a)

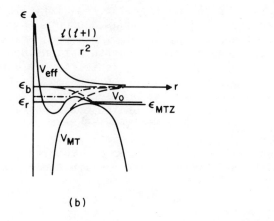

(b)

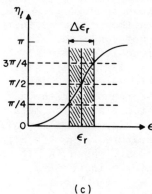

(c)

Figure 2: (a) Construction of the muffin-tin potential. (b) The addition of the centrifugal energy $\ell(\ell+1)/r^2$ to an atomic potential v_a gives rise to an effective potential with a bound state at ε_b. Overlapping to produce muffin-tin potentials v_{MT} turns this into a resonance at ε_r, above the muffin-tin zero. (c) Variation of phase shift with energy at a resonance.

solving the Schrödinger's equation at positive energies and it is natural to use the language of scattering theory. Indeed it turns out that the energy bands only depend on the scattering phase shifts $\delta_\ell(\varepsilon)$ of the individual wells and the arrangement of these wells. This is due to the fact that the potential wells are separated by a constant plateau.

In the K.K.R. method of solving the Schrödinger's equation the energy bands, $\varepsilon_{k,n}$, are found by searching for the zeros of the determinant

$$||[\tau^{-1}(\underline{k},\varepsilon)]_{LL'}|| \equiv ||\sqrt{\varepsilon}(\cot_\ell(\varepsilon)-i)\delta_{LL'}-G_{LL'}(\underline{k},\varepsilon)|| = 0 \quad (13)$$

where L stands for ℓ and m: the polar and azimuthal quantum numbers, respectively, and the $G_{LL'}(\underline{k},\varepsilon)$ are the structure constants. They are the lattice Fourier transforms of free electron Green's function coefficients which describe the propagation of free spherical waves from site to site. One may fix $\underline{k}$ and search in ε to find $\varepsilon_{k,n}$ where $||[\tau^{-1}(\underline{k},\varepsilon)]_{LL'}|| = 0$. Alternatively, one may fix ε and search in $\underline{k}$ space for the corresponding wave vector.

In this language the characteristic feature of transition metals is the fact that their $\ell = 2$ phase shift varies rapidly through $\pi/2$ at an energy close to the energy of the d-bound state in the atom. Such behavior is called a resonance and is due to a metastable, almost bound, state in the potential $v(\underline{r}) + [\ell(\ell+1)/r]$ as shown in Fig. 2b. The energy where the resonance occurs is called the resonance energy ε_r and at this energy the Born series does not converge. This is why one cannot use perturbation theory for transition metals. Since the scattering cross section is given by $\sigma = (1/\varepsilon)\Sigma (2\ell+1)\sin^2\delta_\ell(\varepsilon)$ we see that at $\varepsilon = \varepsilon_r$ the muffin-tin well for a ℓ transition metal is a very strong scatterer. This is the reason for the narrow d-band and most of the other characteristic transition metal properties such as large cohesion, large resistivity and as I shall show, large superconducting transition temperature. (The $\ell = 2$ phase shifts for a number of transition metals are shown in Fig. 3.) To do this let me rewrite the expression for η in Eq. (9) as

$$\eta = \frac{1}{\pi^2 n(\varepsilon_F)} \int d^3r \int d^3r' \; \text{Im} \; G(\underline{r},\underline{r}';\varepsilon_F) \underline{\nabla} v(\underline{r})$$

$$\cdot \underline{\nabla} v(\underline{r}') \; \text{Im} \; G(\underline{r}',\underline{r};\varepsilon_F) \tag{14}$$

where $G(\underline{r},\underline{r}';\varepsilon_F)$ is the one electron Green's function. In terms of Bloch states $\text{Im } G(\underline{r},\underline{r}';\varepsilon_F) = \sum_{k,n} \psi^*_{k,n}(\underline{r})\psi_{k,n}(\underline{r}') \delta(\varepsilon_F-\varepsilon_{k,n})$. Then, use the multiple scattering result due to Gyorffy and Stott [20] that for $\underline{r}$ and $\underline{r}'$ within the same muffin-tin

$$\text{Im } G(\underline{r},\underline{r}';\varepsilon) = \sum_{LL'} \Delta_L(\underline{r},\varepsilon)\Delta_{L'}(\underline{r}',\varepsilon) \; \text{Im } \tau^{oo}_{LL'}(\varepsilon) \tag{15}$$

where $\tau^{oo}_{LL'}(\varepsilon)$ may be written in terms of the inverse KKR matrix defined in Eq. (13), as

$$\tau^{oo}_{LL'}(\varepsilon) = \int_{Bz} d^3k \; \tau_{LL'}(\underline{k},\varepsilon) \tag{16}$$

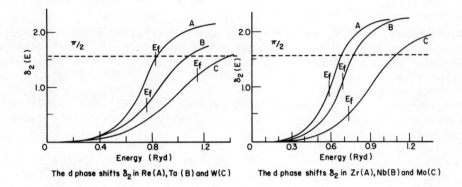
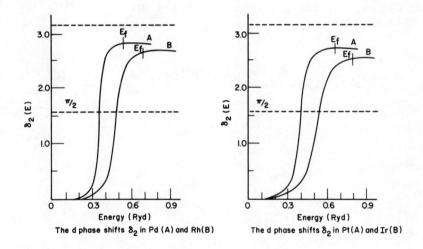

Figure 3

and, in terms of the radial solution of the Schrödinger equation for a single muffin-tin well $R_\ell(r,\varepsilon)$ and the spherical harmonics $Y_L(\hat{r})$,

$$\Delta_L(\underset{\sim}{r},\varepsilon) = - \frac{\sqrt{\varepsilon}}{\sin\delta_\ell(\varepsilon)} R_\ell(r,\varepsilon) Y_L(\hat{r}) \; . \tag{17}$$

Furthermore, $R_\ell(r,\varepsilon)$ satisfies the usual scattering boundary conditions i.e., for $r \geq r_{MT}$, the muffin-tin radius, $R_\ell = \cos\delta_\ell j_\ell - \sin\delta_\ell n_\ell$ where j_ℓ and n_ℓ are the spherical Bessel and Neuman functions, respectively.

When Eq. (15) is substituted into Eq. (14) all the spatial integrations can be carried out analytically. In doing the radial integrals one needs the relation

$$\int_0^{r_{MT}^+} dr \, r^2 \, R_\ell \, \frac{dv}{dr} \, R_{\ell+1} = \sin(\delta_\ell - \delta_{\ell+1})^*$$

which was proved by Gaspari and Gyorffy [11]. The algebra necessary to do the angular integrations was worked out by Jewsbury [21]. One then obtains

$$\eta = \frac{1}{\pi^2 n(\varepsilon_F)} \sum_\alpha \sum_{L_1 L_2 L_3 L_4} F^\alpha_{L_1 L_4}(\varepsilon_F) F^\alpha_{L_2 L_3}(\varepsilon_F) \; \operatorname{Im} \tau^{oo}_{L_1 L_2}(\varepsilon_F)$$

$$\cdot \; \operatorname{Im} \tau^{oo}_{L_3, L_4}(\varepsilon_F) \tag{18}$$

where

$$F^\alpha_{L,L'}(\varepsilon) = \frac{\varepsilon}{\sin\delta_\ell(\varepsilon)\sin\delta_{\ell'}(\varepsilon)} A^\alpha_{L,L'} \sin(\delta_{\ell_<}(\varepsilon) - \delta_{\ell_>}(\varepsilon)) \tag{19}$$

---

* r_{MT}^+ is infinitesimally greater than r_{MT}. Consequently, the range of integration includes the discontinuity in the potential at r_{MT}. In Ref. 11 this formula is given with the wrong sign. I thank B. Harmon and S.K. Sinha for bringing this error to my attention. Fortunately, it has no effect on any subsequent calculations since these always involve bilinear products of such matrix elements.

ELECTRON-PHONON INTERACTION

and the Clebsch-Gordon like coefficients $A^{\alpha}_{L,L'}$ are given in the Appendix. Note that α is x,y,z respectively, and in Eq. (19) $\ell_>$ means the greater and $\ell_<$ the lesser of ℓ and ℓ'. Equations (18) and (19) is the exact form of an earlier approximate theory by Gaspari and Gyorffy [11] and in its present form was given by Gomersall and Gyorffy [18].

Thus it appears to be the case that, though η depends on the wave functions, nevertheless, like the band structure, it is determined solely by the phase shifts and the structure constants. For the rest of the paper I will concentrate on evaluating Eqs. (18) and (19) and show how it gives a simple, physically appealing picture of the electron phonon coupling in transition metals.

The One Scatterer Approximation

Note that in Eq. (14) the potentials refer to the muffin-tin well on the same site. In fact it can be shown by examining the sum rule given in Eq. (8) that η is related to the site diagonal part of the pairing interaction given in Eq. (1). To put it in another way if I did not let the ion response propagate to other sites from where the first electron-ion interaction took place by keeping only the site diagonal part of $v^{eff}(\underline{r},\underline{r}';\omega)$ then I would have obtained $\lambda = 1/M<\omega^{-2}>\eta$ without the necessity of the approximation involved in Eq. (11). This is the approximation used by Appel and Kohn [5]. Thus η and λ are very much local quantities and it makes sense to inquire what happens if we neglect the interaction of the electrons with all the ions except the one which mediates the electron-electron interaction. To do this we merely have to drop $G_{LL'}(\underline{k},\varepsilon)$ in Eq. (13) and use $\tau^{oo}_{L,L'}(\varepsilon) = -(\sin\delta_\ell/\sqrt{\varepsilon})e^{i\delta_\ell}$ in Eq. (18). Using the properties of $A^{\alpha}_{LL'}$ given in the Appendix this yields

$$\eta = \frac{\varepsilon_F}{\pi^2 n(\varepsilon_F)} \sum_\ell 2(\ell+1)\sin^2(\delta_\ell - \delta_{\ell+1}) . \qquad (20)$$

There are several comments to be made about this obvious oversimplified formula;

(a) For weak potentials the phase shifts are small and can be calculated in the Born approximation. Substituting $\delta_\ell = \int dr\, r^2 v(r) j^2_\ell(\sqrt{\varepsilon}\, r)$ in Eq. (20) gives, after some changes of variables,

$$\eta = \frac{3\varepsilon_F}{8k_F^6} \int_0^{2k_F} dq\, q^3 v(q)$$

which is the same as the simple pseudo-potential formula of McMillan [16]. Thus Eq. (20) is the generalization of the pseudopotential formula for the case where there is an arbitrary strong scatterer at the mediating site.

(b) As mentioned, the characteristic feature of transition metals is the resonant behavior of their $\ell = 2$ phase shift. Using the Friedel sum which for one scatterer gives $n(\varepsilon_F) = (\sqrt{\varepsilon_F}/2\pi^2) + 2\Sigma_\ell(2\ell+1)\delta'_\ell(\varepsilon_F)$ and neglecting all but the strongly scattering $\delta_2(\varepsilon_F)$, Eq. (20) takes the following simple form

$$\eta = \frac{2\varepsilon_F \sin^2 \delta_2(\varepsilon_F)}{\pi \delta'_2(\varepsilon_F)} \quad . \tag{21}$$

Clearly, for ε_F near the resonance energy ε_r η will be particularly large. To see how large let us parameterize the energy dependence of η by assuming that it is of the Breit-Wigner form $\tan\delta_2 = \Gamma/\varepsilon-\varepsilon_r$ where Γ is the width of the resonance. After some simple manipulations this leads to $\eta = \varepsilon_F \Gamma/\pi$. For Nb, to take an example, $\varepsilon_F \sim .7$ Ry, $\Gamma \sim .1$ Ry and hence $\eta \simeq 3(\text{ev}/\text{Å}^2)$. This explains why transition metals are such good superconductors in spite of their large Debye temperatures which tend to make λ small.

(c) Note that $\Sigma_\ell 2(\ell+1)\sin^2(\delta_\ell - \delta_{\ell+1})$ is just the transport cross section σ_{tr}^{ex} $\int d\theta(1-\cos\theta)\sigma(\theta)$, where $\sigma(\theta)$ is the differential scattering cross section. Since σ_{tr} is the dominant factor in determining the phonon contribution to the resistivity [10] we can thus understand the source of the empirical correlation between the resistivity and T_c.

(d) It is easy to understand the factor $1/M\langle\omega^2\rangle$ in physical terms. It is roughly the mean square amplitude of the vibration of the mediating atom $\langle U^2 \rangle$ $([1/M\langle\omega\rangle]\langle 1/\omega\rangle = 2[\langle U^2\rangle/h\langle\omega\rangle])$. From Eq. (20) we can see that η also has a simple physical interpretation. It is, roughly, the momentum transfer cross section of the electrons off the rattling ions. That the strength of the Cooper force should be determined by these two factors is eminently plausible. Note that for high λ one needs both large amplitude vibrations (soft modes) and large scattering cross sections (resonant scatterers).

A Simple Theory of the Band Structure Effects on η

In the previous section I have calculated η without taking into account the scattering of the electrons at sites other than the one which mediated the interaction between two electrons. I now want to put back into the theory these multiple scattering effects. As a consequence instead of a plane wave I shall now have Bloch waves incident on that site and one expects that the density of states would enter into even an approximate formula. Of course, Eqs. (17) and (18) are exact. The problem is to find some simple, tractable approximation to these formulae that nevertheless take into some account the band structure.

To do this note that $\tau_{L,L'}^{oo}(\varepsilon)$ is diagonal [22] for cubic systems for $\ell < 2$ due to the fact that the cubic harmonics are the same as the spherical harmonics Y_L to this order. Assuming that it is diagonal for all L leads to the following simple formula

$$\eta = \frac{\varepsilon_F}{\pi^2 n(\varepsilon_F)} \sum_\ell 2(\ell+1) \sin^2(\delta_\ell(\varepsilon_F) - \delta_{\ell+1}(\varepsilon_F)) \frac{n_\ell(\varepsilon_F) n_{\ell+1}(\varepsilon_F)}{n_\ell^{(1)}(\varepsilon_F) n_{\ell+1}^{(1)}(\varepsilon_F)} \quad (22)$$

where $n_\ell(\varepsilon_F)$ is the angular momentum decomposition of the density of states per muffin tin. This latter quantity is defined as

$$-\frac{1}{\pi} \int_{\Omega_{MT}} d^3 r \; \text{Im} \; G(\underline{r},\underline{r};\varepsilon_F)$$

with the integral over the volume within the muffin-tin radius. Then, from Eq. (15), we have

$$n_\ell(\varepsilon_F) = -\frac{2}{\pi} \frac{\varepsilon_F}{\sin^2 \delta_\ell(\varepsilon_F)} \sum_m \int_0^{r_{MT}} dr \; r^2 R_\ell^2(r,\varepsilon_F) \; \text{Im} \; \tau_{\ell m, \ell m}^{oo}(\varepsilon_F) , \quad (23)$$

Also, $n_\ell^{(1)}$ is the same as n_ℓ except it is for a single muffin-tin well with a constant potential everywhere outside. Using $\tau_{LL'}^{oo}(\varepsilon) = -\frac{\sin\delta_\ell}{\sqrt{\varepsilon}} e^{i\delta_\ell}$ from Eq. (23) we have that

$$n_\ell^{(1)}(\varepsilon_F) = \frac{2}{\pi}(2\ell+1) \int_0^{} dr \; r^2 R_\ell^2(r;\varepsilon) . \quad (24)$$

The relatively simple expression in Eq. (22) was first derived by Gaspari and Gyorffy [11] and since then has been used extensively for estimating η in many strong scattering systems. Before describing these applications I want to make two general comments regarding Eq. (22).

Firstly, I note that if $n_2(\varepsilon_F)$ is much bigger than the other components and these are not too different from their one scatterer value $n_\ell^{(1)}$ the term in Eq. (22) reduces to the one scatterer formula given in Eq. (21). Thus the remarks I made in connection with this simple formula apply with more force than its derivation would allow us to suspect.

Secondly, observe that while $n_\ell(\varepsilon_F)$ is not usually calculated during the course of a standard band structure calculation nevertheless, it can be obtained with relative case. Thus the full evaluation of Eq. (22) is an eminently tractable calculation with currently available computing. To show that this is a worthwhile exercise and to gain familiarity with the ways of the theory for particular systems I will now describe some simple minded attempts for estimating η without purpose built programs for calculating n_ℓ.

Rough Estimates of η

In this section I want to show how η can be estimated using existing band structure calculations.

(a) Pure metals: assuming that $n_2(\varepsilon_F)$ is dominant, $n_\ell = n_\ell^{(1)}$ for all ℓ other than $\ell = 2$ and neglecting all terms with $\ell > 3$ from Eq. (20) we obtain

$$\eta = \frac{\varepsilon_F}{\pi^2}\left[\frac{2\sin^2(\delta_0-\delta_1)}{n(\varepsilon_F)} + \frac{4\sin^2(\delta_1-\delta_2)+6\sin^2\delta_2}{n_2^{(1)}(\varepsilon_F)}\frac{n_2(\varepsilon_F)}{n(\varepsilon_F)}\right] \quad (25)$$

If a band structure calculation has been carried to the point where the density of states is calculated and the Fermi energy ε_F is determined, the above formula may be used as follows: take $n(\varepsilon_F)$ and ε_F as given, calculate δ_ℓ and $n_\ell^{(1)}$ from the muffin-tin potential used in the band structure calculation (this is a trivial computational job) and, as appropriate, estimate n_2/n as 1 or if $n(\varepsilon_F)$ is very low, as in the case of W, substract from $n(\varepsilon_F)$ the free electron contributions $n_0^o(\varepsilon_F) = .35/k_F^2$; $n_1^o(\varepsilon_F) = .32/k_F^2$ to obtain $n_2(\varepsilon_F)$. Estimates of η obtained in this way by Evans [23] et al., for eleven transition metals are shown in Table 1. Estimates of $1/M<\omega^2>$ by Foulkes and Gomersall [24] are also shown. These are in reasonable agreement with the more sophisticated calculations of Allen and Dynes [15]. The calculated and empirical values for λ are shown in the last two lines in Fig. 4. Both the trends and the absolute values are in good agreement. In view of the fact that there are no adjustable parameters in the theory, this is highly gratifying.

More accurate calculations starting from Eqs. (18) and (19) seem to confirm these rough estimates except for Nb, V, Ta. In

Table 1

Rough estimates of η and λ for a number of transition metals using Eq. (25) in the text.

	Zr	V	Nb	Ta	Mo	W	Re	Rh	Ir	Pd	Pt
(in $\frac{ev}{Å^2}$)	2.2	2.4	3.8	4.2	4.2	4.3	5.2	2.5	4.5	1.3	2.8
$<\omega^2>^{\frac{1}{2}}$(in °K)	202	216	148	149	252	204	218	244	212	150	127
λ calculated	.34	.57	1.02	.64	.39	.31	.33	.23	.27	.31	.50
λ empirical	.41	.59	.82	.65	.41	.29	.46	.44	.37	.66	.63

these cases the above agreement is fortuitous. η_{Nb} turns out to be ~ 7 ev/Å². I will discuss this later.

(b) Similarly the pressure dependence of η was analyzed by Ratti, Evans and Gyorffy [25]. They found that for reduced lattice parameters the resonance shifted and broadened as shown in Fig. 4a. This volume dependence togther with some knowledge about the way the band structure and phonon spectra changes under pressure was sufficient to understand why dT_c/dP is always negative for simple metals but can be positive for the early transition metals. The relevant quantity $d \ln \eta/d \ln \Omega$ turned out to ve roughly -3, in reasonable agreement with experiments. Stocks et al. considered Cs under pressure [26]. They found that with pressure ε_F moves into the d-resonance as shown in Fig. 4b and predicted $T_c \sim .5$ °K at 35 kbar.

(c) The theory was generalized to two component systems by Gomersall and Gyorffy [19]. They found that η as defined in Eq. (8) separates rigorously into a sum of two terms $\eta_A + \eta_B$ each of which is given to a good approximation by an expression like Eq. (22) with all quantities evaluated at the relevant site. They considered NbN and NbC and obtained surprisingly large values for η_N and η_C on account of a p-resonance at the carbon and nitrogen sites. Unfortunately one can write

$$\lambda = \frac{1}{M_A <\omega^2>_A} \eta_A + \frac{1}{M_B <\omega^2>_B} \eta_B \qquad (26)$$

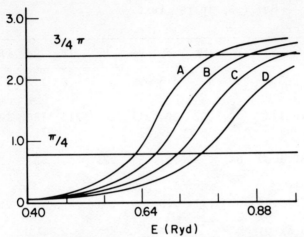

The d phase shifts δ_2 in vanadium as a function of the reduced volume V/V_0. A, $V/V_0=1$; B, $V/V_0=0.95$; C, $V/V_0=0.9$; D, $V/V_0=0.85$. The energy E is measured with respect to the muffin tin zero calculated at each volume.

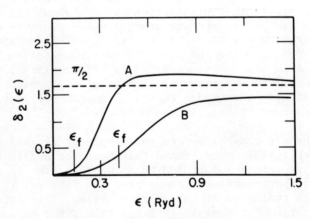

The d wave phase shift for bcc Cs at normal pressure (A) and fcc Cs at 35 kbar (B).

Figure 4

only approximately and even then it is not clear what to take for $<\omega^2>_{A,B}$. Gomersall and Gyorffy [10] took $<\omega^2>_A \sim <\omega^2>_B \sim \theta_D^2$ and found large contributions to λ from the light atoms. A more realistic procedure is probably that of Klein and Papaconstantopoulos [27] which prescribes $<\omega^2>_C = <\omega^2>_{opt}$ and $<\omega^2>_{Nb} = <\omega^2>_{ac}$, where $<\ >_{ac}$ and $<\ >_{opt}$ means averages with respect to the acoustic and optic phonons, respectively, and this gave much smaller contributions from the C and N sites. Apparently, in these systems the optic modes are too high and therefore the light element oscillates with too small an amplitude to make a sizable contribution to the pairing potential in spite of the fact that it is a strong scatterer. On the other hand in PdH where the optic modes are low (40 MeV compared to 70 MeV in NbC) the strong scattering proton contributes. Tunneling experiment measuring $\alpha^2(\omega)F(\omega)$ for frequencies up to 70 MeV would help to decide whether N and C contribute to λ in NbN and NbC, respectively.

The Electron per Atom Ratio and T_c

From Eq. (22) it is clear that if one component of the density of state dominates η will not vary as much as $n(\varepsilon_F)$ (about a factor of 3) across a transition series. In fact η remains remarkably constant according to both theroy [23] and experiment [16]. Thus, comparing Eq. (10) and Eq. (7) we have to conclude that $n(\varepsilon_F)$ in the BCS like form $n(\varepsilon_F)v_{oo}^{eff}$ cannot be used to correlate high T_c with high density of states [28]. This correlation must come through $<\omega^2>$ and possibly μ^*. How $<\omega^2>$ might depend on $n(\varepsilon_F)$ was worked out by Gomersall and Gyorffy [29]. They constructed a rough microscopic theory of $<\omega^2>$ and found that

$$<\omega^2> = \Omega^2(1 - \varepsilon_F n(\varepsilon_F)\frac{\eta}{M\Omega^2}) \qquad (27)$$

where Ω is a bare frequency which was assumed not to vary too much within a family of metals and the second term on the right hand side represents the softening of the phonon frequencies by the electron mediated ion-ion attraction. It was obtained from the bubble correction to the phonon self energy in the rigid muffin-tin approximation (hence η). As can be seen from Eq. (27), for a strong scattering system with large η, large density of states means soft phonons and therefore large T_c. Using the rigid band idea and Eq. (27) a good account can be given of the variations of T_c with the electron per atom ratio e/a for the 4-d, 5-d transition metals and the A-15 [30] compounds.

More Accurate Calculations

Klein and Papaconstantopoulos[32,33] have evaluated Eq. (22) fully, up to and including the term with $\ell = 3$ for a series of very interesting systems. They used self-consistent x-α potentials and the APW method to calculate the energy bands. To find $n_\ell(\varepsilon_F)$ they employed a Slater-Koster interpolation Hamiltonian and the Brillouin zone integrals were carried out with the QUAD scheme. Some of their more impressive results are discussed elsewhere in this volume.

Obviously, each of these would merit separate and detailed discussion. This is particularly the case with V_3Si which, significantly, shows no sign of having bands which can easily be related to linear chains. However, this would take me too far from my main concern in this section which is to catalog the evidence that Eq. (22) is a useful guide to η and, with judicious choices for $1/M<\omega^2>$, for λ and T_c as well.

From this point of view the calculation for PdH is most interesting. It is the first example where in a compound the non-transitional metal element is clearly contributed strongly to λ. It indicates that there is nothing magic about a d-resonance. Any strongly scattering phase shift will do just as long as it is close to $\pi/2$ at the Fermi energy. The large amplitude of vibrations ($<\omega_{opt}> \sim 40$ MeV) and the strong s-scattering by the proton combine to give a substantial contribution to the pairing force. This picture is well substantiated by the tunneling data of Buckel [34] et al. Note also that since ε_F is well outside the d-band there is no question of a large density of state playing a role.

Of course many details of the band structure may not be quite right. Hydrogen in a transition metal and particularly in Pd constitutes a very complicated system. But, as the non-self-consistent calculation of Switendick [35] shows, the strong s-scattering by the proton and the rough position of ε_F appears not to depend too strongly on the details.

As was shown by Appel [36] a reasonable description of λ for a disordered two component system can be derived in the form

$$<\lambda> = C<\lambda_A>_A + (1-C)<\lambda_B>_B$$

where $<\>_A$, $<\>_B$ denote partial averages with either an A or a B atom being fixed at a site. This implies a formula like Eq. (22) for $<\lambda_A>_A$ with the phase shifts and $n_\ell^{(1)}$ referring to the atom fixed at the site and n_ℓ being replaced by the partially averaged, partial density of states on that site. It would be most interesting if the CPA calculation of Faulkner [37] could be extended to yield $<\lambda>$ for PdH off-stoichiometry.

ELECTRON-PHONON INTERACTION

The Exact η for Rigid Muffin Tins

In deriving Eq. (22) from Eq. (18) I have assumed that $\tau_{LL'}^{oo}(\varepsilon_F)$ was diagonal for all ℓ and m. In a recent calculation by Butler et al. [38] Eq. (22) was evaluated for all diagonal and off-diagonal terms with $\ell \leq 3$.

An attractive feature of using the language of scattering theory for this problem is that the angular momentum summations in Eq. (18) converge rapidly because the phase shifts in the matrix-element factor became practically zero for $\ell > 2$. Thus from an $\ell \leq 3$ calculation one obtains η virtually exactly within the rigid muffin-tin approximation. Note also that to calculate η we need quantities at the Fermi energy only. This makes the constant energy search KKR of Faulkner et al. [39] a powerful tool for this problem.

As is shown in ref. 38

$$\eta = \frac{2}{n(\varepsilon_F)} [\sum_\ell (\ell+1) F^2_{\ell,\ell+1} \overline{T}_{\ell\ell} \overline{T}_{\ell+1,\ell+1} + \frac{3}{35} F^2_{2,3} (T^{(25')}_{22} - T^{(1,2)}_{22})(T^{(25)}_{33} - 3T^{(15)}_{33} + 2T^{(2')}_{33}) + \frac{12}{5}\sqrt{\frac{3}{7}} F_{12} F_{32} \cdot (T^{(25')}_{22} - T^{(12)}_{22}) T^{(15)}_{13}]$$

where

$$F_{\ell,\ell+1} = \sin(\delta_\ell(\varepsilon_F) - \delta_{\ell+1}(\varepsilon_F))$$

$$\overline{T}_{oo} = T^{(1)}_{oo}, \quad \overline{T}_{11} = T^{(15)}_{11}, \quad \overline{T}_{22} = 1/5(3T^{(25')}_{22} + 2T^{(12)}_{22})$$

$$\overline{T}_{33} = 1/7(3T^{(25)}_{33} + 3T^{(15)}_{33} + T^{(2')}_{33}) \quad (28)$$

and $T^{(t)}_{\ell\ell}$ is a Brillouin zone average of products of wave function coefficients, $C^t_{\ell,\mu}(\underline{k})$ which are solutions to the symmetrized KKR equations, i.e.,

$$T^{(t)}_{\ell,\ell'} = \frac{1}{(2\pi)^3} \sum_n \int_{FS} k_n^2 \, d\Omega_k \frac{C^t_{\ell,\mu}(\underline{k}) C^t_{\ell,\mu}(\underline{k})}{\underline{k}_n \cdot \nabla \varepsilon_{k,n}} \quad (29)$$

where k_n are the wave vectors for which the KKR determinant vanishes at $\varepsilon = \varepsilon_F$. In the cubic harmonics representation, whose elements are denoted by ℓ and μ, $T^t_{\ell,\ell'}$ is diagonal in μ and independent of μ for a given irreducible representation t. Since the integral in Eq. (29) is over the Fermi surface, in the constant energy scheme there is no need to calculate the bands at any other energy and hence the necessity of using interpolation Hamiltonians is eliminated. The result is a relatively rapid and very accurate calculation.

Instead of $\tau^{oo}_{L,L'}(\varepsilon_F)$ being diagonal in the spherical harmonics representation Eq. (27) has all the elements allowed by the cubic symmetry of the crystal. Noting that $\overline{T}_{\ell,\ell} = (\sqrt{\varepsilon_F}/\pi)(n_\ell/n_\ell^{(1)})$ we see that the first sum on the right hand side is just the formula given in Eq. (22). The rest are all the symmetry allowed correction up to $\ell = 3$.

Butler et al. illustrate their method by evaluating η for Nb. They found $\eta = 7.06$ ev/Å^2 for an $\alpha = 1$ Mattheiss prescription potential and $\eta = 7.6$ ev/Å^2 for an x-α self-consistent potential. This agrees with the APW result of Klein et al. [29]. Moreover, both sets of authors find that η is much larger than the rough estimate in Section III(5) because of the unusually large $n_3(\varepsilon_F)/n_3^{(1)}(\varepsilon_F)$. Presumably this is due to the fact that the loosely bound d-orbitals in Nb have a substantial f component about a neighboring site and therefore enhance the f density of states there over and above the one scattering value. Though in principle the correction terms in Eq. (28) can be significant, for Nb they cancel against each other to give essentially the same result as Eq. (22).

For most reasonable values of $1/M<\omega^2>$ this large η gives rise to $\lambda > 1$ (for $<\omega^2>^{\frac{1}{2}} = 183^0$K, $\lambda = 1.28$ for $\eta = 7.1$ ev/Å^2) in disagreement with $\lambda = .82$ obtained from measurements of T_c and the McMillan formula [16]. One of the main points of my talk is to stress that except for the rigid muffin-tin approximation we are now dealing with a very accurate theory indeed. Therefore, if tunneling experiments measuring η directly ($\eta = 2 \int_0^\infty d\omega\, \omega\, \alpha^2(\omega)F(\omega)$) found it to be as small as the empirical T_c suggests we would be forced to conclude that the rigid muffin-tin approximation can serve only as a rough guide to estimating η. To have this conclusion firmly established would be very important from the point of view of our fundamental understanding of the electron phonon interaction. However, current calculations of $n(\varepsilon_F)$ [38] and the measured specific heat could accommodate a $\lambda > 1$ (for precise agreement one would need $\lambda = 1.19 \pm 0.01$). Moreover, the very first tunneling data on Nb by Bostock et al. [40] is confusing to say the least. Thus, a definite conclusion cannot be drawn as yet.

IV. P-WAVE SUPERCONDUCTIVITY [41]

If metallic electrons could form Cooper pairs with a relative angular momentum $\ell = 1$ we would have a superconductor which is the charged particle analogue of superfluid He^3. Clearly, it would have many fascinating properties.

To find the transition temperature $T_{c,1}$ for a p-wave superconductor we could again start from Eq. (2) and look for an instability in the p-channel. On this basis Foulkes and Gyorffy [43] suggested the following approximate strong coupling formula

$$T_{c,1} = \frac{\theta_D}{1.45} e^{(1+\lambda_0/\lambda_1)} \tag{30}$$

where λ_0 is the s-component of the pairing potential and is given in Eq. (5) and the p-component, λ_1, is defined as

$$\lambda_1 = 2 \frac{\sum_{n,n',\nu} \int \frac{d^2k}{v_F} \int \frac{d^2k'}{v_{F'}} \frac{|\langle \psi_{k,n}|\nabla v|\psi_{k',n'}\rangle|^2}{(2\pi)^3 2M_N \hbar^2 \omega_{k-k';\nu}} \cos(\hat{k}\cdot\hat{k}')}{\int (d^2k/v_F)} \tag{31}$$

In contrast to Hayszenau and Appel [39] we have $1+\lambda_0$ appearing in the exponent instead of $1+\lambda_1$. This is because this factor arises from the self-energy of the intermediate Green's functions in Eq. (2) and this is the same for the s and p channel. In agreement with these quthors there is no residual Coulomb repulsion factor (μ^* in case of s-wave superconductors) because Cooper pairs in odd relative angular momentum state have antisymmetric relative wave functions and therefore cannot overlap. This makes a contact repulsive interaction ineffective [44].

Clearly it would be very interesting to evaluate Eq. (31) for transition metals in order to know where to look for this strange phenomena. We have made some preliminary estimates of λ_1 and $T_{c,1}$ using high temperature resistivity data by noting that roughly $\lambda_1 = \lambda_0 - (1/4k_B)\hbar\omega_p^2(dR/dT)_{T>\theta_D}$ where ω_p^2 is the plasma frequency and R is the resistivity. Our estimates are shown in Table II. We also note that even paramagnetic impurities would tend to suppress p-wave superconductivity. In agreement with Larkin [45] we found that

$$T_{c,1} = T^o_{c,1} - \alpha(R_{4^\circ K}/R_{293^\circ K}) \tag{32}$$

Table II

Values of parameters involved in estimating λ_1, and results for λ_1, $T_{c,1}$ and α for a number of transition metals. For explanation and discussion see text.

	λ_0	ω_p in eV	$(\frac{dR}{dT})_{T>\theta_0}$ in $\mu\Omega/K^{02}$	λ_1	$T_{c,1}$ (in °K)	$T_{c,0}$ (in °K)	α
Nb	.82	9.1	.04	.409	1.94	8.00	3×10^3
Mo	.41	7.4	.027	.226	.61	.84	1.3×10^3
W	.29	5.04	.029	.199	.41	.01	1.3×10^3
Pd	.31	5.63	.030	.192	.20	.028	4.7×10^3
		2.80		.280	1.70		
Ir	.37	3.6	.028	.325	4.32	.33	1.8×10^3
Rh	.44	5.2	.025	.356	5.80	1.42	2.1×10^3

where $T_{c,1}^o$ is the transition temperature without impurities and $\frac{R_{293°K}}{R_{4°K}}$ is the usual resistivity ratio. Estimates of α are also in Fig. 6. Clearly, p-wave superconductivity could be seen only in very pure samples. The samples also have to be larger than the size of the Cooper pairs [45]. Nevertheless, metals at the end of the transition series are attractive candidates because for these the large μ^* tend to suppress ordinary superconductivity but the electron-phonon interaction is still strong. In Pd paramagnetic fluctuations might enhance the $\ell = 1$ pairing force but the purity required $R_{293°K}/R_{4°K} = 10^4$ may be too difficult to achieve.

V. CONCLUSION

In summary, I argued that (a) Eq. (22) is a good starting point for estimating η. (b) In the rigid muffin-tin approximation η can be calculated exactly [Eq. (28)] and therefore it would be very important to be able to do tunneling experiments in transition metals to test this fundamental approximation. As I mentioned in the introduction the theory of no other electron-phonon interaction effect is sufficiently well developed to make such quantitative comparison between first principles theory and experiment in the case of transition metals. (c) Evaluation of λ_1 as given in Eq. (20) should be a considerable help in assessing whether p-wave pairing is likely to occur in metals.

ACKNOWLEDGMENT

It is a pleasure to thank G.D. Gaspari for introducing me to this problem and Drs. W.H. Butler, J.S. Faulkner, R. Evans, G.D. Gaspari, R.I. Gomersall, V. Ratti and G.M. Stocks for stimulating collaborations. Helpful conversations with P. Allen and D. Papaconstantopoulos and B. Klein are also gratefully acknowledged.

Also, I would like to thank Dr. J.S. Faulkner for the hospitality in the Metals and Ceramics Division of the Oak Ridge National Laboratory during which most of the above paper was written.

REFERENCES

1. L.N. Cooper, Phys. Rev. 104, 1189 (1956).
2. J. Bardeen, L.N. Cooper and J.R. Schrieffer, Phys. Rev. 106, 162 (1957).
3. A. Abrikosov, L. Gorkov, I. Dzyloshinskii, "Quantum Field Theoretical Methods in Statistical Physics" (Pergamon Press 1965) Sec. 33.1 p. 283.
4. J.M. Luttinger "Many Body Theory" (Syokabo Tokyo and Benjamin 1966) p. 86.
5. J. Appel and W. Kohn, Phys. Rev. B4, 2162 (1971).
6. A.B. Migdal, Sov. Phys. JETP 7, 996.
7. D.J. Scalapino, "Superconductivity" Ed. R.D. Parks (Marcel Dekker, Inc. New York 1969) Vol. 1, p. 449.
8. L.J. Sham, "Modern Solid State Physics" Vol. II (Eds. R.H. Evans and R.R. Haering, Gordon and Breach, London, 1969).
9. W. Kohn and L.J. Sham, Phys. Rev. 140, A1133.
10. R. Evans, B.L. Gyorffy, N. Szabo and J.M. Ziman, Proc. 2nd Int. Conf. on Properties of Liquid Metals, Tokyo, Ed. Takeuchi, p. 319, 1973.
11. J.D. Gaspari and B.L. Gyorffy, Phys. Rev. Letters 28, 801.
12. P.B. Allen, Phys. Rev. B13, 1416 (1976).
13. G.M. Eliashberg, Sov. Phys. JETP 11, 696.
14. G. Bergmann and D. Rainer, Z. Phys. 263, 59 (1973).
15. P.B. Allen and R.C. Dynes, Phys. Rev. B12, 905 (1975).
16. W.L. McMillan, Phys. Rev. 167, 331 (1968).
17. P.B. Allen and M.L. Cohen, Phys. Rev. 187, 525.
18. I.R. Gomersall and B.L. Gyorffy, J. Phys. F.: Metal Physics 4, 1 (1974).
19. B.L. Gyorffy and M.J. Scott,"Band Structure Spectroscopy of

Metals and Alloys" Ed. D.J. Fabian (Academic Press, London) (1971).
20. P. Jewsbury, Ph.D. Thesis (Bristol University, 1973).
21. W. Kohn, J. Phys. F: Metal Physics $\underline{3}$, L231 (1973).
22. R. Evans, G.D. Gaspari and B.L. Gyorffy, J. Phys. R: Metal Physics $\underline{3}$, 39 (1973).
23. I.F. Foulkes and I.R. Gomersall, J. Phys. F: Metal Phys. $\underline{5}$, 153 (1975).
24. V.K. Ratti, R. Evans and B.L. Gyorffy, J. Phys. F: Metal Phys. $\underline{3}$, L238 (1973).
25. G.M. Stocks, G.D. Gaspari and B.L. Gyorffy, J. Phys. F: Metal Phys. $\underline{2}$, L123 (1972).
26. B.M. Klein and D.A. Papaconstantopoulos (to be published).
27. D. Pines, Phys. Rev. $\underline{109}$, 280 (1958).
28. I.R. Gomersall and B.L. Gyorffy, Phys. Rev. Lett. $\underline{33}$, 1286 (1974).
29. I.R. Gomersall and B.L. Gyorffy, "Low Temperature Physics LT 14" Vol. 5, p. 449, Ed. M. Krusius and M. Vuorio (North Holland, 1975).
30. B.M. Klein and D.A. Papaconstantopoulos, Phys. Rev. Letters $\underline{32}$, 1193 (1974).
31. D.A. Papaconstantopoulos and B.M. Klein, Phys. Rev. Letters $\underline{35}$, 110 (1975).
32. B.M. Klein and D.A. Papaconstantopoulos (to be published).
33. A.C. Switendick and Ber. Bunsenger, Phys. Chem. $\underline{76}$, 535 (1972).
34. A. Eichler, H. Wühl, and B. Stritzker, "Low Temperature Physics LT-14" Ed. M. Krusius and M. Vuorio (North Holland, 1975).
35. J. Appel, Solid State Comm. $\underline{15}$, 1043 (1974).
36. J.S. Faulkner (to be published).
37. W.H. Butler, J. Olson, J.S. Faulkner and B.L. Gyorffy (to be published).
38. J.S. Faulkner, H.L. Davis and H.W. Joy, Phys. Rev. $\underline{161}$, 556 (1967).
39. J. Bostock, V. Diadiuk, W.N. Cheung, K.H. Lo, R.M. Rose and M.L.A. MacVicar, Phys. Rev. Letters $\underline{36}$, 603 (1976).
40. R. Balian and N.R. Werthamer, Phys. Rev. $\underline{131}$, 1553 (1963).
41. A.J. Legett, Rev. Mod. Phys. $\underline{47}$, 331 (1975).
42. I.F. Foulkes and B.L. Gyorffy (to be published).
43. J. Appel and H. Heyszenau, Phys. Rev. $\underline{188}$, 755 (1969).
44. V. Larkin, Sov. Phys. JETP Letters $\underline{2}$, 130 (1965).

APPENDIX

The coefficient $A^{\alpha}_{L,L'}$ are zero unless ℓ and ℓ' differ by ± 1, and are given by

$$A^{x}_{L_1,L_2} = \tfrac{1}{2}\sqrt{\frac{(\ell_1-m_1)!(\ell_2-m_2)!}{(2\ell_1+1)(2\ell_2+1)(\ell_1+m_1)!(\ell_2+m_2)!}}$$

ELECTRON-PHONON INTERACTION

$$\cdot \ [\delta_{m_1,m_2+1}(\delta_{\ell_1,\ell_2-1} - \delta_{\ell_1,\ell_2+1}) \frac{(\ell_1+m_1)!}{(\ell_1-m_1)!} +$$

$$\delta_{m_2,m_1+1}(\delta_{\ell_2,\ell_1-1} - \delta_{\ell_2,\ell_1+1}) \frac{(\ell_2+m_2)!}{(\ell_2-m_2)!} \ ]$$

$$A^y_{L_1 L_2} = \tfrac{1}{2} \frac{(-1)^{m_1+1}}{\sqrt{(2\ell_1+1)(2\ell_2+1)}} \ [\delta_{m_1,1-m_2} \sqrt{\frac{(\ell_2+m_2)!(\ell_1+m_1)!}{(\ell_2-m_2)!(\ell_1-m_1)!}} \cdot$$

$$(\delta_{\ell_1,\ell_2-1} - \delta_{\ell_1,\ell_2+1}) + \delta_{-m_1,m_2+1} \sqrt{\frac{(\ell_2-m_2)!(\ell_1-m_1)!}{(\ell_2+m_2)!(\ell_1+m_1)!}}$$

$$\cdot \ (\delta_{\ell_1,\ell_2-1} - \delta_{\ell_1,\ell_2+1})$$

$$A^z_{L_1 L_2} = \sqrt{\frac{(\ell_1+m_1)!(\ell_2-m_2)!}{(2\ell_1+1)(2\ell_2+1)(\ell_1-m_1)!(\ell_2+m_2)!}} \ [(\ell_1-m_1)\delta_{\ell_1,\ell_2+1} +$$

$$(\ell_2+m_2)\delta_{\ell_1,\ell_2+1}]\delta_{m_1,m_2} \quad . \tag{A.2.7}$$

We note here that these coefficients satisfy the relation

$$\sum_{\alpha} (A^{\alpha}_{LL'})^2 = \frac{4\pi}{3} \sum_{m_1} (C^{1\ell\ell'}_{m,mm'})^2 \tag{A.2.8}$$

and the sum rule

$$\sum_{mm'} A^{\alpha}_{LL'} A^{\beta}_{LL'} = \frac{\delta_{\alpha\beta}}{3} [(\ell+1)\delta_{\ell;\ell+1} + (\ell'+1)\delta_{\ell,\ell'+1}] \tag{A.2.9}$$

where $C^{\ell\ell'\ell''}_{mm'm''}$ is a real Gaunt number.

QUESTIONS AND COMMENTS

M. Cohen: Do you use 183 K for niobium? There is some talk about it being 195 K. Do you have the latest word on it?

B. Gyorffy: Well, I used all of them and λ just goes up and down. I can not say the calculation gives only η and this is why I think it is important, from the point of view of the fundamental theory of electron phonon interaction to measure η directly. Then one is not involved with the phonon part of the problem. Of course, the phonon part is important for finding T_c, but if you want to check the theory which is supposed to be very good (there are only a few approximations) the best way to make the comparison is to measure η directly.

P. Allen: In connection with your p-wave superconductivity, I have been doing calculations in collaboration with B. Chakraborty and W.E. Pickett [Phys. Rev. B, in press] in which we are able to get the empirical λ from dc resistivity at room temperature. For example, in niobium we find the number is 1.15. Now in your language that is approximately equal to the superconducting λ minus the p-wave channel λ which would indicate that the p-wave channel λ is very small or possibly negative (if you believe the McMillan λ for niobium) or zero (if you believe your λ).

B. Gyorffy: We are using different relationships between λ_1 and the resistivity. Besides the s-wave T_c is very large for Nb and therefore is not an interesting case.

P. Allen: We find for four bcc transition metals a rather remarkable similarity between the λ that comes from resistance and the λ from superconductivity which would tend to indicate that coupling constants in other than s-wave channels are surprisingly small.

B. Gyorffy: From this theory, the difference is λ_1 and indeed, I do find that they are always smaller. The average of $\cos^2 \theta$ is 1/2, so it has to be.

J. Bostock: From your point of view, the λ's that you calculated with the Shapiro Theorem or s-wave theory are very, very high. Now when we go to p-wave---they seem very small. How would making your estimates of the scattering smaller, effect your p-wave?

B. Gyorffy: Well, it just pushes everything down, and you have it.

The point I am making is that for p-wave pairing you do not mind working with a T_c of millidegrees, while for s-wave superconductivity you want to have a high T_c.

D. Papaconstantopoulos: I would like to offer a possible explanation as to why the η's are rather high. These calculations are done within the muffin-tin spheres. We have tried to extend these calculations to the real world; i.e., the Wigner-Seitz cell. Then the values of the η's actually come down and are very close to the empirical values. Our calculation is not yet rigorous but it points out the importance of the interstitial region.

B. Gyorffy: Yes, it could be. It is a technical question. One needs to discuss it.

UNEXPECTED SUPERCONDUCTIVITY IN rf. SPUTTERED Nb/Ge SAMPLES*

A.K. Ghosh and D.H. Douglass

University of Rochester

Rochester, New York 14627

ABSTRACT

Samples of Nb/Ge scanning the concentration range of ~ 5 a% Nb to ~ 80 a% Nb, were prepared by the technique of rf. sputtering. It was found that the Ge-NbGe$_2$ two phase region exhibited superconductivity to 16°K, in addition to the superconductivity of the A-15 Nb$_3$Ge phase. We show that the presence of Ge and NbGe$_2$ are necessary, and that the superconductivity is associated with NbGe$_2$, although by itself it is not superconducting above 2.5°K. Various explanations are considered in explaining the observed superconductivity in mixtures of metallic NbGe$_2$ and semiconducting Ge.

INTRODUCTION

Niobium digermanide (NbGe$_2$) has been examined for superconductivity by Hardy and Hulm [1] and found to remain normal to 1.2°K. However in this study, a preliminary report of which is being published [2], anomalously high superconducting critical temperatures are observed for sputtered samples of niobium-germanium in the 95-65 a% germanium concentration range. In this region, the sample consists of a mixture of metallic NbGe$_2$ and semiconducting Ge. Our results are concerned with these Ge-NbGe$_2$ samples.

---

* Supported in part by the National Science Foundation.

The superconducting transition temperature, T_c, was obtained by resistive methods for sample specimens of various germanium concentrations. The T_c measurements were further complemented by critical current, I_c, measurements, and by x-ray analysis, in an effort to determine the nature of the superconducting phase. Based on these experiments, we arrive at certain conclusions which are discussed and compared with the predictions of existing theoretical models.

EXPERIMENTAL METHOD

The samples for this study were prepared by the dual target rf. sputtering technique. The utilization of two targets of different chemical composition, and the simultaneous co-sputtering of these targets permit the deposition of samples with varying compositions. In this case, arc-melted targets of high purity germanium and niobium were co-sputtered onto heated sapphire substrates. The sputtering module used is similar to that described by Cadieu [3]. The apparatus and procedures used have been described in previous publications [4,5]. The substrates were held in place in a stainless steel holder with a 5 mil thick, photoetched, molybdenum mask, which divided the deposit into 27 strips, with each strip forming an alloy specimen of a particular concentration. Before sputtering, the system was evacuated to 10^{-8}-10^{-7} Torr with the substrate held at the desired temperature of deposition. The system was then valved off from the ion pump and high purity argon gas was introduced at pressures ~ 50 - 100 mTorr. A typical bias voltage of 2.0 Kv and sputtering times of ~ 250 mins. yielded samples ranging in thickness from 2,000 Å to ~ 15,000 Å. A removable shutter permitted the substrates to be shielded during the surface cleaning of the electrodes; and titanium sublimation pumping and a continuous flow of argon was usually used to reduce the background level of gaseous contamination present in the system.

In the latter part of the study, samples were made in a sputtering chamber which enabled the preparation of several (typically four) samples under different sputtering conditions without breaking vacuum, thereby minimizing those changes in sputtering conditions which occur from one pump down to the next.

Resistance measurements were carried out in a special probe utilizing the four probe technique. Temperatures were determined by measurements using germanium thermometers calibrated in the range of 2^0 - 40^0K. The structural properties of the sample were obtained by x-ray diffractometer techniques scanned at room temperature. The nominal compositions of the samples were calculated from a knowledge of the sputtering radiation pattern as has been demonstrated by Johnson [6]. Sample thickness was estimated from the sputtering times and weight loss of targets used during the sputtering process.

RESULTS

Figure 1 shows resistance measurements of a sample spanning the range 95 to 19 a% Ge prepared on a single crystal sapphire substrate maintained at a temperature of ~ 810°C. Curves representing a number of different concentrations are shown. The sample and target geometry is schematically shown along with the Mo mask defining the various sample strips, numbered as indicated. Beginning at the Ge rich end of the sample, strips 1 and 2 (not indicated) show a negative temperature derivative of the resistance indicating a predominant semiconducting behavior down to 4°K with no evidence of superconductivity. In strip 3, we see an abrupt decrease in the resistance curve at 16.0°K. This abrupt drop becomes more pronounced for subsequent strips. We interpret this abrupt drop in resistance as the onset of a superconducting phase. The onset temperature T_{co} is defined as the temperature at which the deviation from the normal state resistance is observed. Strip 5 is the first to show a complete superconducting transition. The temperature at which the sample becomes completely superconducting is denoted by $T_{c\ell}$. As the germanium concentration is reduced (i.e., going to higher strip numbers), T_{co} begins to drop while $T_{c\ell}$ rises to ~ 6°K at strip 13. At this point there is a dramatic change in the superconducting behavior, with strips 14 to 16 showing no evidence of superconductivity down to 2°K. Starting with strip 17, a superconducting transition (incomplete) is observed; T_{co} rises with strip 19 showing values of T_{co} of 18.5°K and $T_{c\ell} \simeq 14$°K. These critical temperatures are shown in the top diagram of Figure 2. The range from T_{co} to $T_{c\ell}$ is indicated.

Two things are to be noted from the T_c diagram: (1) two separate regions of superconducting behavior are observed and (2) these are separated by a region of no apparent superconductivity.

The resistance ratio Γ, defined as $R_{300°K}/R_{20°K}$, is shown in the next diagram. A peak in the Γ curve is observed at strip 12. The critical current I_c, defined as the current required to induce a small voltage drop (~ 2 - 5 µv) across the sample, was measured at 4.2°K in zero field and is shown in Figure 2. It is noted that the I_c curve peaks at strip 13 and that for strips 18 through 27, I_c is greater than 1.0 amp.

X-ray measurements were carried out to identify the various phases in each strip. The phases were established by comparing x-ray lines with data available in the literature, and the phase boundaries were estimated by observing the profile of the peak height of the diffraction line characteristic of a particular phase. The range of the various phases as a function of strip number is shown in Figure 2. The thick bar indicates the position of maximum x-ray line intensity of the respective phases. Line

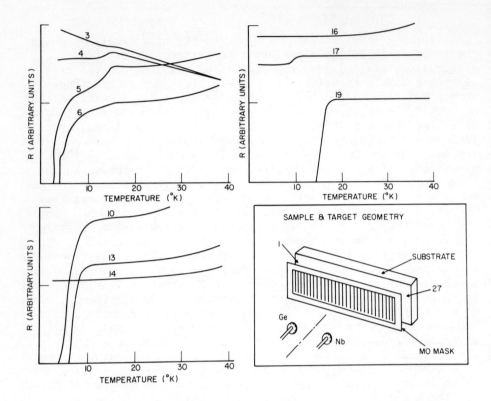

Figure 1: Resistance versus temperature plots of several strips of a Ge-Nb sample prepared at $T_s \sim 810°C$. The numbers of the curves indicate the sample strips of different composition, with low numbers corresponding to high Ge concentration. The fourth figure is a schematic of the two-electrode sputtering, mask and substrate.

width measurements show that for our samples, the average crystallite size > 500 Å for both the Ge and the $NbGe_2$ phases.

X-ray measurements show that: (1) The T_c observed in the low number strips is in the Ge-$NbGe_2$ part of the phase diagram, with superconductivity being depressed below $2°K$ after the Ge-$NbGe_2$ phase boundary is crossed going towards the Nb rich end. (2) The T_c peak in the high number strips is due to A-15 Nb_3Ge and (3) the sharp T_c's $\sim 7°K$ observed for strips 23 - 26 is due to "dirty" α-Nb. In the rest of this paper we will only be concerned with the superconducting region in the Ge-rich end of the phase diagram unless stated otherwise.

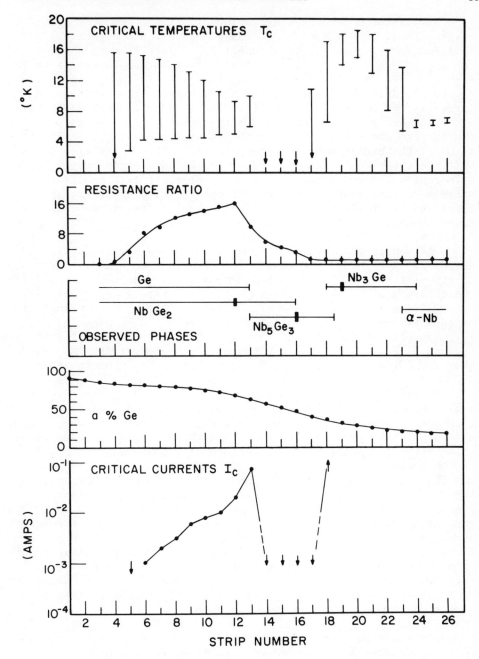

Figure 2: Diagrams from top to bottom: Superconducting transition temperature, resistance ratio $R_{300°K}/R_{20°K}$, phases present, Ge concentration, and critical currents. The abscissa (sample number) is the same for all curves.

Samples were deposited at various substrate temperatures, T_S, ranging from $690°C$ - $870°C$. It was found that the higher T_{co} was obtained for $T_S > 800°C$, the highest $T_{co} = 16.0°K$ being observed for a sample made at $\sim 870°C$. Results of this sample are shown in Figure 3. It is noted that the behavior of this sample is similar to the one described earlier, but has a higher $T_{c\ell} > 10°K$ and a larger $I_c \sim 0.5$ amps. The shift in the T_c profile towards higher strip numbers is due to changes in the sputtering efficiency of the targets and the decrease in the relative sticking coefficient of Ge at the higher substrate temperature. This shift also shows that the observed T_c is independent of the sample geometry. The A-15 region always had an $I_c > 1.0$ amp at $4.2°K$ (1.0 amp is the maximum current that can be put through the electrical leads). The essential features of most of the samples are the same as for those shown in Figures 2 and 3.

One sample made at room temperature showed a maximum $T_{co} \sim 4°K$, having an "amorphous" structure with $\Gamma < 1.0$ (the behavior of "dirty amorphous" Nb). On subsequent annealing at $750°C$, the A-15 phase formed at $\sim 25\%$ Ge with $T_c = 15.6°K$, while a weak peak with $T_c = 2.7°K$ was observed at ~ 67 a% Ge. We note that the behavior of this sample is quite similar to that reported by Hanak et al. [7], who also studied the Nb-Ge system by sputtering onto substrates held at $77°K$ and which was subsequently annealed at $750\ °C$. They show a $T_{co} \sim 6°K$ in the Ge-rich region but do not comment on this.

To eliminate any possible effect of the mask on our results, two samples, one with the mask and one without, were made under identical sputtering conditions in a multi-substrate holder. It was found that the T_c behavior of both these samples was identical, eliminating any possible "mask" effect. In addition, the edges of the sample strips of one of the high T_c samples was ground off and its T_c was remeasured and found to be unchanged. However, the critical current I_c was reduced by $\sim 20\%$, which is almost proportional to the decrease in the width of the sample strip, suggesting that the superconducting phase is homogeneous through the sample strip. To show whether germanium is necessary to observe the high T_c, a sample was prepared under essentially similar conditions as the two dual target experiments using a single alloy target of composition $Nb_{.35}Ge_{.65}$ (bulk T_c measured = $1.85°K$). No T_c above $2°K$ was observed, and x-ray analysis showed almost single phase $NbGe_2$ with no observable Ge.

The anomalous superconductivity in the Ge-rich end of the Nb-Ge phase diagram was observed on a total of 17 samples in 13 separate runs. Considerable variation in the magnitude of the maximum T_c, which ranged from $6.0°K$ to $16°K$, was observed. Although

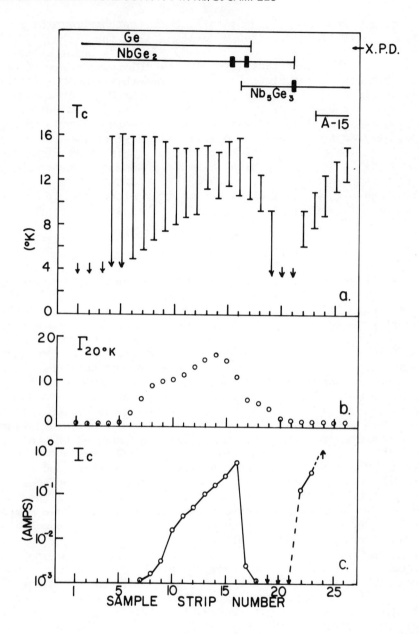

Figure 3: Data for sample deposited at $T_S \sim 870°C$. (a) Superconducting transition temperatures and phases observed in x-ray diffraction (X.P.D.) (b) Resistance ratio $R_{300°K}/R_{20°K}$ and (c) Critical current I_c. The abscissa (sample number) is the same for all curves.

it is not yet known what causes this, there is some correlation of high T_c with large resistance ratio.

Based on the experimental results obtained so far, we arrive at the following conclusions concerning anomalous superconductivity in the Ge rich end of the Ge-Nb system: (1) Both Ge and $NbGe_2$ are present and both appear to be necessary in order to observe high T_c. (2) From the critical current measurements and its obvious similarity to resistance ratio and also to the $NbGe_2$ x-ray line intensity profile, it is concluded that the enhanced superconductivity is associated with the metallic $NbGe_2$ phase. (3) High substrate temperatures (700^0 - 850^0C) are necessary to obtain the high T_c, and (4) the maximum T_c is observed at the concentration showing the highest $T_{c\ell}$.

INTERPRETATION OF DATA

In this section we consider a number of possible explanations for the occurrence of this anomalous superconductivity in Ge-rich Nb/Ge system. (i) The possibility of a third undetected metastable or impurity phase could explain the enhanced T_c. If such a phase is present, then its volume must be less than 10% of the sample volume, since x-ray analysis showed only Ge and $NbGe_2$. From the I_c measurements it is inferred that the amount of this phase would have to be proportional to the quantity of $NbGe_2$ present. Furthermore, since Ge and $NbGe_2$ are necessary, the most likely location would be the grain boundaries between the two phases. If this phase were a metastable phase of $NbGe_2$, it's most likely crystal structure would be related to either $MoSi_2$ or $TiSi_2$, a class of crystal structures in which superconductivity above 2.5^0K has not been observed. However, as has been pointed out by Andrews and Phillips [8] in their classification of semiconductor-metal interfaces, metastable phases can be formed at the interfaces. Phillips has suggested [9] that such an unknown, metastable Nb-Ge phase may be present at the crystallite boundaries. The exact nature of this interfacial phase is however not known.

The only candidate among known compounds that could account for the high T_c must be the Nb compounds. Although free Nb is ruled out, the reactive formation of NbN is possible. Gavaler et al. [10] have extensively studied the reactive sputter formation of NbN and have shown that the high T_c NbN ($\sim 15.5^0K$) depends strongly on the N_2 partial pressures. Since in our system N_2 partial pressure $\sim 10^{-9}$ Torr, which may rise somewhat during the sputtering process, we believe the formation of stochiometric NbN is quite remote under these conditions. In support of this inference, one of the high T_c samples was analyzed for nitrogen by the spark source mass spectrographic method; nitrogen was found at a concentration level of 0.2 ± 0.1 a% [11]. This value is consistent with zero, since an unrelated niobium specimen measured at the same time gave the same reading.

Microprobe analysis [12] of the sample described in Fig. 2 showed that a major contamination was O occurring as NbO. Although NbN as a species was not observed, a weak N signal was detected, the order of ten times less than the O signal. However, quantitative values could not be assigned to the contamination levels observed.

Nb_3Ge is not likely to be formed at the Ge rich end of the phase diagram. However it could conceivably have formed under the mask at the edges of the sample strip, due to the depletion of the germanium. This apparently is not the cause of the observed superconductivity since no change in T_c was observed after the sample strip edges were ground off. Additionally, no difference in T_c behavior was observed for samples made with the mask and without the mask.

(ii) The surface plasmon enhancement of Ngai and Economou [13] is ruled out as a possibility, principally because such an enhancement requires a fairly high carrier density $\sim 10^{22}$ e/cm^3, which is not satisfied for the semiconductor.

(iii) The alternate layer scheme of Cohen and Douglass [14] as a possible mechanism for the T_c enhancement is ruled out because the barrier thickness (in this case the crystallite size) is too large.

(iv) A fourth possibility is that of the so-called "exciton mechanism" for superconductivity, originally proposed by Ginzburg [15]. Such a mechanism was put on a theoretical footing by Allender, Bray and Bardeen [16] (ABB theory) who considered the following configuration: a thin metallic film in contact with a narrow band semiconductor like PbTe or Ge. Many of the experimental facts observed in the $NbGe_2$-Ge system seem to agree with the features of the "exciton mechanism". Large T_c enhancement is observed in the region of metal-semiconductor mixtures of $NbGe_2$-Ge, both being a necessary constituent. Using equation 4.10 of ABB, and using reasonable parameters, it is calculated that a T_c increase from 2.5^0K to 16^0K would require λ_{ex} (the electron-exciton interaction strength) to be $\simeq 0.37$, which is not unreasonable. However, there is a serious problem in using the ABB theory to describe the results. If the sample geometry is assumed to be one of alternating metal and semiconductor crystallites, then from the average particle size estimation, the metal layers ($NbGe_2$) have to be taken to be ~ 500 Å. Since the "exciton" exchange is only possible within the semiconductor region, then to achieve any appreciable enhancement, one has to assume a tunneling depth of the metal electrons of the

~ 150 Å. This distance is an order of magnitude too large, because the metal electron wave function tunneling into the band gap region of the semiconductor will decay by ~ 5-10 Å [16]. Hence the applicability of the ABB theory in explaining the results is questionable.

In conclusion, we have to say that up till now a satisfactory explanation for the T_c enhancement in the NbGe$_2$-Ge system has not been found. Additional experiments are planned to elucidate the various possibilities.

ACKNOWLEDGEMENTS

We wish to acknowledge the kind assistance of R. Gram and C. Manheimer and the experimental suggestions of M.H. Cohen and M. Strongin. Also, conversations with J. Bardeen and R.S. Knox were most helpful.

REFERENCES

1. G.F. Hardy and J.K. Hulm, Phys. Rev. <u>93</u>, 1004 (1954).
2. A.K. Ghosh and D.H. Douglass, Phys. Rev. Lett. (to be published).
3. F.J. Cadieu in <u>Superconductivity in d- and f-Band Metals</u>, ed. by D.H. Douglass (AIP, New York, 1972).
4. G.R. Johnson and D.H. Douglass, J. of Low Temp. Phys. <u>14</u>, 575 (1974).
5. L.A. Pendrys and D.H. Douglass, J. of Low Temp. Phys. <u>23</u>, 367 (1976).
6. G.R. Johnson, <u>Superconductivity in New A-15 Compounds of Niobium</u>, thesis, University of Rochester.
7. J.J. Hanak, J.I. Gittleman, J.P. Pellicane and S. Bozowski, J. Appl. Phys. Vol. <u>41</u>, 4958 (1970).
8. J.M. Andrews and J.C. Phillips, Phys. Rev. Lett. <u>35</u>, 56 (1976).
9. J.C. Phillips, private conversation.
10. J.R. Gavaler, D.W. Deis, J.K. Hulm and C.K. Jones, Appl. Phys. Letts. <u>15</u>, 329 (1969).
11. We are grateful to R. Dynes of Bell Telephone Laboratories for arranging to have this sample analyzed.
12. The microprobe analysis was done at the Brookhaven National Laboratory through the courtesy of Ming Yu and Myron Strongin, for which we are grateful.
13. E.N. Economou and K.L. Ngai, Solid State Comm. <u>17</u>, 1155 (1975).
14. M.H. Cohen and D.H. Douglass, Phys. Rev. Lett. <u>19</u>, 537 (1967).
15. V.L. Ginzburg, Sov. Phys. USP, <u>13</u>, 335 (1971).
16. D. Allender, J. Bray and J. Bardeen, Phys. Rev. <u>B7</u>, 1021 (1973).

QUESTIONS AND COMMENTS

J. Phillips: I would like to elaborate a little on one of those viewgraphs which had a reference to a paper [Phys. Rev. Lett. 35, 56 (1975)] John Andrews and I wrote, which is really just one of many papers that are being written these days on transition metal silicides in particular, which are of interest to the integrated circuit industry. But in any case, the studies of these systems have shown that it is very easy to prepare interfacial phases in this system. My own guess on this system is that barring the existence of Nb_3Ge precipitates what you have here is an interfacial phase of NbGe which would be cubic and would have the NbN structure and would be metastable. Of course, you know that the transition metal non-transition metal compounds with the niobium-nitride structure are made up of the second element from the frist period, not a larger element, but that is presumably because it would just become too soft if you had a larger element. Suppose z is the coordinate normal to the interface and to the left ($z < 0$) we have, say, germanium, which in Nb_xGe_{1-x} means x equals zero and to the right ($z > 0$) we get the $NbGe_2$. So that is x equals one-third and what I think happens is that at the interface there is an overshoot and x equals one half at $z = 0$. And this sort of looks like a Gibbs oscillation. I am very fond of these things because they happen in lots of physical problems; whenever you try to go through something abruptly you tend to overshoot. People never expect that; they always think in terms of smooth (overdamped) interpolations. But in a soft lattice $x(z)$ could oscillate up and this would happen in a region probably as thick as 100 Å. Now you might say how could you have a metastable interfacial phase which is 100 Å thick. The answer is that you are dealing with a soft lattice and the strain fields in a soft lattice propagate long distances and that is well known, not so much to physicists but to metallurgists and they call it the shape memory effect. It is used in a lot of metallurgical applications and in this particular case the fact that this might be NbGe is a very natural thing. For example, many phases are seen in the transition metal silicides systems of the form transition metal two, non-transition metal (T_2M) or transition metal (TM) or TM_2; many of these things are seen and they all occur in the temperature range similar to the substrate temperature you are using, namely around 800°C [see figure 1 next page]. So that is why I would prefer this for instance to NbN since

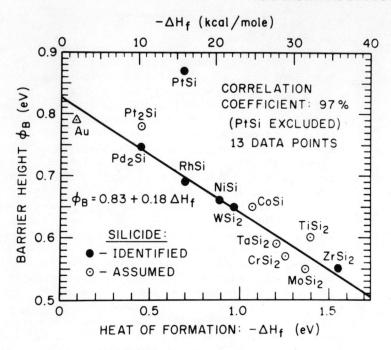

Figure 1: Barrier heights of tm Si – Si interfaces plotted against heats of formation ΔH_f of the tm Si compound.

I think NbN would form basically at any substrate deposition temperature. So if you were making NbN you would not have to operate in this particular range of temperature, or at only Ge-NbGe$_2$ interfaces.

A. Ghosh: Yes. We do observe that T_c drops rapidly once the substrate temperature is below 650°C.

J. Phillips: Yes, that strongly suggests that you are forming an interfacial phase.

M. Cohen: Is is fair to say [assuming the Phillips hypothesis of an interfacial phase] that since the NbGe is being formed along grain boundaries, then once you go down to cryogenic temperatures, the material is under pressure?

A. Ghosh: It is quite possible.

M. Cohen: If you are looking for something metastable which becomes stable under pressure then you have pressure available.

B. Matthias: Well, Jim's [Phillips] intriguing picture unfortunately from a metallurgical point of view does not make any sense at all, because if you want to have NbGe, one to one, you should go to the right (toward the Nb-rich end of the phase diagram) not to the left.

J. Phillips: That is what I just said. Everybody expects that it [the change in concentration] goes up smoothly but it does not happen. It can oscillate.

B. Matthias: I assure you it is not going to be NbGe, one to one.

J. Phillips: Guaranteed?

B. Matthias: Yes.

J. Phillips: All right.

F. Cadieu: Did you have any liquid nitrogen cooled surfaces in your sputtering system?

A. Ghosh: No.

F. Cadieu: When we [at Queen's College] sputter we have a lot of liquid nitrogen cooled surfaces.

A. Ghosh: That is one of the things we are going to do.

F. Cadieu: Our pressure drops when we sputter, does your pressure rise?

A. Ghosh: No, we do it under dynamic conditions---that is, we throttle the argon gas while we sputter.

F. Cadieu: It seems you may have a lot of impurity gases here if under static conditions there is a pressure rise over four hours.

A. Ghosh: No. We do not observe any. On static runs, we did not observe anything on the millitorr scale.

GAP ANISOTROPY AND T_c ENHANCEMENT: GENERAL THEORY, AND CALCULATIONS FOR Nb, USING FERMI SURFACE HARMONICS

W. H. Butler

Oak Ridge National Laboratory[*]

Oak Ridge, Tennessee 37830

and

P. B. Allen [†]

State University of New York

Stony Brook, New York 11794

ABSTRACT

A general theory is given of the anisotropy of the energy gap and the resulting transition temperature (T_c) enhancement of pure superconductors. The frequency dependence of the gap $\Delta(k,\omega)$ is approximated by the two square well form of McMillan, but otherwise an exact algebraic solution of the strong-coupling Eliashberg equations is given, valid for arbitrarily large anisotropy. In the limit of weak anisotropy a simple perturbative formula is also derived. The method of solution relies on the use of expansion functions called Fermi surface harmonics (FSH's) which are velocity polynomials orthonormalized on the Fermi surface. Methods for explicit construction of these functions are described. As an application of these techniques the mass enhancement and gap anisotropy are calculated for Nb, in an approximation which includes all electronic anisotropy but neglects the contribution to the anisotropy which arises from phonons. The rms gap anisotropy in this model is 6% which is not inconsistent with most of the current experimental data. The resulting T_c enhancement is predicted to be 0.7% or .06°K.

[*] Research supported by ERDA under contract with Union Carbide Corp.
[†] Supported in part by NSF Grant No. DMR 73-07578A01

I. INTRODUCTION

Microscopic theory[1,2,3] gives a four-dimensional integral equation for the energy gap $\Delta(k,\omega)$. Experiments such as tunneling[4] are especially good at probing the dependence of Δ on the external frequency ω (given by voltage eV in tunneling), and can also yield information about the variation of Δ as k varies on the Fermi surface. To make theoretical calculations possible it is helpful to approximate the four dimensional integral equation by a lower dimensional equation. The usual procedure is to replace $\Delta(k,\omega)$ by an average value $\Delta_o(\omega)$

$$\Delta_o(\omega) = \frac{1}{\nu} \sum_k \Delta(k,\omega)\delta(\varepsilon_k) \tag{1}$$

$$\nu = \sum_k \delta(\varepsilon_k) \tag{2}$$

which is the isotropic average of $\Delta(k,\omega)$ over the Fermi surface. In these equations k is shorthand for wavevector k, band index n, and spin σ. Thus the density of states at the Fermi surface, ν, is for both spin orientations.

Most of our current understanding of the interactions and thus transition temperature (T_c) of superconductors derives from such an isotropically averaged theory. In particular, McMillan's inversion program[4] and T_c equation[5] both make this approximation. An excellent justification for the isotropic approximation exists in dirty superconductors where impurity scattering causes the actual gap to be isotropically averaged[6]. However, it has long been known that in situations where $\Delta(k,\omega)$ has a significant variation on the Fermi surface, T_c is higher than in the dirty or isotropic limit. The earliest theoretical model of this was the two band model proposed by Suhl et al.[7] and by Moskalenko[8], in which the Fermi surface is presumed to have two unrelated pieces and the gap is allowed to take a different value on each piece. The case of more general anisotropy has been treated in the weak coupling and weak anisotropy approximation by many authors, perhaps first by Pokrovskii[9]. The exactly soluble model of a separable potential was introduced and solved as a function of impurity scattering by Markowitz and Kadanoff[10], Tsuneto[11], and Caroli et al.[12]. Numerous other authors have extended this theory[13-15].

Motivated by the conjecture that anisotropy enhancement might contribute significantly to the T_c of high temperature superconductors, we have reexamined the theory of anisotropic superconductors without making the weak coupling or weak anisotropy approximations. Similar work has been reported by other authors[16,17] but we believe to have found an approach which is particularly simple and general, and well adapted for both analytic arguments and numerical calculations. In this paper we present detailed discussion of a calculation of the gap anisotropy and T_c enhancement of pure niobium. Other problems such as impurity effects[18], critical fields, and normal state transport properties[19] can be profitably discussed in the same language, and such work will be published at a later date. The calculations presented here take full account of the anisotropy which derives from the energy bands and Fermi surface, but neglects anisotropy which arises from the phonon dispersion. We hope to present more complete calculations including phonon-induced anisotropy at a later date.

The new mathematical method which enables us to find simple solutions to anisotropic problems is the technique of Fermi Surface Harmonics[20] (FSH's). The general properties of these functions are discussed in a recent paper[20]. In brief, the idea is that an integral equation such as the Eliashberg equation will become a simple matrix equation if we expand the kernel and the gap in a basis set which is orthonormal <u>on the actual Fermi surface</u> under consideration. Such functions are labeled $F_J(k)$, and must be constructed separately for each metal. The appropriate orthonormality relation is

$$\frac{1}{\nu} \sum_k F_J(k) F_{J'}(k) \delta(\varepsilon_k) = \delta_{JJ'} . \qquad (3)$$

There are five additional properties which it would be desirable for $F_J(k)$ to have:

(i) $F_J(k)$ should be cell periodic in k-space (that is $F_J(k+G) = F_J(k)$ for all reciprocal lattice vectors G). This is accomplished by choosing F_J to depend only on the velocity $v_k = \partial_k \varepsilon_k$, where ε_k is the band-theoretic electron dispersion relation.

(ii) The isotropic case should have a simple representation. This is accomplished by choosing the first function $F_0 = 1$.

(iii) The point symmetry of the crystal should be fully exploited This is done by requiring F_J to transform as basis functions for irreducible representations of the point group.

(iv) The functions $F_J(k)$ should reduce to spherical harmonics $Y_{\ell m}(\hat{k})$ if the Fermi surface is a sphere. This is accomplished by choosing $F_J(k)$ to be polynomials in $(v_{k_x}, v_{k_y}, v_{k_z})$. In the spherical case, v_k is proportional to $\hat{k}$ and $Y_{\ell m}$ is a polynomial in k_x, k_y, k_z.

(v) The gap $\Delta(k,\omega)$ and other properties should have rapidly convergent expansions in FSH's. This is not possible to guarantee <u>a priori</u>, and will be tested in this paper for Nb.

The plan of this paper is as follows. In Section II the Eliashberg equations are written in the FSH representation, and a T_c equation is found in the two-square well approximation. In Section III, perturbation theory is used to simplify the T_c equation in the limit of weak anisotropy. The relation between T_c enhancement and rms gap anisotropy is given for strong-coupled superconductors for the first time. In Section IV the equations of Section II are examined in the two-band approximation, and the T_c equation of Suhl et al[7] is derived with strong-coupling corrections included. In Section V we enumerate the low order Fermi surface harmonics for Nb and explain how they are constructed. In Section VI we show how the quantities entering the equations of Section II can be obtained from energy band calculations. In Section VII the completeness of the FSH's tested for Nb by explicit calculations of the expansion coefficients of various components of the density matrix. Typically 95% completeness has been achieved with an average of six functions on each of the three sheets of Fermi surface. In Section VIII the calculated coefficients of the anisotropic gap and mass enhancement are presented and discussed and compared with experiment.

II. DERIVATION OF T_c EQUATION

Our starting point is the Eliashberg theory[2,3] for the frequency-dependent complex gap $\Delta(\omega)$ as a function of the real external frequency ω. The equations are given by McMillan[5] as eq. (2). Our treatment will be exactly parallel to McMillan's except that we generalize to the case of an anisotropic gap

$$\Delta(k,\omega) = \sum_J \Delta_J(\omega) F_J(k) \qquad (4)$$

and we also make minor modifications to give an improved[21] prefactor, ω_{log}, in the final T_c equation. The anisotropic equations are eqs. (37–39) of ref. 20

$$\phi_J(\omega) = \sum_{J'} [\delta_{JJ'} + \Lambda_{JJ'}(\omega)] \Delta_{J'}(\omega)$$

$$= -\sum_{J'} \int_{-\infty}^{\infty} \frac{d\omega'}{\omega'} \int_0^\infty d\Omega \alpha_{JJ'}{}^2 F(\Omega) \operatorname{Re}[\Delta_{J'}(\omega')] K(\omega,\omega',\Omega)$$

$$- \sum_J \mu_{JJ'} \int_0^{\omega_p} \frac{\ell d\omega'}{\omega'} \operatorname{Re}[\Delta_{J'}(\omega')] \tanh \frac{\beta\omega'}{2} \ . \qquad (5)$$

GAP ANISOTROPY AND T_c ENHANCEMENT

The pair field ϕ is equal to $Z\Delta$ in the isotropic limit, where Z is $1 + \Lambda_{oo}(\omega)$ and is called the renormalization function. The anisotropic theory gives a slightly more complicated relation, and we introduce a tensor $\Lambda_{JJ'}(\omega)$ defined by

$$\omega \Lambda_{JJ'}(\omega) = - \sum_{J''} C_{JJ'J''} \int_{-\infty}^{\infty} d\omega' \int_0^{\infty} d\Omega \, \alpha_{J''0}^2 F(\Omega) K(\omega,\omega',\Omega) \qquad (6)$$

where $C_{JJ'J''}$ is a Clebsh-Gordan coefficient defined in ref. 20. The interaction parameters $\alpha_{JJ'}^2 F(\Omega)$ and $\mu_{JJ'}$ are matrix elements in the FSH basis of the phonon and electron interactions. An explicit formula for $\alpha_{JJ'}^2 F(\Omega)$ is given in eq. (33) of ref. 20. For our purposes, the most important parameters are the integrated strength $\lambda_{JJ'}$,

$$\lambda_{JJ'} = 2 \int_0^{\infty} \frac{d\Omega}{\Omega} \alpha_{JJ'}^2 F(\Omega)$$

$$= \frac{1}{\nu} \sum_{kk'i} (|M_{kk'}^i|^2/\omega_{k-k'}^i) F_J(k) F_{J'}(k') \delta(\varepsilon_k) \delta(\varepsilon_{k'}) \qquad (7)$$

and the zero frequency, zero temperature renormalization parameter $\Lambda_{JJ'}$

$$\Lambda_{JJ'} = \sum_{J''} C_{JJ'J''} \lambda_{J''0}$$

$$= \frac{1}{\nu} \sum_{kk'i} (|M_{kk'}^i|^2/\omega_{k-k'}^i) F_J(k) F_{J'}(k) \delta(\varepsilon_k) \delta(\varepsilon_{k'}) \qquad (8)$$

In these equations $M_{kk'}^i$ is the electron-phonon matrix element, $\omega_{k-k'}^i$ is the phonon frequency, and i is the phonon polarization index. The difference between $\lambda_{JJ'}$ and $\Lambda_{JJ'}$ is that both FSH's have the same argument, k, in eq. (8), whereas in (7) the arguments are different. The Coulomb coupling matrix $\mu_{JJ'}$ is defined as

$$\mu_{JJ'} = \frac{1}{\nu} \sum_{kk'} V_{kk'} F_J(k) F_{J'}(k') \delta(\varepsilon_k) \delta(\varepsilon_{k'}) \qquad (9)$$

where $V_{kk'}$ is the matrix element of the (screened and vertex-corrected) Coulomb potential which scatters a spin up electron from k to k' and a spin down electron from $-k$ to $-k'$. The plasma frequency $\omega_{p\ell}$ is the natural cutoff of the Coulomb interaction.

The equations of an isotropic or dirty superconductor are obtained from (5) and (6) by assuming that $\alpha^2_{JJ'}F(\Omega)$ is diagonal in JJ'. This guarantees that $\Lambda_{JJ'}(\omega)$ is diagonal because of the relation[20] $C_{JJ'0} = \delta_{JJ'}$. In this approximation equation (5) becomes a set of uncoupled equations, one for each value of J, with the $J=0$ equation being eq. (2) of McMillan's paper[5]. The usual parameters λ, μ and $\alpha^2 F$ are identified λ_{oo}, μ_{oo}, $\alpha_{oo}^2 F$. The effect of impurities is to introduce a strong repulsive term which will drastically reduce or eliminate all coupling constants except for the $J=0$, $J'=0$ component where the impurity term is absent in accordance with Anderson's theorem[6].

In an attempt to solve these equations, we now follow McMillan[5] in making the two-square-well approximation,

$$\Delta_J(\omega) = \begin{cases} \Delta_J(0) & \text{if } |\omega| < \omega_{ph} \\ \Delta_J(\infty) & \text{if } \omega_{ph} < |\omega| < \omega_{p\ell} \\ 0 & \text{otherwise} \end{cases}$$

$$\Lambda_{JJ'}(\omega) = \begin{cases} \Lambda_{JJ'} & \text{if } |\omega| < \omega_{ph} \\ 0 & \text{otherwise} \end{cases} \quad (10)$$

where ω_{ph} is the maximum phonon energy and $\Delta_J(0)$, $\Delta_J(\infty)$ are assumed real. The kernel $K(\omega,\omega',\Omega)$ of eqs. (5) and (6) is given by

$$K(\omega,\omega',\Omega) = \frac{f(\omega') + N(\Omega)}{\omega-\omega' + \Omega} + \frac{f(-\omega') + N(\Omega)}{\omega - \omega' - \Omega}. \quad (11)$$

Observing that $|\Omega|$ is confined to frequencies less than ω_{ph}, we make the approximation that K is zero unless both $|\omega|$ and $|\omega'|$ are less than ω_{ph}. Then using these approximations, we evaluate eq. (5) at $\omega=0$ and $\omega=\infty$ (actually $\omega \lesssim \omega_{p\ell}$ is meant rather than ∞) to get self-consistent equations for $\Delta_J(0)$ and $\Delta_J(\infty)$,

$$\sum_{J'} (\delta_{JJ'} + \Lambda_{JJ'})\Delta_{J'}(0) =$$

$$\sum_{J'} (A_{JJ'} - \mu_{JJ'} \ln(1.13\beta\omega_{ph}))\Delta_{J'}(0)$$

$$- \sum_{J'} \mu_{JJ'} \ln(\omega_{p\ell}/\omega_{ph})\Delta_{J'}(\infty) \qquad (12)$$

$$\Delta_J(\infty) = -\sum_{J'} \mu_{JJ'} \ln(1.13\beta\omega_{ph})\Delta_{J'}(0)$$

$$- \sum_{J'} \mu_{JJ'} \ln(\omega_{p\ell}/\omega_{ph})\Delta_{J'}(\infty) . \qquad (13)$$

Here $A_{JJ'}$ is defined as

$$A_{JJ'} = \int_0^\infty d\Omega \alpha^2_{JJ'} F(\Omega) \int_{-\omega_{ph}}^{\omega_{ph}} d\omega' \, K(0,\omega',\Omega)/\omega' \qquad (14)$$

$$\approx 2 \int_0^\infty \frac{d\Omega}{\Omega} \alpha^2_{JJ'} F(\Omega) \ln(1.13\beta\Omega) \qquad (15)$$

$$\equiv \lambda_{JJ'} \ln 1.13\beta\omega_{\log} + \delta\lambda_{JJ'} \qquad (16)$$

and the logarithmic average frequency, $\omega_{\log}$, is defined as in ref. 21

$$\ln(\omega_{\log}) \equiv \frac{2}{\lambda} \int_0^\infty \frac{d\Omega}{\Omega} \alpha^2 F(\Omega) \ln\Omega \qquad (17)$$

$$\delta\lambda_{JJ'} \equiv 2 \int_0^\infty \frac{d\Omega}{\Omega} \alpha_{JJ'}^2 F(\Omega) \ln(\Omega/\omega_{\log}) \qquad (18)$$

The isotropic component $\delta\lambda_{oo}$ vanishes by the definition (17). In going from (14) to (15) we have extracted the leading term in weak coupling (i.e. treating $\beta\Omega$ as large). It has been found that treating corrections to (15) of higher order in $(\beta\Omega)^{-1}$ leads to incorrect conclusions. The problem is that the two square well model fails if the coupling is very strong[22]; T_c has been shown[21] both experimentally and theoretically to increase more strongly as λ increases than is allowed in this model. However, it has also been shown[21] that this model is highly accurate for materials with $\lambda < 1.5$. Therefore in the present paper, we work in the two square-well approximation.

Equations (12), (13), and (16) are now easily solved to find T_c. First we solve (13) for $\Delta(\infty)$,

$$\Delta_J(\infty) = - \sum_{J'} \ln(1.13\beta\omega_{ph}) \mu^*_{JJ'} \Delta_{J'}(0) \qquad (19)$$

where the matrix $\underset{\sim}{\mu^*}$ is related to $\underset{\sim}{\mu}$ by

$$\underset{\sim}{\mu^*} = (\underset{\sim}{1} + \underset{\sim}{\mu} \ln(\omega_{p\ell}/\omega_{ph}))^{-1} \underset{\sim}{\mu}. \qquad (20)$$

Then we use (18) to eliminate $\Delta(\infty)$ from (12), giving

$$\rho(\underset{\sim}{1} + \underset{\sim}{\lambda} - \underset{\sim}{\delta\lambda} + \underset{\sim}{\delta\mu^*}) \cdot \underset{\sim}{\Delta}(0) = (\underset{\sim}{\lambda} - \underset{\sim}{\mu^*}) \cdot \underset{\sim}{\Delta}(0) \qquad (21)$$

where we have defined

$$\underset{\sim}{\delta\mu^*} = \ln(\omega_{ph}/\omega_{log}) \underset{\sim}{\mu^*} \qquad (22)$$

$$1/\rho = \ln(1.13\beta\omega_{log}) \qquad (23)$$

Equation (21) is the final equation which determines the anisotropic zero frequency gap $\Delta(k,0)$ at $T = T_c$. This equation has non-trivial solutions $\Delta \neq 0$ only for discrete eigenvalues ρ. The actual transition temperature corresponds to the maximum eigenvalue ρ which we denote λ_{eff}.

$$T_c = 1.13\omega_{log} \, e^{-1/\lambda_{eff}} \qquad (24)$$

$$\lambda_{eff} = \text{Max. Eigenvalue of } (\underset{\sim}{1} + \underset{\sim}{\lambda} - \underset{\sim}{\delta\lambda} + \underset{\sim}{\delta\mu^*})^{-1} (\underset{\sim}{\lambda} - \underset{\sim}{\mu^*}) \qquad (25)$$

Equations (24) and (25) are the T_c equation we have been seeking. They take full account of strong coupling corrections and are valid for arbitrarily large anisotropy. However, the use of the two square-well model limits their validity to materials with $\lambda \lesssim 1.5$. Furthermore, detailed numerical solutions[21] in the isotropic approximation suggest that the prefactor 1.13 should be replaced by 1/1.20.

In terms of practical applications, it may appear disturbing that eq. (25) involves a large number of microscopic coupling constants, namely five matrices λ, μ^*, Λ, $\delta\lambda$, and $\delta\mu^*$. However, things are not as complicated as they may seem. For example, the matrix $\delta\mu^*$ is proportional to μ^*. The upper left hand element, $\mu^*_{oo} = \mu^*$ is of order 0.1, and the other elements in all likelihood are considerably smaller (and therefore negligible). The matrix $\delta\lambda$ has a vanishing upper left element. The other elements are non-vanishing only to the extent that the shape of $\alpha_{JJ'}^2 F(\Omega)$ differs from the shape of $\alpha^2 F(\Omega)$. Each element of $\delta\lambda$ is small compared to the corresponding element of λ and therefore, in all likelihood, also negligible. The Matrix Λ has elements which can all be expressed in terms of the elements of the matrix λ and the Clebsh-Gordan coefficients. These can be calculated if a reliable Fermi surface is known, and most of the important Clebsh-Gordan coefficients are either 0 or 1. Thus to a good approximation we seek the maximum eigenvalue of $(1 + \Lambda)^{-1}(\lambda - \mu^* \delta_{J0}\delta_{J0'})$ which involves only one matrix of coupling constants $\lambda_{JJ'}$. Group theory has been used in constructing the functions $F_J(\hat{k})$. This has the helpful consequence that all five matrices are block-diagonal, i.e. the only non-vanishing off-diagonal elements are those connecting functions which belong to the same row of the same irreducible representation. In cubic symmetry, there are ten irreducible representations, and a total of twenty rows. Thus the matrix $(1 + \Lambda - \delta\lambda + \delta\mu^*)^{-1}(\lambda - \mu^*)$ consists of twenty submatrices, i.e. a Γ_1 submatrix which determines the s-wave transition temperature, three (identical appearing) Γ_{15} submatrices corresponding to the three rows of the Γ_{15} representation which determine the p-wave transition temperature, and so forth. Ordinarily we expect the s-wave transition temperature to be highest, although it has been suggested[23] that because μ^* may be small in the p-wave channel, some metals with small values of λ may have p-wave instabilities at very low temperatures. At any rate, the problem of determining the s-wave transition temperature involves examining only the Γ_1 submatrix, which in favorable cases can probably be accurately represented with a fairly small number of terms.

Finally it is important to note that the matrices λ and Λ describe material properties relevant in other problems besides superconductivity. For example, the low temperature normal state specific heat is enhanced by $1 + \lambda_{oo} = 1 + \lambda$. The high temperature electrical resistance is determined by the p-wave submatrix[20] of $\Lambda-\lambda$. Thus the zoo of new coefficients introduced here is by no means so large and arbitrary as may at first sight appear.

III. THE LIMIT OF WEAK ANISOTROPY

In the previous section we have derived equations for the anisotropic gap parameter and the transition temperature. We repeat here a simplified version of these equation

$$\lambda_{eff}(\underline{1} + \underline{\Lambda}) \cdot \underline{\Delta} = (\underline{\lambda} - \underline{\mu}^*) \cdot \underline{\Delta} \tag{26}$$

$$T_c = (\omega_{log}/1.2) \exp(-1/\lambda_{eff}). \tag{27}$$

We have neglected the small matrices $\delta\underline{\lambda}$, $\delta\underline{\mu}^*$, and introduced an empirical adjustment of the prefactor. These equations have the same form in any orthonormal basis set. In particular, Entel and Peter[17] have recently presented these equations, working in the "locally constant" representation - that is, their orthonormal functions are constants in isolated (i.e. non-overlapping) regions and zero elsewhere. Clearly such functions can be normalized so as to satisfy the condition (1). However, the FSH basis set has several advantages over that of Entel and Peter, one being that the equations take a simple form if the anisotropy is weak. Specifically, the matrices $\underline{\Lambda}$, $\underline{\lambda}$, and $\underline{\mu}^*$ become nearly diagonal, and the maximum eigenvalue is closely approxiamted by the upper left hand term, i.e.

$$\lambda_{eff}(iso)\ (1 + \lambda) = \lambda - \mu^*$$

$$T_c(iso) = (\omega_{log}/1.2) \exp(-(1 + \lambda)/(\lambda - \mu^*)). \tag{28}$$

Apart from small adjustments, this is the familiar McMillan equation[5]. If the anisotropy is not too strong, it is a simple matter to find corrections to the isotropic equation by perturbation theory, and also to find the gap anisotropy perturbatively. These tasks would be harder in alternative basis sets such as that of Entel and Peter.

Equation (26) allows us to calculate the gap anisotropy only at $T = T_c$ where Δ is vanishingly small. At lower temperatures where Δ is no longer negligibly small, non-linear terms occur in the original integral equation which determine the magnitude of Δ. The non-linearity consists of putting the J' component of $Re[\Delta/\sqrt{\omega'^2-\Delta^2}]$ in place of $Re[\Delta_{J'}]/\omega'$ in eq. (5). If we make the approximation

$$[\Delta/\sqrt{\omega'^2-\Delta^2}]_{J'} \approx \Delta_{J'}/\sqrt{\omega'^2-\Delta_o^2} \tag{29}$$

GAP ANISOTROPY AND T_c ENHANCEMENT

then the gap equation is soluble by methods similar to those we used to find T_c. Within this approximation we find that the relative anisotropy (i.e. Δ_J/Δ_o) is independent of temperature. Thus it is sufficient to solve the linear equation (21) or (26) to learn the relative anisotropy at all temperatures. This approximation was first described by Anderson and Morel[24] and Pokrovskii[9], and has been tested by Leavens and Carbotte[25] who found that the corrections were small, using an iterative procedure. In the weak anisotropy approximation, we seek solutions of the form

$$\Delta(k,o) = \Delta_o [1 + \sum_{J \neq 0} (\Delta_J/\Delta_o) F_J(k)] \tag{30}$$

$$\lambda_{eff} = (\lambda - \mu^*)/(1 + \lambda) + \delta\lambda_{eff} \tag{31}$$

where Δ_J and $\delta\lambda_{eff}$ are assumed small compared with Δ_o or $\lambda_{eff}(iso)$ respectively. In the weak coupling limit where $\underset{\sim}{\lambda}$ is neglected relative to $\underset{\sim}{1}$, eq. (26) is a Hermitean eigenvalue problem, and the Rayleigh-Ritz perturbation series for ground state wavefunction and energy in quantum mechanics. The results in lowest order are:

$$\left(\frac{\Delta_J}{\Delta_o}\right) = \frac{(\underset{\sim}{\lambda}-\underset{\sim}{\mu}^*)_{0J}}{(\underset{\sim}{\lambda}-\underset{\sim}{\mu}^*) - (\underset{\sim}{\lambda}-\underset{\sim}{\mu}^*)_{JJ}} \approx \frac{\lambda_{0J}}{\lambda-\mu^*}$$

(J $\neq$ 0, weak coupling)

$$\delta\lambda_{eff} = \sum_{J\neq 0} \frac{(\underset{\sim}{\lambda}-\underset{\sim}{\mu}^*)^2_{0J}}{(\underset{\sim}{\lambda}-\underset{\sim}{\mu}^*)_{oo} - (\underset{\sim}{\lambda}-\underset{\sim}{\mu}^*)_{JJ}} \approx \sum_{J\neq 0} \frac{\lambda_{0J}^2}{\lambda-\mu^*} \tag{32}$$

(weak coupling)

The approximate forms on the right invoke the assumption that $\underset{\sim}{\mu}^*$ has only an upper left term μ^* which is non-zero. The weak anisotropy approximation assumes only that off-diagonal elements of $\underset{\sim}{\lambda}$ are small. It is probably reasonable to assume that the diagonal elements for $J\neq 0$ are also small, and thus to omit λ_{JJ} from the denominators as a higher order correction. This approximation has also been introduced on the right in eq. (32). We introduce the root-mean-square (rms) fractional anisotropy α, defined as

$$\alpha \equiv \sqrt{\langle\Delta^2\rangle - \langle\Delta\rangle^2}\,/\langle\Delta\rangle$$

$$= [\sum_{J\neq 0} (\Delta_J/\Delta_0)^2]^{\frac{1}{2}} \tag{33}$$

where the last expression is a general formula derived from eq. (1), and the angular brackets $\langle\;\rangle$ are used to denote a Fermi surface average,

$$\langle P_k\rangle \equiv \nu^{-1} \sum_k P_k \delta(\varepsilon_k). \tag{34}$$

In the weak coupling approximation (where $\lambda_{eff} = \lambda-\mu^*$) we find from (32) and (33)

$$\alpha^2 = \delta\lambda_{eff}/\lambda_{eff} = \lambda_{eff}\,\delta T_c/T_c(iso) \tag{35}$$

This result has been derived many times, usually within the factorizable interaction model[10-12], and is derived here for weak coupling and weak anisotropy (but with no restriction as to the form of the anisotropic interactions). This result shows that the fractional T_c enhancement due to anisotropy scales as the square of the rms gap anisotropy. Thus a 10% gap anisotropy (as is commonly seen by tunneling in s-p metals) gives a T_c enhancement of order 1%. The actual size of the fractional T_c enhancement is larger for low T_c materials where λ_{eff} is small. For example, the alloy data of Farrell et al[26] for Zn, a weak coupling material ($\lambda_{eff} \sim 0.20$), indicate a T_c enhancement of 20%; eq. (35) then implies a 20% rms gap anisotropy.

Let us now derive the strong coupling corrections to eqs. (32) and (35). If we made the assumption that the renormalization matrix $(\underset{\sim}{1} + \underset{\sim}{\Lambda})$ were diagonal and equal to $(1 + \lambda)\underset{\sim}{1}$ (as it is in the isotropic limit) then the derivation would be simple. Each term $\lambda_{JJ'}$ and $\mu^*_{JJ'}$ of eq. (32) would simply be renormalized (i.e. reduced) by a factor $(1 + \lambda)$ leaving the gap anisotropy Δ_J/Δ_0 unaffected, and reducing $\delta\lambda_{eff}$ by $(1 + \lambda)$. The form of eq. (35) would be unaltered (because the formula for λ_{eff} is altered to $(\lambda-\mu^*)/(1+\lambda)$. However, not all of these conclusions are correct, because the renormalization $\underset{\sim}{\Lambda}$ has anisotropic corrections which need to be handled in the same order of approximation as the anisotropic corrections to $\underset{\sim}{\lambda}$. The derivations are somewhat tricky and are given below, but the final results are gratifyingly simple:

$$\frac{\Delta_J}{\Delta_0} = (\frac{1+\mu^*}{1+\lambda}) \frac{\lambda_{0J} - \mu^*_{0J}(1+\lambda)/(1+\mu^*)}{(\lambda-\mu^*) - (\lambda-\mu^*)_{JJ}} \approx (\frac{1+\mu^*}{1+\lambda}) \frac{\lambda_{0J}}{\lambda-\mu^*}$$

($J \neq 0$, strong coupling)

$$\delta\lambda_{eff} = \frac{(1+\mu^*)^2}{(1+\lambda)^3} \sum_{J \neq 0} \frac{[\lambda_{0J} - \mu^*_{0J}(1+\lambda)/(1+\mu^*)]^2}{(\lambda-\mu^*) - (\lambda-\mu^*)_{JJ}}$$

$$\approx \frac{(1+\mu^*)^2}{(1+\lambda)^3} \sum_{J \neq 0} \frac{\lambda_{0J}^2}{\lambda-\mu^*} \approx \frac{(1+\mu^*)^2}{(1+\lambda)^3(\lambda-\mu^*)} [<\lambda_k^2> - \bar{\lambda}^2] \qquad (36)$$

(strong coupling, weak anisotropy)

The last form of eq. (36) is written in a form which can be calculated in a conventional way without the use of Fermi surface harmonics. The function λ_k is the mass enhancement parameter at point k on the Fermi surface

$$\lambda_k = \sum_{k',i} (|M^i_{kk'}|^2/\omega^i_{k-k'}) \, \delta(\epsilon_{k'})$$

$$= \sum_J \lambda_{0J} F_J(k) \qquad (36a)$$

$$<\lambda_k^2> = \frac{1}{\nu} \sum_k \lambda_k^2 \, \delta(\epsilon_k) = \sum_J \lambda_{0J}^2 . \qquad (36b)$$

This function is one which in certain metals can be extracted from normal state data such as cyclotron resonance, and has been calculated theoretically in non-transition elements by many authors. We make use of the last form of eq. (36) at the end of our paper as an independent check on the accuracy of our expansions.

The effect of anisotropy in Λ is to reduce the gap anisotropy Δ_J/Δ_0 (for fixed coupling constants λ_{0J}) by a factor $(1+\lambda)/(1+\mu^*)$, and to reduce $\delta\lambda_{eff}$ by the square of this quantity. The third $(1+\lambda)^{-1}$ term in eq. (36) for $\delta\lambda_{eff}$ is the strong-coupling correction from the isotropic part of the renormalization Λ. Remarkably, all the strong coupling corrections cancel in eq. (35) leaving this equation correct provided λ_{eff} is written as $(\lambda-\mu^*)/(1+\lambda)$, the coupling constant of an isotropic superconductor. Thus we can rewrite eq. (35),

$$\delta T_c/T_c(iso) = \alpha^2 \ln(\omega_{log}/1.2T_c(iso)). \qquad (37)$$

This formula has been derived allowing for strong coupling provided $\lambda \lesssim 1.5$ and arbitrary from of (weak) anisotropy. A few additional comments should be made about these perturbative results. We see that anisotropy always increases T_c. This was well known for the weak coupling case ignoring Coulomb repulsion, but is less obvious in the general case. In particular, suppose the electron-phonon interaction is isotropic but the Coulomb interaction is not. Then the gap will be locally depressed at values of k where the Coulomb coupling strength is strong and enhanced where it is weak. This gives rise to a T_c enhancement that is just as strong as if the signs were reversed and it had been the phonon interaction which was anisotropic. If Coulomb and phonon interactions have the same type of anisotropy, cancellations occur which weaken the gap anisotropy and the resulting T_c enhancement. The role of Coulomb and phonon interactions is not completely parallel in this problem (i.e. note the factor $(1+\lambda)/(1+\mu^*)$ which strengthens the role of Coulomb anisotropy in eqs. (36). The reason for the lack of parallelism is that Coulomb effects do not occur in the renormalization matrix $\underline{1} + \underline{\Lambda}$. It is assumed that Coulomb alterations of the normal metal spectrum are included in the band structure ε_k.

The rest of this section gives an outline of the somewhat tedious derivation of eqs. (36). We seek the maximum eigenvalue and the corresponding <u>right</u> eigenvector of the matrix $(1+\underline{\Lambda})^{-1}(\underline{\lambda}-\underline{\mu}^*)$. Although the matrices involved are all individually symmetric, the product of two symmetric matrices is no longer symmetric, and this gives rise to an important alteration in the algebra. The matrix $\underline{\Lambda}$ becomes $\lambda\underline{1}$ in the isotropic limit. Thus it is convenient to make some changes in notation

$$\underline{1} + \underline{\Lambda} \equiv (1 + \lambda)(\underline{1} + \underline{\gamma}) \qquad (37)$$

$$(\underline{1} + \underline{\gamma})^{-1}(\underline{\lambda} - \underline{\mu}^*) \equiv \underline{B} = \underline{B}^{(0)} + \underline{B}^{(1)} + \underline{B}^{(2)} + \ldots \qquad (38)$$

We now seek the maximum eigenvalue and corresponding right eigenvector of the matrix B which differs from $(1+\underline{\Lambda})^{-1}(\underline{\lambda}-\underline{\mu}^*)$ by the constant factor $(1+\lambda)^{-1}$. Factoring out $(1+\lambda)$ does not affect eigenvectors of $\underline{B}$, but we must remember when we are all through to divide the eigenvalue by $(1+\lambda)$. The matrix elements of $\underline{\gamma}$ and the off-diagonal elements of $\underline{\lambda}$ and $\underline{\mu}^*$ all vanish in the isotropic limit and are therefore the small parameters of our perturbation theory. The matrix $\underline{B}$ is then expanded in powers of these small terms. The leading term, $\underline{B}^{(0)}$, is diagonal, and the first and second order terms are denoted $\underline{B}^{(1)}$ and $\underline{B}^{(2)}$. The formulas of the Rayleigh-Ritz perturbation series are

$$\Delta_J/\Delta_0 = B_{JO}^{(1)}/(B_{oo}^{(0)} - B_{JJ}^{(0)}) \quad (J \neq 0) \qquad (39)$$

$$\delta\lambda_{eff} = (1+\lambda)^{-1} \{B_{oo}^{(1)} + B_{oo}^{(2)} + \sum_{J \neq 0} B_{OJ}^{(1)} B_{JO}^{(1)}/(B_{oo}^{(0)} - B_{JJ}^{(0)})\} \qquad (40)$$

where it is important to remember that B_{OJ} and B_{JO} are not identical. We can construct the matrix $(\underline{1}+\underline{\chi})^{-1}$ perturbatively and find the necessary formulas for the elements of $\underline{B}$,

$$\underline{B} = (\underline{1} - \underline{\chi} + \underline{\chi}^2 - \ldots)(\underline{\lambda} - \underline{\mu}^*)$$

$$B_{JJ}^{(0)} = (\underline{\lambda} - \underline{\mu}^*)_{JJ}$$

$$B_{oo}^{(1)} = 0$$

$$B_{oo}^{(2)} = -\sum_{J \neq 0} [\gamma_{OJ}(\underline{\lambda}-\underline{\mu}^*)_{JO} - \gamma_{OJ}\gamma_{JO}(\underline{\lambda}-\underline{\mu}^*)_{oo}]$$

$$B_{OJ}^{(1)} = (\underline{\lambda} - \underline{\mu}^*)_{OJ} - \gamma_{OJ}(\lambda - \mu^*)_{JJ}$$

$$B_{JO}^{(1)} = (\lambda - \mu^*)_{JO} - \gamma_{JO}(\lambda - \mu^*)_{oo} \qquad (41)$$

Finally, we need the formula for $\gamma_{JO} = \gamma_{OJ}$,

$$\gamma_{OJ} = (1+\lambda)^{-1}\Lambda_{OJ} = (1+\lambda)^{-1}\sum_{J'} C_{OJJ'}\lambda_{J'o}$$

$$= (1+\lambda)^{-1}\lambda_{JO} \qquad (42)$$

where we have used the definition (8) of $\Lambda_{JJ'}$, and the fact that $C_{OJJ'} = \delta_{JJ'}$. Combining eqs. (39), (41), (42), we easily get the formula (36) for the eigenvector Δ_J/Δ_0. Deriving the formula (36) for $\delta\lambda_{eff}$ is similarly straightforward but involves a delicate cancellation of the terms in eq. (40).

IV. THE TWO BAND MODEL

In this section we specialize the results of section II to the case of a two band model[7,8]. This model is exact if the interaction $V_{ph}(k,k') = |M_{kk'}|^2/\hbar\omega_{k-k'}$ has the form

$$V_{ph}(k,k') = \sum_{\alpha,\beta} V_{\alpha\beta}\, \delta_{k\alpha}\, \delta_{k'\beta} \tag{43}$$

where $V_{\alpha\beta}$ are arbitrary positive numbers, α,β run over two sheets of Fermi surface (which we label a and b), and the delta's δ_{ka}, δ_{kb} are 1 if k is on sheet a, b respectively, and zero otherwise. The reason for exhibiting this case explicitly are: first, it is the simplest example of the application of our equations to a situation of some intrinsic interest; and second, we can anticipate in a simple context some of the complexities that will arise when we apply these methods to an actual Fermi surface, that of Nb, which has many different sheets.

Within the two band model, the only band structure information required besides the interaction parameters $V_{\alpha\beta}$ is the partial densities of states ν_α

$$\nu_\alpha = \sum_k \delta_{k\alpha}\, \delta(\varepsilon_k) \tag{44}$$

where the total density of states ν equals $\nu_a + \nu_b$. As described in ref. 20, one way to construct Fermi surface harmonics is to construct two different sets, F_{Ja} and F_{Jb}, each of which vanishes on the opposite sheet. Thus F_{Ja} is automatically orthogonal to $F_{J'b}$ for all J and J'. With the two-band potential (43), it is only necessary to keep the constant function on each sheet, and we denote them F_a and F_b. Using the orthonormality relation (3), we get explicit formulas for the normalized FSH's

$$F_a = \sqrt{\nu/\nu_a}\, \delta_{ka} \equiv |a\rangle$$

$$F_b = \sqrt{\nu/\nu_b}\, \delta_{kb} \equiv |b\rangle \tag{45}$$

The λ matrix has elements defined in eq. (7) which can be written using an inner product notation

$$\lambda_{\alpha\beta} = \nu\langle\alpha|V_{kk'}|\beta\rangle = \sqrt{\nu_\alpha \nu_\beta}\, V_{\alpha\beta}\,. \tag{46}$$

GAP ANISOTROPY AND T_c ENHANCEMENT

In the weak-coupling approximation with Coulomb coupling ignored (which is the model solved by Suhl et al[7]), T_c is determined by eq. (27) with λ_{eff} the maximum eigenvalue of the 2×2 matrix $\underset{\sim}{\lambda}$. This gives the answer found by Suhl et al[7]. We now examine the strong-coupling corrections.

The functions (45) differ slightly from the choice of FSH's described in section I and implicitly used up to now. The difference is that previously we have assumed that there was a function $F_0=1$, whereas with two distinct sheets it is more natural to use the choice (45). However, it is still possible to revert back to our former convention by making a unitary transformation, and there are certain advantages to doing so. The advantages are that the previous choice correctly simplifies to a 1×1 problem in the isotropic limit, whereas the present choice remains 2×2. (The isotropic limit here is obtained if $V_{\alpha\beta} = V$.) In the case of weak anisotropy it is much easier to do perturbation theory in the former picture. However, the present picture is more natural for doing calculations of the coefficients. These two pictures have been given the names[20] symmetric (S) (for the previous one) and disjoint (D) representation. The transformation equations

$$\begin{pmatrix} F_o \\ F_u \end{pmatrix} = \underset{\sim}{U} \begin{pmatrix} F_a \\ F_b \end{pmatrix}$$

$$\underset{\sim}{U} = \begin{pmatrix} \cos\theta & \sin\theta \\ -\sin\theta & \cos\theta \end{pmatrix}$$

$$\theta = \tan^{-1}\sqrt{\nu_b/\nu_a}, \quad \cos\theta = \sqrt{\nu_a/\nu}, \quad \sin\theta = \sqrt{\nu_b/\nu}$$

$$(\underset{\sim}{\lambda})_{sym} = \underset{\sim}{U} (\underset{\sim}{\lambda})_{disj} \underset{\sim}{U}^{-1} \tag{47}$$

define a unitary transformation from the disjoint to the symmetric representation. Note that F_o is 1 as required, and F_u orthogonal to F_o.

Another advantage of the symmetric representation is that construction of the matrix $\underset{\sim}{\Lambda}$ is conceptually somewhat simpler. A rigorous definition of $\Lambda_{JJ'}$, (in any orthonormal basis set) is eq. (8) which can be written

$$\Lambda_{JJ'} = \nu^{-1} \sum_{kk'} F_J(k) F_{J'}(k) V_{ph}(k,k') \delta(\varepsilon_k) \delta(\varepsilon_{k'}) \tag{47}$$

where F_J and $F_{J'}$ both have the argument k. Now V_{ph} has by definition the expansion

$$V_{ph}(k,k') = \nu^{-1} \sum_{JJ'} \lambda_{JJ'} F_J(k) F_{J'}(k') \qquad (48)$$

We also define coefficients $C_{JJ'J''}$ and θ_J by

$$\nu^{-1} \sum_k F_J(k) F_{J'}(k) F_{J''}(k) \delta(\varepsilon_k) \equiv C_{JJ'J''} \qquad (49)$$

$$\nu^{-1} \sum_k F_J(k) \delta(\varepsilon_k) \equiv \theta_J. \qquad (50)$$

Using eqs. (48-50), eq. (47) becomes

$$\Lambda_{JJ'} = \sum_{J''J'''} C_{JJ'J''} \theta_{J'''} \lambda_{J''J'''}. \qquad (51)$$

In the symmetric representation, θ_J equals δ_{J0} and we recover eq. (8) which is valid in the symmetric representation only. In the disjoint representation, however, the coefficients θ are less simple, namely $\theta_\alpha = \sqrt{\nu_\alpha/\nu}$. Thus the more complicated equation (51) must be used. This complication is offset by the simplicity of the resulting formulas for $\Lambda_{\alpha\beta}$. The only non-zero Clebsh-Gordon coefficients are of the type $C_{\alpha\alpha\alpha}^{\alpha\beta} = \sqrt{\nu/\nu_\alpha}$ which implies that Λ is diagonal in the disjoint representation of the 2-band model,

$$(\Lambda_{\alpha\beta})_{disj} = \delta_{\alpha\beta} \sum_\gamma \sqrt{\frac{\nu_\gamma}{\nu_\alpha}} \lambda_{\alpha\gamma} \qquad (52)$$

$$(\underset{\sim}{\Lambda})_{sym} = \underset{\sim}{U} (\underset{\sim}{\Lambda}_{disj}) \underset{\sim}{U}^{-1}. \qquad (53)$$

We can now solve for T_c in either representation. The disjoint representation proves simpler because $\underset{\sim}{\Lambda}$ is diagonal. We can factor $(\underset{\sim}{1} + \underset{\sim}{\Lambda})$ into $[(\underset{\sim}{1} + \underset{\sim}{\Lambda})^{\frac{1}{2}}]^2$ and then seek the maximum eigenvalue of

$$\underset{\sim}{\lambda}^* \equiv (\underset{\sim}{1}+\underset{\sim}{\Lambda})^{-\frac{1}{2}}(\underset{\sim}{\lambda}-\underset{\sim}{\mu}^*)(\underset{\sim}{1}+\underset{\sim}{\Lambda})^{-\frac{1}{2}}$$

$$(\lambda^*_{\alpha\beta})_{disj} = (\lambda_{\alpha\beta}-\mu^*_{\alpha\beta})/\sqrt{(1+\Lambda_{\alpha\alpha})(1+\Lambda_{\beta\beta})} \ . \tag{54}$$

The final formula for λ_{eff} is

$$\lambda_{eff} = \frac{1}{2}(\lambda^*_{aa} + \lambda^*_{bb}) + \sqrt{[\frac{1}{2}(\lambda^*_{aa}-\lambda^*_{bb})]^2 + \lambda^{*2}_{ab}} \tag{55}$$

We believe that this is the first time the complete solution of the strongly coupled two-band model has been given. A correct result was given by Geilikman and Masharov[27], except they have folded the renormalization Λ into the coupling constant λ, and have not provided the connection (52) between the mass renormalization and the coupling constant matrix.

V. CONSTRUCTING FERMI SURFACE HARMONICS

There are two parts to the problem of constructing FSH's: first, enumerating a maximal set of linearly independent polynomials of a given symmetry type, and second, orthogonalizing them. Both of these involve standard mathematical procedures, but it seems appropriate to give a summary of the procedures we have found useful.

First we work out the case where there is a single sheet of Fermi surface. It does not matter whether this sheet is open as in the Fermi surface of Cu or the "jungle-gym" of Nb, or closed as the "jack" of Nb; the form of the functions is the same. The task is to construct orthonormal polynomials starting with zeroth order and increasing in order until convergence is obtained for the problem at hand. In N^{th} order there are in general $(N+1)(N+2)/2$ linearly independent functions of the form $v_x^\ell v_y^m v_z^n$ with $\ell, m, n > 0$ and $\ell + m + n = N$. The operation R of the point group G of the crystal transforms points in $\underset{\sim}{k}$ space into new points $R^{-1}\underset{\sim}{k}$. When the operations R are performed on the polynomials, they will transform into other polynomials of the same order. A trivial example is the zero order polynomial, 1, which is invariant under all operations R, and thus transforms according to the identity representation Γ_1, of G. The first order polynomials are the three components of the velocity. The velocity vector $\underset{\sim}{v}_k$ transforms into $\underset{\sim}{v}_{R^{-1}k} = \underset{\sim}{\Gamma}(R)\underset{\sim}{v}_k$, and the matrices $\underset{\sim}{\Gamma}(R)$ form a representation of the group G. For cubic symmetry this is the irreducible representation Γ_{15} (in the notation of Bouchaert, Smoluchowski, and Wigner[28]). For lower symmetry the representation Γ thus defined is a reducible representation.

From now on we will assume the crystal has cubic symmetry. The four functions (1, v_x, v_y, v_z) so far discussed are automatically orthogonal because they belong either to different representations or to different rows of the same irreducible representation. Normalizing the functions is easy: 1 is already normalized by the definitions (3), and is given the name F_0. The remaining three are normalized by dividing by $\langle v_x^2\rangle^{\frac{1}{2}} = (\frac{1}{3}\langle v^2\rangle)^{\frac{1}{2}}$ where $\langle v^2\rangle^{\frac{1}{2}}$ is the rms Fermi velocity. These functions are named F_x, F_y, F_z. In second order there are six polynomials, v_x^2, v_y^2, v_z^2, v_xv_y, v_yv_z, v_zv_x. These form a reducible six-dimensional representation of G. It is clear that in cubic symmetry the first three do not mix with the other three, so the representation can be immediately reduced to two three-dimensional representations. The general process of reduction is illustrated in Table I, where the character table of the cubic group is given. Only even representations and proper rotations are displayed explicitly. It is all that is necessary because in the examples worked below, it is always obvious whether the representations are even or odd. The characters (traces) of the 6×6 matrices for the second order polynomials are easily worked out and are displayed in the row labled Γ_{vv} in Table I. Using the orthogonality relation for characters, it is found that Γ_{vv} reduces to $\Gamma_1 + \Gamma_{12} + \Gamma_{25'}$. Thus the three functions v_xv_y, v_yv_z, v_zv_x are basis functions for the irreducible representation $\Gamma_{25'}$, while the functions v_x^2, v_y^2, v_z^2 can be further reduced to $\Gamma_1 + \Gamma_{12}$. It is also clear that $v^2 = v_x^2 + v_y^2 + v_z^2$ is a totally symmetric (Γ_1) function, and Γ_{12} functions can easily be found ($v_x^2 - v_y^2$ and $3v_z^2 - v^2$ are conventional choices, but there is no rule that specifie a priori which linear combinations or signs should be chosen. Once a choice has been made, however, the higher order Γ_{12} functions are constrained except for signs.). Of the six new functions, five are automatically orthogonal to the lower ones but v^2 is not orthogonal to F_0. Our specific conventions and proceduces for doing the orthogonalization will be discussed below.

The reduction of higher order functions proceeds in a similar fashion. The results of the character analysis for N=3, 4 and 6 are in Table I. The explicit construction of linearly independent functions with the proper transformation properties can usually be accomplished by intuition assisted by simple rules, but projection operator techniques are also available if intuition fails. Usually only the Γ_1 and Γ_{15} representations will be of interest (for use in superconductivity and transport, respectively). There are two Γ_{15} sets in third order. One is trivial - the product of the first order Γ_{15} functions and the second order Γ_1 function, namely v^2v_x, v^2v_y, v^2v_z. The other is a new set, v_x^3, v_y^3, v_z^3. Similarly the fourth order Γ_1 functions are trivial ones; $(v^2)^2$, and a new one, $v_x^4 + v_y^4 + v_z^4$. If we stop at fourth order, we have a total of four Γ_1 functions which we can put into orthonormal form, and group theory assures us that there are no more to be found without going to higher order.

TABLE I

Character table of cubic group, and characters of nth order polynomials for $n \leq 4$ on a single sheet and on a set of 6 equivalent sheets centered at the N point of the bcc Brillouin zone.

label of representation / class	E	$3C_4^2$	$6C_4$	$6C_2$	$8C_3$	decomposition of reducible representation
Γ_1	1	1	1	1	1	
Γ_2	1	1	-1	-1	1	
Γ_{12}	2	2	0	0	-1	
$\Gamma_{15'}$	3	-1	1	-1	0	
$\Gamma_{25'}$	3	-1	-1	1	0	
single sheet Γ_c	1	1	1	1	1	Γ_1
Γ_x	3	3	1	1	0	Γ_{15}
Γ_{xx}	6	2	0	2	0	$\Gamma_1 + \Gamma_{12} + \Gamma_{25'}$
Γ_{xxx}	10	-2	0	-2	1	$\Gamma_{2'} + 2\Gamma_{15} + \Gamma_{25}$
Γ_{xxxx}	15	3	1	3	0	$2\Gamma_1 + 2\Gamma_{12} + \Gamma_{15'} + 2\Gamma_{25'}$
Γ_v^6	28	4	0	4	1	$3\Gamma_1 + \Gamma_2 + 3\Gamma_{12} + 2\Gamma_{15'} + 4\Gamma_{25'}$
N point ellipsoids Γ_c	6	2	0	2	0	$\Gamma_1 + \Gamma_{12} + \Gamma_{25'}$
Γ_x	18	-2	0	-2	0	$\Gamma_{2'} + \Gamma_{12'} + 3\Gamma_{15} + 2\Gamma_{25}$
Γ_{xx}	36	4	0	4	0	$3\Gamma_1 + \Gamma_2 + 4\Gamma_{12} + 3\Gamma_{15'} + 5\Gamma_{25'}$
Γ_{xxx}	60	-4	0	0	0	$2\Gamma_{1'} + 2\Gamma_{2'} + 4\Gamma_{12'} + 8\Gamma_{15} + 8\Gamma_{25}$
Γ_{xxxx}	90	6	0	10	0	$7\Gamma_1 + 2\Gamma_2 + 9\Gamma_{12} + 8\Gamma_{15'} + 13\Gamma_{25'}$

The next Γ_1 functions are sixth order. The intuitive method suggests four basis functions, two "old ones", v^6 and $v^2(v_x^4 + v_y^4 + v_z^4)$, and two "new ones", $v_x^6 + v_y^6 + v_z^6$ and $v_x^2 v_y^2 v_z^2$. However, group theory shows that only three Γ_1 functions occur in sixth order. The four we have listed are actually linearly dependent, because $v^6 = 3v^2(v_x^4 + v_y^4 + v_z^4) + 6v_x^2 v_y^2 v_z^2 - 2(v_x^6 + v_y^6 + v_z^6)$. We can choose any three of these four functions as the maximal linearly independent set.

The second example is a case where the Fermi surface has multiple sheets which are related to each other by the symmetry operations of the crystal. Specifically, Nb has surfaces at the N or $(\pi/a)(110)$ points. There are twelve half-surfaces or 6 whole surfaces in the first Brillouin zone, and we refer to them as "potatoes" because of their distorted ellipsoidal shape. Each potato individually has orthorhombic symmetry, but the collection of six potatoes has full cubic symmetry. Here we pretend no other sheets occur besides the six potatoes. In actual application to Nb, the normalizations will need to be altered because two additional sheets occur.

We now have six times as many linearly independent functions in each order as we had before. For example in zeroth order we can choose F_a to be $\sqrt{\nu/\nu_a}$ on potato "a" in the [110] direction and zero elsewhere. Similarly there are functions F_b, F_c, F_d, F_e, F_f, corresponding to potatoes "b" in the $[1\bar{1}0]$ direction, "c" in the [011] direction, "d" in the $[01\bar{1}]$ direction, "e" in the [101], and "f" in the $[\bar{1}01]$ direction. These six functions are clearly linearly independent and orthonormal in the sense of eq. (3). Under the operations R these functions transform into each other, forming a six-dimensional reducible representation. From Table I we see that this reduces to $\Gamma_1 + \Gamma_{12} + \Gamma_{25'}$. The Γ_1 function will clearly be $F_o = 1 = \sqrt{\nu_a/\nu} \; (F_a + F_b + F_c + F_d + F_e + F_f)$ which is totally symmetric. The function $F_a - F_b$ is clearly orthogonal to F_o and along with its partners $F_c - F_d$, $F_e - F_f$, form the basis of the $\Gamma_{25'}$ representation. Finally, $F_c + F_d - F_e - F_f$ and $3(F_a + F_b) - F_o$ from the two Γ_{12} partners.

In doing character analysis for multipy-sheeted Fermi surfaces, it is not necessarily true that even polynomials will belong to even representations. For example, if two sheet transform into each other under inversion, then the constant polynomial which is 1 on the first sheet, -1 on the second, is odd. This is not possible on the Nb potatoes because each sheet transforms into itself under inversion, and even or odd polynomials always belong to even or odd representations respectively.

We need explicit formulas for the 3 linearly independent second order Γ_1 polynomials on the potatoes. These are most simply found by examining the subgroup that leaves a single potato invariant. There are three orthogonal axes with C_2 operations, in the [110],

GAP ANISOTROPY AND T_c ENHANCEMENT

[110] and [001] directions respectively for the "a" potato. Correspondingly the polynomials $(v_x + v_y)^2$, $(v_x-v_y)^2$, and v_z^2 are invariants under this subgroup. Each of these generates a Γ_1 function of the full group if we operate on it by the six operations which generates the six potatoes and add the results in phase. Thus for example, v_z^2 generates a functions $v_z^2(\delta_{ka}+\delta_{kb})+v_x^2(\delta_{kc}+\delta_{kd})+v_y^2(\delta_{ke}+\delta_{kf})$ which has Γ_1 symmetry. We have chosen v^2, $v_z^2 k_c$ and $v_x v_y$ as a convenient choice of three independent Γ_1 functions on potato "a" which generate Γ_1 functions on the full serface.

This concludes the outline of how the maximal set of linealy independent functions is found. We now outline the procedures we followed in orthornormalizing them. We consider only the Γ_1 subset, and list the linearly independent functions in order starting with polynomials of lowest degree. The ordering of functions of the same degree can be chosen arbitrarily and must be clearly specified because it affects the form of the function. We label the linearly independent set f_1, f_2.... Our task is to construct an orthonormal set F_1, F_2.... This is done by the usual Gram-Schmidt procedure

$$F_1 = f_1/\text{norm}_1$$

$$F_2 = [f_2 - <F_1|f_2>F_1]/\text{norm}_2$$

$$F_n = [f_n - \sum_{j=1}^{n-1} <F_j|f_n> F_j]/\text{norm}_n. \qquad (56)$$

We find it convenient to keep track of the transformation matrix T, from the f_j's to the F_i's.

$$F_i = \sum_{j \leq i} T_{ij} f_j \qquad (57)$$

$$A_{ij} \equiv <f_i|f_j>. \qquad (58)$$

The matrix T has positive diagonal elements and zeros on every location above the diagonal. It is relatively easy to express T_{ij} in terms of the matrix A_{ij} of the overlaps of the original functions. Finally, we need to know expansion coefficients p_J and $Q_{JJ'}$ of functions $p(k)$ and operators $Q(k,k')$ in the FSH basis. We find it more convenient to calculate first in the non-orthogonal basis of f_i's and then transform

$$P_J = \sum_i T_{ji} \langle f_i | p \rangle$$

$$Q_{JJ'} = \sum_{ii'} T_{ji} \langle f_i | Q | f_{i'} \rangle (T^{tr})_{i'J'}. \qquad (59)$$

There are two reasons for doing all the calculations first in the nonorthogonal basis and then transforming. One is that often integrals like $\langle f_i | p \rangle$ are interesting in their own right, or more easily interpreted than the more abstract $\langle F_J | p \rangle$. The other reason is that this enables us to do all the Fermi surface integrals simultaneously. That is, our computer programs do Fermi surface integrals such as $\langle f_i | f_j \rangle$ which are needed to construct orthonormal functions, and at the same time integrals such as $\langle f_i | p \rangle$ are done which are later used to find p_J. After all the integrals are done, a separate program is written which constructs the matrix $\underset{\sim}{T}$ from the coefficients A_{ij} and then constructs p_J and $Q_{JJ'}$.

Finally, since Nb has three separate types of Fermi surface, it is necessary to decide which representation to work in, disjoint (D) or symmetric (S). There are several reasons for preferring the D-representation in the numerical calculations. The most compelling one is that our programs do integrals over the three types of surface by somewhat different procedures and at different places in the program. This makes it difficult to do integrations over functions which are non-zero on more than one type of surface. When the calculations are all done, it is desirable to transform to the S-representation so that perturbation theory for weak anisotropy can be used. For this purpose, only the constant functions need to be in the S-representation - the functions derived from higher order polynomials can be left untouched. Thus the matrix U which transforms D → S is the unit matrix except for a 3×3 sub-block which describes how the three constant functions transform. This sub-block was chosen in the form

$$U = \begin{pmatrix} \cos\theta & \sin\theta \cos\phi & \sin\theta \sin\phi \\ 0 & \sin\phi & -\cos\phi \\ \sin\theta & -\cos\theta \cos\phi & -\cos\theta \sin\phi \end{pmatrix}$$

$$\cos\theta = \sqrt{\nu_a/\nu}, \quad \sin\theta \cos\phi = \sqrt{\nu_b/\nu}, \quad \sin\theta \sin\phi = \sqrt{\nu_c/\nu}$$

$$F_J(S) = \sum_{J'} U_{JJ'} F_{J'}(D) \qquad (60)$$

where ν_α is the partial density of states of the band α. The functions $F_J(D)$ are the form $\sqrt{\nu/\nu_\alpha} \, \delta_{k\alpha}$, ($\alpha$ = a,b,c). Applying the transformation

(60), the first of the S functions has the form $\Sigma \delta_{k\alpha} = 1$, and is the function we have previously labeled F_0.

We have made no effort to calculate the μ^* matrix from first principles. Therefore we make the simplest possible assumptions, namely that μ^* is isotropic. This means that in the S-representation we get $\mu^*_{JJ'}(S) = \mu^* \delta_{J0}\delta_{J'0}$. The matrix μ^* in the D-representation can be found using the inverse of the transformation (60) (equal to the transpose of the unitary matrix $\underline{U}$), namely $\underline{\mu}^*(D) = \underline{U}^{-1} \underline{\mu}^*(S) \underline{U}$.

VI. CALCULATION OF THE MATRIX $\lambda_{JJ'}$ FROM FIRST PRINCIPLES BAND THEORY

In this section we show that if one neglects anisotropy arising from the phonons, the matrix $\lambda_{JJ'}$ can be obtained rather simply from a sufficiently detailed energy band calculation. Let us define a new matrix, $\eta_{JJ'}$, by

$$\eta_{JJ'} = 2M \int_0^\infty d\Omega\, \Omega\, \alpha^2_{JJ'} F(\Omega)$$

$$= \frac{M}{\nu} \sum_{kk';i} |M^i_{kk'}|^2 \omega^i_{k-k'} F_J(k) F_{J'}(k') \delta(\epsilon_k) \delta(\epsilon_{k'}) .$$

(61)

$\eta_{JJ'}$ is a generalization of the quantity η introduced by McMillan[5] and Hopfield[29], related to $\lambda_{JJ'}$ through the identity

$$\lambda_{JJ'} = \eta_{JJ'}/M \langle\Omega^2\rangle_{JJ'} ,$$

(62)

where the average square frequency, $\langle\Omega^2\rangle_{JJ'}$, is given by

$$\langle\Omega^2\rangle_{JJ'} = \int_0^\infty d\Omega\, \Omega\, \alpha^2_{JJ'} F(\Omega) / \int_0^\infty d\Omega/\Omega\, \alpha^2_{JJ'} F(\Omega)$$

(63)

For $J = J' = 0$ it is usually a good approximation to replace $\langle\Omega^2\rangle_{JJ'}$ by $\langle\Omega^2\rangle$ where $\langle\Omega^2\rangle$ is obtained by neglecting the frequency dependence of $\alpha^2_{JJ'}$ in Eq. (63). We shall assume $\langle\Omega^2\rangle_{JJ'} = \langle\Omega^2\rangle$ in the following for all J and J'. This has the effect of omitting the phonon contribution to the gap anisotropy.

The computational advantage of working with $\eta_{JJ'}$ as opposed to $\lambda_{JJ'}$ is that $\eta_{JJ'}$ may be written in terms of single (rather than double) integrals over the Fermi surface. In addition $\eta_{JJ'}$ depends only upon the electronic structure of the system. The phonons enter $\lambda_{JJ'}$ in this approximation only through the denominator $M\Omega^2$. These simplifications result from the fact that the factor $M\omega_{k-k'}^i$ in Eq. (61) cancels against a similar factor coming from the matrix elements,

$$M_{kk'}^i = \int d^3r\, \psi_k^*(r)\, \hat{e}_{k-k'}^i \cdot \nabla V\, \psi_k(r)/\sqrt{M\omega_{k-k'}^i}. \quad (64)$$

The matrix element contains electronic wave function $\psi_k(r)$, a phonon polarization vector $\hat{e}_{k-k'}^i$, a potential gradient ∇V, the ionic mass, M, and phonon frequency $\omega_{k-k'}^i$. It is difficult to calculate ∇V accurately for a transition metal. We shall use the rigid muffin-tin approximation of Gaspari and Gyorffy[30-32] which seems to give reasonable agreement with empirical values of η.

Substituting the matrix elements (63) into our expression (61) for $\eta_{JJ'}$ we obtain (using the completeness of the polarization vectors)

$$\eta_{JJ'} = \frac{1}{\nu} \sum_{kk'} \int d^3r \int d^3r'\, \psi_k^*(r)\, \psi_{k'}(r)\, \psi_k(r')\, \psi_{k'}^*(r')\, \nabla V(r)$$

$$\cdot \nabla' V(r') \times F_J(k)\, F_{J'}(k')\, \delta(\varepsilon_k)\, \delta(\varepsilon_{k'}). \quad (65)$$

One easily recognizes in (64) the Fermi-surface density matrix,

$$\rho_k(r,r') = \psi_k^*(r)\, \psi_k(r')\, \delta(\varepsilon_k), \quad (66)$$

and its expansion coefficient in FSH's,

$$\rho_J(r,r') = \frac{1}{\nu} \sum_k \rho_k(r,r')\, F_J(k). \quad (67)$$

Thus, in terms of $\rho_J(r,r')$ we have

$$\eta_{JJ'} = \nu \int d^3r \int d^3r'\, \rho_J(r,r')\, \rho_{J'}(r',r)\, \nabla V(r) \cdot \nabla' V(r'). \quad (68)$$

The quantities $\rho_J(r,r')$ must be expanded in coordinate space cubic harmonics in order to make contact with quantities available from band theory.

GAP ANISOTROPY AND T_c ENHANCEMENT

$$\rho_J(r,r') = \frac{1}{\nu} \sum_k \psi_k^*(r) \psi_k(r') F_J(k) \delta(\varepsilon_k)$$

$$= \frac{1}{\nu} \sum_k \sum_{\ell\mu t} \sum_{\ell'\mu't'} i^{\ell'-\ell} c_{\ell\mu}^t(k) c_{\ell'\mu'}^{t'}(k) R_\ell(r) R_{\ell'}(r')$$

$$\times K_{\ell\mu}^t(\hat{r}) K_{\ell'\mu'}^{t'}(\hat{r}') \delta(\varepsilon_k) F_J(k) \tag{69}$$

where we have used the Bloch wave expansion

$$\psi_k(r) = \sum_{\ell\mu t} i^\ell c_{\ell\mu}^t(k) R_\ell(r) K_{\ell\mu}^t(\hat{r}). \tag{70}$$

$R_\ell(r)$ is a regular solution to the radial Schrödinger equation, $K_{\ell\mu}^t$ is a cubic harmonic of row μ of irreducible representation t and orbital quantum number ℓ, and $c_{\ell\mu}^t$ is an expansion coefficient which is obtained from an energy band calculation. It may be noted that all quantities which enter (68) and (69) are available if a detailed Fermi surface calculation has been performed.

Additional simplifications can be achieved by exploiting the cubic symmetry of Nb. Equation (69) contains the Fermi surface integral

$$T_{\ell\mu,\ell'\mu',J}^{tt'} = \frac{1}{\nu} \sum_k c_{\ell\mu}^t(k) c_{\ell'\mu'}^{t'}(k) F_J(k) \delta(\varepsilon_k). \tag{71}$$

For s-wave superconductitity we are only interested in the Γ_1 submatrix of $\eta_{JJ'}$. For this reason only FSH's with full cubic symmetry will be considered in Eq. (71), and for this case one can show that $T_{\ell\mu,\ell'\mu',J}^{tt'}$ is diagonal in t and t' and also in μ and μ', and independent of μ.

$$T_{\ell\mu,\ell'\mu',J}^{tt'} = T_{\ell\ell',J}^t \delta_{tt'} \delta_{\mu\mu'}. \tag{72}$$

Substituting (72) and (71) into (69) we have

$$\rho_J(r,r') = \sum_{\ell\ell',t} T_{\ell\ell',J}^t R_\ell(r) R_{\ell'}(r') F_{\ell\ell'}^t(\hat{r},\hat{r}') \tag{73}$$

where

$$F_{\ell\ell'}^t(\hat{r},\hat{r}') = \sum_\mu K_{\ell\mu}^t(\hat{r}) K_{\ell'\mu}^t(\hat{r}'). \tag{74}$$

The functions $F^t_{\ell\ell'}$ are listed in ref. 32.

At this point it is helpful to condense the notation slightly. One can show that for most values of (ℓ,ℓ',t), $T^t_{\ell\ell'J}$ is zero. In fact for $\ell,\ell' \leq 3$ there are only 8 non-vanishing coefficients for a given value of J. Thus we write α for the set of indices (ℓ,ℓ',t), i.e.

$$T^t_{\ell\ell'J} \equiv T^\alpha_J \qquad (\alpha = 1\ldots 8) \; . \tag{75}$$

The translation $\alpha \leftrightarrow (\ell\ell't)$ can be found in Table II.

Using this notation (73) becomes

$$\rho_J(r,r') = \sum_\alpha T^\alpha_J R_\ell(r) R_{\ell'}(r') F_\alpha(\hat{r},\hat{r}') \tag{76}$$

and (68) becomes

$$\eta_{JJ'} = \nu \sum_{\alpha\beta} T^\alpha_J T^\beta_{J'} g_{\alpha\beta} \tag{77}$$

where the coupling matrix is given by

$$g_{\alpha\beta} = V'_{\ell_1\ell_4} V'_{\ell_2\ell_3} \int d\hat{r} \int d\hat{r}' F^t_{\ell_1\ell_2}(\hat{r},\hat{r}') F^{t'}_{\ell_3\ell_4}(\hat{r}',\hat{r})$$

$$\frac{(xx' + yy' + zz')}{rr'} \qquad \alpha = (\ell_1\ell_2 t) \qquad \beta = (\ell_3\ell_4 t') \; . \tag{78}$$

$V'_{\ell_1\ell_4}$ is a radial integral over the derivative of the band theory potential in the rigid muffin-tin approximation.

$$V'_{\ell_1\ell_4} = \int r^2 R_{\ell_1}(r) R_{\ell_4}(r) \frac{\partial V}{\partial r} dr \tag{79}$$

$g_{\alpha\beta}$ is displayed explicitly in Table II.

To summarize: $\eta_{JJ'}$ is obtained from (77) where $g_{\alpha\beta}$ is obtained from Table II, and $T^\alpha_J = T^t_{\ell\ell',J}$ is obtained from (71) which we may write as

TABLE II

Coupling constant matrix, $g_{\alpha\beta}$ for Nb. Radial wave functions are normalized to 1 at the muffin-tin radius. The constants are dimensionless.

$\ell\ell'$t		$00\Gamma_1$	$11\Gamma_{15}$	$22\Gamma_{25'}$	$33\Gamma_{12}$	$33\Gamma_{25}$	$33\Gamma_{15}$	$33\Gamma_{2'}$	$13\Gamma_{15}$
	α \ β	1	2	3	4	5	6	7	8
$00\Gamma_1$	1	0	A	0	0	0	0	0	0
$11\Gamma_{15}$	2	A	0	6/5 B	4/5 B	0	0	0	0
$22\Gamma_{25'}$	3	0	6/5 B	0	0	6/7 C	18/35 C	3/7 C	$\frac{12}{5}\sqrt{3/7}$ D
$22\Gamma_{12}$	4	0	4/5 B	0	0	3/7 C	27/35 C	0	$-\frac{12}{5}\sqrt{3/7}$ D
$33\Gamma_{12}$	5	0	0	6/7 C	3/7 C	0	0	0	0
$33\Gamma_{15}$	6	0	0	18/35 C	27/35 C	0	0	0	0
$33\Gamma_{2'}$	7	0	0	3/7 C	0	0	0	0	0
$13\Gamma_{15}$	8	0	0	$\frac{12}{5}\sqrt{3/7}$ D	$-\frac{12}{5}\sqrt{3/7}$ D	0	0	0	0

$A = V_{01}'^2 = 11.999$

$B = V_{12}'^2 = 19.787$

$C = V_{23}'^2 = 306.290$

$D = V_{12}' \, V_{23}' = -77.850$

$$T^t_{\ell\ell',J} = \frac{1}{\nu} \sum_k c^t_{\ell\mu}(k) \, c^t_{\ell'\mu}(k) \, F_J(k) \, \delta(\varepsilon_k) \ . \tag{80}$$

These Fermi surface integrals are the heart of the numerical calculation. They can be restricted to the irreducible 1/48th of the Brillouin zone by virtue of the fact that the quantity,

$$T^t_{\ell\ell'}(k) \equiv \frac{1}{g} \sum_\mu c^t_{\ell\mu}(k) \, c^t_{\ell'\mu}(k) \ , \tag{81}$$

(where g is the dimensionality of t) has full cubic symmetry. Thus the main numerical effort involves calculation of integrals of the form

$$T^t_{\ell\ell',J} = 48 \frac{1}{\nu} \sum_{k \in (1/48)} T^t_{\ell\ell'}(k) \, F_J(k) \, \delta(\varepsilon_k) \ . \tag{82}$$

The Fermi surface of Nb is shown in the computer generated perspective drawings Figs. 1-3. For each mesh point we have calculated the magnitude of k, the group velocity $\nabla_k \varepsilon$, and the wave functions $c^t_{\ell\mu}(k)$. The details of the calculation are described in ref. 32. All of the points shown were generated from first principles without the use of interpolation schemes, using the KKR constant-energy-search[33] method. This calculation agrees well with the earlier APW calculation by Mattheiss[34].

VII. COMPLETENESS

The question of whether or not the FSH's for an arbitrary Fermi surface are complete in the mathematical sense was addressed in ref. 20, but the question was not laid to rest. In this section we are concerned with the more practical (and perhaps more important) question of whether or not the functions $T^\alpha(k)$ defined in Eq. (81) can be expanded as a rapidly convergent series of FSH's for Nb. In order to answer this question quantitatively we define a quantity $p^\alpha_{n,s}$ which is the percentage of completeness of the expansion of $T^\alpha(k)$ through nth order FSH's in the disjoint representation on surface s. Thus

$$(p^\alpha_{n,s})^2 = \sum_{J=0}^n (T^\alpha_{J,s})^2 / \langle (T^\alpha(k))^2 \rangle_s \tag{83}$$

where

$$\langle (T^\alpha(k))^2 \rangle_s = \frac{1}{\nu} \sum_{k \text{ on } s} \delta(\varepsilon_k) \, (T^\alpha(k))^2 \ . \tag{84}$$

Figure 1: The central sheet or "jack" of Nb containing holes from the second band. Points of discontinuous slope can be seen corresponding to points where this surface touches the surface of fig. 2

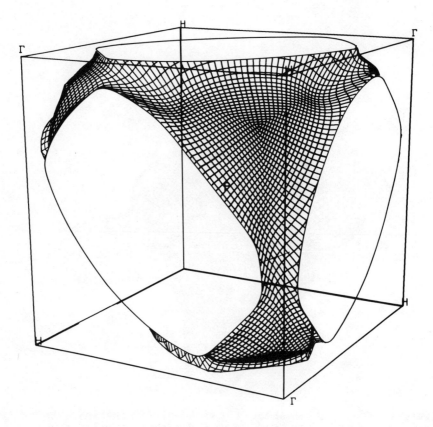

Figure 2: The third band open surface of Nb. In the earlier work by Mattheiss this sheet was drawn centered about Γ, and appears as a jungle-gym containing hole states. From the present perspective centered around P the surface contains electrons and is more reminiscent of a bikini.

GAP ANISOTROPY AND T$_C$ ENHANCEMENT 105

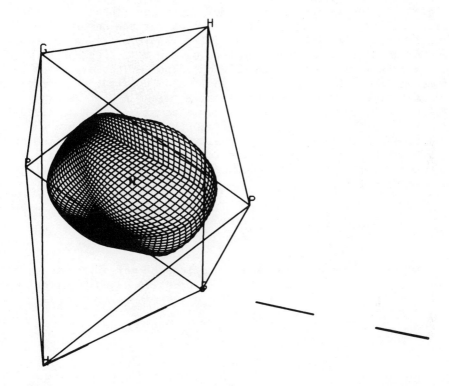

Figure 3: The third band hole surface of Nb centered at the N point. This is one of six such surfaces, called "potatoes", each of which has rhombohedral symmetry.

and where $T_{J,s}^{\alpha}$ is the expansion coefficient of $T^{\alpha}(k)$ on sheet s. If the FSH's were known to be complete, this would guarantee (by the definition of completeness) that the sequence $p_0^2, p_1^2, ...p_n^2$ converge to 1. Since the FSH's are not known to be complete, the empirical convergence of these sequences constitutes an important test of completeness.

We have carried out calculations for Nb using the seven lowest order FSH's on sheets 1 (the jack(fig. 1)) and 2 (the jungle-gym (fig. 2)), and the four lowest order functions (sheet 3) on the potatoes, (fig. 3). Our specific choices for the unnormalized functions f_{J_s} are listed in Table III. The rules given in sec. V provide an unambiguous specification of the resulting FSH's $F_{J_s}(k)$. In table IV we present our calculated values of the expansion coefficients $T_{J_s}^{\alpha}$ for Nb, and our final value p_{ns}^{α} for the percentage of completeness, where only the final term $n = n_{max}$ is shown explicitly. Typically, about 95% completeness has been achieved. We also give the partial densities of states ν_s^{α} which are the contributions to the Fermi energy density of states arising from states of type α on sheet s, and I_{α}, the integral of the radial wave functions $R_{\ell}(r)$ and $R_{\ell'}(r)$ with appropriate weighting functions over the Wigner Seitz cell.

$$I_{\alpha} = \int d^3r\, R_{\ell}(r)\, R_{\ell'}(r)\, F_{\alpha}(\hat{r},\hat{r}) \quad . \tag{85}$$

The condition that the wave functions be normalized to unity over the Wigner-Seitz cell for each k point leads to the requirement that

$$\sum_{\alpha} T^{\alpha}(k)\, I_{\alpha} = 1 \quad . \tag{86}$$

The integral over a sheet of Eq. (86) leads to

$$\sum_{\alpha} \sum_{k\text{ on }s} T^{\alpha}(k)\, \delta(\varepsilon_k)\, I_{\alpha} = \nu_s \quad . \tag{87}$$

Inclusion of a FSH in the k integral of (87) leads to the following sum rules,

$$\sum_{\alpha} T_{J,s}^{\alpha}\, I_{\alpha} = \sqrt{\nu_s/\nu}\, \delta_{J0} \quad . \tag{88}$$

TABLE III

Fermi Surface Harmonics with full cubic symmetry for Nb (non-orthogoanl basis).

J	sheets 1 & 2	sheet 3* (potatoes)
0	1	$\delta_{ka} + \delta_{kb} + \delta_{kc} + \delta_{kd} + \delta_{ke} + \delta_{kf}$
1	v^2	$v^2(\delta_{ka} + \delta_{kb} + \delta_{kc} + \delta_{kd} + \delta_{ke} + \delta_{kf})$
2	v^4	$v_z^2(\delta_{ka} + \delta_{kb}) + v_x^2(\delta_{kc} + \delta_{kd}) + v_y^2(\delta_{ke} + \delta_{kf})$
3	$v_x^4 + v_y^4 + v_z^4$	$v_x v_y(\delta_{ka} - \delta_{kb}) + v_x v_z(\delta_{kc} - \delta_{kd}) + v_y v_z(\delta_{ke} - \delta_{kf})$
4	v^6	
5	$v^2(v_x^4 + v_y^4 + v_z^4)$	
6	$v_x^2 v_y^2 v_z^2$	

* k_i is unity on potato i, zero elsewhere.

TABLE IV

Expansion coefficients and percentage of completeness for the Fermi surface harmonic expansion of the density matrix coefficients of Nb. T_J^α, S, ν_S^α, and I_α are given in dimensionless program units. To convert to atomic units multiply νT_J^α by $(2\pi/a)^2$, ν_S^α by $(a/2\pi)^2$ and I by $(a/2\pi)^3$. a for Nb is 6.2294 a.u.

(a) Sheet 1 (Jack)

$T_{J,1}^\alpha (\times \nu)$

n	J	$00\Gamma_1$	$11\Gamma_{15}$	$22\Gamma_{25'}$	$22\Gamma_{12}$	$33\Gamma_{25}$	$33\Gamma_{15}$	$33\Gamma_{2'}$	$13\Gamma_{15}$
0	0	.000239	.001673	.050642	.004502	.003262	.000537	.000541	-.000909
2	1	.000092	.001114	-.008878	.010720	.002543	.000287	-.000048	-.000551
4	2	-.000095	-.001038	-.00027	.001135	-.000194	-.000332	-.000177	.000569
4	3	.000044	.000127	.001538	-.002168	-.000318	.000108	.000248	-.000121
6	4	-.000098	.000020	.000228	-.000376	.000015	.000087	.000178	-.000097
6	5	.000043	.000230	-.001353	.001660	.000361	-.000017	-.000222	-.000047
6	6	-.000047	-.000096	.000719	-.000873	-.000201	-.000000	.000034	-.000022
$p_{6,1}^\alpha$		67.4%	97.3%	99.9%	99.1%	99.5%	96.0%	91.1%	96.6%
ν_S^α		.0007	.0139	1.1854	.0736	.0284	.0047	.00009	-.0025
I_α		8.1707	23.552	66.200	46.222	24.647	24.529	4.666	7.690

TABLE IV (continued)

$T^{\alpha}_{J,2}(\chi\nu)$

(b) Sheet 2 (Jungle Gym)

n	J	$00\Gamma_1$	$11\Gamma_{15}$	$22\Gamma_{25'}$	$22\Gamma_{12}$	$33\Gamma_{25}$	$33\Gamma_{15}$	$33\Gamma_{2'}$	$13\Gamma_{15}$
0	0	.009766	.017143	.086291	.009875	.005834	.005202	.008104	-.005678
2	1	.008841	-.000225	.000876	-.003412	-.002848	.004249	.004715	-.003021
4	2	.007407	-.002116	-.001656	.001832	.001379	-.000309	-.003000	.000245
4	3	-.001329	-.001435	.001474	-.001054	.000494	-.000371	-.000933	-.000375
6	4	-.000152	.002832	.000216	-.001577	-.000889	.000265	.001442	.000234
6	5	.001404	.001130	-.000546	-.000049	-.000482	.000289	.000533	.000338
6	6	-.001482	-.000241	.001267	-.001184	-.000027	-.000440	-.000075	.000056
$p^{\alpha}_{6,2}$		94.6%	90.2%	99.7%	88.0%	97.1%	99.0%	95.3%	98.3%
ν^{α}_2		.0529	.2676	3.7856	.3025	.0953	.0846	.0251	-.0289
I_α		8.1707	23.552	66.200	46.222	24.647	24.529	4.666	7.690

TABLE IV (continued)

(c) Sheet (3)

$T^{\alpha}_{J,3}(\times\nu)$

n	J	$00\Gamma_1$	$11\Gamma_{15}$	$22\Gamma_{25'}$	22_{12}	$33\Gamma_{25}$	$33\Gamma_{15}$	$33\Gamma_{2'}$	$13\Gamma_{15}$
0	0	.057976	.100722	.029356	.038861	.007540	.001321	.002854	.009696
2	1	-.005141	.001638	-.006819	.009288	-.000522	.000701	-.000181	.002865
2	2	-.017362	-.003379	-.003942	.010091	.000061	.000403	-.000859	.001120
2	3	-.009829	-.007728	-.009658	.018299	.001763	.000446	-.001137	.000882
$P^{\alpha}_{2,3}$		86.9%	99.4%	97.1%	97.2%	88.5%	95.4%	86.9%	93.4%
ν^{α}_3		.3127	1.5661	1.2829	1.1858	.1227	.0214	.0088	.0492
I_α		8.1707	23.552	66.200	46.222	24.647	24.529	4.6663	7.6900

Equation (86) implies that if a single function $T^\alpha(k)$ dominates the sum (86) for a particular sheet, then, $T(k)$ must be approximately independent of k over this sheet and can be represented quite well by the FSH which is constant on that sheet. This circumstance obtains on the jack and jungle gym surfaces where the 22 $\Gamma_{25'}$ contribution dominates. Correspondingly we see that the expansion of $T_{22}^{\Gamma_{25'}}(k)$ is more than 99% complete for both of these surfaces taking only one FSH. Most of the other functions are also satisfactorily complete with the basis we have taken.

The few functions which are not well converged deserve closer scrutiny. The worst offender is the $00\Gamma_1$ function on the jack. Fortunately this contribution is of negligible magnitude ($v_s^\alpha/v < 0.0001$) and will not affect our calculation of $\lambda_{JJ'}$. These states are concentrated near the sharp cusp near the tips of the jack. The wave functions can be discontinuous in this region due to a band crossing in the ΓNH plane which would be removed by spin-orbit coupling (not included here). It is not yet clear whether the lack of convergence is due to unsuitability of the FSH's for describing this function or to numerical noise in the wavefunctions which are almost zero for these states.

The largest error in our estimation of the gap anisotropy and T_c enhancement probably arises from the relatively poor convergence of the ($33\Gamma_{25}$) states on the potato. These states have a fairly high weight for f states and the coupling matrix $g_{\alpha\beta}$ is very large when α is a d state and β is an f state. An examination of the distribution of the ($33\Gamma_{25}$) states over the potatoes does not reveal any pathological behavior and we speculate that only a few more polynomials of higher order than two will be required to complete the expansion. In any event our overall convergence is quite good as we shall show below.

VIII. RESULTS AND COMPARISON WITH EXPERIMENT

Calculations of λ for d-band metals based on reliable band structure information have only very recently been reported. Calculations of η have been done for a number of metals[30-32], but the only detailed calculation of λ for a transition element to our knowledge is that of Yamashita and Asano[35] for Nb and Mo. These authors use a very coarse mesh of points on the Fermi surface so that (in our opinion) the finer details of their results cannot be trusted. However, it is noteworthy that these authors have calculated the anisotropic function λ_k as well as λ for both elements. Extensive calculations of λ_k and λ have been reported for Cu by Nowak[36] and Das[37]. However, Cu has a much simpler Fermi surface than any true transition element. We report here the first calculation of the anisotropy of Δ for any transition element. However, our calculations have not included the part of the anisotropy arising from the phonons. Our work is an extension to the anisotropic problem of previous calculations[30-32] of η.

The results of our calculation are two 18×18 real symmetric matrices, $\eta_{JJ'}$ and $H_{JJ'}$, calculated for Nb as described in previous sections. These matrices are then divided by $M\langle\omega^2\rangle = 5.527$ eV/Å to yield the matrices $\lambda_{JJ'}$ and $\Lambda_{JJ'}$. There would be little point to presenting all 342 numbers-most of them are small at any rate. Instead we present the 18 numbers λ_{0J} and the 18 numbers Δ_J in Table V These coefficients can be used to construct the anisotropic mass renormalization λ_k and gap Δ_k. In the symmetric representation, all except the coefficients λ_{00} and Δ_0 are small. The calculated rms gap anisotropy α is 6%. Approximately half of this anisotropy is already obtained in a 3×3 calculation using only the constant polynomials - i.e. a 3 band model. The value of λ is 1.27, in agreement with the calculation of ref.32. This value is significantly larger than the empirical[5] value of 0.82; probable reasons for the discrepancy are discussed in ref. 32. We believe that the electronic contribution to the relative anisotropy of λ and Δ is correctly accounted for in our model even though an overall uncertainty in magnitude still exists. We obtained $\lambda_{eff}=0.51822$ which may be compared with the isotropic value $(\lambda-\mu^*)/(1+\lambda) = .51625$ yielding $\delta\lambda_{eff} = .00197$. This corresponds to a T_c enhancement of 0.74% or 0.076K. The matrix μ^* was taken to be $0.1\,\delta_{J0}\delta_{J'0}$ in the symmetric representation. These results are found by solution of the eigenvalue equation (26), and are in good agreement with the perturbative eq. (36). As a check on our calculations we have also calculated $\langle\lambda_k^2\rangle$ directly, without using the Fermi surface harmonic expansion. It is easy to verify that $\langle\lambda_k^2\rangle$ is given in our model by

$$\langle\lambda_k^2\rangle = (\nu/M\langle\omega^2\rangle)^2 \sum_{\alpha\beta} \sum_{\alpha'\beta'} T_0^\alpha T_0^{\alpha'} g_{\alpha\beta} g_{\alpha'\beta'} I_{\beta\beta'} \qquad (90)$$

where $I_{\beta\beta'}$ can either be expanded in FSH's or calculated directly,

$$I_{\beta\beta'} = \frac{1}{\nu} \sum_k \delta(\varepsilon_k) T^\beta(k) T^{\beta'}(k) = \sum_J T_J^\beta T_J^{\beta'} \ .$$

From direct calculation not using FSH's we obtain $\langle\lambda_k^2\rangle - \lambda^2 = .02307$ which compares very well with our eighteen term expansion $\sum_{J\neq 0}\lambda_{0J}^2 = .02271$. This gives us considerable extra confidence that the FSH basis set is complete.

It is somewhat disappointing to discover that the theoretically predicted anisotropy is so low. This is in accord with numerous experiments which we cite below. However, it violates an intuitive notion that Nb with its complicated non-spherical Fermi surface should be more anisotropic than a metal like Zn with a Fermi surface not too badly distorted from a sphere. Possibly d-band elements have an intrinsically more isotropic coupling constant than s-p-band elements, but if this is so, we do not yet understand why. To make the situation more confusing, the noble metals, which should occupy a middle ground, are known both from experiment[38,39] and theory[36,37] to have a relatively large anisotropy. Since the transition temperatures of the noble metals (if they in fact are finite) are very low, it seems likely that anisotropy enhancement will give a large increase in T_c in these materials. This in turn suggests that sample purity may play a larger role than usually expected since impurity scattering strongly suppresses anisotropy enhancement[10-15] when $\hbar/\tau \gtrsim k_B T_c$.

The current experimental situation is not inconsistent with our theoretical estimate of 6% rms anisotropy of Δ in Nb. However, the experimental situation is far from unambiguous so we indulge here in a brief review. A variety of experiments, tunneling, ultrasonic attenuation, heat capacity, thermal conductivity, T_c reduction in dilute alloys, and critical field anisotropy, can all be interpreted in terms of gap anisotropy. In niobium most of the experiments so far seem to have a somewhat ambiguous interpretation. Critical field anisotropy[40] is perhaps the least ambiguous experiment, but we have not yet made the kind of calculations which permit comparison. There are several sources of critical field anisotropy, some of which do not require Δ to be anisotropic. Williamson[40] measured a variation of 10% in $H_{c2}(\theta)$ as θ was varied in the $(1\bar{1}0)$ plane. This variation can be accounted for in a non-local model taking into account the known Fermi surface geometry and not necessarily requiring any anisotropy in Δ.

Heat capacity at low temperatures is most sensitive to the minimum energy gap, whereas nearer to T_c, a more nearly isotropic average of the gap is measured. Recent measurements of C_V in Nb by Sellers et al[41] show no evidence for strong anisotropy and indicate that the anomalous results of earlier studies may have arisen from H impurities. A similar conclusion was reached by Anderson and collaborators in studies of the thermal conductivity[42,43].

The behavior of T_c in dilute alloys of Nb has apparently not been much studied, although we may have missed some literature. The most systematic study known to us is by Ronami and Berezina[44] who found a sharp minimum in T_c at concentrations less than 0.6% of column IV and VI transition metal impurities. Unfortunately they were unable to establish the precise concentration of the minimum because of sample difficulties, and they did not measure

TABLE V

Expansion coefficients for the energy gap and mass enhancement in Fermi Surface Harmonics. These coefficients are for the symmetric-orthonormal basis where F_0, $F_{0'}$, and $F_{0''}$ are symmetrized combinations of the functions which are constants on a single sheet.

J	sheet	Δ_J/Δ_0	λ_{0J}
0	1,2,3	1.0000	1.2739
0'	1,2,3	-.0390	-.0883
0''	1,2,3	-.0196	-.0435
1	1	.0273	.0677
2	1	-.0093	-.0178
3	1	.0007	-.0016
4	1	.0027	.0045
5	1	.0017	.0062
6	1	-.0016	-.0045
1	2	.0290	.0821
2	2	-.0043	-.0073
3	2	-.0010	-.0037
4	2	.0004	.0005
5	2	-.0008	-.0015
6	2	-.0034	-.0101
1	3	-.0107	-.0227
2	3	-.0080	-.0169
3	3	.0065	.0181

the residual resistances. It is difficult to learn the magnitude
of the depression at the minimum from their paper, but it is
apparently around $0.2^\circ K$ and species dependent. We urge that further
studies be made of this effect.

In principle the ideal experiment to probe gap anisotropy is
single crystal tunneling, which can measure Δ in specific directions
in $\underset{\sim}{k}$ space. Also polycrystalline tunneling can determine the rms
anisotropy from the broadening of the onset of normal resistance at
$eV \sim \Delta$. Unfortunately the application of tunneling to transition
elements has proved to be extremely difficult, probably because of
surface contamination. A recent report[45] finds $2\Delta \sim 3.93kT_c$ in
single crystal samples, with no evidence of any anisotropy greater
than 2%. Earlier reports had been interpreted in terms of very
large anisotropy, but this is now ascribed[45] to sample problems.

Ultrasonic attenuation has also been interpreted in terms of
very large anisotropy, but recent experiments[46] on very pure samples
are more consistent with an interpretation of very weak anisotropy.
In the quantum regime $q\ell \gg 1$ which is achieved only in very pure
samples and high ultrasonic frequencies, the attenuation is caused
by "belts" of electrons in the fermi surface with $\underset{\sim}{\nu_k}$ perpendicular
to the $\underset{\sim}{q}$ of sound. At low temperatures the attenuation is principally
caused by the states on this belt with the minimum gap. By analyzing
low T attenuation data, Carsey et al[47] have found $2\Delta/k_B T_c = 3.7$ for
$\underset{\sim}{q}$ in the [111] direction, 3.64 in the [100] direction, and 3.84 in
the [110] direction. A later experiment[48] with a purer sample and
larger $q\ell$ revised the value on the [100] direction from 3.64 to 3.56.
These experiments seem at the moment to provide the clearest evidence
that rms gap anisotropy in Nb is in the range 5-10%. It is also
noteworthy that a new technique which amounts to very high frequency
ultrasonic attenuation has been developed using neutrons[49] and that
within the resolution of this experiment, phonons in the [100] and
[110] directions see the same energy gap in niobium.

At the moment it seems premature to make a very detailed comparison between theory and experiment. We can hope that in the near
future both will improve and allow a more critical test of this
interesting lapse in our microscopic understanding of d-band superconductors.

ACKNOWLEDGEMENTS

P.B.A. thanks R.C. Dynes for much instruction and stimulation in this
field. W.H.B. thanks J.S. Faulkner and J.J. Olson for assistance in
the development of KKR constant energy search and plotting routines.

REFERENCES

1. J. Bardeen, L.N. Cooper and J.R. Schrieffer, Phys. Rev. $\underline{108}$, 1175 (1957).
2. G.M. Eliashberg, Zh. Eksp. Teor. Fiz. $\underline{38}$, 966 (1960); $\underline{39}$, 1437 (1960)[Sov. Phys. -JETP $\underline{11}$, 696 (1960); $\underline{12}$, 1000 (1961)].
3. D.J. Scalapino, J.R. Schrieffer, and J.W. Wilkins, Phys. Rev. $\underline{148}$, 263 (1966).
4. W.L. McMillan and J.M. Rowell, in Superconductivity, edited by R.D. Parks (Marcel Dekker, New York, 1969).
5. W.L. McMillan, Phys. Rev. $\underline{167}$, 331 (1968).
6. P.W. Anderson, J. Phys. Chem. Solids $\underline{11}$, 26 (1959); Proc. VII Int. Conf. on Low Temperature Physics, edited by G.M. Graham, Plenum, 1961, p. 298.
7. H. Suhl, B.T. Matthias, and L.R. Walker, Phys. Rev. Lett. $\underline{3}$, 552 (1959).
8. V.A. Moskalenko, Fiz. Met. Metalloved. $\underline{8}$, 503 (1959).
9. V.L. Pokrovskii, Zh. Eksp. Teor. Fiz. $\underline{40}$, 641 (1961). [Sov. Phys. - JETP $\underline{13}$, 447 (1961)].
10. D. Markowitz and L.P. Kadanoff, Phys. Rev. $\underline{131}$, 563 (1963).
11. T. Tsuneto, Prog. Theor. Phys. $\underline{28}$, 857 (1962).
12. C. Caroli, P.G. DeGennes, and J. Matricon, J. Phys. Rad. $\underline{23}$, 707 (1962).
13. P. Hohenberg, Zh. Eksp. Teor. Fiz. $\underline{45}$, 1208 (1963); [Sov. Phys. - JETP $\underline{18}$, 834 (1964)].
14. D.M. Brink and M.J. Zuckermann, Proc. Phys. Soc. (London) $\underline{85}$, 329 (1965).
15. J.R. Clem, Phys. Rev. $\underline{148}$, 392 (1966).
16. H. Teichler, Phil. Mag. $\underline{31}$, 775 (1975), and references therein.
17. P. Entel and M. Peter, J. Low. Temp. Phys. $\underline{22}$, 613 (1976).
18. P.B. Allen, Bull. Am. Phys. Soc. Ser. II, $\underline{21}$, 259 (1976).
19. P.B. Allen, Bull. Am. Phys. Soc. Ser. II, $\underline{21}$, 102 (1976).
20. P.B. Allen, Phys. Rev. B$\underline{13}$, 1416 (1976).
21. P.B. Allen and R.C. Dynes, Phys. Rev. B$\underline{12}$, 905 (1975).
22. P.B. Allen, Sol. State Commun. $\underline{14}$, 937 (1974).
23. J. Appel and H. Heyszenau, Phys. Rev. $\underline{188}$, 755 (1969); I. Foulkes and B.L. Gyorffy, preprint.
24. P.W. Anderson and P. Morel, Physica (Utr.) $\underline{26}$, 671 (1960), Phys. Rev. $\underline{123}$, 1911 (1961).
25. C.R. Leavens and J.P. Carbotte, Ann. Phys. (N.Y.) $\underline{70}$, 338 (1972).
26. D. Farrell, J.G. Park and B.R. Coles, Phys. Rev. Lett. $\underline{13}$, 328 (1964).
27. B.T. Geilikman and N.F. Masharov, Fiz. Met. Metalloved, $\underline{32}$, 492 (1971). B.T. Geilikmann, R.O. Zaitsev, and V.Z. Kresin, Fiz. Trend. Tela. (1967)[Sov. Phys. - Solid State $\underline{9}$, 642 (1967)].
28. L.P. Bouchaert, R. Smoluchowski, and E. Wigner, Phys. Rev. $\underline{50}$, 58 (1936).
29. J.J. Hopfield, Phys. Rev. $\underline{186}$, 443 (1969).
30. G.D. Gaspari and B.L. Gyorffy, Phys. Rev. Lett. $\underline{28}$, 801 (1972).

31. B.M. Klein and D.A. Papaconstantopoulos, Phys. Rev. Lett. $\underline{32}$, 1193 (1974).
32. W.H. Butler, J.J. Olson, J.S. Faulkner, and B.L. Gyorffy, Phys. Rev. B (in press).
33. J.S. Faulkner, H.L. Davis, and H.W. Joy, Phys.Rev. $\underline{161}$, 656 (1967).
34. L.F. Mattheiss, Phys. Rev. B$\underline{1}$, 373 (1970).
35. J. Yamashita and S. Asano, Prog. Theor. Phys. $\underline{51}$, 317 (1974).
36. D. Nowak, Phys. Rev. B6, 3691 (1972).
37. S.G. Das, Phys. Rev. B$\underline{7}$, 2238 (1973).
38. M.J.G. Lee, Phys. Rev. B$\underline{2}$, 250 (1970).
39. N.E. Christensen, Sol. State Commun. $\underline{9}$, 749 (1971).
40. S.J. Williamson, Phys. Rev. B$\underline{2}$, 3545 (1970).
41. G.J. Sellers, A.C. Anderson, and H.K. Birnbaum, Phys. Rev. B$\underline{10}$, 2771 (1974).
42. A.C. Anderson, C.B. Satterthwaite, and S.C. Smith, Phys. Rev. B$\underline{3}$, 3762 (1971).
43. S.G. O'Hara, G.J. Sellers and A.C. Anderson, Phys. Rev. B$\underline{10}$, 2777 (1974).
44. G.N. Ronami and V.P. Berezina, Fiz. Met. Metalloved. $\underline{37}$, 872 (1974).
45. J. Bostock, K. Agyeman, M.H. Frommer and M.L.A. MacVicar, J. Appl. Phys. $\underline{44}$, 5567 (1973).
46. D.P. Almond, M.J. Lea and E.R. Dobbs, Phys. Rev. Lett. $\underline{29}$, 764 (1972).
47. F. Carsey, R. Kagiwada, M. Levy and K. Maki, Phys. Rev. B$\underline{3}$, 854 (1971).
48. F. Carsey and M. Levy, Phys. Rev. B$\underline{7}$, 4123 (1973).
49. S.M. Shapiro, G. Shirane, and J.D. Axe, Phys. Rev. B$\underline{12}$, 4899 (1975).

QUESTIONS AND COMMENTS

E. Fawcett: This is a Fermi surface question. You have shown drawings of the Fermi surface of Nb which do not seem to agree with my recollection of the Fermi surface given by Mattheiss' calculation.

W. Butler: The Fermi surface which we obtained is essentially the same as Mattheiss. I think you recognize the distorted ellipsoid, and you recognize the jack. The third sheet which we have shown as a surface containing electrons centered at P will look quite different if it is drawn as a surface containing holes centered at Γ.

B. Gyorffy: Can you tell if there were any cancellations? The final result might come out small, but if there were cancellations during the calculation one should conclude that this is not a general rule and, indeed, in other systems you might have a bigger anisotropy.

W. Butler: Yes, there were cancellations. The small value of the gap anisotropy is due not only to the fact that the density matrix expansion coefficients T_J^α are small in absolute value for $J \neq 0$, but also to the fact that they can be positive or negative. I would guess that the rms anisotropy might be doubled or tripled if one were to replace the T_J^α's by their absolute values in calculating the gap. Other systems would be expected to have "cancellations" too, however, and it is not clear that niobium is an exceptional system with regard to the importance of this effect.

J. Phillips: This is a cubic crystal. The example you took before is zinc which is hexagonal. Do you expect much bigger anisotropy?

W. Butler: Generally we expect bigger anisotropy in hexagonal systems than in cubic. Zinc, however, is generally considered to be a simple metal with a Fermi surface not too much distorted from a sphere. Our initial expectation was that the complicated Fermi surfaces of transition metals should lead to large anisotropies compared to the simple metals.

J. Bostock: You get a mean square anisotropy of 0.004 which is presumably what your approximation allows you to calculate to some accuracy. What is that accuracy?

W. Butler: There are various sources of error. The most important source is our neglect of anisotropy arising from the phonons. Our initial expectation was that the electronic contribution to the gap anisotropy would be at least as important as the phonon contribution. Now that the electronic contribution has turned out to be so small we can no longer be confident that this is so. We plan to treat the phonons on an equal footing with the electrons in a later calculation. As for the other sources of error---Fermi surface harmonic expansion convergence, band theory convergence, matrix element uncertainties, etc.---we estimate the net effect of these errors on the <u>relative</u> gap anisotropy to be on the order of 10%.

F. Mueller: Would you care to speculate on the anisotropies in the A-15's---in terms of similar kinds of calculations?

W. Butler: No. Except to say that our calculation for Nb indicates that one can have a very anisotropic Fermi surface which does not lead to a large anisotropy of the gap function. This is partly due to the fact that in Nb the density of states is fairly uniformly distributed over the Fermi surface. This might happen in the A-15's also.

F. Mueller: Do you think the anisotropy in the A-15's is larger, for example?

W. Butler: You obviously have a very strange Fermi surface there. A very anisotropic Fermi surface.

F. Mueller: We are not talking about doing real calculations. We all know that they are very difficult.

W. Butler: The only thing I am going to say, Fred [Mueller], is how I use my intuition here. I say 'well, I see a very anisotropic Fermi surface and maybe there is a big anisotropy in the gap. There is a big anisotropy in the Fermi surface obviously of niobium and it did not lead to such a big anisotropy of the gap'.

R. Dynes: I think the point is that niobium is a surprise.

W. Butler: Yes, that is all I am saying.

F. Mueller: I did not find niobium so surprising.

W. Butler: You have better intuition, that's all.

C. Varma: It is quite possible, the fact that the gap function involves a convolution integral over the Fermi surface implies that anisotropy in the Fermi surface will not reflect equally in the gap function. I have a different question. There are several weak-coupling calculations in the literature that suggest that if the mean free path due to impurity scattering is less than the coherence length, the gap will be isotropic. Are there reasons to believe that this result will not hold in a strong-coupling calculation?

W. Butler: No. Our calculation, however, neglected impurity scattering. It should be applicable in the limit of very long mean free path.

P. Allen: It is true that the anisotropy enhancement of T_c is diminished when the mean free path (ℓ) is as short as the coherence length (ξ). However, the enhancement by no means vanishes at that point. It washes out rather slowly and continues to be measureable even when ℓ is much shorter than ξ.

J. Bostock: One more experimental comment: If you look at ultrasonic data, or thermal conductivity, or specific heat there is zero anisotropy observed for niobium.

W. Butler: I thought it was more than zero.

J. Bostock: Zero.

P. Allen: Ultrasonic attenuation measurements by Carsey et al. [Refs. 47 and 48] suggest 5 or 10% anisotropy.

J. Bostock: That is not al all clear; see C. Gough's paper.

M. Strongin: I am a little confused about Varma's remarks because in niobium if I remember the coherence length is something like 400 Å and at room temperature the mean free path is something like 50 Å, but if you have a resistance ratio of 100 then the mean free path at low temperatures would be much larger than the coherence length.

C. Varma: I am talking about the impurity mean free path and I am really concerned about the A-15's.

W. Butler: That depends on how pure you get your niobium, doesn't it?

B. Klein: We know that anisotropies in k space are just something like Fourier transforms of anisotropies in real space. Since it is well known from the band structure point of view that using spherically averaged charge densities and potentials give very good values of macroscopic quantities, maybe it isn't such a surprise that anisotropies in k space do not show up very strongly.

ANISOTROPY PHENOMENA IN CUBIC d-BAND SUPERCONDUCTORS

H. Teichler

Institut für Physik am Max-Planck-Institut für Metallforschung

Stuttgart, Germany

and

Institut für theoretische und angewandte Physik der Universität Stuttgart

Stuttgart, Germany

ABSTRACT

Anisotropy phenomena of monocrystalline superconductors are reviewed. A theory of H_{c2} anisotropy in cubic metals with low impurity content at arbitrary temperatures below T_c is presented. According to this theory the anisotropy of the energy gap of the pure material and details of the electron band-structure anisotropy may be deduced from investigations of the temperature and impurity dependence of H_{c2} anisotropy. By means of this theory microscopic anisotropy parameters for Nb are determined from recent H_{c2} anisotropy data. Estimates of preferential orientations of the fluxline lattice relative to the crystals lattice resulting from these parameters are in agreement with experimental observations.

INTRODUCTION

According to our present understanding of superconductivity the physical properties of a particular superconductor, e.g., the value of the transition temperature, may be calculated by taking

carefully into account all details of the microscopic structure of
the electron-phonon system (such as the electron band structure,
phonon dispersion curves, electron-phonon interaction matrix elements, etc.) and solving an appropriate set of equations like
Eliashberg's equations [1]. However, nowadays this idea appears
not to be practical in general: difficulties arise even when calculating the normal state quantities and, moreoever, when trying
to include the very details of the normal state in the evaluation
of the superconducting properties. That is to say, some important
problems in the theory of superconductivity, in particular in the
area of d- and f-band superconductors, are related not to the
foundations of theory but to the determination and incorporation of
various details of the microscopic electron-phonon system in the
description of the superconducting state.

In order to develop mathematical methods which take into account
the microscopic structure of the system in all details it would be
worthwhile to study systematically such properties of the superconductors that depend on the microscopic structure in a very
direct and simple manner. Physical properties well suited for this
purpose may be found in the anisotropy phenomena: the microscopic
anisotropies of the electron-phonon system lead to anisotropies in
the energy gap of the superconductor which thus becomes a function
of $\underline{k}$-vector on the Fermi surface, $\phi \equiv \phi(\underline{k}_F)$. The anisotropy of the
energy gap now may be detected either directly, e.g., by tunneling
measurements [2], indirectly in 'isotropic' quantities such as the
specific heat [3] or the T_c-shift due to impurities [4], or indirectly in macroscopic anisotropy phenomena like the anisotropy of
H_{c_2} [5,6].

All these phenomena may be used to study the effects of microscopic anisotropies on the superconducting state. In particular
superconducting tunneling measurements offer a direct way for
studying energy gap anisotropies. However, as reported recently
by Bostock and MacVicar [2], at least in the case of Nb the gap
anisotropies deduced from tunneling experiments are inconsistent
with each other and with those obtained form specific heat data.
Due to that it seems to be even more important to investigate the
macroscopic anisotropy phenomena and to discover up to what extent
these macroscopic anisotropies may be used to study microscopic
details of the superconducting electron-phonon system.

In the following, we give a brief survey of macroscopic
anisotropy phenomena known at present for d-band superconductors,
in particular for Nb and V. Then we discuss some recent developments in the theory of H_{c_2} anisotropy which make it possible to
extract informations about the anisotropy of the energy gap from
measurements of H_{c_2} anisotropy. Finally we describe the relation
between the experimentally observed preferential orientations of
the flux-line lattice relative to the crystal lattice and the micro-

scopic anisotropies of the system.

MACROSCOPIC ANISOTROPY PHENOMENA IN CUBIC SUPERCONDUCTORS

Macroscopic anisotropy phenomena are observed in the magnetization curve of monocrystalline superconductors and in the structure of the flux-line lattice (FLL) of the mixed state. Examples for anisotropies in the magnetization curve are:

the anisotropy of the upper critical field H_{c_2} (i.e., the dependence of the value of H_{c_2} on the orientation of the crystal axes relative to the direction of the external magnetic field) [5, 6],
  the anisotropy of κ_2 [7,8],
  the anisotropy of H_{c_1} [9,10],
  the anisotropy of the induction jump B_0 at H_{c_1} [10].

In addition to these anisotropies further anisotropies may be detected by investigating the structure and orientation of the FLL of the mixed state either by means of the decoration technique of Träuble and Essmann [11] or by means of neutron diffraction experiments [12]. By these methods one finds:
  correlation between the symmetry of the crystal lattice (CL) and the structure of the FLL: e.g., Nb single crystals with magnetic field parallel to the fourfold [100] crystallographic direction display a square FLL [13] (as shown schematically in Fig. 1) whereas the FLL is triangular for magnetic fields parallel to the threefold [111] axes,
  preferential orientations of the basis vectors of the FLL relative to the CL [12].

Among all these macroscopic anisotropy phenomena up to now only the anisotropy of H_{c_2} has been investigated extensively by theory [14-19] and experiment (e.g., [4,5,9]). The complications which hinder the extensive use of the other macroscopic anisotropy phenomena are various and differ from case to case: Sample shape and interactions between flux lines and CL imperfections (in particular interactions of flux lines with crystallographic oriented dislocations [20]) affect the anisotropy of H_{c_1} (but see, e.g., [9]). Experimental data concerning the anisotropy of κ_2 in Nb [7,8] are available but the theory [15] of κ_2 anisotropy is valid only near T_c and, moreover, it attributes this anisotropy entirely to the band structure anisotropy of the normal state. Last but not least, the correlations between FLL and CL yield only qualitative information about the microscopic anisotropies of the system but no quantitative results. Because of this we consider in the following mainly the problem of H_{c_2} anisotropy. But, in addition, we shall have a short glance at the problem of preferential orientations of the FLL relative to the CL, too.

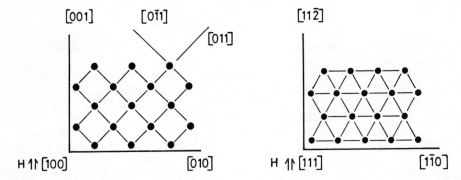

Figure 1: Schematic plot of flux-line arrangements in monocrystalline Nb samples with different crystallographic orientations.

H_{c2} ANISOTROPY IN CUBIC SUPERCONDUCTORS

In cubic metals the dependence of H_{c2} on the orientation of the field direction $\underline{e}_H$ relative to the crystal axes obviously has to reflect the cubic symmetry. Hence the orientation dependence of $H_{c2}(\underline{e}_H)$ (or of the relative anisotropy $\Delta H_{c2}(\underline{e}_H)/\overline{H}_{c2}$) may be expanded in terms of orthonormalized cubic harmonics $H_\ell(\underline{e}_H)$*[21]

$$\frac{\Delta H_{c2}(\underline{e}_H)}{\overline{H}_{c2}} \equiv \frac{H_{c2}(\underline{e}_H) - \overline{H}_{c2}}{\overline{H}_{c2}} = \sum_{\ell=4,6,\ldots} H_\ell(\underline{e}_H) \cdot a_\ell(T,\tau). \quad (1)$$

($\overline{H}_{c2}$ denotes the average over all space directions.) In our notation of cubic harmonics the subscript ℓ refers to the angular momentum value of the spherical harmonics from which the particular $H_\ell(\underline{e}_H)$ is constructed.

The expansion coefficients $a_\ell(T,\tau)$ reveal the microscopic anisotropies of the superconducting electron system. These coefficients depend on the temperature T and on the impurities in the sample which may be characterized by their s-wave scattering time τ. In 1967 Hohenberg and Werthamer [14] proved for the so-called case of small nonlocality, i.e., for temperatures near T_c or for samples with sufficient high impurity content, that the leading nonlocal corrections of Gorkow's theory [22] to the Ginzburg-Landau equations [23] account for the H_{c2} anisotropy in cubic superconductors and that these leading corrections result in a nonvanishing $a_{\ell=4}(T,\tau)$. Attributing the H_{c2} anisotropy entirely to electron band structure anisotropies they showed that $a_{\ell=4}$ factorizes into a universal weight function depending on T and τ and into one single material parameter which describes the anisotropy of Fermi surface and Fermi velocity. This theory for the regime of small nonlocality has been generalized as to include higher angular momentum components [15-17]. However, since the magnitude of relative H_{c2} anisotropy decreases with increasing temperature and impurity content experimental investigations of H_{c2} anisotropy are usually carried out in metals with low or moderate impurity content and at temperatures clearly below T_c, that means in the so-called regime of strong

---

* An additional subscript j should be introduced in the notation of cubic harmonics to label the different irreducible representations for a particular ℓ. But since for $\ell \leq 10$ at most one irreducible representation exists for each ℓ and since higher angular momentum contributions are far beyond the resolution of present H_{c2} anisotropy measurements we shall suppress this subscript for convenience.

nonlocality. Because of this a theory of H_{c_2} anisotropy for the regime of strong nonlocality will be presented in the following which takes into account as sources of H_{c_2} anisotropy the attractive electron-electron interaction as well as the electron band structure.

THEORY OF H_{c_2} ANISOTROPY IN THE REGIME OF STRONG NONLOCALITY

The nonlocal theory [18,19] of H_{c_2} anisotropy is based on a formulation of the superconducting electron-phonon system in terms of anisotropic transport-like equation [24] which mean generalizations of the transport-like equations of, e.g., Eilenberger [25] or Betbeder-Matibet and Nozieres [26]. In order to derive explicit results on the temperature and impurity dependence of $\Delta H_{c_2}/\bar{H}_{c_2}$ some simplifications have been introduced in the theory. In particular, the anisotropic attractive electron-electron interaction which depends in general on the wave vectors of the interacting electrons and on their energies is replaced by an energy independent and separable model interaction as introduced by Markowitz and Kadanoff [4]

$$V(\underline{k}_F, \underline{k}'_F; \omega, \omega') \rightarrow \psi(\underline{k}_F) V(o) \psi(\underline{k}'_F) . \qquad (2)$$

Application of this model interaction to the homogeneous Meissner state yields that $V(o)$ denotes the effective coupling strength which determines T_c and that $\psi(k_F)$ describes the reduced anisotropy of the energy gap of the pure metals, i.e.,

$$\psi(\underline{k}_F) = \phi(\underline{k}_F)/\sqrt{<\phi^2>} . \qquad (3)$$

Here the brackets denote an average over the Fermi surface weighted by the anisotropic density of states at the Fermi surface:

$$<\ldots> = \int_{F.S.} \frac{d^2\underline{k}_F}{v(\underline{k}_F)} \ldots \bigg/ \int_{F.S.} \frac{d^2\underline{k}_F}{v(\underline{k}_F)} . \qquad (4)$$

For impure metals the anisotropy incorporated in $\psi(\underline{k}_F)$ must be distinguished from the anisotropy of $\omega_0(k_F)$, the energy gap in the renormalized excitation spectrum of the superconductor as measured by tunneling experiments [27]. Whereas impurities strongly reduce the anisotropy of $\omega_0(\underline{k}_F)$ [27,28] $\psi(\underline{k}_F)$ retains its clean metal structure. (Nevertheless the effect of $\psi(\underline{k}_F)$ on the anisotropy of H_{c2} also is reduced by impurities.) As additional approximation only linear effects of microscopic anisotropies are taken into account, that means only linear terms in

$$\delta\psi^2 = \psi^2(\underline{k}_F) - 1 \equiv \left. \frac{\phi^2(\underline{k}_F) - <\phi^2>}{<\phi^2>} \right| \text{clean limit,} \quad (5)$$

$$\frac{\delta v^2}{<v_F^2>} = \frac{v^2(\underline{k}_F) - <v_F^2>}{<v_F^2>}, \quad (6)$$

are considered. In the framework of these approximations it follows that for each angular momentum value ℓ the coefficient $a_\ell(T,\tau)$ may be represented by [19]

$$a_\ell(T,\tau) = c_\ell \{ h_\ell^{(1F)}(t,\alpha)\gamma_\ell^{(1F)} + h_\ell^{(1\psi)}(t,\alpha) \cdot \gamma_\ell^{(1\psi)} + h_\ell^{(2)}(t,\alpha)\gamma_\ell^{(2)} \}. \quad (7)$$

$t = T/T_c$ denotes the reduced temperature and

$$\alpha = \frac{h}{2\pi k_B T_c \cdot \tau}$$

measures the impurity scattering in the sample. The quantities

$$\gamma_\ell^{(1F)} = <H_\ell(\underline{v}_F)>, \quad (8)$$

$$\gamma_\ell^{(1\psi)} = <\delta\psi^2 H_\ell(\underline{v}_F)>, \quad (9)$$

$$\gamma_\ell^{(2)} = <\frac{\delta v^2}{<v_F^2>} H_\ell(\underline{v}_F)>, \quad (10)$$

are material parameters of the host metal: The parameters $\gamma_\ell^{(1\psi)}$ describe the anisotropy of the attractive electron-electron coupling; the $\gamma_\ell^{(1F)}$ measure the anisotropy of the direction of the Fermi velocity, the density of states at the Fermi surface, and the shape of the Fermi surface; the $\gamma_\ell^{(2)}$ characterize the anisotropy of $v^2(k_F)$. The $h_\ell^{(i)}(t,\alpha)$ are weight functions which give the weights by which the various microscopic anisotropies contribute to the anisotropy of H_{c2}.

In (7) the coefficients c_ℓ are introduced as to yield in the clean limit ($\alpha = 0$) for temperatures near T_c the asymptotic forms

$$h_\ell^{(1F)}(t,\alpha=0) \equiv h_\ell^{(1\psi)}(t,\alpha=0) \underset{t\to 1}{=} (1-t)^{\ell/2-1} \qquad (11)$$

$$h_\ell^{(2)}(t,\alpha=0) \underset{t\to 1}{=} \tfrac{1}{2}\ell(1-t)^{\ell/2-1}. \qquad (12)$$

For the lowest angular momentum values $\ell = 4$ and $\ell = 6$ the conditions (11), (12) give in particular

$$c_{\ell=4} = \frac{3}{35} \frac{\lambda(5)}{\lambda(3)^2},$$

$$c_{\ell=6} = \frac{135}{2002} \frac{\lambda(7)}{\lambda(3)^3},$$

$$\lambda(n) = \sum_{k=0}^{\infty} (2k+1)^{-n}.$$

The detailed (t,α)-dependence of the $h_\ell^{(i)}$ has to be determined numerically. Figure 2 presents plots of the weight functions for $\ell=4$ and $\ell=6$. (For convenience only the functions with $\alpha=0$, 0.1, 0.5, and 1.5 are displayed, more complete data will be given in [19].) Comparison of Figure 2a with Figure 2b shows that $h_\ell^{(1F)}$ and $h_\ell^{(1\psi)}$ are identical in the clean limit and that $h_\ell^{(1\psi)}$ decreases more rapidly with increasing impurity content than $h_\ell^{(1F)}$ does. From Fig. 2c it is obvious that $h_\ell^{(2)}$ has a completely different temperature dependence even in the pure metal. On account of that value of the parameters $\gamma_{\ell=4}^{(2)}$ and $\gamma_{\ell=4}^{(1)} = \gamma_{\ell=4}^{(1F)} + \gamma_{\ell=4}^{(1\psi)}$ may be extracted from measurements of the H_{c2} anisotropy in clean metals whereas the further decomposition of $\gamma_{\ell=4}^{(1)}$ in $\gamma_{\ell=4}^{(1F)}$ and $\gamma_{\ell=4}^{(1\psi)}$ (that is to say, the decomposition into Fermi surface and electron-electron coupling contributions) may be achieved by investigation of the

ANISOTROPY PHENOMENA

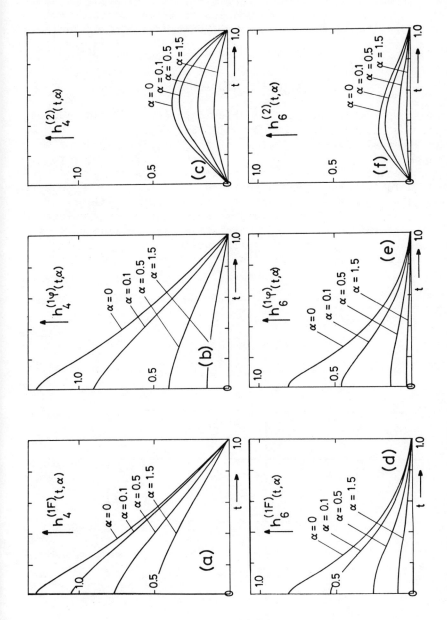

Figure 2: Weight functions $h_\ell^{(i)}(t,\alpha)$ describing the effects of microscopic anisotropies on $\overline{\Delta H_{c2}/\overline{H}_{c2}}$.

impurity dependence of H_{c2} anisotropy [18,19]. As may be seen from Fig. 2d-f the weight functions with $\ell=6$ display similar impurity dependences than those with $\ell=4$. Therefore, also material parameters $\gamma_{\ell=6}^{(1F)}$, $\gamma_{\ell=6}^{(1\psi)}$, and $\gamma_{\ell=6}^{(2)}$ may be deduced from experimental investigations of the temperature and impurity dependence of H_{c2} anisotropy.

APPLICATION TO NIOBIUM*

For Nb the temperature and impurity dependence of the reduced H_{c2} anisotropy has been measured by Seidl and Weber [29] in the system Nb-N for six different values of the impurity parameter α in the range $0.028 \leq \alpha \leq 1.35$. By least squares fits of (7) to their experimental data the values of material parameters $\gamma_\ell^{(i)}$ displayed in Table I have been determined. According to these data the $\ell=4$ angular momentum component of the reduced H_{c2} anisotropy mainly originates from electron band structure and Fermi surface anisotropies of the normal state whereas the overwhelming part of the $\ell=6$ component of H_{c2} anisotropy is due to anisotropies of the attractive electron-electron coupling.

Table I

Microscopic anisotropy parameters of Nb

	$\gamma_\ell^{(1F)}$	$\gamma_\ell^{(1\psi)}$	$\gamma_\ell^{(2)}$
$\ell=4$	-0.177	-0.062	-0.103
$\ell=6$	-0.042	0.251	-0.036

---

* For vanadium estimates of the $\gamma_\ell^{(i)}$ have been determined [19] from Williamson's [5] H_{c2} anisotropy measurements. However, these experimental data are not sufficient to evaluate values of the $\gamma_\ell^{(1)}$ for V unambiguously.

PREFERENTIAL ORIENTATIONS OF THE FLUX LINE LATTICE RELATIVE TO THE CRYSTAL LATTICE IN Nb

Up to now the microscopic anisotropy parameters $\gamma_\ell^{(i)}$ have not yet been determined by other methods or calculated from first principles. Thus, we cannot compare their values derived from H_{c2} anisotropy measurements with other estimates. However, in order to prove the reliability of the deduced $\gamma_\ell^{(i)}$ we may use these values to make predictions about further macroscopic anisotropy phenomena. Well suited for this purpose is the phenomenon of preferential orientations of the FLL relative to CL, the theory [30] of which we shall describe briefly in the following.

The preferential orientations of the FLL indicate an orientation dependent interaction between flux lines. Moreover, the induction jump B_0 observed in Nb [31,32] means that there exists an attractive interaction between flux lines in Nb [31,32]. (According to the theory of Kramer [33] and Leung [34] every superconductor with κ slightly above $1/\sqrt{2}$ has to have an attractive flux-line interaction.) Thus we may conclude that the flux lines in Nb affect each other by an anisotropic attractive interaction. This conclusion is confirmed by a recent theory of K. Fischer [35] who proved for anisotropic superconductors with κ slightly above $1/\sqrt{2}$ that there exists an attractive interaction energy between flux lines which for large flux-line distances r_{ij} has the asymptotic form

$$W_{ij} \sim -\exp[-r_{ij}/\xi(\underline{e}_{ij})] \ . \qquad (13)$$

($\underline{r}_{ij} = r_{ij} \cdot \underline{e}_{ij}$.) The decay length $\xi(\underline{e}_{ij})$ in (13) is identical to the anisotropic coherence length derived in [36].

In order to evaluate the influence of the microscopic anisotropies on the orientation of the FLL we assume that the structure of the FLL is known. If we restrict ourselves to nearest neighbor interactions in the FLL the leading term in the orientation dependent part of the FLL energy reads for small anisotropies of the coherence length

$$\Delta W_{FLL} \sim -(b/\overline{\xi}) \exp(-b/\overline{\xi}) \cdot \tfrac{1}{2} \sum_{ij} \frac{\Delta\xi(\underline{e}_{ij})}{\overline{\xi}} \ .$$

(b: nearest neighbor distance). $\Delta\xi(\underline{e}_{ij})/\overline{\xi}$ denotes the relative anisotropy of the coherence length. For impurity-free superconductors $\Delta\xi/\overline{\xi}$ is given in linear order of microscopic anisotropies by

$$\frac{\Delta\xi(e)}{\overline{\xi}} = \sum_{\ell=4,6,\ldots} H_\ell(e) \cdot \{ g_\ell^{(1F)}(t)\gamma_\ell^{(1F)} + g_\ell^{(1\psi)}(t)\gamma_\ell^{(1\psi)}$$
$$+ g_\ell^{(2)}(t)\gamma_\ell^{(2)} \}, \qquad (15)$$

where the $\gamma_\ell^{(i)}$ are identical to those entering the expression (7) for the relative H_{c2} anisotropy. The $g_\ell^{(i)}(t)$ denote the weights by which the various $\gamma_\ell^{(i)}$ contribute to $\Delta\xi/\overline{\xi}$ and thus to the orientation dependent part of the FLL energy. Numerical calculated plots of these functions are presented in Figure 3 for angular momentum $\ell = 4$ and $\ell = 6$. Obviously the $g_\ell^{(i)}$ are completely different from the $h_\ell^{(i)}$. In particular, the $h_\ell^{(i)}$ are always positive whereas the $g_\ell^{(1\psi)}$ change their signs at intermediate temperatures. (Because of this at low temperatures metals with similar H_{c2} anisotropy may show different preferential orientations of the FLL relative to the CL.) Nevertheless, for $t \to 1$ the asymptotic proportionalities $g_\ell^{(i)} \sim (1-t)^{\ell/2-1}$ hold.

Using the $\gamma_\ell^{(i)}$ values of Table I for Nb and assuming a triangular FLL eq. (14) yields for magnetic fields parallel to the [111], [110], or [112] crystal axes that at low temperatures one of the basis vectors of the FLL should be parallel to the [110] direction. This theoretical result agrees with experimental observations on Nb single crystals [12,37]. In addition, if for magnetic fields parallel to the [001] cyrstal axis a square FLL is realized Eq. (14) predicts that the basis vectors of the FLL are parallel to the crystallographic <110> directions. Even this results agrees with experimental observations [12,13]. Thus we may conclude that the microscopic anisotropy parameters presented in Table I for Nb account for the temperature and impurity dependence of H_{c2} anisotropy as well as for the preferential orientations of the FLL relative to the CL.

SUMMARY

It has been pointed out that there exists a large variety of anisotropy phenomena which contain informations about details of the superconducting electron-phonon system. In particular the anisotropy of H_{c2} in cubic superconductors with low or moderate impurity content at temperatures well below T_c offers an interesting tool to investigate the microscopic anisotropies of the system. Application of the corresponding theory of H_{c2} anisotropy to experimental measurements of Nb yielded explicit data about energy-gap and band-structure anisotropies in this metal. These data may be used to test the validity of theoretical techniques developed for

ANISOTROPY PHENOMENA

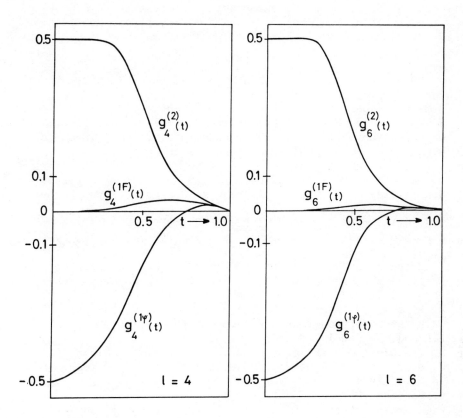

Figure 3: Weight functions $g_\ell^{(i)}(t)$ describing the effects of microscopic anisotropies on $\Delta\xi/\bar{\xi}$.

incorporating into the description of the superconducting state details of the electron band structure, phonon-dispersion curves or electron-phonon interaction.

ACKNOWLEDGEMENTS

The author gratefully acknowledges stimulating remarks of Prof. Dr. A. Seeger and his encouraging interest in this problem. He wishes to thank Drs. E. Seidl and H.W. Weber for submitting their experimental results prior to publication.

REFERENCES

1. G.M. Eliashberg, S.P. -JETP $\underline{11}$, 696 (1960).
2. J. Bostock and M.L.A. MacVicar in "Proc. Internat. Disc. Meeting on Anisotropy Effects in Superconductors, Vienna, 1976" (ed. H.W. Weber) to be published Plenum Publ. Comp.
3. J.R. Clem, Ann. of Phys. $\underline{40}$, 268 (1966).
4. D. Markowitz and L.P. Kadanoff, Phys. Rev. $\underline{131}$, 563 (1963).
5. D.R. Tilley, G.J. van Gurp, and C.W. Berghout, Phys. Letters $\underline{12}$, 305 (1964).
6. S.J. Williamson, Phys. Rev. $\underline{B2}$, 3545 (1970).
7. C.E. Gough, Sol. State Commun. $\underline{6}$, 215 (1968).
8. I. Williams and A.M. Court, Sol. State Commun. $\underline{7}$, 169 (1969).
9. T. Ohtsuka in "Proc. Internat. Disc. Meeting on Anisotropy Effects in Superconductors, Vienna, 1976" (ed. H. W. Weber), to be publ. by Plenum Publ. Comp.
10. H. Kiessig, U. Essmann, H. Teichler, and W. Wiethaup in "Proc. 14th Internat. Conf. on Low Temperature Physics, Otaniemi, Finland, August 1975" Vol. 2 p. 317.
11. H. Träuble and U. Essmann, phys. stat. sol. $\underline{18}$, 813 (1966).
12. J. Schelten, G. Lippmann, and H. Ullmaier, J. Low Temp. Phys. $\underline{14}$, 213 (1974).
13. U. Essmann, Physica $\underline{55}$, 83 (1971).
14. P.C. Hohenberg and N.R. Werthamer, Phys. Rev. $\underline{153}$, 493 (1967).
15. K. Takanaka and T. Nagashima, Progr. Theor. Phys. $\underline{43}$, 18 (1970).
16. K.-D. Harms, Z. Naturforsch. $\underline{25a}$, 1161 (1970).
17. T. Nagashima, Progr. Theor. Phys. $\underline{47}$, 37 (1972).
18. H. Teichler, phys. stat. sol. (b) $\underline{69}$, 501 (1975).
19. H.-W. Pohl and H. Teichler, phys. stat. sol. (b), in press.
20. R. Ihle, Diploma-Work, Stuttgart 1975.
21. F.C. von der Lage and H.A. Bethe, Phys. Rev. $\underline{71}$, 612 (1947).
22. L.P. Gorkov, S.P.-JETP $\underline{7}$, 505 (1958).
23. N.R. Werthamer in "Superconductivity" ed. R.D. Parks (Marcel Dekker Inc., New York, 1969) vol. I.
24. H. Teichler, Phil. Mag. $\underline{31}$, 775 (1975).
25. G. Eilenberger, Z. Phys. $\underline{214}$, 195 (1968).

26. O. Betbeder-Matibet and P. Nozieres, Ann. of Physics 51, 392 (1969).
27. P. Hohenberg, S.P.-JETP 18, 834 (1964).
28. J.R. Clem, Phys. Rev. 148, 392 (1966).
29. E. Seidl and H.W. Weber in "Proc. Internat. Disc. Meeting on Anisotropy Effects in Superconductors, Vienna, 1976" (ed. H.W. Weber) to be published by Plenum Publ. Comp.
30. K. Fischer and H. Teichler, submitted to Phys. Letters.
31. U. Kumpf, phys. stat. sol. (b) 44, 829 (1971).
32. J. Auer and H. Ullmaier, Phys. Rev. B7, 136 (1973).
33. L. Kramer, Z. Physik 258, 367 (1973).
34. M.C. Leung, J. Low Temp. Phys. 12, 215 (1973).
35. K. Fischer, Thesis, Stuttgart 1975.
36. H. Teichler, Phil. Mag. 31, 789 (1975).
37. R. Kahn and G. Parette, Solid State Commun. 13, 1839 (1973).

QUESTIONS AND COMMENTS

J. Phillips: Experimentally, how clean is clean?

H. Teichler: Well, I should say that the niobium data of Dr. Williamson which I included for comparison had an impurity residual resistivity ratio of about 750 or so, whereas in the data of Seidl and Weber the cleanest sample corresponds to a residual resistivity ratio of about 600.

J. Phillips: As it gets dirtier, what happens?

H. Teichler: If it gets dirtier for instance, for α equal 0.7, which corresponds to a residual resistivity ratio of about 15, then the anisotropies are appreciably produced.

J. Phillips: Is this theory or experiment?

H. Teichler: Theory and experiment. The experimental points and the theoretical fittings to these points, i.e., theory and experiment, indicate the same behavior. In particular, a strong impurity dependence in the ℓ equals 6 component indicates that this anisotropy cannot be due to Fermi surface effects. It must be due to other effects, e.g., due to anisotropic coupling.

THE EFFECT OF HIGH PRESSURE ON SUPERCONDUCTING TERNARY

MOLYBDENUM CHALCOGENIDES

R.N. Shelton

Institute for Pure and Applied Physical Sciences
University of California at San Diego

La Jolla, California 92093

ABSTRACT

The variation of the superconducting transition temperature as a function of hydrostatic pressure up to 22 kbar is determined for over fifty Chevrel phase molybdenum chalcogenides. These materials are found to be more sensitive to pressure than any other class of superconductors.

INTRODUCTION

The ternary molybdenum chalcogenides comprise an intriguing class of superconducting materials with exceptional properties. Subsequent to the initial paper by Chevrel et al. [1] describing the synthesis of the ternary molybdenum sulfides, Matthias and coworkers reported superconductivity in some of these compounds [2]. Transition temperatures (T_c's) may reach as high as 15 K. Progress toward a more complete understanding of the crystal structure of these ternaries was initiated in 1973 when Bars and coworkers solved the structure of Mo_3Se_4 [3]. This compound was known to be superconducting at 6.4 K [4] and crystallizes in the structure that forms the prototype for the ternary molybdenum chalcogenides. In the same year, similar work was reported on the structure of some ternary materials [5-7].

Exceptionally high upper critical magnetic fields have been measured for these ternary compounds [8-12]. In fact the highest critical fields reported for any superconductor are those for lead molybdenum sulfide with zero temperature values in excess of 500 kGauss. These data lend importance to further research on this family of materials.

The experimental techniques employed for measuring T_c at high pressures have been detailed elsewhere [13]. However it is important to note that all pressures are generated hydrostatically at room temperature using a one to one mixture of n-pentane and isoamyl alcohol as the pressure transmitting fluid [14]. As a means of verifying the reversibility of the data, the zero pressure value of T_c was redetermined after the release of pressure for every sample exhibiting nonlinear behavior. In each instance $T_c(0)$ reproduced identically within experimental error. All compounds studied in this work were synthesized directly from the elements or from the elements plus the appropriate sulfide or selenide of the tertiary element. Specific methods of sample preparation have been thoroughly discussed in other papers [1,15,16,17].

DATA

The Binary System Mo_3S_4 - Mo_3Se_4 - Mo_3Te_4

The compound Mo_3Se_4 becomes superconducting at 6.3 K in contrast to the isostructural binaries Mo_3S_4 and Mo_3Te_4 which are normal to 1.1 K and 0.05 K respectively. The synthesis of superconducting pseudobinary compounds is possible within certain solubility limits [18] and the variation of T_c as a function of composition is presented in Figure 1. The solubility limit in the system $Mo_3(Se_{1-x}S_x)_4$ terminates at approximately 60 at.% S. Compounds with greater amounts of sulfur do not synthesize by direct combination of the elements, but must be acid reduced from a suitable ternary molybdenum sulfide-selenide [19]. To insure a self-consistent approach, we have measured only pseudobinary compounds prepared by direct synthesis. The sold solubility is complete for $0 \leq x \leq 1.0$ in the system $Mo_3(Se_{1-x}Te_x)_4$; however, no superconductivity was detected in any sample with greater than 70 at.% Te.

Several salient features of superconductivity in this binary system are evident in Figure 1. Clearly T_c reaches a maximum at the composition Mo_3Se_4. The rate of change of T_c with composition near this maximum dT_c/dx, is radically different depending on whether sulfur or tellurium is substituted for selenium. Upon substitution of sulfur, the decline of T_c is immediate and rapid. This drop is almost discontinuous when considered with respect to the gentle slope immediately to the tellurium side of Mo_3Se_4.

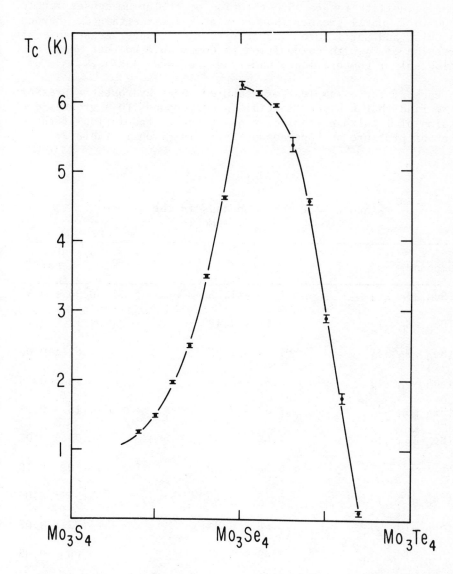

Figure 1: Superconducting transition temperature as a function of composition for the binary system Mo_3S_4-Mo_3Se_4-Mo_3Te_4. Error bars indicate transition widths.

A value of $dT_c/dx = -0.15$ K/at.% correctly describes this slope up to 20 at.% S. This precipitous fall slows near the edge of the direct solubility range. Substitution of tellurium causes little change in T_c until greater than ~ 25 at.% Te is attained. Beyond this point and until superconductivity is no longer measurable, the depression of T_c with composition is comparable to that seen for the low sulfur compositions, namely $dT_c/dx = -0.14$ K/at.%.

Each of these samples was measured under hydrostatic pressure and we find that a higher initial T_c corresponds to a greater depression of T_c with pressure. A quantitative tabulation of the effect of pressure on superconductivity is given in Table 1.

Table 1

T_c and dT_c/dP for compounds in the system
$Mo_3S_4 - Mo_3Se_4 - Mo_3Te_4$

Compound	$T_c(0)$ (K)	dT_c/dP (10^{-5} K-bar^{-1})
$Mo_3(Se_{0.4}S_{0.6})_4$	1.27 - 1.23	-0.96 ± 0.5
$Mo_3(Se_{0.5}S_{0.5})_4$	1.51 - 1.47	-1.12 ± 0.05
$Mo_3(Se_{0.6}S_{0.4})_4$	1.99 - 1.95	-1.62 ± 0.05
$Mo_3(Se_{0.7}S_{0.3})_4$	2.52 - 2.48	-2.12 ± 0.05
$Mo_3(Se_{0.8}S_{0.2})_4$	3.52 - 3.49	-4.12 ± 0.05
$Mo_3(Se_{0.9}S_{0.1})_4$	4.65 - 4.60	-6.75 ± 0.05
Mo_3Se_4	6.30 - 6.19	-13.10 ± 0.10
$Mo_3(Se_{0.9}Te_{0.1})_4$	6.16 - 6.10	-5.95 ± 0.05
$Mo_3(Se_{0.8}Te_{0.2})_4$	5.98 - 5.94	-5.22 ± 0.05
$Mo_3(Se_{0.7}Te_{0.3})_4$	5.49 - 5.34	-4.06 ± 0.05
$Mo_3(Se_{0.6}Te_{0.4})_4$	4.62 - 4.54	-2.29 ± 0/05
$Mo_3(Se_{0.5}Te_{0.5})_4$	2.95 - 2.89	non-linear
$Mo_3(Se_{0.4}Te_{0.6})_4$	1.83 - 1.68	non-linear
$Mo_3(Se_{0.3}Te_{0.7})_4$	0.09 - 0.04	5.96 ± 0.05

The specific behavior of the selenium-sulfur pseudobinaries is illustrated in Figure 2. Immediately evident from the data in Figure 2 is the fact that pressure depresses T_c for all of the selenium-sulfur compounds. For $x \geq 0.4$ a slight deviation from a linear change in T_c occurs between zero and five kbar. In this region, T_c for these samples decreases less rapidly than in the higher pressure range. In fact the 60 at.% S sample exhibits a slight but definite enhancement of T_c (22 mK over 6 kbar). For these compounds the high pressure data have been used to obtain the numerical values of dT_c/dP listed in Table 1. This is consistent with previous treatment of the ternaries [20].

The dependence of T_c on pressure for the compounds $Mo_3(Se_{1-x}Te_x)_4$ with $0 \leq x \leq 0.7$ is presented in Figure 3. The pure selenide remains most sensitive to pressure. As greater amounts of tellurium are substituted for selenium, dT_c/dP becomes progressively less negative until with 50 at.% Te a significant enhancement of T_c with pressure is realized. This increase in T_c occurs for pressures up to ~ 12 kbar, beyond which T_c begins a gradual decline. The compound with 60 at.% Te shows a stronger initial rise in T_c and this enhancement remains for pressures up to ~ 20 kbar although the rate dT_c/dP lessens at these higher pressures. Although the zero pressure value of T_c for $Mo_3(Se_{0.3}Te_{0.7})_4$ lies below the 1.04 K limit of the He^4 cryostat used in this investigation, the effect of pressure on this sample is positive enough to measure the superconducting transition temperature for pressures in excess of ~ 16 kbar. The data in this range reflect a strong linear increase in T_c. The long extrapolation to zero pressure yields an approximate value of 0.1 K for $T_c(0)$ of this sample in excellent agreement with the measured value. So unlike the selenium-sulfur system, some of these compounds exhibit pronounced nonlinear behavior and enhancements of T_c under pressure.

The overall results of the effect of pressure on the superconducting transition temperatures of these pseudobinary compounds can be seen most readily in Figure 4. The value of the slope dT_c/dP is plotted against composition. For those samples with nonlinear pressure effects, the value of dT_c/dP at high pressures is shown in the graph. A tremendous peak in dT_c/dP is evident for Mo_3Se_4. This compound has the highest zero pressure T_c and certain similarities are present between this graph and Figure 1 where $T_c(0)$ is plotted against composition. Values of dT_c/dP for all of the mixed selenium-tellurium samples fall along a straight line reflecting the rapid change that the effect pressure has on T_c as the tellurium content rises. To the sulfur side of Mo_3Se_4 the change in dT_c/dP is just as pronounced; however, this rate of change does not persist so that dT_c/dP values for the samples with 30 at.% to 60 at.% S are very similar. The exceptional sensitivity of these materials to pressure is also evident in the ternary compounds.

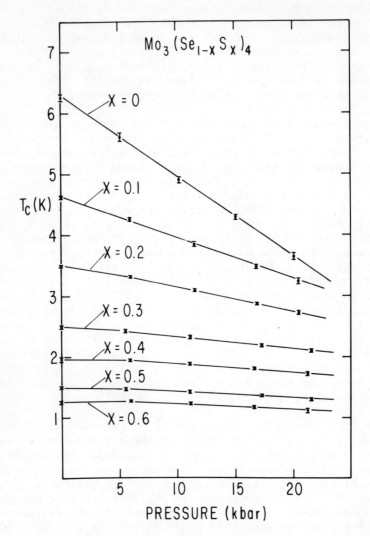

Figure 2: Superconducting transition temperatures as a function of hydrostatic pressure for the pseudobinary compounds $Mo_3(Se_{1-x}S_x)_4$ with $0 \leq x \leq 0.6$.

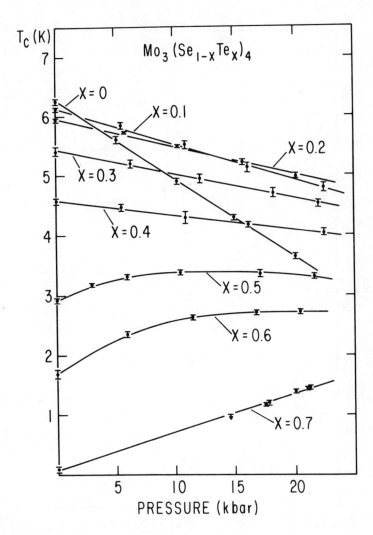

Figure 3: Superconducting transition temperatures as a function of hydrostatic pressure for the pseudobinary compounds $Mo_3(Se_{1-x}Te_x)_4$ with $0 \leq x \leq 0.7$. Horizontal bar with arrow beneath indicates the transition was incomplete to the lowest temperature attained.

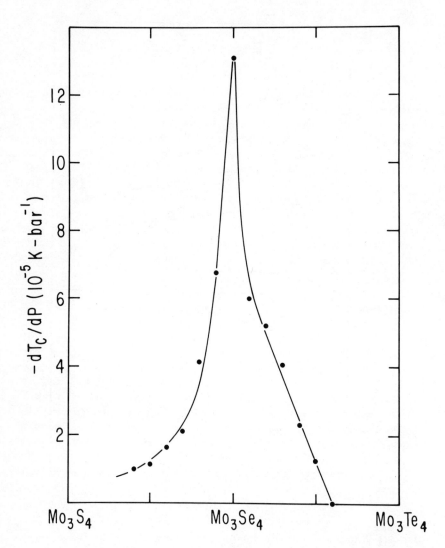

Figure 4: Pressure dependence of the superconducting transition temperature as a function of composition for the binary system Mo_3S_4 - Mo_3Se_4 - Mo_3Te_4.

Ternary Molybdenum Sulfides

An earlier paper presented the effects of hydrostatic pressure on T_c for seven ternary molybdenum sulfides [20]. The extension of this work to include a dozen compounds is displayed in Figure 5. Values for the change in T_c with pressure can be determined from the high pressure, linear portion of the data for each sample. These numerical values are listed in Table II along with the initial transition temperatures. The magnitude of dT_c/dP for the five highest T_c compounds is comparable to the large depression of T_c observed in the binary system. In general one also observes a greater reduction of T_c for samples with higher zero pressure transition temperatures. The relative behavior of the two tin samples as well as the two distinct lead compounds indicates that the pressure effects can be sample dependent.

A significant new feature is the strong initial enhancement of T_c and resulting peak at approximately 6 kbar for $CuMo_3S_4$. Similar behavior is also observed for the zinc and first lead samples. Low temperature x-ray powder diffraction diagrams indicate both the copper and zinc ternaries undergo lattice transformations below room temperature. The nature of this crystallographic change was believed to have been rhombohedral to base centered orthorhombic; [21] however, more detailed x-ray analysis indicates the low temperature structure is even more complex. A triclinic distortion similar to those found in other ternary molybdenum sulfides is suspected [22].

The large enhancement of T_c by the application of pressure for $ScMo_6S_7$ stands in marked contrast to the behavior of the other compounds. This sample is also unique because it does not form with the rhombohedral symmetry common to these materials, but has a slight triclinic distortion at room temperature. These pressure data combined with the copper and zinc ternaries make it tempting to associate positive values of dT_c/dP with the presence of this triclinic distortion. The first lead sample and to some extent the cadmium molybdenum sulfide defy this trend.

Ternary Molybdenum Selenides

The dependence of T_c on hydrostatic pressure for five selenium ternaries is shown in Figure 6. Accompanying numerical data are listed in Table II and indicate that all T_c values for these selenides are lower than those for the corresponding sulfides except for YMo_6Se_8. Two samples exhibit definite nonlinear effects under pressure. The behavior of $Cu_{0.7}Mo_3Se_4$ directly parallels that for $CuMo_3S_4$. The magnitude of both the initial rise in T_c at low pressures and the high pressure depression of T_c are comparable for the

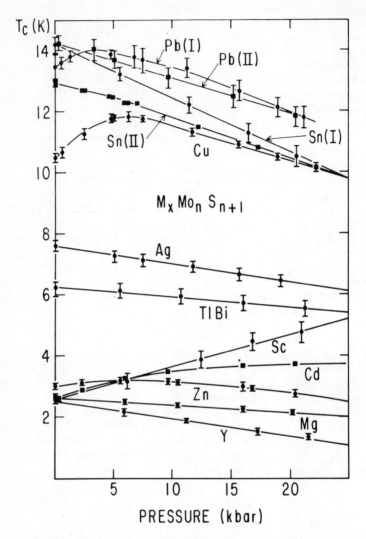

Figure 5: Variation of the superconducting transition temperature with hydrostatic pressure for the ternary molybdenum sulfides.

Table II

Tc and dT_c/dP for ternary molybdenum sulfides and selenides

Compound	$T_c(0)$ (K)	dT_c/dP (10^{-5} K-bar^{-1})
$Pb_{0.92}Mo_6S_{7.5}$(I)	13.98 - 13.09	-16.00 ± 0.10[a]
$PbMo_{5.1}S_6$(II)	14.51 - 14.01	-10.50 ± 0.10
$Sn_{0.6}Mo_3S_4$(I)	14.45 - 13.93	-17.80 ± 0.10
$Sn_{0.6}Mo_3S_4$(II)	13.05 - 12.85	-13.82 ± 0.05[a]
$CuMo_3S_4$	10.66 - 10.35	-10.72 ± 0.10[a]
$AgMo_4S_5$	7.77 - 7.41	-6.00 ± 0.10
$(Tl_{0.46}Bi_{0.46})Mo_6S_{7.5}$	6.41 - 5.95	-3.40 ± 0.10
$ZnMo_5S_6$	3.12 - 2.91	non-linear
$CdMo_5S_6$	2.67 - 2.60	non-linear
$ScMo_6S_7$	2.67 - 2.33	10.93 ± 0.10
YMo_6S_7	2.66 - 2.42	-5.81 ± 0.05
$MgMo_6S_7$	2.72 - 2.60	-2.50 ± 0.05
$Cu_{0.7}Mo_3Se_4$	5.90 - 5.44	-11.20 ± 0.10[a]
$Ag_{0.7}Mo_4Se_5$	5.96 - 5.60	non-linear
$Sn_{0.7}Mo_3Se_4$	4.85 - 4.74	-4.03 ± 0.05
$Pb_{1.2}Mo_6Se_8$	4.32 - 3.97	-6.45 ± 0.10
YMo_6Se_8	6.38 - 6.06	-10.52 ± 0.10

[a] Slope determined from high pressure, linear portion of the data.

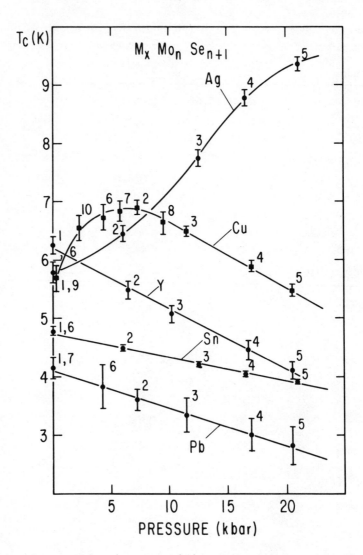

Figure 6: Superconducting transition temperatures as a function of hydrostatic pressure for five ternary molybdenum selenides.

ternary sulfide and selenide. As expected from these data, $Cu_{0.7}Mo_3Se_4$ also undergoes a crystallographic transformation below room temperature. This rise in T_c to 6 kbar followed by a linear decrease at higher pressures may be considered a necessary, although not sufficient, condition for occurrence of a lattice instability in these ternary molybdenum chalcogenides. The silver compound is strikingly different with a positive as well as a nonlinear pressure dependence of T_c over the entire range of pressures. Low temperature x-ray patterns were taken down to 6 K on this sample and no change in lattice symmetry was detected. The peculiar effect of hydrostatic pressure on the T_c of this silver ternary cannot be attributed to the presence of a crystallographic transformation.

Rare Earth Molybdenum Sulfides

Superconductivity in the rare earth molybdenum sulfides was first reported by Fischer and coworkers in 1975 [15]. The pressure dependence of T_c for all twelve superconducting rare earth ternaries is shown in Figures 7 and 8. Two graphs are employed to help distinguish the data below 2 K where the majority of these compounds becomes superconducting. The five heavy rare earth ternary compounds illustrated in Figure 8 have comparable initial transition temperatures and exhibit quite similar behavior under pressure. The slope of the lutetium sample is representative of these five materials and is repeated in Figure 7 to provide an overall comparison. Actual numerical data for transition temperatures and pressure dependencies are presented in Table III.

With the exception of $YbMo_6S_8$, all samples may be characterized by linear changes in T_c with pressure. The ytterbium sample stands out due to its high T_c, large negative slope and distinctive bend in the pressure dependence at approximately 10 kbar. No enhancement of T_c is observed for any of these rare earth compounds. This is in marked contrast to the behavior of those ternaries with low temperature lattice instabilities.

Two of the samples have such low initial transition temperatures that dT_c/dP is computed from only two data points. The T_c's of $Gd_{1.2}Mo_6S_8$ and $Tb_{1.2}Mo_6S_8$ drop below the measurable range of this investigation when pressures exceed 2 kbar. The resulting numerical determination of dT_c/dP must be considered valid only over this limited pressure region. Judging from the other compounds, one would not expect any nonlinear behavior at higher pressures. A continuation of this linear depression would cause superconductivity to vanish at a pressure of roughly 10 kbar for both samples.

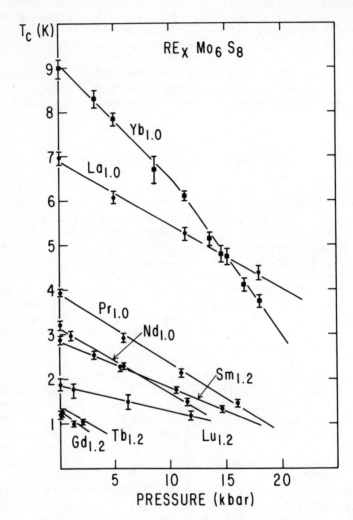

Figure 7: Variation of the superconducting transition temperature with hydrostatic pressure for eight rare earth molybdenum sulfides.

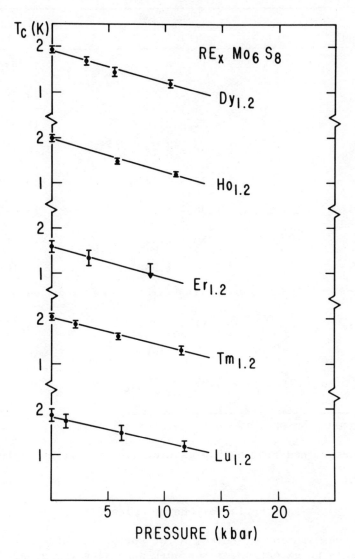

Figure 8: Effect of hydrostatic pressure on T_c for five heavy rare earth molybdenum sulfides. Horizontal bar with arrow beneath indicates the transition was incomplete to the lowest temperature attained.

Table III

T_c and dT_c/dP for rare earth molybdenum sufides

Compound	$T_c(0)$ (K)	dT_c/dP (10^{-5} K-bar^{-1})	
$LaMo_6S_8$	7.12 - 6.83	-14.22	0.10
$PrMo_6S_8$	4.02 - 3.85	-14.78	0.10
$NdMo_6S_8$	3.32 - 3.19	-15.60	0.10
$Sm_{1.2}Mo_6S_8$	2.97 - 2.77	-10.42	0.10
$Gd_{1.2}Mo_6S_8$	1.32 - 1.14	-13.80[a]	
$Tb_{1.2}Mo_6S_8$	1.36 - 1.17	-13.48[a]	
$Dy_{1.2}Mo_6S_8$	1.99 - 1.85	-7.12	0.10
$Ho_{1.2}Mo_6S_8$	2.06 - 1.93	-7.10	0.10
$Er_{1.2}Mo_6S_8$	1.72 - 1.46	-7.32	0.10
$Tm_{1.2}Mo_6S_8$	2.11 - 1.95	-6.50	0.10
$YbMo_6S_8$	9.19 - 8.81	-36.90	0.10[b]
$Lu_{1.2}Mo_6S_8$	2.05 - 1.79	-6.26	0.10

[a] Calculated from only two data points; see text.

[b] Slope determined from high pressure, linear portion of the data.

Rare Earth Molybdenum Selenides

The substitution of selenium for sulfur in these rare earth ternaries is an extension of the work done with the binary system and the other ternary molybdenum chalcogenides. The same twelve rare earth elements form superconducting ternaries when added to the Mo_6Se_8 basis; however, the transition temperatures are generally higher in the selenium compounds. From Figures 9 and 10 and Table IV, it can be seen that the effect of pressure on the selenides is comparable to that of the sulfides. The heavy rare earth ternaries featured in Figure 10 are again very similar to each other

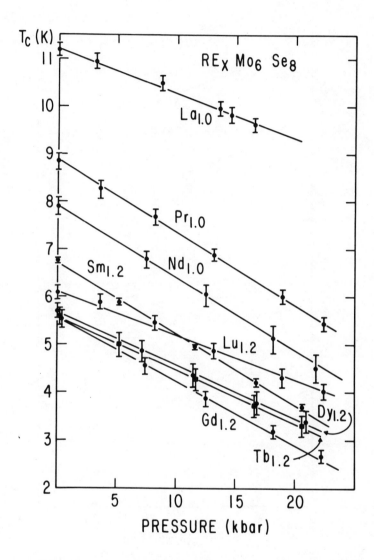

Figure 9: Effect of hydrostatic pressure on T_c for eight rare earth molybdenum selenides.

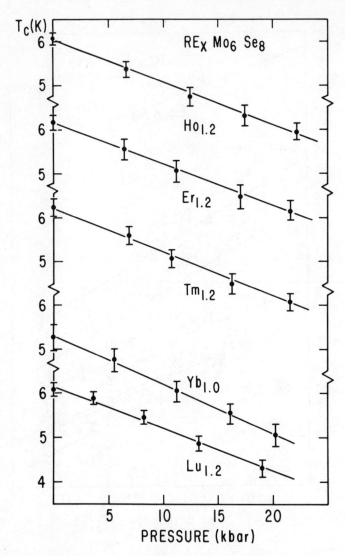

Figure 10: Variation of the superconducting transition temperature with hydrostatic pressure for five heavy rare earth molybdenum selenides.

Table IV

T_c and dT_c/dP for rare earth molybdenum selenides

Compound	$T_c(0)$ (K)	dT_c/dP (10^{-5} K-bar^{-1})
$LaMo_6Se_8$	11.39 - 11.05	-9.40 ± 0.10
$PrMo_6Se_8$	9.16 - 8.65	-15.00 ± 0.10
$NdMo_6Se_8$	8.22 - 7.65	-15.91 ± 0.10
$Sm_{1.2}Mo_6Se_8$	6.83 - 6.35	-14.90 ± 0.10
$Gd_{1.2}Mo_6Se_8$	5.59 - 5.24	-13.15 ± 0.10
$Tb_{1.2}Mo_6Se_8$	5.70 - 5.34	-11.02 ± 0.10
$Dy_{1.2}Mo_6Se_8$	5.77 - 5.30	-11.03 ± 0.10
$Ho_{1.2}Mo_6Se_8$	6.10 - 5.54	-9.71 ± 0.10
$Er_{1.2}Mo_6Se_8$	6.17 - 5.70	-9.56 ± 0.10
$Tm_{1.2}Mo_6Se_8$	6.33 - 5.68	-10.24 ± 0.10
$YbMo_6Se_8$	5.80 - 4.70	-11.10 ± 0.10
$Lu_{1.2}Mo_6Se_8$	6.20 - 5.47	-9.30 ± 0.10

in T_c and dT_c/dP values. The $YbMo_6Se_8$ compound does exhibit a slightly greater depression of T_c than other heavy rare earth samples; however, it shows no nonlinear behavior of the type observed for $YbMo_6S_8$. The "kink" in the T_c versus pressure data for $YbMo_6S_8$ is found to be unique among these rare earth ternaries. For comparison with the other materials, data for the lutetium compound are duplicated in Figure 9.

The existence of superconductivity with transition temperatures as high as 11.3 K in these rare earth molybdenum sulfides and selenides is quite remarkable. Although there is essentially a regular lattice of magnetic ions in a superconducting material, the interaction between the rare earth ions is extremely weak. This is due to the unique crystal structure of these materials which locates the rare earth ions far from each other and far from the molybdenum

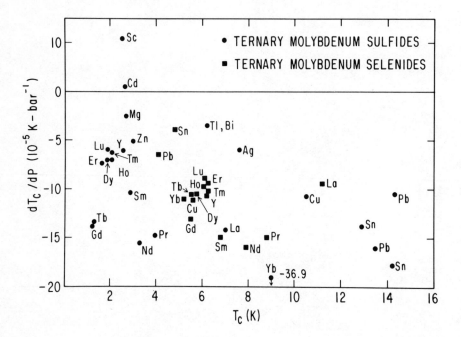

Figure 11: dT_c/dP for some ternary molybdenum sulfides (●) and selenides (■).

octahedron. Hydrostatic pressure measurements are ideally suited to test the effect on T_c of reducing these interatomic distances. The result that dT_c/dP is always strongly negative could be interpreted as the effect of increasing the interaction between rare earth ions due to the decreased interatomic spacing. This hypothesis must be qualified with the observation that the lanthanum and lutetium compounds, which possess no magnetic ions, also exhibit strong depressions of T_c with pressure. These depressions may simply be characteristic of these ternary molybdenum chalcogenides.

SUMMARY

In an attempt to conveniently summarize the effect of pressure on these ternary compounds, the sensitivity to pressure dT_c/dP is plotted against the transition temperature in Figure 11. The use of T_c as a plotting parameter should not be taken as significant. It is utilized simply as a means of displaying the large amount of data accumulated on these materials. For those samples with non-linear pressure effects, the high pressure slope is taken as representative of the rhombohedral Chevrel phase.

The data is quite varied, but certain observations may be made. First, the overwhelming effect of pressure is to depress the superconducting transition temperature of these materials. The magnitude of the depression is exceptionally large since values of $|dT_c/dP|$ in excess of 5×10^{-5} K-bar^{-1} are rare [13,23]. Second, two clusters of points are evident. The sulfide group includes at least a half dozen ternary molybdenum sulfides while the selenide cluster contains an even larger number of compounds. Third, the only ternary with a significant positive slope also crystallizes with a slight triclinic distortion, namely $ScMo_6S_7$.

Considering both binary and ternary compounds, the general behavior of the superconducting properties of these Chevrel phase materials is one of remarkable sensitivity to hydrostatic pressure. Large changes in the transition temperature occur for both binary and ternary materials. These large shifts in T_c with pressure must be related to the unique crystal structure of these materials as opposed to the presence or absence of a third element.

ACKNOWLEDGMENTS

The author wishes to thank B.T. Matthias, D.C. Johnston, A.C. Lawson, M.B. Maple and R.W. McCallum for numerous discussions concerning the properties of these materials. A sample of $PbMo_{5.1}S_6$ was generously provided by S. Foner and E.J. McNiff, Jr. The superb technical assistance of B. Ricks is gratefully acknowledged.

REFERENCES

1. R. Chevrel, M. Sergent and J. Prigent, J. Sol. State Chem. $\underline{3}$, 515 (1971).
2. B.T. Matthias, M. Marezio, E. Corezwit, A.S. Cooper and H.E. Barz, Science $\underline{175}$, 1465 (1972).
3. O. Bars, J. Guillevic and D. Grandjean, J. Sol. State Cheml $\underline{6}$, 48 (1973).
4. F. Hulliger, Structure and Bonding, Vol. 4, p. 189 (Springer-Verlag, Berlin (1968)).
5. J. Guillevic, O. Bars and D. Grandjean, J. Sol. State Chem. $\underline{7}$, 158 (1973).
6. O. Bars, J. Guillevic and D. Grandjean, J. Sol. State Chem. $\underline{6}$, 335 (1973).
7. M. Marezio, P.D. Dernier, J.P. Remeika, E. Corenzwit and B.T. Matthias, Mater. Res. Bull. $\underline{8}$, 657 (1973).
8. O. Fischer, R. Odermatt, G. Bongi, H. Jones, R. Chevrel and M. Sergent, Phys. Lett. $\underline{45A}$, 87 (1973).
9. R. Odermatt, O. Fischer, H. Jones and G. Bongi, J. Phys. C: Solid State Phys. $\underline{7}$, L13 (1974).
10. N.E. Alekseevskii, N.M. Dobrovol'skii and V.I. Tsebro, Zh. Eksp. Teor. Fiz. Pis. Red. $\underline{20}$, 59 (1974); (Sov. Phys. - JETP Lett. $\underline{20}$, 25 (1974)).
11. S. Foner, E.J. McNiff, Jr. and E.J. Alexander, Phys. Lett. $\underline{49A}$, 269 (1974).
12. O. Fischer, Proceedings of "Physics in High Magnetic Fields", Conference held at Grenoble, Sept. 1974, to be published.
13. T.F. Smith, J. Low Temp. Phys. $\underline{6}$, 171 (1972).
14. A. Jayaraman, A.R. Hutson, J.H. McFee, A.S. Coriell and R.G. Maines, Rev. Sci. Inst. $\underline{38}$, 44 (1967).
15. O. Fischer, A. Treyvaud, R. Chevrel and M. Sergent, Solid State Commun. $\underline{17}$, 721 (1975).
16. R.N. Shelton, R.W. McCallum and H. Adrian, Phys. Lett. $\underline{56A}$, 213 (1976).
17. R.N. Shelton, Ph.D. Thesis, University of California, San Diego (1975).
18. R. Chevrel, M. Sergent and O. Fischer, Mater. Res. Bull. $\underline{10}$, 1169 (1975).
19. R. Chevrel, M. Sergent and J. Prigent, Mater. Res. Bull. $\underline{9}$, 1487 (1974).
20. R.N. Shelton, A.C. Lawson and D.C. Johston, Mater. Res. Bull. $\underline{10}$, 297 (1975).
21. A.C. Lawson, Mater. Res. Bull. $\underline{7}$, 773 (1972).
22. A.C. Lawson, private communication.
23. T.F. Smith, Superconductivity in d- and f-Band Metals, ed. D.H. Douglass (New York, AIP), pp. 293-319 (1972).

QUESTIONS AND COMMENTS

M. Ashkin: Do you pressurize at room temperature and then cool down?

R. Shelton: Yes, the technique used is to place the samples in a teflon capsule with a pressure transmitting fluid which remains fluid up to 40 kbar at room temperature. So it is as hydrostatic a technique available unless you utilize helium gas to apply the pressure.

B. Gyorffy: We could try to understand these trends if we knew something about what happens to the phonons, the Gruneisen constant, the melting temperature, or anything like that. Do you have any data on these?

R. Shelton: Not yet; the one parameter we would like to measure is the compressibility of these compounds. Because they come in a powdered form, it is a little difficult but we are working on that. Unfortunately we have no data yet. I should emphasize that these are relatively new compounds as far as superconductors go. They were found to be superconducting in the early 1970's.

D. Ginsberg: How was the transition temperature measured? Do the error bars indicate the width of the transition?

R. Shelton: Yes, all T_c's were measured inductively and as you can see the width of the transition does vary from a hundred millidegrees to half a degree.

P. Schmidt: How do you synthesize these materials?

R. Shelton: The general synthesis is to first react binary sulfides such as lead sulfide. Then this material is thoroughly ground with Mo powder and sulfur powder. All the reaction is done in a quartz tube and goes through about three or four stages.

B. Ganguly: I think if I went through the data correctly, you find that the T_c under pressure is relatively independent of whether or not the third element is magnetic.

R. Shelton: Yes, there do seem to be many mechanisms involved in these compounds. For the rare earth molybdenum selenides we do see a larger depression of T_c for the non-lanthanum, non-lutetium compounds, that is, the magnetic compounds; however, the difference is very slight. Now to understand why the difference is slight, I think the clue lies in the crystal structure. These rare earths are positioned very far apart from each other as well as very far apart from the molybdenum octahedron. It is not completely clear how pressure affects these separation distances because the crystal structure is so complex.

HIGH CRITICAL FIELD SUPERCONDUCTORS

S. Foner

Francis Bitter National Magnet Laboratory[†]
Massachusetts Institute of Technology

Cambridge, Massachusetts   02139

ABSTRACT

A summary of recent studies in high field superconductors is presented including high T_c, β-W compounds, ternary lead molybdenum chalcogenides, high T_c oxides, and anisotropic superconductors.

I. INTRODUCTION

High field superconductors have continued to be of interest since the last d- and f-band Superconductors Conference. I will limit my discussion to the high field properties of superconducting materials which have been chosen because they have the highest T_c or highest H_{c2} of their class, or because they are representative of new classes of superconducting materials. The summary presented here involves work at the Francis Bitter National Magnet Laboratory, often in collaborative programs with other research groups. The references give further details and discuss related work.

The most extensively explored materials in the last decade are the β-W compounds which have some of the highest transition temperatures of any superconductors. One reason for interest in this class is their promise for practical application. The major new development since the last d- and f-band Superconductivity Conference in this class of materials was the discovery by Gavaler et al.[1] that Nb_3Ge can be made with T_c approaching 23 K. With this impetus, a number of investigators then showed that high T_c, Nb_3Ge could be made in

[†]Supported by the National Science Foundation.

a variety of ways including chemical vapor deposition. A newer class of materials, the ternary molybdenum chalcogenides, have been examined recently. Although these have lower values of T_c, at present they hold the record as the highest critical field superconductors yet discovered. Only a few aspects will be presented here in order to indicate some areas of agreement and disagreement with work of Fischer and coworkers. Some superconducting oxide compounds have been discovered with $T_c \sim 13$ K, e.g., Li-Ti-O compounds,[2] and the high field characteristics of the Li-Ti-O compounds are mentioned. Finally, some of the high field properties of some anisotropic superconductors also are mentioned.

II. EXPERIMENTAL PROBLEMS

There are some unique experimental problems in studying the very high H_{c2} superconductors. The main difficulty is that dc field measurements are limited to the range of 230 kG at our Laboratory, and materials with higher H_{c2}'s must be examined with pulsed magnetic fields. At best one achieves thermal equilibrium during the pulse; if thermal equilibrium is not achieved, the observed value of H_{c2} is expected to be too low. We have checked results with pulsed fields for cases where H_{c2} was less than 230 kG. In general we have found good agreement between measurements with dc and pulsed fields when they could be compared in liquid helium. In addition, very often the superconducting-to-normal transition has a finite width which varies from ~ 10 kG for some of the sharper transitions to possibly ~ 60 kG for some of the very broad transitions. Thus, even if a consistent criterion for H_{c2} is used to define the transition field, systematic errors in the determination of H_{c2} and $(dH_{c2}/dT)_{T=T_c}$ can arise. Finally, when $(dH_{c2}/dT)_{T=T_c}$ is large, measurements in the dc fields are restricted to a narrow interval near T_c, even for dc fields of ~ 200 kG. Thus extrapolations to low temperature based on data very close to T_c may involve a considerable error, assuming that the experiments follow theory in detail. At best, we have found that extrapolations to low T are hazardous.

Measurements at low temperature (and consequently at high fields) are essential to clarify the properties of very high field superconductors for several reasons:

1) theory often involves several assumptions which are not readily tested near T_c;

2) experimental uncertainties are significant factors for extrapolations far from T_c;

3) $H_{c2}(T)$ generally becomes a slowly varying function as T approaches 0 K, so that experimental results are less susceptible to small changes in T;

HIGH CRITICAL FIELD SUPERCONDUCTORS

4) the dependence of $H_{c2}(T)$ at low T often permits more unambiguous evaluation of theory; and

5) effects such as Pauli paramagnetic limiting are much more prominent at highest field (and lowest temperature) because the paramagnetic contribution to the free energy varies as H^2.

Examples presented here and in the references will illustrate these aspects.

For most of the studies it is convenient to compare results with theoretical predictions for a "dirty" type II superconductor.[3] For no Pauli paramagnetic limiting (PPL) the upper critical field at 0 K, $H^*_{c2}(0)$ for a dirty type II superconductor is given by

$$H^*_{c2}(0) = 0.69\, T_c [dH_{c2}/dT]_{T=T_c} \qquad (1)$$

where T_c and $[dH_{c2}/dT_2]_{T=T_c}$ are measured quantities. If the "dirty" type II superconducting theory holds then relatively few parameters are involved. The assumption that there is no Pauli paramagnetic limiting is to be proven in each case by measurements. Usually, Pauli paramagnetic limiting is suppressed by very large spin-orbit scattering, which is expected if atoms with large atomic numbers, Z, are present. Theory suggests that the spin-orbit scattering parameter λ_{so} varies as Z^4, a very strong function of Z. Thus, one expects that superconducting materials with low Z components would show PPL effects, and indeed this is often observed. The behavior of $H^*_{c2}(T)$ can be determined by computer[3] for various values of $\alpha = H^*_{c2}(0)/H_p$ and spin-orbit scattering parameter λ_{so}, where $H_p = 18.4\, T_c$(kG) is the paramagnetic limiting field. Generally, both α and λ_{so} are required for fits of experiment to theory. When λ_{so} becomes very large ($\lambda_{so} = \infty$), the effects of PPL are completely suppressed and the maximum value of $H^*_{c2}(0)$ is achieved as given in Eq. (1).

The general behavior of $H^*_{c2}(T)$ is best illustrated by normalized plots. Fig. 1 shows the normalized field $h = H^*_{c2}(\alpha,\lambda_{so})/T_c(dH_{c2}/dT)_{T=T_c}$ as a function of normalized $t = T/T_c$ for various values of λ_{so} assuming $\alpha = 3$ [corresponding to $(dH_{c2}/dT)_{T=T_c} = 57$ kG/K observed for one of our best ternary lead molybdenum sulfide material[4]]. Note that: 1) In order to estimate a lower limit for λ_{so}, data at low t are essential because all the calculations converge as t approaches 1. 2) The dc field range of 230 kG only permitted measurements for $t > 0.85$ for this very high H_{c2} material. If α is small, it is still difficult to obtain an estimate of λ_{so} even at low T. This is illustrated by the normalized plots in Fig. 2 for $t = 0.3$, where $h_1/h_2 = H^*_{c2}(\alpha,\lambda_{so})/H^*_{c2}(\alpha,\lambda_{so} = \infty)$. The dashed line, corresponding to $h_1/h_2 = 0.95$, shows that if experiment is within 5% of h_1, for $\alpha = 0.7$ the lower limit of $\lambda_{so} \simeq 1$ whereas for $\alpha = 3$ the lower limit of $\lambda_{so} \simeq 50$ (see also Fig. 1). Thus extremely precise data are required to estimate λ_{so} when α, which is proportional to $(dH_{c2}/dT)_{T=T_c}$, is small. The highest H_{c2} materials have large α's so that the paramagnetic contribution and λ_{so} can be evaluated with reasonable accuracy.

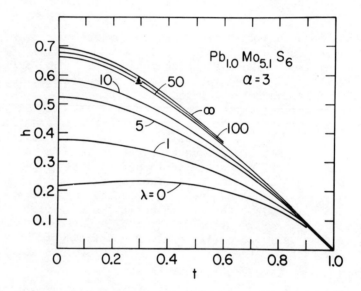

Fig. 1. Comparison of data for ternary lead molybdenum sulfide with nominal composition of $Pb_{1.0}Mo_{5.1}S_6$ as a function of h and t with calculations for a dirty type II superconductor. Normalized $\underline{h} = H^*_{c2}(\alpha,\lambda_{so})/T_c(dH_{c2}/dT)_{T=T_c}$ and $\underline{t} = (T/T_c)$. The heavy line near T_c and the bold arrow correspond to the data in Fig. 4. The solid curves are calculated for various values of $\lambda = \lambda_{so}$. Maximum paramagnetic limiting corresponds to $\lambda = 0$ and complete suppression of paramagnetic limiting corresponds to $\lambda = \infty$. The results indicate $\lambda_{so} \geq 50$ for this material. (After Phys. Letters, Ref. 4).

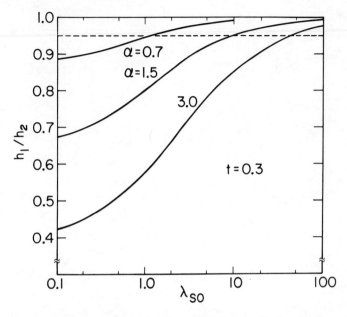

Fig. 2. Ratios of normalized upper critical field h_1/h_2 versus spin-orbit scattering parameter λ_{so} and paramagnetic limiting parameter α. Here $h_1/h_2 = H^*_{c2}(\alpha,\lambda_{so})/H^*_{c2}(\alpha,\lambda_{so}=\infty)$ and $t = T/T_c = 0.3$. The values of $\alpha = 0.7$, 1.5 and 3.0 corresponds to $(dH_{c2}/dT)_{T=T_c} \simeq 13$, 29 and 57 kG/K respectively.

III. Nb_3X COMPOUNDS

A composite of $H_{c2}(T)$ for many of the Nb_3X materials is shown in Fig. 3. All are very good superconductors. The theory for a dirty type II superconductor with no PPL ($\lambda_{so} = \infty$), shown by the dashed lines, is in good agreement with the measurements. This is consistent with the high Z of Nb, and the large concentration of Nb in these β-W materials.[5]

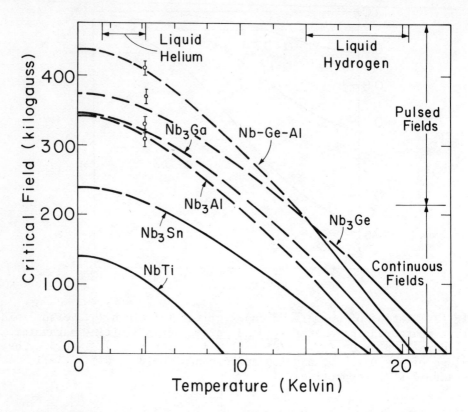

Fig. 3. Comparison of critical field versus temperature for several Nb_3X materials. The solid curves for each material correspond to data taken in dc magnetic fields. The data points at 4.2 K are taken with pulsed magnetic fields. The dashed curves are calculated assuming a dirty type II superconductor with no paramagnetic limiting. The dashed curves are normalized with the measured T_c and $(dH_{c2}/dT)_{T=T_c}$. (After Colloq. Int. CNRS, Ref. 5).

Because the Nb_3X compounds are excellent candidates for high field superconducting materials, a number of related experiments have been carried out on these systems. In particular, Sweedler et al.[6] have examined the properties of these materials with high flux neutron irradiation. They found a large decrease in T_c with irradiation, suggesting that use of such materials in fusion reactors would require substantial shielding. We have been examining

HIGH CRITICAL FIELD SUPERCONDUCTORS

the high field properties of selected irradiated Nb_3Al in collaboration with Sweedler et al. to examine the effects of atomic order on $H_{c2}(T)$. A brief discussion of some of these results has been presented[6]. The data showed that $(dH_{c2}/dT)_{T=T_c}$ increased slightly with initial irradiation, and further irradiation decreased T_c, but the value of $(dH_{c2}/dT)_{T=T_c}$ remained relatively constant. Measurement with pulsed fields have not yet been completed.

Because the β-W structure has chains of atoms along the faces of the cubic cell, a linear chain model (LCM) for describing the unusual superconducting properties has been suggested. One of the predictions of the quasi-onedimensional LCM is that the Fermi surface would be strongly anisotropic, and this is expected to be reflected in the anisotropy of H_{c2} in the tetragonal phase. Measurements of H_{c2} in high dc fields in selected single crystal Nb_3Sn and V_3Si indicate that the anisotropy of H_{c2} is quite small, and that any evidence of a quasi-onedimensional band structure is not reflected in the upper critical field of these high field superconductors.[7] In order to examine the anisotropic features of H_{c2} with sufficient resolution, measurements at low $\underline{t}$ (and high fields) were essential.

IV. TERNARY MOLYBDENUM CHALCOGENIDES

In the last few years a new crystal structure was discovered by Chevrel et al.[8] Shortly thereafter, Matthias et al.[9] showed that many of these materials were superconducting with high transition temperatures. Our interest in these materials was kindled by a publication of Odermatt et al.[10] which indicated that the ternary molybdenum sulfides were very high field superconductors.

We have made a number of the ternary lead molybdenum sulfide compounds in order to maximize T_c and to study the critical fields. In contrast with early work[10] we found that the tin molybdenum sulfide materials had a relatively low H_{c2}, but that the PbMoS compounds were the highest field superconductors we found in this class. Results of some of the early measurements are shown in Fig.4. The general features of our data are that the H_{c2} appears to fit the theory for a dirty type II superconductor with no PPL with a lower limit[4] of $\lambda_{so} \gtrsim 50$. To date, we have not been able to improve on T_c of approximately 14.5 K by variation of composition, reaction temperature, or pressure. We have observed a slightly higher T_c in some materials, but the high T_c fraction appears to have a much lower critical field.[4] In general the transition widths for the PbMoS compounds range from about 40 to 60 kG, so that the transition is not as sharply defined as other materials. However, $(dH_{c2}/dT)_{T=T_c} \simeq$ 60 kG/K which is extremely large and nearly twice that of the β-W materials.

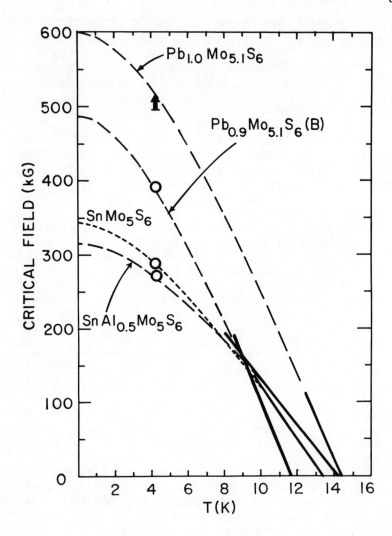

Figure 4: H_{c2} versus T for selected ternary molybdenum sulfide materials. Nominal compositions are given; the general formula for these compounds is $A_xMo_6S_8$, where A is a metal. The solid lines extending to $T=T_c$ are the data taken in dc magnetic fields. The dashed curve corresponds to calculations for a dirty type II superconductor assuming complete suppression of paramagnetic limiting. The open circles at 4.2 K are the pulsed field data. The bold arrow for the $Pb_{1.0}Mo_{5.1}S_6$ compound indicates that the transition is not complete at 495 kG. From data on related Pb compounds with lower T_c, we estimate $H_{c2}(4.2\ K) \simeq 510$ kG which is the tip of the arrow. (After Ref. 4)

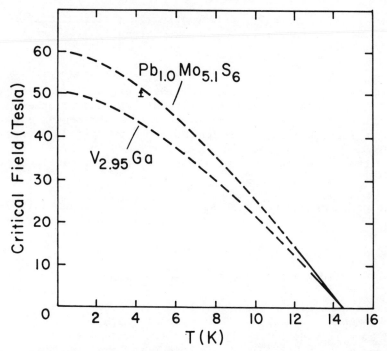

Fig. 5. Comparison of H_{c2} versus T for ternary lead molybdenum sulfide of Fig. 4 and $V_{2.95}Ga$ (Ref. 11). The solid lines near T_c correspond to dc field measurements, and the dashed curves are calculated assuming a dirty type II superconductor with no paramagnetic limiting. Measurements of $H_{c2}(T)$ for $V_{2.95}Ga$ at low temperatures shows strong paramagnetic limiting and $H_{c2}(4.2K) <$ 20 T(200 kG) (After Ref. 12).

Whether the unusually high critical fields of these compounds require new physical mechanisms is open for discussion. The characteristic $(dH_{c2}/dT)_{T=T_c}$ is not much higher than in V_3Ga. The characteristics of the ternary lead molybdenum sulfides are compared with V_3Ga in Fig. 5 based on data of more than a decade ago by Wernick et al.[11] Note that the characteristics of V_3Ga are plotted[12] assuming no PPL whereas it is well known that V_3Ga is strongly paramagnetically limited ($H_{c2}(4.2) \simeq 180$ kG is actually measured). Clearly if sufficient scattering could be introduced into V_3Ga without reducing T_c appreciably, one might expect a very high critical field for this material. The particular V_3Ga chosen here has a T_c equal to that of the ternary compound. Despite the fact that V_3Ga is strongly paramagnetically limited, it still is a useful commercial material with an exceptionally high critical current.

Recently Shelton et al.[13] have developed $REMo_6Se_8$ compounds (where RE is a rare earth) with relatively high T_c's. Although the T_c's for these materials have not yet matched those of the lead molybdenum sulfide compounds, we have made measurements in collaboration with Shelton, McCallum and Maple which indicate that the $(dH_{c2}/dT)_{T=T_c} \simeq 70$ kG/K for RE = La. If T_c can be improved, it is possible that this class of materials will have extremely high critical fields.[14]

The paper by Fischer discusses ternary molybdenum sulfide results of his group, and references to their work will be found in that paper.

V. OTHER HIGH FIELD SUPERCONDUCTORS

Johnston et al.[2] recently discovered superconductivity in the Li-Ti-O compounds with $T_c \sim 13$ K. Because high T_c oxides have been studied to a limited extent, the question arose whether this class shows promise as a high field superconductor. Although T_c is not as high as that of the best β-W materials, we have seen that the ternary molybdenum sulfides with comparable T_c's can have exceptionally high values of H_{c2}. We made a number of these oxides by arc melting and find that these materials show effects of the large PPL, so that the upper critical fields are limited to $H_{c2}(4.2K)$ of less than 200 kG.[15] This is expected for the low Z values of the major constituents.

Another class of materials which have been studied in the last few years are anisotropic superconductors. Intercalated layered superconductors have been examined extensively by Geballe, Thompson and Gamble. We studied a number of their intercalated single crystal materials over a wide range of field. All have low T_c's ($\gtrsim$ 4K) and values of $(dH_{c2}/dT)_{T=T_c}$ often well above 100 kG/K.[16] Thus, $H_{c2}(1.5K)$ often exceeds available dc fields. Because these intercalated materials are strongly anisotropic they are very difficult to study. Careful alignment of the layered planes parallel to the applied field is essential in order to avoid suppression of the superconductivity. One of the major problems we observed was that the angular dependence of H_{c2} was not as sharp as expected for uniformly intercalated systems. If the intercalation is not perfectly uniform, then there are regions of the layer-planes which are misaligned with respect to the applied field. In addition to an artificially broadened transition, it is difficult to examine the detailed functional dependence of $H_{c2}(T)$, partly because of the extremely large $(dH_{c2}/dT)_{T=T_c}$. Our results could best be fit by a linear H_{c2} versus T over the range of available field.

Nonintercalated superconducting single crystals are interesting for studies of moderate anisotropy. Measurements of $H_{c2}(T)$ in $NbSe_2$, for example, showed that $H_{c2}(T)$ was well above that expected for dirty (or clean) type II superconductors.[17] The anisotropy $(H\|_{c2}/H^\perp_{c2}) \simeq 3.2$ for $NbSe_2$. The effects of an anisotropic Fermi surface are evident in the H_{c2} vs T behavior. Studies in cubic systems such as Nb and V by Williamson et al.[18] have shown that in the clean limit, $H_{c2}(T)$ was influenced by Fermi surface effects so that the upper critical field is well above that calculated for clean type II superconductors. Layered materials and very clean superconductors present a fertile area for studies of Fermi surface effects.

VI. SUMMARY

This brief review presents some of our recent work with high critical field superconducting materials. It is clear that the upper critical field some of the recent new superconductors are now reaching the limits of conventional pulsed field technology. It is also apparent that if these higher field superconducting materials can be developed as practical wires, they would be likely candidates for means to extend available dc fields so that further progress can be made.

VII. ACKNOWLEDGMENTS

I am grateful to E.J. McNiff, Jr. who has been involved in all aspects of the research summarized here. In addition, we have been involved in active collaboration with E.J. Alexander for fabrication of all of the ternary molybdenum sulfides and x-ray analyses at our Laboratory. Numerous collaborative efforts included: studies of the rare earth molybdenum selenides, R.N. Shelton, R.W. McCallum, and M.B. Maple of the University of California, San Diego; the Nb_3X materials, B.T. Matthias of Bell Laboratories and the University of California, San Diego, T.H. Geballe of Bell Laboratories and Stanford University, E. Corenzwit and R.H. Willens of Bell Laboratories, G.W. Webb and L.J. Vieland, R.E. Miller and A. Wicklund of RCA Laboratories, J.R. Gavaler and M.A. Janocko of Westinghouse Research Laboratories; and studies in layered compounds, T.H. Geballe, A.H. Thompson, F.R. Gamble, and R.E. Schwall of Stanford University, and F.J. DiSalvo of Bell Laboratories. Several V_3Ga materials were furnished to us by J.H. Wernick of Bell Laboratories.

REFERENCES

1. J.R. Gavaler, Appl. Phys. Letters 23, 480 (1973).
2. D.C. Johnston, H. Prakash, W.H. Zachariasen, R. Viswanathan, Mat. Res. Bull. 8, 777 (1973).
3. K. Maki, Phys. 1, 127 (1964); N.R. Werthamer, E. Helfand, and P.C. Hohenberg, Phys. Rev. 147, 295 (1966).
4. S. Foner, E.J. McNiff, Jr., and E.J. Alexander, Applied Superconductivity Conference, Oakbrook, Ill., Sept. 30-Oct. 2, 1974, Proc. I.E.E.E. Trans. Magnetics, MAG 11, Mar. 1975, p. 155; Phys. Letters 49A, 269 (1974); Bull. Am. Phys. Soc. Series II, 20, 343 (1975).
5. See S. Foner, E.J. McNiff, Jr., J.R. Gavaler and M.A. Janocko, Phys. Letters 47A, 485 (1974) and references cited therein. Our criterion for determination of H_{c2} is also compared to dc resistive measurements in this reference. See also, S. Foner, Colloq. Int. CNRS No. 242 (1975), p. 423.
6. A.R. Sweedler, D. Schweitzer, and G. Webb, Phys. Rev. Letters 33, 168 (1974); A.R. Sweedler, D.E. Cox, S. Foner, and E.J. McNiff, Jr., Bull. Am. Phys. Soc., Series II, 20, 461 (1975).
7. S. Foner and E.J. McNiff, Jr. (in press).
8. R. Chevrel, M. Sergent and J. Prigent, J. Solid State Chem. 3, 515 (1971).
9. B.T. Matthias, M. Marezio, E. Corenzwit, A.S. Cooper, and H.E. Barz, Science 175, 1465 (1972).
10. R. Odermatt, Ø, Fischer, H. Jones, and G. Bongi, Phys. C., Solid State Phys. 7, L13 (1974).
11. J.H. Wernick, F.J. Morin, F.S.L. Hsu, D. Dorsi, J.P. Maita, and J.E. Kunzler, High Magnetic Fields (Eds. H.H. Kolm, B. Lax, F. Bitter, and R. Mills) MIT Press and John Wiley, New York (1962), p. 609. The highest $(dH_{c2}/dT)_{T=T_c}$ material was chosen.
12. See Physics Today, Feb. 1976, p. 77 and references cited therein.
13. R.N. Shelton, R.W. McCallum and H. Adrian, Phys. Letters 56A, 213 (1976).
14. S. Foner, E.J. McNiff, Jr., R.N. Shelton, R.W. McCallum, and M.B. Maple, Phys. Letters (in press).
15. S. Foner and E.J. McNiff, Jr., Bull. Am. Phys. Soc. Series II, 21, 384 (1976), and in press.
16. S. Foner, E.J. McNiff, Jr., F.R. Gamble, T.H. Geballe, and F.J. DiSalvo, Bull. Am. Phys. Soc. Series II, 17, 293 (1972).
17. S. Foner and E.J. McNiff, Jr., Phys. Letters 45A, 429 (1973).
18. S.J. Williamson, Phys. Rev. 12, 2161 (1975) and references cited therein.

QUESTIONS AND COMMENTS

J. Bardeen: What is known about the mechanical properties?

S. Foner: The question was: 'have the mechanical properties of any of these materials been measured?'. Indirectly, yes. Generally, working has reduced T_c. In most of our experiments we look at the powder or the conglomerate as prepared. I do not know of any systematic studies of the mechanical properties. Some discussions on the materials aspects of the ternary molybdenum sulfides were given in Letters [see Physics Today December 1975 Letter by Foner, p. 12; by Strongin, p. 13; Physics Today, Feb. 1976, p. 77; and article in Physics Today, Jan. 1975, p. 17].

Ø. Fischer: Just one question concerning your Nb-Ge-Al material. The composition of the highest field Nb-Al-Ge was not marked here, but in your publication it is marked $Nb_{79}(Al_{73}Ge_{27})_{21}$.

S. Foner: That is the composition of the Nb-Ge-Al of the material for which we published the data in Physics Letters 31A, 349 (1970).

Ø. Fischer: Have you tried to change the composition?

S. Foner: We chose the best materials (highest T_c) that Matthias' group had fabricated. We looked at a series of samples in this general class with small differences in composition. Examples are discussed by S. Foner, E.J. McNiff, Jr., B.T. Matthias, and E. Corenzwit, 11th Int. Conf. on Low Temp. Phys. (1968), Eds. J.F. Allen, D.M. Finlayson and D.M. McCall, p. 1025, for $Nb_{1-x}(Al_{1-y}Ge_y)_x$ where x = 0.20 - 0.21 and y = 0.29 - 0.25. Additions of Ga as a fourth component also gave very similar results.

B. Gyorffy: You said the effects of paramagnetic limiting are suppressed by large spin orbit scattering effects. Does that correlate with high Z quantitatively?

S. Foner: Well, certainly all the lead molybdenum sulfides have high Z components and seem to show essentially complete suppression of paramagnetic limiting effects. Generally, when you have a spin-orbit scattering parameter of 15 or 20 or so, like in Nb_3X compounds, the deviation from the theoretical curve is very small. In the case of the lanthanum molybdenum selenides the only reason we can observe a λ_{so} of 20 to 50, is that $(dH_{c2}/dT)_{T=T_c}$ is very large and any paramagnetic limiting would have a significant effect at low T. The general effects appear to be qualitatively correct. Quantitatively, I do not think we can be certain within the experimental limitations. Whether or not the spin orbit scattering varies as Z^2 or Z^4 is open to discussion. Meservey and Tedrow, Phys. Letters 51A, 57 (1975) have used polarized electronspin tunneling experiments with gallium, and find that λ_{so} appears to vary as Z^2 rather than Z^4.

ON THE UPPER CRITICAL FIELD OF THE TERNARY MOLYBDENUM CHALCOGENIDES

Ø. Fischer, M. Decroux, R. Chevrel

Départment de Physique de la Matière Condensée

Université de Genève, Switzerland

and

M. Sergent

Laboratoire de Chimie Minérale B, Equipe Associée

au C.N.R.S., Université de Rennes, France

ABSTRACT

A simple analysis of the critical field of ternary molybdenum chalcogenides with and without magnetic impurities is presented.

I. INTRODUCTION

Ternary molybdenum chalcogenides, first described by Chevrel et al.[1] are characterized by relatively high critical temperatures[2,3] and by extremely high critical fields[4,5,6]. It has been found that the superconducting properties are very sensitive to pressure[7], they are however rather insensitive to Rare Earth magnetic impurities[8,9,10].

The general formula for these compounds may be written $M_y Mo_6 X_8$, where $0 \leq y \leq 4$, X = S, Se, Te and M is a metallic element like Pb, Sn, Cu, Ag, La, Y, Mo, etc. From the study of the superconducting properties of many of these compounds, we have suggested that superconductivity is mainly due to the 4d electrons of the Mo-atoms[8,9]. Since these Mo-atoms are grouped together in octahedrons ("clusters") which are relatively well separated in space we think that as a zero-th order approximation the electronic properties can be described as the one of weakly coupled Mo_6-clusters. Such a picture could then account for the extremely high critical fields found in these compounds[8,11].

The Mo-Mo intracluster distances are comparable to those found in Mo-metal, typically 2.7 - 2.8 Å [12]. The Mo-intercluster distances are generally much larger, ranging from 3,08 Å to 3,66 Å. [12,13]. When comparing different compounds of this class we find that the size of the Mo_6 cluster is roughly the same in all compounds, what changes is the relative position of the clusters. To illustrate this, we show in Table 1 the mean Mo-Mo intracluster distance a and the shortest intercluster distance d.[12]

Table 1

Mean value of the intracluster distances (a) and shortest intercluster distance (d) in some $M_yMo_6X_8$ compounds.

Compound	$\bar{a}$(Å)	d(Å)
Mo_3S_4	2.78	3.08
$Ni_2Mo_6S_8$	2.73	3.18
$Fe_2Mo_6S_8$	2.71	3.22
$PbMo_6S_8$	2.70	3.27
Mo_3Se_4	2.76	3.27
$PbMo_6Se_8$	2.72	3.49
Mo_3Te_4	2.73	3.66

II. A SIMPLE ANALYSIS OF THE CRITICAL FIELDS

As we pointed out in the Introduction, one of the most striking properties of these compounds is their extremely high critical field. We have suggested that this high critical field is due to a certain localization of the conduction-electron wave function on the Mo_6 clusters. There are two ways that such a localization could produce a high critical field.

1. The localization could lead to a low fermi velocity which would in turn lead to a small B.C.S. coherence length ξ_o and a high critical field. This would then be an intrinsic high field superconductor.

2. The localization may produce an extreme short mean free path ℓ, and in this way give a high critical field.

We have earlier pointed out that nearly all compounds of this type are high field superconductors. However, not all of them are extreme high field superconductors. In Table 2, we give the initial slope $(dH_{c2}/dT)_{T=T_c}$ and H_{c2} for some compounds. The critical field of $PbMo_6S_8$ is particularly low taking into account the large Mo-intercluster distance. These results need therefore a somewhat more detailed analysis. What we will do is to estimate values for ξ_o and ℓ, assuming that these systems are dirty superconductors. ξ_o can be estimated from the density of states $N(0)$ and the electron concentration n

$$\xi_o = 1.68 \; 10^{-21} \; \frac{M}{\rho} \; \frac{n^{2/3}}{N(0)(1+\lambda)T_c} \quad .$$

M is the mean atomic weight, ρ the density and λ is the electron-phonon coupling parameter. The mean free path can be estimated from dH_{c2}/dT [14]:

$$\ell = \frac{6.60 \; 10^{-8}}{\xi_o \cdot T_c \cdot (dH_{c2}/dT)_{T=T_c}} \quad .$$

To estimate n, we have assumed that the 4d electrons remaining on the MO-atoms after charge transfer participate on the conduction band. This charge transfer has been estimated from the radius of the X-atoms as compared to the ionic radius. Considering what we said above, this value might be somewhat too high. However, we are only interested in an order of magnitude and since n enters in the power 2/3, an error of a factor of two in n will not be important in what follows. The density of states for some of these compounds has been given by Morton et al.[15] who calculated it from the susceptibility, arguing that other contributions to χ will cancel each other. Specific heat measurements [16,17] on $PbMo_6S_8$, $SnMo_6S_8$ and $Cu_2Mo_6S_8$ confirm this and we have used room temperature susceptibility [18] to estimate the density of states of the other compounds given in Table 2. The electron-phonon coupling parameter λ has been found to be of the order 1 in these compounds. For our estimate we use $\lambda = 1$ for all compounds.

Table 2

Comparison of superconducting and electronic parameters for some $M_y Mo_6 X_8$ compounds.

	T_c °K	dH_{c_2}/dT kGauss/°K	$H_{c_2}(0)$ kGauss	n 10^{22} cm^{-3}	X 10^{-6} emu/gram	N(0) eV^{-1}atom^{-1}spin^{-1}	ξ_0 Å	ℓ Å
$Cu_2 Mo_6 S_8$	10.6	35	130	10.2	0.36	0.35	505	3.5
$Pb Mo_6 S_8$	12.6	58	500	9.1	0.56	0.63	245	3.7
$Sn Mo_6 S_8$	11.8	47	340	8.9	0.54	0.55	300	4.0
$La Mo_6 S_8$	7.0	15	54	9.8	0.19	0.20	1450	4.3
$Pb Mo_6 Se_8$	3.8	22	38	8.4	0.25	0.37	2190	3.6

The obtained values for ξ_0 and ℓ are shown in Table 2. We see that the variation in H_{c_2} from compound to compound corresponds to a variation in the density of states and thus in ξ_0. The mean free path turns out to be about the same for all compounds, about 4 Å. This distance happens to be about the diameter of the Mo_6 cluster, a correspondence which should not be taken too seriously. What we retain from this estimate is that the mean free path is very short probably less than the rhombohedral lattice parameter (a ≈ 6,5 Å) and that the different values of (dH_{c_2}/dT) are brought about by changes in N(0). These compounds turn out to be rather low density state material which would have been low field superconductors if they were pure ($\ell = \infty$). From the estimated values of ℓ we may estimate the resistivity to be about 0,2 to 0,3 mΩcm. This is in reasonable agreement with the values of 1 to 2 mΩcm measured on sintered samples.

The critical temperatures given in Table 2 are not the highest ones that can be obtained. The exact value of T_c depends on the preparation conditions. This variation of T_c is probably a result of defects. The highest reported value for $PbMo_6S_8$ is about 15°K, a value which can be obtained also under less favorable preparation conditions by adding a trivalent rare earth.

Concerning the upper critical field itself, we note that for $PbMo_6S_8$, H_{c_2} is more than twice the Clogston limit. This may be understood as a result of strong spin-orbit scattering. In this case, the paramagnetic limiting field may be written [19]

$$H_p \approx 1.33\sqrt{\lambda_{so}}\, H_{po} \qquad (H_{po} = 18.4\, T_c)$$

where λ_{so} is the spin-orbit scattering parameter introduced by Werthamer et al.[14]

Since H_p has to be larger than H_{c2}, we see that $\lambda_{so} > 3$ for $PbMo_6S_8$. In principle λ_{so} may be determined by fitting the temperature dependence of H_{c2}. In fig. 1, is shown the critical field for some of the ternary systems. We get a reasonable fit to the WHH theory for $SnMo_6S_8$, $Cu_2Mo_6S_8$ and apparently also $PbMo_6S_8$ and $PbGd_{.2}Mo_6S_8$. The values of λ_{so} are very large indicating no paramagnetic limitation for $PbMo_6S_8$. However, to be sure of these values, one should measure H_{c2} for $PbMo_6S_8$ down to low temperatures to check that the extrapolation for H_{c2} shown in fig. 1 is valid. We have done such a measurement for a $PbMo_6S_8$ sample with low T_c ($\sim 12°$) and we did find values below the theoretical extrapolation. This shows that this $PbMo_6S_8$ sample cannot be fitted to the WHH theory. We have not yet checked if this is connected with the low value of T_c or if it is also true for the higher T_c $PbMo_6S_8$ samples.

III. RARE EARTH MOLYBDENUM CHALCOGENIDES

The fact which most clearly suggests that the d-electrons of molybdenum are responsible for superconductivity in these compounds is the existence of superconductivity in the $(RE)_xMo_6S_8$ [9] and $(RE)_xMo_6Se_8$ [10]. The RE-atoms are situated far away from the Mo-atoms and we expect a very weak exchange interaction between the 4f-electrons and the 4d-electrons of Mo.

The exchange interaction in these compounds is so weak that T_c in $Gd_{1.2}Mo_6S_8$ is practically not influenced by it. This can be seen by plotting T_c as a function of volume for $Pb_{1-x}Gd_{1.2x}Mo_6S_8$ and $Pb_{1-x}Lu_{1.2x}Mo_6S_8$. The two curves fall one on the top of each other[20], although Lu is nonmagnetic and Gd has a spin of 7/2. The decrease of T_c from $LaMo_6S_8$ to $Lu_{1.2}Mo_6S_8$ [9] seems essentially to be a result of the contraction of the lanthanides, thus a result of decreasing volume. A similar situation is found in $(RE)_xMo_6Se_8$ [10] with the difference, however, that the critical temperature is roughly a factor 2 - 3 higher. This difference has clearly a nonmagnetic origin.

Let us now look at the critical fields for the $(RE)_xMo_6S_8$ compounds. From the estimated density of states for $LaMo_6S_8$ (0.2 states/eV atom spin) and for $Lu_{1.2}Mo_6S_8$ (0.14 states/eV atom spin)

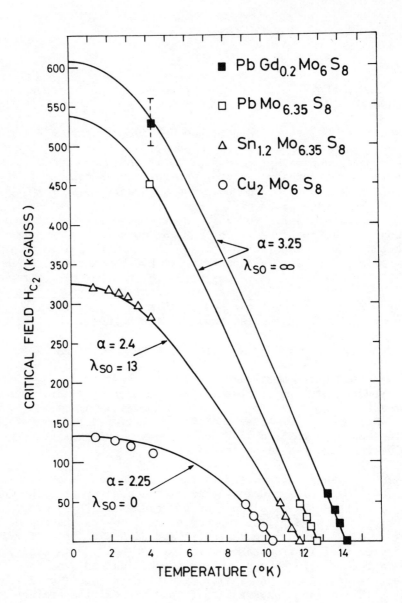

Figure 1: Critical field for some ternary molybdenum sulfides.

we expect relatively low critical fields. This is also what is
found. However, there is a strong depression of the critical field
at low temperatures, characteristic of the effect of an exchange
field acting on the superconducting electrons. In figure 2 is
shown the critical field of $Pb_{.4}Gd_{.6}Mo_6S_8$ and $Gd_{1.2}Mo_6Se_8$. The
critical field was determined as the midpoint of the resistive
transition. The total width of the transition was typically 6
kGauss. For $Gd_{1.2}Mo_6Se_8$ the transition width decreased at lower
temperature to about 4 kGauss. Here the effect of the exchange
field is clearly seen. The estimated orbital critical field
(without the effect of the exchange field) is above 30 kGauss and
60 kGauss respectively, and it is clear that the critical field is
determined only by the effect of the magnetic ions. The exchange
field is therefore equal to the paramagnetic critical field (for
$T \ll T_c$) and we have

$$H_J \approx 1.3 \sqrt{\lambda_{so}} \, H_{po} \quad .$$

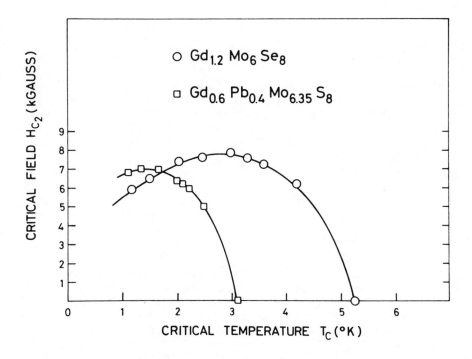

Figure 2: Critical field of $Gd_{.6}Pb_{.4}Mo_6S_8$ and $Gd_{1.2}Mo_6Se_8$ as a function of temperature.

We see that the larger λ_{so}, the larger has the exchange interaction to be to produce the observed H_{c2}. Using $\lambda_{so} = 8$, a value determined from the measurements on the samples containing Eu, gives $H_J \approx 200$ kGauss for $Pb_{.4}Gd_{.6}Mo_6S_8$ and $H_J \approx 340$ kGauss for $Gd_{1.2}Mo_6Se_8$. These values correspond however, only to partly aligned spins. From susceptibility measurements we estimate the magnetization to be about 50% of the saturation magnetization at $1,2°K$ and 6 kGauss. Thus for fully aligned spins we expect an exchange field of about 700 kG in $Gd_{1.2}Mo_6Se_8$ and 800 kGauss in $Gd_{1.2}Mo_6S_8$.

It may sound as a contradiction that the exchange field is so high and yet there seems to be no magnetic effect on T_c. However, one should not forget that the exchange field is a first order effect and that the decrease of T_c results from a second order effect. The above calculated exchange field corresponds to an exchange parameter $J(q = 0) \approx 0.04$ eV. In the expression for the exchange depression of T_c enters $\overline{J(q)}$ which generally is smaller than $J(0)$. If we assume $\overline{J(q)} = (1/3)J(0)$ we get for a system with $N(0) = 0.2$ states/eV atom spin $dT_c/dC = 0.08°K/at\%$. This is small enough that we would not detect it, due to fluctuations in T_c from sample to sample.

IV. CRITICAL FIELD IN THE SERIES $Sn_{1-x}Eu_xMo_6S_8$ AND $Pb_{1-x}Eu_xMo_6S_8$

Eu is a special case among the Rare Earths. In the divalent state, it has the same ionic radius as Sn^{+2}. We therefore suspect Eu to be divalent in these compounds with a spin of 7/2. This is confirmed by susceptibility measurements. However, if it is correct that there is no exchange depression of T_c, we expect T_c to remain unchanged upon substitution of Sn or Pb by Eu in $SnMo_6S_8$ and $PbMo_6S_8$. In figure 3, we show T_c in the series $Sn_{1-x}Eu_xMo_6S_8$. We see that in fact T_c remains unchanged, at least up to $x = 0,5$. At higher concentrations there is a rapid drop in T_c. We should note that all samples in these series had relatively low T_c values, compared with $Sn_{1.2}Mo_6S_8$ showed in fig. 1. This depends as noted above, on the exact conditions of preparation.

Let us now turn to the critical fields. From the results obtained on the other $(RE)_xMo_6S_8$ systems and $Gd_{1.2}Mo_6Se_8$, we expect the critical field to be strongly diminished due to the action of an exchange field. The experimental results shown in figure 4, show however, the contrary; an anomalous increase of H_{c2} at low temperatures and high Eu-concentration [21]. The width of the transitions was typically 60 kGauss at low Eu-concentration and 100 kGauss at higher Eu-concentration. In the sample with $x = .8$, we found narrow transitions near T_c and at low temperatures (~ 60 kG) and very wide transitions in the region where H_{c2} increases rapidly (> 100 kG). The points given in Fig. 3 are the midpoints of the

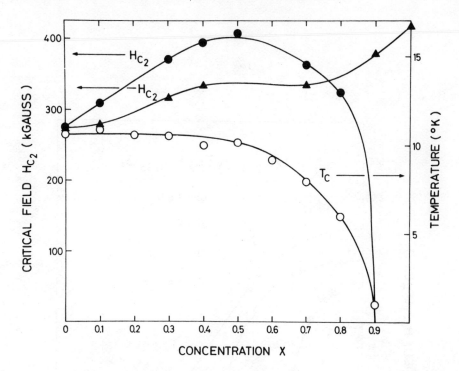

Figure 3: Critical field at T = 0 (●) and critical temperature (○) as functions of concentration χ for $Sn_{1.2(1-x)}Eu_xMo_{6.35}S_8$. For comparison, the critical field of $Sn_{1.2(1-x)}Pb_xMo_{6.35}S_8$ is also shown (▲).

transition. We have interpreted this as the result of a negative exchange field compensating the effect of the external field on the conduction electron spins, an effect first proposed by Jaccarino and Peter[22]. In fig. 3, is shown H_{c2} (T =0) as a function of concentration x. We see that this curve goes through a maximum as we expect for such a compensation effect. At the maximum of H_{c2} versus x, we should have $H_{c2} = |H_J|$, thus $H_J \approx -400$ kGauss for $Sn_{0.6}Eu_{0.5}Mo_6S_8$ which is the right order of magnitude. To analyze the rest of the curve in figure 3, we proceed as indicated in ref. 23, using the series $Sn_{1-x}Pb_xMo_6S_8$ as a nonmagnetic matrix. The value of the spin-orbit scattering parameter λ_{so} can be determined from the height of the maximum. We find a value of about 8 in these series. Knowing this value, it is possible to calculate H_J for all concentrations. The result is a linear dependence of H_J with x as expected. Thus assuming a negative exchange field, the upper critical field can be explained with an exchange field which has the right magnitude and the right concentration dependence. Also the anomalous temperature dependence shown in fig. 4 can be

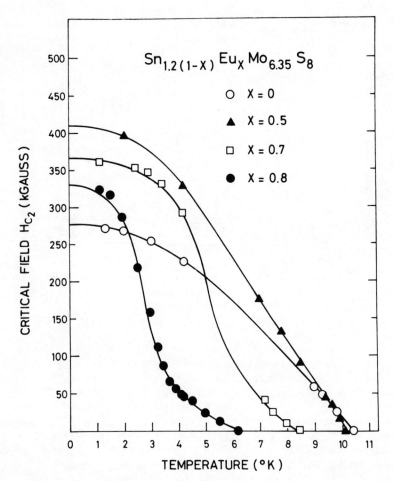

Figure 4: Critical field as a function of temperature for $Sn_{1.2(1-x)}Eu_xMo_{6.35}S_8$ for four values of x:

explained by this assumption. In fact, such a positive curvature as observed in the samples with x = .7 and x = .8 is expected in the case of a negative exchange interaction and in samples where $|H_J| > H_{c_2}$. [23,24]

The value of $\lambda_{so} = 8$ disagrees with the value given in fig.1. This either means that $H_{c_2}(T)$ does not exactly correspond to the WHH-theory or that our interpretation is not the right one. This second possibility can of course not be completely excluded since we have no other evidence for the negative J yet. In this case, one has however to explain the apparent absence of exchange fields in these compounds in addition to finding an explanation for the anomalous increase in H_{c_2}. The $Pb_{1-x}Eu_xMo_6S_8$ series shows a similar behavior as $Sn_{1.2(1-x)}Eu_xMo_6S_8$ and also there the behavior can be analyzed and explained in terms of an exchange-compensation.[24]

CONCLUSION

The $M_yMo_6S_8$ compounds are extremely dirty superconductors with very short yeman free paths. They fit reasonably well the WHH-theory with only a weak reduction due to paramagnetic limitation. If the interpretation in terms of a compensation effect of H_{c_2} in the samples containing Eu is correct, the depression of H_{c_2} due to paramagnetic effects is about 50 - 100 kGauss.

ACKNOWLEDGEMENTS

We thank Dr. R. Shelton, A. Paoli and M. Pellizone for communicating their results prior to publication.

REFERENCES

1. R. Chevrel, M. Sergent and J. Prigent, J. Sol. State Chem. 3, 515 (1971).
2. B.T. Matthias, M. Marezio, E. Corenzwit, A.S. Cooper, and H.E. Barz, Science 175, 1465 (1972).
3. Ø. Fischer, R. Odermatt, G. Bongi, H. Jones, R. Chevrel and M. Sergent, Phys. Letters 45A, 87 (1973).
4. R. Odermatt, Ø. Fischer, H. Jones and G. Bongi, Journal Phys. C 7, L13 (1974).
5. S. Foner, E.J. McNiff and E.J. Alexander, Phys. Letters 49A, 269 (1974).
6. Ø. Fischer, H. Jones, G. Bongi, M. Sergent and R. Chevrel, Journ. Phys. C 7, L450 (1974).
7. R.N. Shelton, A.C. Lawson and D.C. Johnston, Mat. Res. Bull. 10, 297 (1975).

8. Ø. Fischer, Coll. internat. CNRS "Physics in high magnetic fields" 242, 79 (1974).
9. Ø. Fischer, A. Treyvaud, R. Chevrel and M. Sergent, Sol. State Comm. 17, 721 (1975).
10. R.N. Shelton, R.W. McCallum and H. Adrian, to be published.
11. Ø. Fischer, Proceedings of LT14 Helsinki 5, 172 (1975).
12. J. Guillevic, Thesis, University of Rennes (1974).
13. R. Chevrel, Thesis, University of Rennes (1974).
14. N.R. Werthamer, E. Helfand and P.C. Hohenberg, Phys. Rev. 147, 295 (1966).
15. N. Morton, J.G. Booth and C.F. Woodhead, J. Less Comm. Metals 34, 125 (1974).
16. S.D. Bader, G.S. Knapp and A.T. Aldred, Puerto Rico Superconductivity Conference, December 1975.
17. A. Paoli, private communication.
18. M. Pellizoni, private communication.
19. Ø. Fischer and M. Peter, in H. Suhl Magnetism Vol. 5, p. 327, 1973. The numerical factor was there erroneously given as 1.2, the correct value is 1.33.
20. M. Sergent, R. Chevrel and Ø. Fischer, to be published.
21. Ø. Fischer, M. Decroux, S. Roth, R. Chevrel, M. Sergent, J. Phys. C 8, L474 (1975).
22. V. Jaccarino and M. Peter, Phys. Rev. Lett. 9, 290 (1962).
23. Ø. Fischer, Helv. Phys. Acta 45, 331 (1972).
24. M. Decroux and Ø. Fischer, to be published.
25. It is possible depending on preparation technique to influence T_c in these compounds. This will be discussed in detail in ref 20.

QUESTIONS AND COMMENTS

B. Matthias: As you probably all have seen, for the gadolinium salts the critical field decreases as the temperature goes down and I do not think this requires any great theoretical explanation as a matter of fact. The explanation you will get tomorrow by McCallum, to whom I am going to give part of my time, is that these compounds do order magnetically below the superconducting transition.

B. Gyorffy: Spin exchange scattering might help you over the paramagnetic limit but it will also act as a pair breaker and will reduce T_c.

Ø. Fischer: Yes, from the determined value of the exchange parameter you can estimate the decrease in T_c. In low density of states materials like these, it is possible to have an exchange field of a few hundred kilogauss without having more than, say, half a degree decrease in T_c due to exchange scattering.

S. Sinha: Is there any evidence as to what the outer electrons associated with the heavy interstitial atom are doing? Does the atom behave as if it is more like a neutral atom?

Ø. Fischer: We believe that these compounds are partly ionic. If you estimate the sulphur radius from the sulphur-sulphur distance you can estimate a negative charge between 1 and 2 on sulphur.

H. Teichler: There is a paper by Entel and Peter in which they show that almost arbitrary shapes of H_{c2} curves may be obtained by taking into account anisotropic coupling between quasi-particles at the Fermi surface. Why do you exclude this explanation?

Ø. Fischer: We do not exclude this explanation. The work of Entel and Peter show that in a real superconductor one very likely gets another temperature dependence than the one predicted from a calculation not taking into account anisotropy effects. This could account for deviations in the temperature dependence of the non-magnetic systems. The anomalous temperature dependence in the Eu-samples is clearly linked with the magnetic properties of Eu, and it seems to me unlikely that anisotropy effects can explain it.

E. Collings: What normal state electrical resistivity values are observed in these materials?

Ø. Fischer: So far no good measurements have been done due to the lack of real bulk samples. Sintered samples show resistivities of the order of 1 mΩcm. Whereas the resistivities estimated from H_{c2} are of the order of 0.2 mΩcm.

PHONON SPECTRA OF A-15 COMPOUNDS AND TERNARY MOLYBDENUM CHALCOGENIDES

B.P. Schweiss, B. Renker, E. Schneider[+] and W. Reichardt

Kernforschungszentrum Karlsruhe, Institut für Angewandte Kernphysik

Federal Republic of Germany

ABSTRACT

A survey is given on studies of the phonon spectra of several A-15 compounds by inelastic neutron scattering on polycrystalline samples. Comparison of the results for V_3Si, V_3Ge, V_3Ga, Nb_3Sn and Nb_3Al at 297 K lead to the conclusion that the interatomic forces are to a good approximation the same for all compounds with 4.75 valence electrons but are reduced by about 20% for those with 4.5 valence electrons. For all compounds investigated we observe a softening of the phonon frequencies on cooling which is most pronounced for those materials with the highest T_c values. From a comparison of our results with the experimentally determined Eliashberg function of Nb_3Sn we derive information about the energy dependence of the electron-phonon coupling function $\alpha^2(\omega)$.

From similar studies on the Chevrel phase superconductors $PbMo_6S_8$, $SnMo_6S_8$, Mo_6Se_8, and $PbMo_6Se_8$ the modes associated with displacements of the Pb(Sn)-atom could be uniquely identified. The results are discussed with respect to the molecular crystal model proposed recently for the Chevrel phases. For $PbMo_6S_8$ strong low temperature phonon softening has been observed which seems to occur mainly in modes with large Pb-displacements.

[+]Present address: Physik Department TU München.

INTRODUCTION

In recent years the A-15 materials and the Ternary Molybdenum Chalcogenides (Chevrel phases) have attracted considerable interest both from the experimental and the theoretical side due to their unusual electronic and mechanical properties. In particular several of the A-15 compounds have the highest superconducting transition temperatures T_c known today whereas among the Chevrel phases critical fields up to 600 KG have been found. In order to understand superconductivity in these materials information about both the band structure and the lattice dynamics is needed.

For a diatomic system an approximate expression to calculate the electron phonon coupling constant λ and hence T_c from electronic and lattice dynamical properties is given by [1,2]

$$\lambda = \frac{\eta_1}{M_1 <\omega^2>_1} + \frac{\eta_2}{M_2 <\omega^2>_2} \quad . \quad (1)$$

Here M_i are the atomic masses of the two components and η_i are the average electron-phonon interactions which can be obtained from band structure calculations. This expression, which has been derived for systems with widely different atomic masses, may be applied also to compounds with masses of comparable magnitude if the frequency moments $<\omega^2>_i$ are determined from the amplitude-weighted phonon densities of states $F_i(\omega)$ of the two components, which are defined by

$$F_i(\omega) = \frac{1}{N} \sum_{j,q<BZ} |\vec{e}_i^{\,j}|^2 \delta(\omega-\omega_j(\vec{q})) \quad , \quad (2)$$

where $\vec{e}_i^{\,j}$ is the polarization vector of atom i associated with the jth phonon branch.

It was a major goal of this work to obtain information about these partial distribution functions $F_i(\omega)$ from neutron scattering experiments on polycrystalline samples by comparing the results of closely related compounds. In such an experiment a generalized phonon density of states is determined, which is given by

$$G(\omega) = \left[\sum_i \frac{\sigma_i}{M_i}\right]^{-1} \cdot \sum_i \frac{\sigma_i}{M_i} F_i(\omega) \quad , \quad (3)$$

with σ_i the bound scattering cross-section of the ith scatterer.

This is strictly true only for compounds with nuclei which are purely incoherent scatterers (e.g. Vanadium). However most elements are predominantly coherent scatterers. In this case the one-phonon cross-section for a polycrystalline sample contains an additional interference term besides the term proportional to expression (3). However, this term can be made sufficiently small if the cross-section is averaged over a large range of momentum transfers Q. In the experiment this is accomplished by using high incident neutron energies and integrating the scattered intensity over a large range of scattering angles. A good elimination of the interference term is obtained if the range of momentum transfers covers a large number of Brillouin zones.

Table 1 lists $\frac{\sigma}{M}$ - values for elements which are constituents of those A-15 compounds and Chevrel phases on which we performed neutron scattering experiments. It can be seen from this table that in many cases the $G(\omega)$ are not very different from the true phonon density of states $F(\omega)$ (e.g. V_3Ga, Nb_3Al).

Table 1

$\frac{\sigma}{M}$ - Values for Elements which are Constituents of High T_c A-15 Compounds and Chevrel Phases.

Element	σ/M [barn/a.u.]
V	.100
Nb	.071
Al	.056
Si	0.77
Ga	.108
Ge	1.24
Sn	.041
Mo	.064
Pb	.056
S	.037
Se	.127

A-15 COMPOUNDS: RESULTS AT ROOM TEMPERATURE

Most of the measurements on the A-15 compounds were performed on a multidetector time-of-flight spectrometer at the Karlsruhe research reactor FR2 using a primary energy E_o = 65 meV. For Nb_3Sn we carried out additional measurements at the HFR Grenoble with E_o = 45 meV. The good agreement between the results obtained with different E_o indicate that the Q-averaging in the experiments was sufficient.

In Figs. 1 and 2 generalized phonon densities of states $G(\hbar\omega)$ are shown for Nb_3Al, Nb_3Sn, V_3Si, V_3Ge and V_3Ga at room temperature. For Nb_3Al and V_3Si we observe a fairly well separated high frequency peak which is primarily due to the vibrations of the light masses Al and Si respectively. Below 30 meV for Nb_3Al and 36 meV for V_3Si the two distributions are very similar. A strong similarity is also found between the spectra of Nb_3Sn, V_3Ge and V_3Ga, for which the mass ratios of the two components are approximately the same (Nb_3Sn: .78, V_3Ge: .70, V_3Ga: .73). (See also Fig. 4.) The comparison between V_3Si and V_3Ge and between Nb_3Al and Nb_3Sn strongly suggests that the two sidepeaks around the central peak can be attributed to vibrations of the non-transition metal (NTM). A more direct proof that this interpretation is correct is obtained

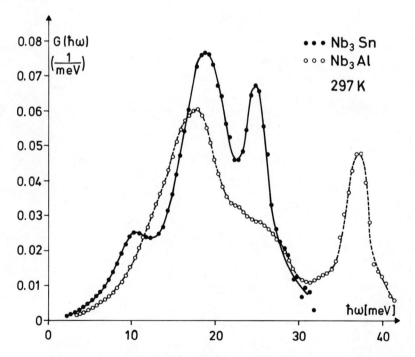

Figure 1: $G(\hbar\omega)$ for Nb_3Sn and Nb_3Al at 297 K.

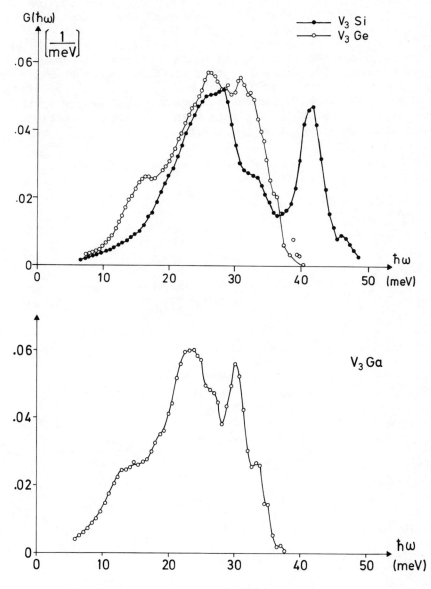

Figure 2: G($\hbar\omega$) for V$_3$Si, V$_3$Ge and V$_3$Ga at 297 K.

from a study of the angular dependence of the inelastic scattering from V_3Ga (Fig. 3). Here we make use of the fact that V is a purely incoherent and Ga a purely coherent scatterer. Strong interference patterns are observed for energies in the regions of the sidepeaks while the scattering distribution is rather smooth in the region of the main peak. Notice that in Fig. 4 the energy scales for V_3Ga and Nb_3Sn differ by a factor 1.25 whereas $\sqrt{M_{Nb}/M_V}$ = 1.35. Furthermore the main peak of Nb_3Al is shifted to lower energies by about 10% compared to that of Nb_3Sn, and the positions of the high frequency peaks for Nb_3Al and V_3Si differ by about the same amount. Thus one is tempted to conclude that generally the interatomic forces in A-15 compounds with 4.5 valence electrons (VE) are about 20% lower than those for members with 4.75 VE.

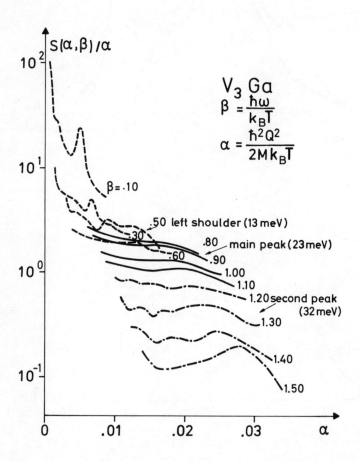

Figure 3: Scattering Law of V_3Ga at 296 K. The energy transfer $\hbar\omega$ and the square of the momentum transfer Q are given in reduced units ($\beta = \hbar\omega/k_BT$, $\alpha = (\hbar^2Q^2)/(2Mk_BT)$).

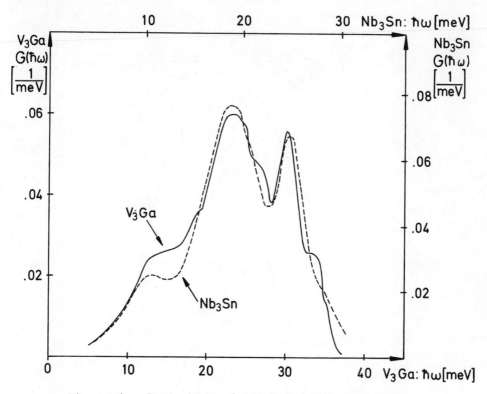

Figure 4: Comparison of $G(\hbar\omega)$ for V_3Ga and Nb_3Sn.

There is some indication that these systematics may be extended to include also compounds with 5 VE. Knapp et al. [3] determined from specific heat measurement the zeroth moment of the phonon spectrum at 300 K of the two closely related compounds Nb_3Sn and Nb_3Sb at 288 K and 333 K respectively, i.e. the value for Nb_3Sb is about 15% larger than that for Nb_3Sn.

The similarity of the phonon spectra of all compounds investigated is demonstrated in Fig. 5 where our results are plotted on a reduced energy scale which takes into account the differences in mass of the TM-component and accounts for the reduction of the interactomic forces in the compounds with 4.5 VE (f_{III} = .91 for 4.5 VE and 1.0 for 4.75 VE). The dashed line represents the amplitude weighted phonon density of states of the TM which according to our analysis is to a good approximation the same for all A-15 compounds when plotted on this reduced scale. If we use this distribution to decompose our results into the two partial densities of states we find that a relation $M_{TM}<\omega^2>_{TM} = M_{NTM}<\omega^2>_{NTM}$ is fairly well fulfilled.

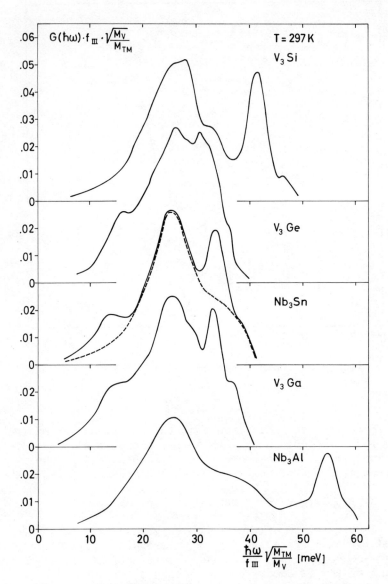

Figure 5: G(ℏω) at 297 K for all A-15 materials investigated plotted on a reduced scale in order to show the close relationship of these spectra. (f_{III} = .91 (1.0) for 4.5 (4.75) VE.)

Our results are summarized in the following three statements:

i. $M_{TM}<\omega^2>_{TM}$ is nearly constant for a given number of valence electrons.
ii. The interatomic forces are increased by about 20% when going from 4.5 to 4.75 valence electrons.
iii. $M<\omega^2>$ is nearly the same for both the TM- and the NTM-component.

Thus the hypothesis of McMillan[4], which assumes that within a role class of materials variations in λ are mainly due to changes in $M<\omega^2>$ whereas the electronic quantity η is essentially constant, is not valid for the A-15 compounds.

The assumption of constant η together with the systematics about the phonon spectra found in our investigations would lead to the following predictions:

- All V-based compounds should have higher T_c values than the Nb-based compounds due to the larger prefactor in McMillan's T_c formula.

- Compounds with 4.5 VE should have higher T_c values than compounds with 4.75 VE.

- V_3Ge and V_3Si should have about the same T_c.

Apparently these predictions are not observed in most cases.

The observations ii and iii suggest that the vibrational spectra of the A-15 compounds are largely determined by the forces between the TM- and the NTM-component. This may explain the failure of McMillan's hypothesis.

Our discussions were based on results at room temperature. In the following section it will be shown that considerable changes of the phonon density of states occur when we cool down to liquid helium temperature. However, these changes will only slightly affect the conclusions drawn from the room temperature data.

A-15 COMPOUNDS: TEMPERATURE DEPENDENCE OF THE PHONON DENSITY OF STATES

Ultrasonic measurements [5] and investigations of the phonon dispersion curves [6] on single crystals of Nb_3Sn and V_3Si have shown that at least at low frequencies a considerable softening of some phonon branches occurs when the temperature is lowered. The question arises whether this effect will show up in the phonon density of states and - if so - which part of the spectrum is affected by this softening. In Fig. 6 the phonon densities of states at room temperature are compared with those at 77 K for the three Vanadium compounds. One observes that up to rather high frequencies a phonon softening occurs on cooling and that this effect is more pronounced for the high T_c compounds. When cooled further to about 4 K only minor changes have been observed. A similar behavior is found for the Nb-compounds (Fig. 7). From Table 2 it is seen that there is a clear correlation between the frequency shifts and the magnitude of T_c, the changes being slightly lower for the Nb-compounds than for the V-compounds.

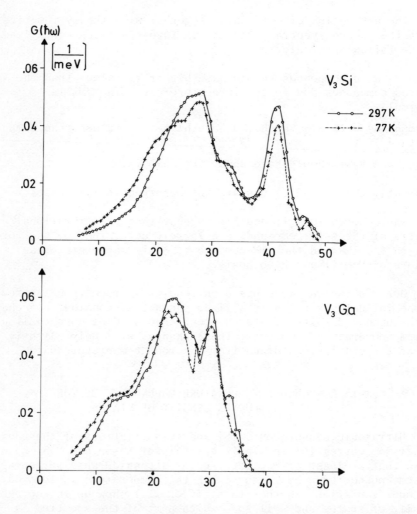

Figure 6: Temperature Dependence of $G(\hbar\omega)$ for V_3Si and V_3Ga.

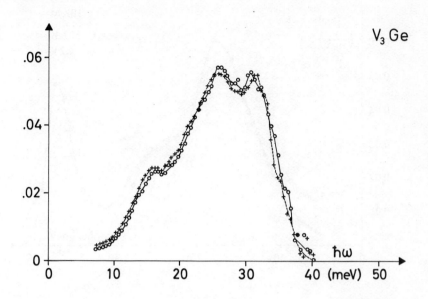

Figure 6 (cont'd): Temperature Dependence of $G(\hbar\omega)$ for V_3Ge.

Table 2

Relative Frequency Shifts between 297 K and 4 K. $\langle\omega^2\rangle$ has been calculated according to $\langle\omega^2\rangle = \langle\omega\rangle/\langle\omega^{-1}\rangle$.

	T_c (K)	$\rho(\langle\omega^2\rangle^{\frac{1}{2}})$ (%)
V_3Ge	6.2	3
V_3Ga	13.9	6*
V_3Si	16.7	9
Nb_3Sn	17.8	7
Nb_3Al	18.4	8

*This value refers to 77 K, as no 4 K data are available.

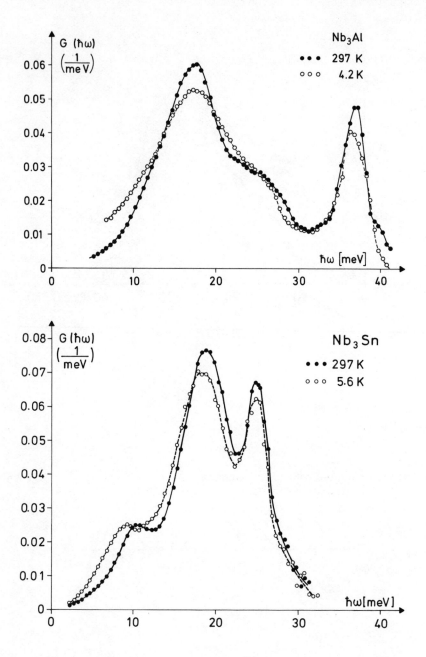

Figure 7: Temperature Dependence of G($\hbar\omega$) for Nb_3Al and Nb_3Sn.

From the measured T_c data the λ-values for V_3Si and V_3Ge are obtained as $\lambda_{V_3Si} = 1.02$ and $\lambda_{V_3Ge} = .68$, i.e. the ratio of the two values is 1.5. Under the assumption that η is the same for both compounds and calculating $M\langle\omega^2\rangle$ from our phonon spectra this ratio will be 1.0 at 297 K and increases to 1.11 at 4 K due to the softening of the phonon modes. The latter value is still far below the value 1.5 obtained from experiment.

This example shows that the differences in T_c cannot be explained solely by differences in the phonon spectra and that the conclusion drawn in the preceding section are essentially unchanged if we use the low temperature results.

All samples have been investigated on a high resolution neutron diffractometer for the occurrence of a tetragonal phase at low temperature. Only for Nb_3Sn and V_3Si a phase transformation was found. For the other compounds a possible tetragonal distortion has to be well below $1°/°°$.

In Fig. 8 our result for Nb_3Sn at 5.4 K is compared with the Eliashberg function $\alpha^2(\omega)F(\omega)$ determined by Shen [7] from tunneling experiments. Both distributions show the same three-peak-structure with very good agreement in the positions of the peaks. The ratio

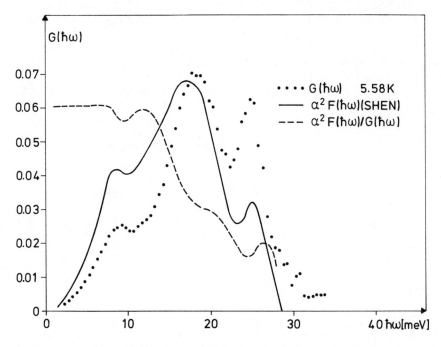

Figure 8: Nb_3Sn: Comparison of $G(\omega)$ with the Eliashberg function $\alpha^2(\omega)F(\omega)$.

of the two distributions, given by the dashed curve, represents a
good approximation to the electron-phonon-coupling function $\alpha^2(\omega)$.
There are dips in this curve at the positions of the side-peaks,
which---according to the preceding discussion---are essentially
due to the Sn vibrations. Taking into account that in the neutron
result the Sn vibrations are weighted too low [$(\sigma/M)_{Sn} = .58(\sigma/M)_{Nb}$]
these dips will be more pronounced in the true $\alpha^2(\omega)$ indicating
that the Sn contributes less than the Nb to λ and T_c. However,
the presence of the two side-peaks in the tunneling data clearly
shows that this contribution is not negligible. Besides the difficulties arising from the differences between $G(\hbar\omega)$ and the true
phonon density of states $F(\hbar\omega)$ a quantitative analysis is further
complicated by possible uncertainties in the tunneling spectrum:
In high T_c transition metals tunneling experiments are very difficult
due to short coherence length. Unless the tunnel junctions are very
perfect one has to expect that the phonon structure in the tunneling
characteristic and hence the experimental $\alpha^2(\omega)F(\omega)$ is reduced at
high energies. Most likely, this applies also to the Nb_3Sn results
as is indicated by the very low value $\mu^* = 0.06$ obtained in the
analysis of the tunneling experiment. Thus the decrease of the
dashed curve in Fig.8 towards high energies may at least partly
be explained by a slightly dirty tunnel junction.

A crude estimation which is affected by these uncertainties
in the experimental $\alpha^2(\omega)F(\omega)$ and by uncertainties in the decomposition of $G(\omega)$ into the two partial densities of states shows
that about 10% of λ is due to the Sn component. This is in contrast to the results of Klein and Papaconstantopoulos [7] who
performed band structure calculations for V_3Si with the aim to
calculate superconducting quantities. They found that the contribution of Si to λ is completely negligible.

STUDIES ON SOME CHEVREL PHASE SUPERCONDUCTORS

Powder samples of $PbMo_{6.35}S_8$ (T_c = 14 K), $SnMo_6S_7$ (13 K),
$Pb_{1.2}Mo_6Se_8$ (3.9 K) and Mo_6Se_8 (6.5 K) were prepared using the
method described by Marezio et al. [9]. The Chevrel phases exist
as defect structures and the quoted formulae give the initial
composition. Neutron diffraction patterns were taken for all of
the samples and did not show any appreciable amount of other phases.
The transition temperatures given in the brackets were measured
by the inductive method. The values indicate the midpoint of the
transition region which was typically of the order of .5 K.

Neutron scattering experiments on the Chevrel phases are
difficult as the phonon spectra range up to about 50 meV and very
good resolution is required in order to resolve the rich structure
in the spectra especially in the low frequency region. Therefore

we performed most of our experiments on a time-of-flight spectrometer at the cold source of the FR2 reactor using 5 meV neutron incident energy and neutron upscattering, which gives good resolution at low energy transfers whereas the resolution is only moderate at large energy transfers. However, under these experimental conditions the range of momentum transfers Q covered by the experiment may be too small to obtain a sufficient Q-averaging for the elimination of the interference scattering especially for small energy transfers. This effect will not be detrimental if we compare results for different compounds under the same experimental conditions.

Fig. 9 shows the room temperature results obtained for $PbMo_{6.35}S_8$ and $SnMo_6S_8$ respectively. The $G(\hbar\omega)$ results for $Pb_{1.2}Mo_6Se_8$ and Mo_6Se_8 are compared in Fig. 10.

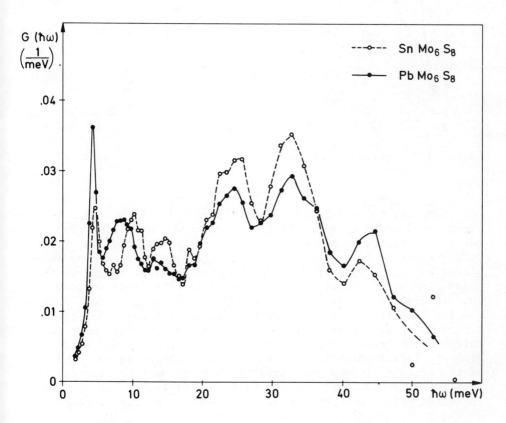

Figure 9: Comparison of $G(\hbar\omega)$ for $SnMo_6S_8$ and $PbMo_{6.35}S_8$ at 296 K.

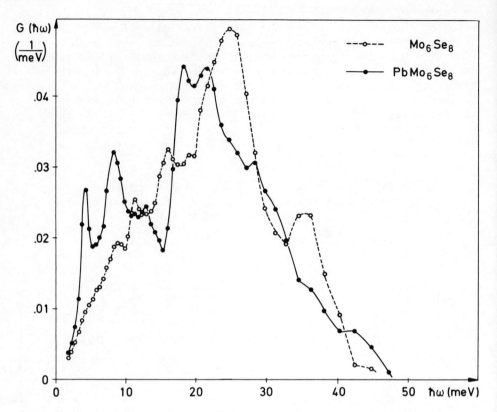

Figure 10: Comparison of $G(\hbar\omega)$ for Mo_6Se_8 and $Pb_{1.2}Mo_6Se_8$ at 296 K.

In order to explain their results of specific heat and Mössbauer experiments Bader et al. [10] have suggested that the Chevrel phases may be considered as molecular crystals with slightly distorted CsCl-structure, where the X-atoms (Pb, Sn, etc.) are at the cation places and the quasi-rigid Mo_6S_8 (Mo_6Se_8) "molecules" are at the anion places. Then the total number of 45 normal modes group into 36 high frequency internal and 9 soft external modes which consist of 6 translational acoustic and optic modes and 3 librational modes of the Mo-chalcogenide cage. Within this picture intersection between high internal and low external modes is probably indicated by the gap at 18 meV shown in Fig. 9. The ratio of the areas is indeed approximately 9:36 as must be expected and this value is even better approximated if we try to correct for the different atomic scattering cross-section of the components.

The most prominent features within the region of the external modes are the extremely sharp peak at 4.3 meV and a broader peak at 8 meV for both $PbMo_{6.35}S_8$ and $Pb_{1.2}Mo_6Se_8$. For $SnMo_6S_8$ these two peaks are shifted to higher frequencies (4.7 meV and 10.5 meV). As for Mo_6Se_8 the corresponding features are absent it seems natural to correlate these excitations with movements of the X-atoms. Comparing $SnMo_6S_8$ and $PbMo_{6.35}S_8$ the shift of the higher peak corresponds very well to ratio $(M_{Pb}/M_{Sn})^{\frac{1}{2}} = 1.32$ whereas for the lower peak the shift is only 10%. Though the low frequency peaks are absent in Mo_6Se_8 there is still a large number of low frequencies modes and therefore we have to assume that also in the other three compounds the peaks due to exitation of the X-atom sit upon a high density of states which is due to translational and librational motions of the Mo-chalcogenide unit.

For the two sulfides the region of the internal vibrations exhibit a very similar three-peak-structure. The peak at about 45 meV corresponds mainly to vibrations of the S-atoms. The shift in frequency and the change in intensity when the S-atoms are replaced by Se with the larger mass and larger scattering crosssection support this conclusion. Taking an average frequency of 7 meV for the Pb-vibrations and 45 meV for the S-vibrations we can estimate that the effective force-constant acting on the S-atom is about 5 times as large as that acting on the Pb-atom.

So far all discussion seem to be in favor of the molecular crystal model. However, by comparing the results of Mo_6Se_8 and $PbMo_6Se_8$ we observe that also the internal vibrations are affected to some extent by the insertion of the Pb-atoms. Thus this model provides a fruitful basis for a general understanding of the phonon spectra, but it may be too simplified for a quantitative description of the lattice dynamics of these compounds.

We have started studies on the temperature dependence of $G(\hbar\omega)$ for $PbMo_6S_8$. Investigations at 4 K are hampered by the necessity that the measurements have to be performed in down scattering. This requires a high incident neutron energy (and thus necessarily poor resolution) if one wants to cover the total frequency range necessary to obtain reliable normalization of the spectra.

Measurements at 296 K and 4.2 K using an incident energy of 65 meV showed that the spectra are essentially unchanged in the region of the internal vibrations above 18 meV whereas a considerable softening of the 4 K spectrum was observed in the low frequency range. However, the resolution was insufficient to resolve the prominent peak at 4.3 meV. Further measurements again at 296 K and 4.2 K using an incident energy of 23.8 meV with improved resolution allowed to determine $G(\hbar\omega)$ up to about 12 meV. The normalization of the spectra was obtained by using the 65 meV results. In Fig. 11 the 4.2 K data are compared with the $G(\hbar\omega)$

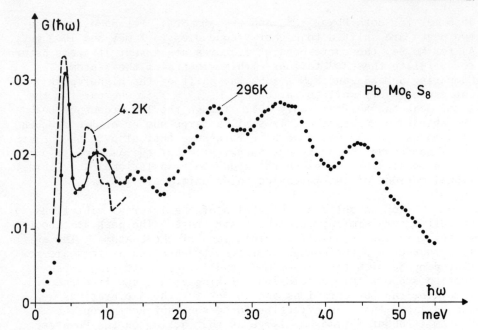

Figure 11: PbMo$_{6.35}$S$_8$: Comparison of G($\hbar\omega$) at 4.2 K and 296 K. The room temperature data have been obtained with E_o = 5 meV in upscattering. The slight differences of this distribution compared to that of Fig. 9 are caused by using a larger range of scattering angles in the analysis. This indicates that the Q-averaging was not sufficient in this experiment. The dashed line represents the 4 K data obtained with E_o = 23.8 meV. An additional measurement with E_o = 65 meV has been used to fix the height of the 296 K curve.

at 296 K obtained with E_o = 5 meV in upscattering. There is both an enhancement of the low frequency modes and a shift of the peak positions when the temperature is lowered, the shift being more pronounced for the peak at 8 meV. This shows that those modes associated with displacements of the Pb-atom soften considerably on cooling. A similar behavior has been found for SnMo$_6$S$_8$ from ^{119}Sn Mössbauer studies by Kimball et al. [11].

From our results it is not possible to conclude uniquely that the softening is restricted to vibrational modes of the Pb-atom only. As discussed before the Pb-peaks sit on top of a high density of modes not associated with the displacements of the Pb-atoms which may also be strongly temperature dependent. A final answer may be possible after similar measurements on Mo$_6$Se$_8$ at low temperatures.

Quite clearly the low lying frequencies of the X-atoms are unimportant for the superconducting properties of the Chevrel phases as the comparison of the low T_c = 3.9 K for $PbMo_6Se_8$ and the higher T_c = 6.5 K for the Mo_6Se_8 shows. Furthermore it is most likely that the low temperature anharmonicity of these modes is of different origin than in the A-15 compounds. In the A-15 compounds phonon softening is caused by the strong electron-phonon interaction which is also responsible for the high superconducting transition temperatures. In the Chevrel phases the X-atoms experience a highly anharmonic potential which results from the large space available in the open channels between the Mo-chalcogenide cages. If this interpretation is correct no correlation between the amount of softening and T_c may be expected.

ACKNOWLEDGEMENT

We would like to thank Drs. S.K. Sinha, S.D. Bader, and G.S. Knapp for valuable collaboration. We also thank Dr. Ø. Fischer for providing a $PbMo_6S_8$ sample and Dr. A. Müller for preparing several Vanadium based A-15 compounds. The help of Dr. N. Nücker in the experiment on Nb_3Sn is gratefully acknowledged.

REFERENCES

1. J.C. Phillips, in Superconductivity in d- and f-Band Metals, ed. D.H. Douglass, p. 339 (1972).
2. B.M. Klein and D.A. Papaconstantopoulos, to be published in J. Phys. F.
3. G.S. Knapp, S.D. Bader and Z. Fisk, Phys. Rev. B13, 3783 (1976).
4. W.L. McMillan, Phys. Rev. 167, 331 (1968).
5. L.R. Testardi and T.B. Bateman, Phys. Rev. 154, 402 (1967).
6. G. Shirane, J.D. Axe and R.J. Birgenau, Sol. State Commun. 9, 397 (1971).
7. L.S.L. Shen, Phys. Rev. Lett. 29 1082 (1972).
8. B.M. Klein and D.A. Papaconstantopoulos, to be published.
9. M. Marezio, P.D. Dernier, J.P. Remeika, E. Corenzwit and B.T. Matthias, Mat. Res. Bull. 8, 657 (1973). B.T. Matthias, M. Marezio, E. Corenzwit, A.S. Cooper and H.E. Barz, Science 175, 1465 (1972).
10. S.D. Bader, G.S. Knapp, and A.T. Aldred, Proc. Int'l. Conf. on Low-lying Lattice Vibrational Modes and their Relationship to Superconductivity and Ferroelectricity, San Juan, Puerto Rico December, 1975). Ferroelectricity, to be published.
11. C.W. Kimball, L. Weber, G. Van Landuyt, F.Y. Fradin, B.D. Dunlap and G.K. Shenoy, Phys. Rev. Lett. 36, 412 (1976).

QUESTIONS AND COMMENTS

J. Bostock: I would like to know how your A-15 samples were made?

W. Reichardt: We got them from several places. That is to say, we made only a few ourselves and this we did by pressing the powder of the components and then arc-melting it.

J. Rowell: If you take a material where neutron work has been done, to obtain the dispersion curves and then $F(\omega)$, say lead or niobium, and then you take the same material and measure $G(\omega)$ using your method, how different are $F(\omega)$ and $G(\omega)$?

W. Reichardt: That depends on what kind of material it is. In the material consisting of just one kind of atom then you should get, if you do the experiment properly, the true phonon density of states. In the other case where you have several different atoms in the material the individual contributions are weighted by the ratio σ/M. For instance, in the case of the A-15's, the σ/M values vary about 40%.

S. Sinha: In regard to the low frequency modes in the moly-sulfides, isn't it reasonable to think of a flat Einstein-like mode which is hybridized with the acoustic modes which are going to push through them, producing that structure which you see. You cannot separate them in any real way.

W. Reichardt: I am certain that this is the explanation. That is to say, all the atoms are moving in phase and therfore you have the rise given by the elastic constants then the branch will drop off. I think you have hybridization here with a very low-lying optic mode.

S. Sinha: We have also measured molybdenum selenides at low temperature. We don't find any appreciable shifts at low temperatures.

W. Reichardt: The molybdenum selenides?

S. Sinha: That's right.

W. Reichardt: Okay, then this would answer the question I raised and I would say it is all due to the lead atom.

INELASTIC NEUTRON SCATTERING STUDIES OF THE PHONON SPECTRA OF

CHEVREL PHASE SUPERCONDUCTORS[*]

S. D. Bader and S. K. Sinha[**]

Argonne National Laboratory

Argonne, Illinois 60438

and

R. N. Shelton[†]

Institute for Pure and Applied Physical Science
University of California, San Diego

La Jolla, California 92093

ABSTRACT

Phonon spectra are obtained using inelastic neutron scattering by polycrystals of the Chevrel-phase superconductors $SnMo_6S_8$, $PbMo_6S_8$, Mo_6Se_8, and $Pb_{1.2}Mo_6Se_8$. Modes associated primarily with Sn (or Pb) atomic displacements are clearly identified. Acoustic softening on cooling is noted for $SnMo_6S_8$. Anharmonicity and the superconductivity are discussed utilizing the molecular-crystal concept.

---

[*] Submitted after the Conference.
[**] Work supported by the U. S. Energy Research and Development Administration.
[†] Research supported by the U. S. Air Force Office of Scientific Research under Contract No. AFOSR/F-44620-72-C-0017.

INTRODUCTION

The recently discovered [1] Chevrel-phase molybdenum chalcogenide superconductors [2] are of interest primarily due to their extremely large upper critical fields [3], and also because of their modestly high T_c values [4], ability to accommodate paramagnetic rare-earth ions [4,5] interesting pressure effects [4], lattice instabilities [6], and soft modes [7]. The inelastic neutron scattering measurements reported here were undertaken as part of an effort to help elucidate the lattice dynamics of this interesting class of superconductors. We find an appreciable spectral density in a frequency range Bergmann and Rainer [8] indicate is optional for high T_c values. We also tentatively identify these modes as being associated with the collective displacements of quasi-rigid groupings of atoms, using a molecular-crystal model [9,10]. Softening with decreasing temperature of particular types of modes is observed. These observations, when properly combined with an understanding of the electronic properties, and an appreciation of the crystal structure (including its imperfections), should lead to deeper insights into the nature of the superconductivity of these interesting materials.

The samples used in this study were prepared from the elements, and found from X-ray powder patterns, to be single phase, except for the binary Mo_6Se_8, which may contain an order 10% second phase ($MoSe_2$). Also, the superconducting transitions, measured calorimetrically in a separate study [11], were quite sharp, which provides confidence that these samples are of appropriate quality to warrant further study. The crystal structure [12] is rhombohedral with space group $R\bar{3}$. There is one formula unit per unit cell. The structure of the ternaries may be envisioned as being based on a distorted CsCl-type arrangement of individual X atoms at the origin, where X is Sn or Pb, and discrete Mo_6S_8 (or Mo_6Se_8) units, each of which consists of a Mo_6 octahedron within a distorted chalcogenide cube that is tilted $\sim$15-20° from the rhombohedral axes. Binary Mo_6Se_8 has the same structure [13], but the origin is not occupied. There may be appreciable numbers of vacancies in these crystals. For instance, Marezio, et al. [12] found that their nominal $PbMo_6S_8$ crystal fit X-ray diffraction intensities better using the composition $Pb_{0.92}Mo_6S_{7.5}$. Also, the sample with nominal composition $Pb_{1.2}Mo_6Se_8$ presumably contains its excess Pb within the channels formed by the chalcogenide network. The T_c values of the samples of $PbMo_6S_8$, $SnMo_6S_8$, Mo_6Se_8, and $Pb_{1.2}Mo_6Se_8$ are 13.2, 12.1, 6.3, and 3.9 K, respectively [11].

We find it useful to regard the Chevrel-phase compounds as molecular crystals composed of quasi-rigid Mo_6S_8 (or Mo_6Se_8) units bound to each other, and for the ternaries, bound also to individual X atoms. The lattice dynamics then simplifies since the

3n normal modes of the unit cell, where n is the number of atoms in the unit cell, group into two categories: the low-lying "external" modes, and the higher-lying "internal" modes, which are internal to the molybdenum chalcogenide units. There are nine external modes for the ternaries and six for the binary. Six of the nine (three acoustic plus three optic) are associated with translational motion of an X atom and a quasi-rigid molybdenum chalcogenide unit, and three with torsional motion of the quasi-rigid units. For the binary compound, the low-lying optic modes are absent.

The technique of determining phonon densities of states from polycrystalline samples by coherent inelastic neutron scattering has been discussed extensively in the literature [14-16]. The one-phonon coherent partial differential cross-section for thermal neutron scattering from a crystal is given by [17]

$$\frac{d^2\sigma}{d\Omega dE} = \frac{(2\pi)^3 k}{\Omega\, k_o} \sum_{qj} \left| \sum_{\kappa} \frac{b_\kappa [\underset{\sim}{K} \cdot \underset{\sim}{e}_\kappa(qj)] e^{i\underset{\sim}{K} \cdot \underset{\sim}{r}_\kappa} e^{-W_\kappa}}{(2M_\kappa)^{\frac{1}{2}}} \right|^2$$

$$\times \left(\frac{1}{\omega_{qj}}\right)\left(\frac{1}{e^{\beta\omega_{qj}} - 1}\right) \delta(\underset{\sim}{K} - \underset{\sim}{q} - \underset{\sim}{G}) \delta(\omega - \omega_{qj}), \qquad (1)$$

where E is the scattered neutron energy, Ω is the unit cell volume, b_κ is the coherent neutron scattering length for the κth atom in the unit cell, $e_\kappa(qj)$ denotes the eigenvector for the κth atom (of mass M_κ and position r_κ) associated with phonon mode (qj), K is the neutron momentum transfer vector, G is the reciprocal lattice vector, k and k_o are the final and incident neutron wavevectors, respectively, ω is the neutron energy gain (in units where $\hbar=1$) and e^{-W_κ} is the Debye-Waller factor for atom κ. In Eq. (1) we have assumed only phonon annihilation processes, corresponding to the conditions of the present series of measurements. For polycrystalline samples, one automatically averages over all orientations of $\underset{\sim}{K}$. However, the spectrum seen by one detector at a fixed scattering angle does not yield enough of an average over q in Eq. (1). Proper averaging is achieved using a large number of detectors grouped at various angles, in such a way that the volume of reciprocal space sampled by $\underset{\sim}{K}$, for a given energy transfer, is large compared to the volume of the Brillouin zone. Then the summed intensity may be related to a "weighted" or "generalized" phonon frequency spectrum given by

$$\left[\frac{k_o}{k} \omega\, (e^{\beta\omega}-1)\, \frac{1}{K^2}\right] \frac{d\sigma}{dE} \propto \sum_j P_j g_j(\omega) \equiv G(\omega) \qquad (2)$$

where the factors on the left mean we have multiplied the spectrum in each detector group by the term in brackets for a given energy transfer before adding the spectra from the detector groups, $g_j(\omega)$ is the partial density of phonon states for branch j,

$$g_j(\omega) = \sum_q \delta(\omega - \omega_{qj}), \qquad (3)$$

and

$$p_j = \frac{1}{3} \sum_\kappa \frac{b_\kappa^2}{2M_\kappa} \langle e^{-2W_\kappa} |e_\kappa(qj)|^2 \rangle, \qquad (4)$$

where the average is over all K sampled by the detectors for the given energy transfer. We have written the above expressions in terms of sums over "phonon branches" because we have in mind the molecular crystal model mentioned above, where the phonon branches would naturally separate into groups of different types of modes, each with their own characteristic pattern of eigenvectors. We have also assumed in Eq. (4) that the K average causes the interference terms between different atoms in the unit cell to sum to zero because of the presence of the phase factor $e^{i K \cdot (r_\kappa - r_{\kappa'})}$ in Eq. (1). Thus, using Eq. (2), we may extract the generalized frequency distribution function $G(\omega)$. We should bear in mind that p_j from Eq. (4) may itself have some frequency dependence, but it is likely to be relatively smooth and the basic structure (including positions of peaks, etc.) of $g(\omega)$ should be obtainable by this method. The generalized frequency spectrum obtained in this way is in fact quite analogous to the "$\alpha^2 F$" spectrum obtained in tunneling measurements on superconductors. In the present case, the electron-phonon matrix elements are replaced by neutron-phonon matrix elements. The experiment must be performed under conditions such that sufficient K-averaging is obtained but K must not be so large as to render multi-phonon scattering processes appreciable.

The present experiments were performed on the thermal neutron time-of-flight (TNTOF) facility at the CP-5 reactor at Argonne National Laboratory. The incident beam had an energy of 4.83 meV, was obtained from a double graphite monochromator, and was passed through a cooled Be filter and a correlation chopper. The measurements were performed with neutron energy gain. Seventy-five detectors were used to detect the scattered neutrons and these were connected for various runs in ten to twelve different groups spanning scattering angles from roughly 30° to 110°. Care was taken to avoid detectors which could observe Bragg scattering for a particular sample to avoid large elastic peaks. Nevertheless, the imperfections of the correlation chopper necessitated a correction for spurious oscillations resulting from the diffuse elastic peak. This was carried out by subtracting a suitable renormalized decorrelated elastic peak from vanadium during the data analysis.

The K values ranged from 1 Å^{-1} to 2.84 Å^{-1} for 3 meV energy transfer, and from 3.2 Å^{-1} to 5.1 Å^{-1} for 35 meV energy transfer. Note that the distance to the boundary of the first Brillouin zone for these materials is of the order of 0.7 Å^{-1}. It appears therefore that K-averaging should be reasonably complete for these spectra. ~The K-averaging in this case is also helped by the considerable lack of dispersion for many of the modes seen. No corrections have been made for multi-phonon processes.

RESULTS

The room-temperature $G(\omega)$ for Mo_6Se_8 appears in Fig. 1. In the molecular-crystal model we expect only two types of external modes for the binary: acoustic and torsional. We assign these designations to the 8 and 12 meV peaks, respectively. The higher frequency peaks are assigned to the internal modes.

The room-temperature $G(\omega)$ for $PbMo_6S_8$ appears in Fig. 2. Note the richness of structure in $G(\omega)$. Six peaks can be identified whose frequencies range from 4.3 to ~47 meV. By analogy with the Mo_6Se_8 spectrum, we can assign 8 and 12.5 meV peaks with the acoustic and torsional modes, and the three higher-frequency peaks to the internal modes. The 4.3 meV peak can then be recognized as being the Einstein-like optic mode peak, which is largely associated with displacements of the Pb atoms. The frequency range for both the internal and external modes and the value of the frequency that separates them (~18 meV) are all in agreement with average values obtained from an analysis of heat-capacity measurements [9,10] between 2 and 400 K. Also, note that the internal modes of $PbMo_6S_8$ contains a higher-frequency peak than Mo_6Se_8. The peak at ~47 meV for $PbMo_6S_8$ is presumably primarily associated with the displacements of the light S atoms. Comparing Figs. 1 and 2, the corresponding chalcogenide spectral density seems to have redistributed itself in the 25 to 30 meV region for Mo_6Se_8, in rough agreement with a $\sqrt{M}$ mass effect between S and Se. (The shift in the average internal-mode frequency between the sulfides and selenides is also noted in the heat-capacity analyses.)[18]

The results for $SnMo_6S_8$ appear in Fig. 3. At room temperature the frequency spread in $G(\omega)$ for $SnMo_6S_8$ is similar to that for $PbMo_6S_8$. The lowest-energy peak, denoted ω_X, is broader for the Sn than the Pb ternary. The 300-K value of ω_{Sn} is 4.7 meV. The

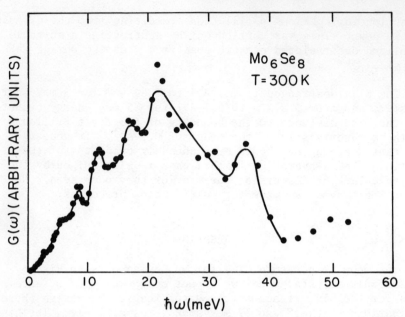

Figure 1: The G(ω) for Mo_6Se_8 at 300°K. The curve is a guide to the eye.

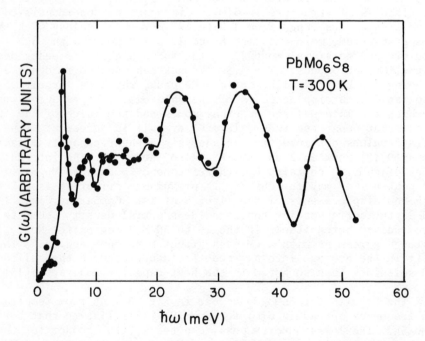

Figure 2: The G(ω) for $PbMo_6S_8$ at 300°K. The curve is a guide to the eye.

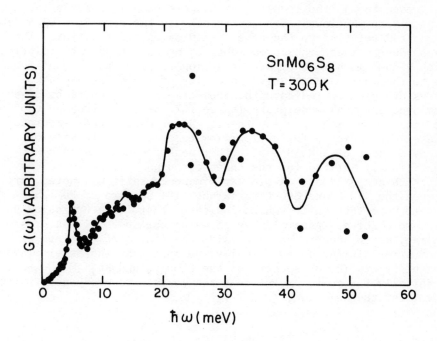

Figure 3: The $G(\omega)$ for $SnMo_6S_8$ at 300 K. The curve is a guide to the eye.

mass effect for the ω_X modes expected from the equal-force-constant harmonic expression $\omega_{Sn}/\omega_{Pb} = (M_{Pb}/M_{Sn})^{\frac{1}{2}}$ is not fully observed, since $\omega_{Sn}/\omega_{Pb} \simeq 1.1$, while $(M_{Pb}/M_{Sn})^{\frac{1}{2}} = 1.32$, or using the appropriate reduced mass ratio $(\mu_{Pb-Mo_6S_8}/\mu_{Sn-Mo_6S_8})^{\frac{1}{2}} \simeq 1.26$. The features corresponding to the remaining modes, both external and internal, are similar for the two ternaries.

We now focus on lower temperature spectra of these three compounds and of $Pb_{1.2}Mo_6Se_8$ (see Fig. 4). Only the low frequency regions ($\lesssim 25$ meV) of these spectra have adequate counting statistics, even though the counting times were much longer than at 300 K. Note the lowest frequency peak in the $PbMo_6S_8$ $G(\omega)$ appears attenuated relative to the other features of the spectrum. This may be due to inaccuracies in subtracting the chopper oscillations, which become much more important at low temperature than at 300 K, due to the increased relative intensity of the elastic peak. Arrows are used to indicate the room-temperature positions of the prominent external mode peaks (three for the ternaries and two for the binary) on each spectrum in Fig. 4 for which room-temperature data exists. Except for a clear low temperature softening of the 10-meV feature of $SnMo_6S_8$, all other peak frequencies have not significantly shifted for the three compounds for which we can make such comparisons. A similar frequency shift was noted earlier between the 4.2-K and 300-K Karlsruhe [10] spectra of $PbMo_6S_8$. We fail to note that shift, which indicates that it occurs below 90 K. Also, we observe a general sharpening of the spectral features on cooling, which appears to be outside the limits of statistical fluctuations in the data. Finally, we note that the external-mode peak positions in the spectra of the $PbMo_6S_8$ and $Pb_{1.2}Mo_6Se_8$ are essentially the same, except that the lowest-frequency peak is at ~ 5 meV for the selenide.

DISCUSSION

We find that the molecular-crystal model does provide a reasonable starting point for a description of the spectra of the Chevrel-phase superconductors. There are two important features we will now discuss. The first is the anharmonic character of the external modes. The second is the relationship of the spectra to the superconducting transition temperatures, and, in particular, the importance of the acoustic and torsional modes.

The Chevrel-phase compounds provide yet another example that high-T_c values are accompanied by soft modes. In general, the presence of anharmonic terms in a crystal's potential energy causes the actual phonon frequencies ω_{qj} to suffer complex shifts

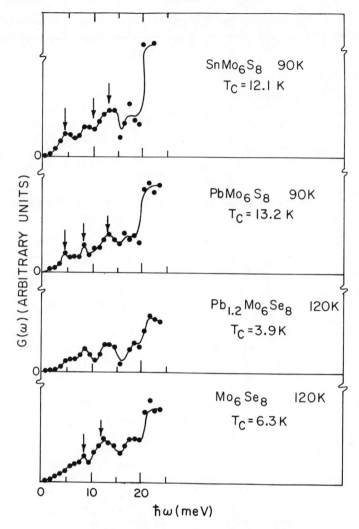

Figure 4: The low-temperature generalized phonon spectra for the indicated compounds plotted using a 1 meV mesh for clarity. The curves were drawn utilizing a finer mesh and serve as guides to the eye. The arrows locate the <u>room-temperature</u> positions of the prominent external-mode peaks on the appropriate spectra. For the ternaries these peaks are tentatively identified as being largely optic-like, acoustic-like, and torsional with increasing frequency, respectively. For the binary Mo_6Se_8 the arrows at 8.5 and 12 meV tentatively locate the acoustic and torsional mode peaks, respectively. Note the acoustic softening for $SnMo_6S_8$, and the sharpening of the spectral features with respect to their character at room temperature. The attenuation of the lowest-frequency peak for the ternaries is believed to be spurious (see text).

$\Delta\omega_{qj} + i\Gamma_{qj}$ with respect to the unperturbed harmonic normal-mode frequencies [19]. The real part $\Delta\omega_{qj}$ gives the frequency shift, while the imaginary part Γ_{qj}, which is the inverse lifetime, causes frequency broadening. Both parts are temperature dependent. The mode softening on cooling noted for high-T_c superconductors is opposite in sign to the more common anharmonicity found in solids, i.e., softening as the melting temperature is approached. Third and fourth order terms in the expansion of a crystal's potential energy give rise to cubic and quartic anharmonicity, respectively, which yield opposite signs for $d\omega_{qj}/dT$ [15]. Ordinarily these contributions cancel and the average $d\omega/dT$ is negative due to dilation. In superconductors, however, the mode softening on cooling is attributed to strong quartic anharmonicity.

Soft mode behavior in the Chevrel-phase superconductors was first noted in a Mössbauer study [7] of $^{119}SnMo_6S_8$. The Sn atoms were found to experience a highly anharmonic potential. Physically, this is because Sn occupies essentially the same volume as a Mo_6 octahedron within the crystal structure, and hence large amplitude displacements are expected for Sn. Furthermore, since the Sn site has inversion symmetry, cubic anharmonicity is identically zero to first order, and quartic anharmonicity then would cause the shifting of the spectral features associated with Sn atom displacements. However, in the present study we clearly identify the lowest-frequency feature in the spectra of the ternaries as being primarily due to the X atom, and this feature does not significantly shift with temperature. We do note that there is a softening on cooling of the second-lowest frequency feature (the acoustic feature) of the $SnMo_6S_8$ spectrum, as was first noted at Karlsruhe for $PbMo_6S_8$. This suggests that the anharmonic X atom modes hybridize with the acoustic modes, and in a very peculiar manner, which requires further theoretical work to understand. The apparent discrepancy between the present study ($d\omega_{Sn}/dT \simeq 0$) and the Mössbauer study ($d\omega_{Sn}/dT \gg 0$) is most likely due to this hybridization, which causes the frequency distribution associated with the X atom to be more complicated than merely a single Einstein peak. A second and completely different possible explanation of the acoustic mode shifting observed in the present study is that it is intimately related to a strong electron-phonon interaction, as, for instance, is the case for the classical high-T_c superconductors, the A-15 compounds. Finally, another manifestation of anharmonicity in the Chevrel-phase compounds is the sharpening of the spectral features on cooling. We tentatively attribute this effect to a loss of dispersion on cooling as the internal-external mode couplings weaken. This would suggest that the displacements of the molybdenum chalcogenide units with respect to each other may also need to be considered beyond the harmonic approximation. Relatively flat phonon branches then become dispersive at high temperatures, but in a manner that preserves the average frequency of each branch.

A primary motivation for this study was to obtain some understanding of the relationship between the phonon spectra and the T_c values of the Chevrel-phase superconductors. The T_c values of the Sn and PbMo$_6$S$_8$ can be understood to result from the external modes being in an optimal frequency range, according to Bergmann and Rainer [8], to maximize T_c ($\omega_{optimal} \sim 2\pi T_c$). We note first that the presence of the low-lying optic feature ω_X does not directly correlate with the magnitude of T_c, i.e., Pb$_{1.2}$Mo$_6$Se$_8$ (as well as PbMo$_6$Se$_8$) has a lower T_c value than binary Mo$_6$Se$_8$. We now check for other possible qualitative correlations between trends in T_c and in average force constants of the remaining spectral features we can readily identify. This approach is motivated by the common practice of approximating the superconductivity parameter λ as being proportional to the frequency-independent factor $N(0)<<I^2>>$ (where $N(0)$ is the density of states at the Fermi energy, and $<<I^2>>$ is the double Fermi-surface average of the matrix elements that connect pairs of electronic states via changes in the lattice potential due to individual ionic displacements), and also inversely proportional to the frequency-dependent average force constant $M<\omega^2>$ [16]. As pointed out earlier, the differences between the internal mode structure of Mo$_6$Se$_8$ compared to PbMo$_6$S$_8$ can be attributed to softening of the highest frequency modes, which are attributed primarily to chalcogenide displacements, on going from the sulfide to the selenide. And this softening appears to occur in a manner that preserves the internal mode force constants. Hence there are no interesting correlations between λ and internal mode force constants. (Although the lower frequency internal modes of the selenides compared to the sulfides could be detrimental to T_c by depressing the prefactor of any of the strong-coupled BCS-like T_c expressions.) [20,21] The frequencies of the acoustic and torsional modes of all four compounds are quite similar (see Fig.4). The corresponding force constants are then larger for the low-T_c selenides since they have both a larger average mass and a larger moment of inertia than the high-T_c sulfides. Thus a qualitative correlation appears to exist here between λ and the inverse of the acoustic and torsional force constants. Both of these types of modes involve relative displacements of Mo$_6$ octahedra with respect to each other. It is physically reasonable that these types of modes should be most important in facilitating Cooper pair formation because of the following argument. The important Fermi-surface states are the d-like states derived from Mo. The intra-octahedral Mo-Mo separations are smaller than the inter-octahedral spacings, which are also significantly larger than the spacings in Mo metal. This geometric arrangement makes the d-electrons quasi-localized within octahedra. The external modes that involve primarily the translations and librations of the Mo_6S_8 (or Mo_6Se_8) groups are more likely to affect the electronic overlap integrals between the Mo_6 octahedra, and thus likely to dominate the electron-phonon interaction. Thus the strong correlation between T_c and an optimal inter-octahedral spacing [22] in the Chevrel-phase compounds may be understood in these terms.

In summary, we characterize the phonon spectra of four Chevrel-phase superconductors. We find that the molecular-crystal model is valuable for purposes of classifying the types of modes present, and for discussing the anharmonicity and the superconductivity. The phonon mode softening with decreasing temperature clearly observed for the $SnMo_6S_8$ spectrum might be caused by the quartic anharmonicity experienced by the Sn atom, whose modes are hybridized to the acoustic branches. The spectral features conjectured to provide the dominant electron-phonon couplings that determine the magnitude of T_c are the acoustic and torsional modes, which involve relative displacements of Mo_6 octahedra.

ACKNOWLEDGMENT

We thank Drs. G. S. Knapp, W. Reichardt, P. Schweiss, and B. Renker for valuable exchanges. We also thank J. W. Downey for preparing the $SnMo_6S_8$, and G. E. Ostrowski for technical assistance. We appreciate the help of Drs. T. A. Postol and C. A. Pelizzari in analyzing the data.

REFERENCES

1. R. Chevrel, M. Sergent, and J. Pringent, J. Sol. State Chem. 3, 515 (1971).
2. B. T. Matthias, M. Marezio, E. Corenzwit, A. S. Cooper, and H. E. Barz, Science 175, 1465 (1972).
3. See S. Foner and also Ø. Fischer in these proceedings.
4. See R. Shelton in these proceedings.
5. See R. W. McCallum, D. C. Johnson, R. N. Shelton and M. B. Maple in these proceedings.
6. A. C. Lawson, Mat. Res. Bull. 7, 773 (1972).
7. C. W. Kimball, L. Weber, G. Van Landuyt, F. Y. Fradin, B. D. Dunlap and G. K. Shenoy, Phys. Rev. Lett. 36, 412 (1976).
8. G. Bergmann and D. Rainer, Z. Phys. 263, 59 (1973).
9. S. D. Bader, G. S. Knapp, and A. T. Aldred, Int. Conf. on Low-Lying Lattice Vibrational Modes and their Relationship to Superconductivity and Ferroelectricity, San Juan, Puerto Rico (December (1975). *Ferroelectricity*, to be published.

10. S. D. Bader, G. S. Knapp, S. K. Sinha, P. Schweiss, and B. Renker, to be published.
11. See F. Y. Fradin, G. S. Knapp, S. D. Bader, G. Cinader, and C. W. Kimball, in these proceedings.
12. M. Marezio, P. D. Dernier, J. P. Remeika, E. Corenzwit, and B. T. Matthias, Mat. Res. Bull. $\underline{8}$, 657 (1973).
13. O. Bars, J. Guillevic, and D. Grandjean, J. Sol. State Chem. $\underline{6}$, 48 (1973).
14. M. M. Bredov, B. A. Kotov, N. M. Okuneva, V. S. Oskotskii, and A. L. Shakh-Bodagov, Sov. Phys. $\underline{9}$, 214 (1967).
15. V. S. Oskotskii, Sov. Phys. $\underline{9}$, 420 (1967).
16. F. Gompf, H. Lau, W. Reichardt, and J. Salgado, Inelastic Scattering of Neutrons, (Proc. Conf. Grenoble, 1972) IAEA, Vienna (1972) p. 137.
17. W. Marshall and S. M. Lovesey, Theory of Thermal Neutron Scattering, (Clarendon Press, Oxford, 1971) p. 83.
18. S. D. Bader, G. S. Knapp, and S. K. Sinha, to be published.
19. A. A. Maradudin and A. E. Fein, Phys. Rev. $\underline{128}$, 2589 (1962).
20. W. L. McMillan, Phys. Rev. $\underline{167}$, 10 (1968).
21. P. B. Allen and R. C. Dynes, Phys. Rev. B $\underline{12}$, 905 (1975).
22. R. Chevrel, M. Sergent, and Ø. Fischer, Mat. Res. Bull. $\underline{10}$, 1169 (1975).

PHONON ANOMALIES IN TRANSITION METALS, ALLOYS AND COMPOUNDS*

H.G. Smith, N. Wakabayashi, and Mark Mostoller

Solid State Division, Oak Ridge National Laboratory

Oak Ridge, Tennessee 37830

ABSTRACT

The anomalous features observed in the dispersion curves of several d-band superconducting materials are reviewed and comparisons are made with related non-superconducting materials. The results are briefly discussed in terms of some theoretical models. New neutron scattering data are presented for Nb with oxygen impurities, the Mo-Re bcc alloy system, and the hcp metals Tc and Ru.

INTRODUCTION

At the first Rochester conference, phonon dispersion curves measured by inelastic neutron scattering were presented for the transition metal carbides [1,2]. The data showed that the high T_c superconducting carbides have anomalous dips in their phonon dispersion curves, whereas, the non-superconducting carbides do not exhibit these anomalous features. It was pointed out that other high T_c superconductors have anomalies in their phonon spectra not seen in neighboring low T_c materials. Among the bcc transition metals, for example, the phonon dispersion curves of Nb [3,4] have anomalous features which are not present in the phonon spectra of Mo [5,6]. A similar relationship is observed for the V [7], Cr [8] and Ta [9], W [10] pairs; however, the dip is largest in Nb, the

---

* Research sponsored by the U.S. Energy Research and Development Administration under contract Union Carbide Corporation.

element with the highest transition temperature. It was also noted that anomalies in the phonon dispersion curves of superconductors are not restricted to the more complex d-band transition metals, but are present as well in non-transition metals and alloys, notably Pb [11] and the Pb-Tl-Bi [12] system, in which the anomalous dips grow more pronounced with increasing T_c. Technetium (Tc) has the highest measured transition temperature of the elemental hcp metals, and recent neutron measurements [13] of its phonon dispersion curves also reveal quite anomalous features compared to those of the low T_c hcp materials.

A number of models have been proposed to explain the phonon anomalies observed in various T_c superconductors, particularly in Nb [14-18] and the transition metal carbides [18-23]. Most of these proposed explanations [15-22] correlate the anomalies with special features of the electronic response function. Ganguly and Wood [17], for example, suggested that acoustical plasmons are responsible for the dip in the longitudinal branch in Nb along the [100] direction. Weber and co-workers [18,20,21], in order to fit the dips in the dispersion curves of Nb and of TaC and NbC, introduced a phenomenological double shell model (DSM). In the DSM, an extra electronic "supershell" is added around the metal atoms (there are also "normal" shells around the metal and carbon atoms), and these additional electronic degrees of freedom produce resonance behavior in the electronic response which gives rise to the phonon anomalies. Sinha and Harmon [19] attribute the anomalies in Nb and NbC to charge fluctuations in the localized d-shells which produce a resonance denominator in the electronic contributions to the dynamical matrix. Gupta and Freeman [22] have performed constant matrix element calculations of the diagonal electronic susceptibility, $\chi(\vec{q})$, of NbC, and they find pronounced structure in the intra-band contributions to $\chi(\vec{q})$ which correlates strongly with the "resonance cube" or soft mode surface proposed by Weber [18] and observed by Smith and Glaser [1]. Here, it should be noted that calculations of $\chi(\vec{q})$ for Nb with matrix elements included [24] do not show any pronounced structure at the wave vector where the dip occurs in the longitudinal mode along the [100] direction.

Most of this presentation will be concerned with the 4-d metals from Y to Pd and their compounds and alloys, excluding Rh, although a few 5-d and 3-d metals will be mentioned. The first part is devoted to compounds, and the latter part to elemental metals and alloys. For most of the materials we shall be concerned with, it is well known that there is an empirical correlation between T_c and the number of valence electrons per atom, e/a, with maxima in T_c occurring at e/a ~ 4.7, and at e/a ~ 6.5.* Most of the phonon

---

*Matthias, at this conference, has emphasized exceptions to this correlation, particularly in the rare earth-molybdenum chalcogenides.

studies to be described fall in the area of the first maximum, but some results will be presented here for the region of the second maximum as well.

The reader is referred to the literature [1-12] for the details of the experimental techniques. In the coherent one-phonon scattering process a neutron of incident energy E_0 and momentum $h\vec{k}_0$ is scattered to a final energy E' and momentum $h\vec{k}'$ by the vibrating lattice. A phonon is created or destroyed in the process. Imposing the conservation of energy and crystal momentum, the energy and wave vector of the phonon can be determined, and the dispersion curves can be mapped out. The neutron energies are the order of the energies associated with the lattice vibrations and the changes in energies on scattering are large and can easily be measured.

TRANSITION METAL COMPOUNDS

TM Carbides

As shown in Fig. 1 for ZrC, Weber [21] was able to fit the dispersion curves of non-superconducting ZrC and HfC rather well with a screened shell model (SSM). It should be noted that the calculated curves in Fig. 1 were fitted only to data for the acoustic modes and the optic modes at q = 0, since measurements of the optic mode frequencies for other wave vectors were not available at the time. As in the simple shell model, used so successfully for the alkali halides [25,26] each atom is described by a core consisting of the nucleus and the inner, tightly bound electrons, and a shell of outer, loosely bound electrons, without specifying in detail the division between inner and outer electrons. Each shell is connected to its own core by a spring, and in the simple shell model all short range interactions act through the shells only. In ionic crystals, the Coulomb interactions among the charged constituents of the model produce the well-known Lyddane-Sachs-Teller splitting of the longitudinal optic (LO) and transverse optic (TO) modes in the long wavelength or $\vec{q} \to 0$ limit. To eliminate this splitting, Weber added a screening term to the effective ion-ion dynamical matrix in a manner similar to the treatment of free carriers in the semiconductors PbTe and SnTe [27,28]. We have modified the screened shell model slightly by making the screening periodic, but this produces little difference in the results. The relatively good fit for ZrC and HfC obtained with the screened shell model should be contrasted with the simple pseudopotential results of Mostoller [29], which were not as successful in reproducing the observed dispersion of the optic modes, although the model was rather successful for UC.

In Weber's double shell model for the high T_c transition metal

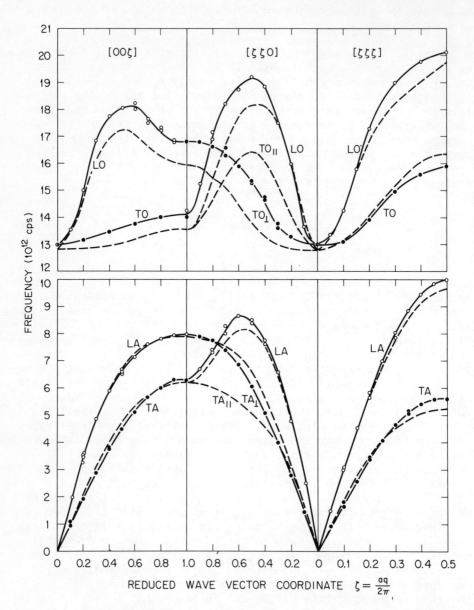

Figure 1: Phonon dispersion curves for ZrC at room temperature. The dotted curves are calculations with a screened shell model based on the parameters of Weber (Ref. 20).

carbides, the added supershells around the metal atoms are bound weakly to the metal shells, and interact <u>attractively</u> with other supershells out to second neighbors in the metal sublattice. The behavior of the supershells is illustrated in Fig. 2, which shows dispersion curves for the supershells as well as the atoms of TaC. These were calculated by diagonalizing the full 15 x 15 dynamical matrix of the DSM. In units of the electronic mass m, the tantalum shell mass was set equal to $|Y(Ta)|$, the magnitude of the shell charge, while the carbon shell and tantalum supershell masses were set equal to unity. These latter are somewhat arbitrary choices, (they were set equal to zero by Weber). All Coulomb interactions between charged constituents of the model were screened, so that no Lyddane-Sachs-Teller splitting occurred for any of the ionic or electronic modes at $\vec{q} = 0$.

Weber's parameter set b for TaC was used for the calculations, and the ionic dispersion curves turned out to be essentially identical to those shown by Weber. The tantalum and carbon shell modes have energies of about 30 and 50 eV, respectively, and exhibit little dispersion. Although there may be some question as to the exact interpretation to be placed on the dispersion curves of the supershells, the dispersion, as shown in Fig. 2, is large, and quite pronounced and rather broad dips occur at the wave vectors at which the phonon anomalies are observed. The term "electronically-driven" lattice instabilities would appear to be appropriate here.

The double shell model is rather complicated with about fifteen adjustable parameters but, it does have the advantage that the components of the model responsible for the phonon anomalies are clearly identified. Verma and Gupta [23] have obtained reasonably good agreement with the measured dispersion curves of both TaC and HfC using their three-body-force shell model. While this model has the advantage of using fewer adjustable parameters, the source of the anomalies in TaC and the absence of anomalies in HfC is not explained.

In the approach of Sinha and Harmon [19], localized charge fluctuations or "incipient charge density waves" produce resonance denominators in the dynamical matrix similar to those arising from the supershells in the DSM. For their numerical calculations for NbC, Sinha and Harmon assumed a constant diagonal electronic susceptibility for purposed of convenience (for Nb, they used a $\chi(\vec{q})$ derived from APW band structure calculations, with matrix elements included). Gupta and Freeman [22] have emphasized the possible importance of the $\vec{q}$-dependence of the diagonal susceptibility $\chi(\vec{q})$; their constant matrix element results for $\chi(\vec{q})$ in NbC exhibit substantial peaks at the wave vectors at which the anomalous dips in the phonon spectra occur. In Sinha and Harmon's treatment, the momentum off-diagonal or local field corrections to the electronic response function play a crucial role, but the diagonal

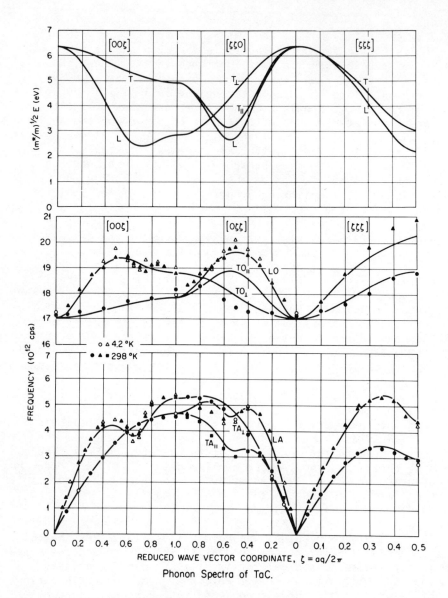

Figure 2: Phonon dispersion curves for TaC. The closed (open) symbols are the experimental data points at room temperature (4.2°K). The curves in the uppermost section represent the dispersion curves of the "supershells" calculated as described in the text. The curves in the middle and lower sections are based on the double shell model parameters of Weber (Ref 20) but with the motions of all the shells included in the dynamical matrix.

PdH(D)

The discovery of superconductivity in the palladium hydrides stimulated further interest in the lattice dynamics of the β-phase Pd-H(D) system. The observation and possible explanation of the inverse isotope effect emphasized the need to know and to understand the detailed behavior of the optic modes (predominantly H-motion) as well as the acoustic modes of the superconducting (H > 75%) and non-superconducting (58% < H < 75%) hydrides. Large single crystals are desirable for inelastic neutron scattering measurements, but for technical reasons it is rather difficult to obtain such samples containing nearly stoichiometric hydrogen concentrations. However, a large single crystal of non-superconducting $PdD_{0.63}$ was prepared and neutron scattering measurements of the phonon dispersion curves were reported by Rowe et al. [30]. More low temperature data have since been obtained, and experimental results along the three major symmetry directions are shown in Fig. 3.

The acoustic modes of $PdD_{0.63}$ are similar in shape to those for pure Pd [31], but are somewhat lower in frequency; this lowering is at least in part due to the observed lattice expansion of about four percent in going from pure Pd to $PdD_{0.63}$. Compared to the transition metal carbides, the optic modes frequencies are low, particularly when the light deuterium mass is taken into account, but the large dispersion of the optic modes is similar. Theoretical calculations [32-34] and tunneling measurements [35-37] indicate that these low-lying optic modes play an important, perhaps dominant role in the observed high T_c values and inverse isotope effect for near-stoichiometric PdH ($T_c \sim 9^0K$) and PdD ($T_c \sim 11^0K$). No pronounced anomalies of the kind typically observed in high T_c superconductors appear to be present in the phonon spectra of $PdD_{0.63}$, which is not surprising since $T_c < 1^0K$ at this deuterium concentration.

Efforts have been made to fit a variety of models to the neutron scattering data for $PdD_{0.63}$, treating the crystal as stoichiometric for the calculations. The solid curves in Fig. 3 give the results of one such fit; these curves were calculated from a simple screened shell model like that used by Weber [21] for the non-superconducting transition metal carbides. Beyond the screened Coulomb interactions between its charged constituents, the model includes only first unlike-neighbor Pd-D and first like-neighbor

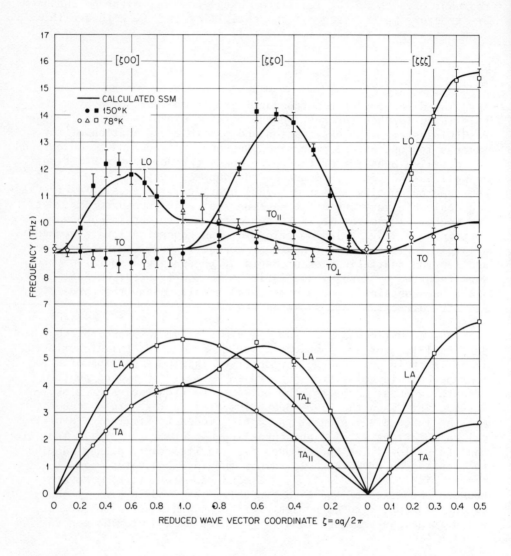

Figure 3: Phonon dispersion curves of PdD$_{0.63}$. The closed (open) symbols are the experimental data points at 150°K (78°K). The solid curves are results of calculations with a screened shell model as described in the text (Ref 30).

Pd-Pd interactions. No short range D-D interactions are included.
The parameters for the model are given in Table I.

Table I

Screened shell model parameters for $PdD_{0.63}$. The ionic and shell
charges Z and Y are in units of $|e|$, and all force constants are
in units of $e^2/2r_o^3$, where $r_o = a/2$ is the nearest neighbor distance.
n_e and m^*/m are the number and dimensionless effective mass of the
screening electrons, and the Fermi wave vector corresponding to
n_e is given in units of $2\pi/a$.

Charges	
$Z(Pd) = -Z(D)$	0.367
$Y(Pd)$	-1.354
$Y(D)$	0
Screening parameters	
n_e	0.050
m^*/m	0.333
k_F	0.288
Core-shell force constants	
$k(Pd)$	90
$k(D)$	200
Short-range force constants	
(Pd-D; 100, xx)	1.187
(Pd-D; 100, yy)	-0.232
(Pd-Pd; 110, xx)	1.030
(Pd-Pd; 110, zz)	0.076
(Pd-Pd; 110, xy)	1.451

A quantitatively better fit to the data can be obtained with
a Born-von Kármán model with second like-neighbor Pd-Pd and D-D
interactions. The latter are necessary in order to fit the large

dispersion of the LO modes, and are indicative of the long range forces provided automatically by something like the screened shell model. Born-von Kármán parameters for a fit with a χ^2 of about 1 are given in Table II. In addition to the screened shell and Born-von Kármán models, fits to the data with screened rigid ion and simple pseudopotential models were investigated. In neither case could the observed dispersion of the LO modes be reproduced. However, it should be emphasized that the parameter values listed in Tables I and II for the screened shell and Born-von Kármán model fits to the data are not unique, so no great physical significance should be ascribed to them.

Table II

Born-von Kármán force constants for $PdD_{0.63}$. Units are dynes/cm.

Pd-D:	(100,xx)	2398
	(100,yy)	1439
Pd-Pd:	(110,xx)	13534
	(110,zz)	-409
	(110,xy)	19722
	(200,xx)	3060
	(200,yy)	-370
D-D:	(110,xx)	241
	(110,zz)	-345
	(110,xy)	1697
	(200,xx)	1938
	(200,yy)	-40

$NbSe_2$

Another interesting class of superconducting compounds in which Nb also plays a dominant role is the transition metal dichalcogenides. For example, $NbSe_2$ is metallic and has a superconducting transition temperature of $\approx 7°K$, whereas MoS_2 is a semiconductor. In addition, the compound exhibits some features which are characteristic of two-dimensional lattices. The interactomic forces between the layers are much weaker than those within the layers and this fact is reflected in the phonon dispersion curves, in which the frequencies of the phonons propagating normal to the layers are much lower than the phonons propagating parallel to the layers.

The structure of 2H-NbSe$_2$ is nearly the same as that of 2H-MoS$_2$. The atomic arrangement within a layer is identical, but the positions of atoms with respect to those of the adjacent layers are slightly different. Figure 4 shows the phonon measurements [38] for a small single crystal (volume = 0.02 cm^3). Only a limited amount of data could be obtained. Smooth curves were calculated from a valence-force-field model, using parameters obtained for MoS$_2$ [39], but with proper masses substituted for NbSe$_2$. In the [001] direction the observed phonon frequencies are reproduced within the experimental uncertainties by the calculations, indicating that the interlayer interactions are not seriously affected by the conduction electrons within the layer. However, in the [100] direction the measured phonon frequencies are considerably softer than the model results. The abrupt change in the [100] LA (Σ_1) branch at q = 0.2 has been attributed by Wakabayashi, et al. [38] to be a Kohn anomaly [40] associated with a two-dimensional Fermi surface. This has been confirmed at BNL by Moncton et al. [41] for NbSe$_2$ and also observed by them in TaSe$_2$ in their studies of charge density waves in these materials.

ELEMENTAL METALS, ALLOYS, AND IMPURITIES

Neutron scattering measurements have been made for a number of transition metals and alloys. The alloy systems are particularly interesting in that one can hope to change the interatomic forces gradually and in a controlled manner, although the presence of disorder can make interpretation more difficult than for the pure materials. We present below experimental data for several transition metals, alloys, and transition metal-impurity systems which illustrate the complex and interesting phonon properties of these materials.

Figure 5 shows dispersion curves taken from the work of Powell, Martel, and Woods [42] on the Nb-Mo system, which they have studied extensively throughout the concentration range. In Nb there are numerous sharp kinks interpreted as Kohn anomalies and many, but not all of these, can be related to features of the Fermi surface. In Nb-rich alloys these kinks shift with increasing Mo concentration but the dips just fade away. At a concentration of 75 at % Mo no anomalous features are apparent. In pure Mo [5,6], a dip and Kohn anomalies appear to be developing at the zone boundary point H, but similar structure is observed in Cr [8]---a magnetic material.

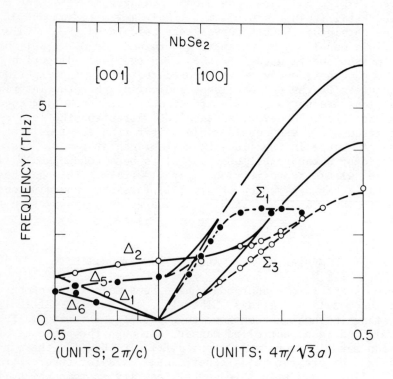

Figure 4: Phonon measurements on a small single crystal of $NbSe_2$ at room temperature (Ref. 38).

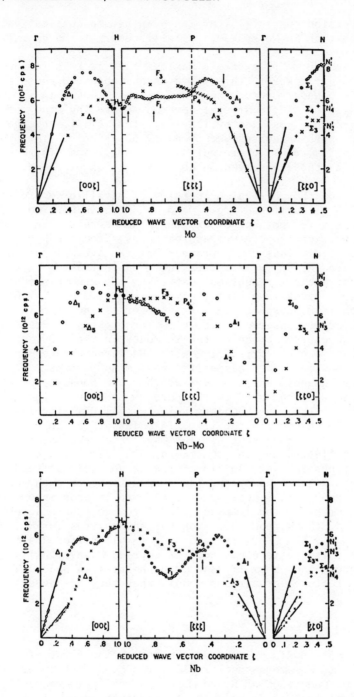

Figure 5: Phonon dispersion curves of Nb, $Nb_{0.75}Mo_{0.25}$, and Mo at room temperature (Ref. 42).

The dispersion curves [43] of $Ta_{0.77}W_{0.23}$ are qualitatively similar to the Nb-rich Nb-Mo system. We will return to Mo shortly.

In the $Nb_{1-x}Zr_x$ system a maximum $T_c = 12°K$ is reached at about $Nb_{0.70}Zr_{0.30}$; however, this is in an unstable two-phase region. The bcc phase boundary occurs near $Nb_{0.85}Zr_{0.15}$ where T_c is about 11°K. Accordingly, a single crystal neutron scattering study of an alloy of $Nb_{0.87}Zr_{0.13}$ was made by Traylor, et al. [44], and the changes in the dispersion curves were readily observable. The phonon frequency shifts relative to pure Nb are small and, in general, opposite to those observed in $Nb_{0.85}Mo_{0.15}$ as expected.

Many of the physical properties of metallic Nb are seriously affected by impurities (e.g., O, N, and H); in particular, it is known that a few percent of oxygen in Nb decreases its superconducting transition temperature appreciably. Specific heat measurements by Koch, et al. [45] demonstrated that the average phonon frequencies (as measured by θ_D) were increasing with increasing oxygen content. Subsequent inelastic neutron scattering measurements [46] of the dispersion curves in a single crystal of Nb (2.6 at. % O) confirmed their conclusions in that an overall increase in most phonon frequencies was observed (Fig. 6). The greatest increases occurred in the low energy region of the TA modes in the [100] and [110] directions, hence, correlating with the large changes in θ_D derived from the low temperature specific heat measurements. The maximum lattice frequency, which occurs at the zone boundary point H, did not change when compared to the pure Nb spectrum. The maximum frequency at the zone boundary point N (LA[100]) increased about 5%, but the changes in the [110] TA modes at the zone boundary were minimal. Changes of similar magnitude were observed in a Nb alloy with 10% Mo and in $NbD_{0.15}$ (at 573 °K)[47], but in each case different portions of the dispersion curves were affected in slightly different ways. Wakabayashi [48] has not observed any deviations from the room temperature spectra of pure Ta upon incorporating 20% deuterium in the lattice. The force constants determined by 7th and 8th neighbor Born-von Kármán analyses for Nb, $NbO_{0.026}$, $Nb_{0.85}Mo_{0.15}$, $Nb_{0.87}Zr_{0.13}$, and $NbD_{0.15}$ all differed slightly, but a definite pattern has not yet emerged. Nevertheless, these differences, though small, should be helpful in understanding the detailed features of various models when they are developed.

Mo-Re

The alloys discussed above belong to the first e/a group. Let us return to Mo--more specifically, the Mo-Re system of the second e/a group. Addition of Re to Mo dramatically increases T_c from less than 1°K to over 12°K; however, the maximum occurs in a two phase region. We have initiated a study [49] of the Mo-Re system, but so far have limited our study to the bcc single phase region with a

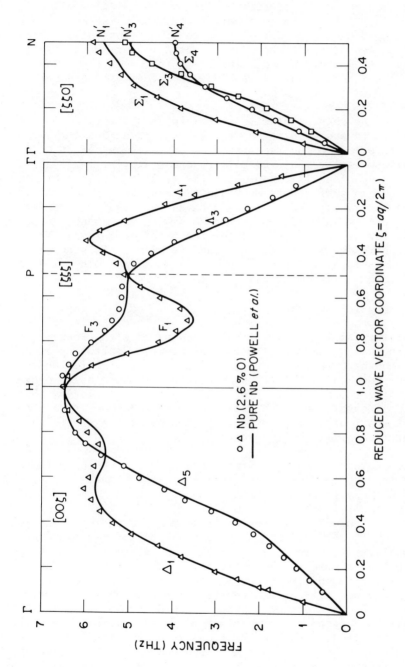

Figure 6: Phonon dispersion curves of Nb with 2.6 at. % oxygen. The solid curves are those of pure Nb (Ref. 2).

maximum Re concentration of 25 at. %, where T_c is about $10.5°K$.
We have concentrated our efforts so far on the 15 at. % alloy in
hopes of reducing short range order effects which may be improtant
near the single phase boundary. Because of the unusual behavior
of the modes near the zone boundary point H in pure Mo, we anticipated that this would be an interesting region in the alloys,
and we will see below that it is, indeed, a region of unusual phonon
behavior.

Because the rhenium atom is almost twice as heavy as Mo, we
also expected to see effects due to resonant mode behavior. The
reader is referred to the recent review articles on defect modes
in crystals by Taylor [50] and Elliott and Leath [51]. Figure 7
shows several phonon line shapes for the [111] transverse branch.
A few phonon groups for pure molybdenum are shown for comparison,
indicating that instrumental resolution has not seriously affected
the observed line shapes. In the Nb-Mo system the phonon groups were
just as narrow as those observed for the pure elements, so the
widths here are due mainly to the heavy Re atoms. As shown in the
figure a quasi-local resonant mode occurs at a frequency of about
3.5 THz (1 THz = 4.14 meV).

The unusual line shapes for $Mo_{0.85}Re_{0.15}$ make it impossible to
draw dispersion curves in the usual way. Instead, any model for the
lattice vibrations in the alloy can be compared with experiment only
by comparing calculated results for the neutron scattering cross
section with the observed neutron groups. We have so far performed
a series of such calculations using the coherent potential approximation (CPA) for mass disorder only [52,53]. Very briefly, the
CPA is a self-consistent mean field theory for elementary excitations
in random substitutional alloys, which in some sense treats the
scattering from single sites exactly.

At the outset, Born-von Kármán model parameters for pure Mo
were used to describe the "host" lattice in the CPA calculations.
The resonant mode frequency for the Re atoms in $Mo_{0.85}Re_{0.15}$ was
found to be about 4.2 THz (17.4 meV). Scattering intensity curves
are shown in Figure 8 for a series of values for the host lattice
frequency, which are indicated at the right of the figure. The
overall qualitative agreement of these first-attempt calculations
with the behavior observed experimentally was encouraging, and
indicated that the observed behavior was not primarily due to short
range order in the crystal. However, two rather obvious conclusions
are: (1) the use of pure Mo parameters to describe the host lattice
is probably not a good approximation, for the electronic properties
change with the addition of appreciable amounts of Re (which has
one more electron per atom that Mo), thus changing the lattice
dynamical properties of the alloy; and (2) there may be appreciable
local force constant changes in the alloy, that is, the force constants between a Re atom and its neighbors may be substantially
different from those between a Mo atom and its neighbor.

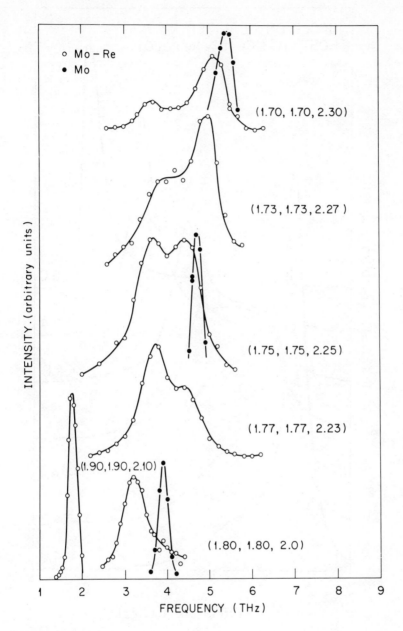

Figure 7: Several transverse acoustic phonon groups in the [111] direction in $Mo_{0.85}Re_{0.15}$ (open symbols) and pure Mo (closed symbols).

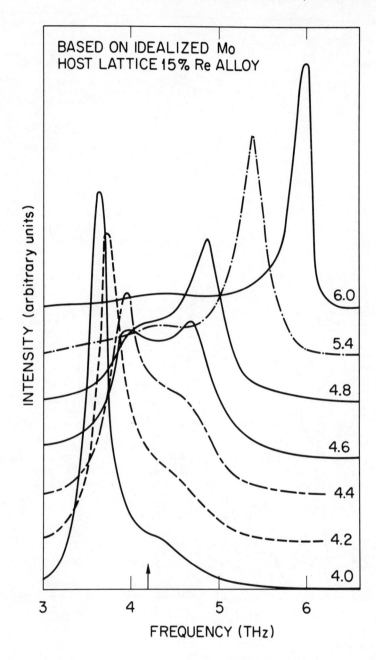

Figure 8: Neutron scattering intensity curves of $Mo_{0.85}Re_{0.15}$ for a series of values for the host lattice frequency, which are indicated at the right of the figure.

To explore the first of these two effects, we compared the observed line shapes with the CPA results in a somewhat self-consistent manner until we obtained what we thought represented approximate host lattice dispersion curves. An eight neighbor Born-von Kármán model was fit to these calculations for mass disorder only. The qualitative features of the CPA intensity curves remained the same, but the calculated resonant mode frequency decreased to 3.9 THz, appreciably closer to the observed value of 3.5 THz. It is possible that the remaining discrepancies between calculated and experimental results can be eliminated by doing CPA calculations which include local force constant changes [54]. Such calculations are computationally more difficult, but the programs are available and the calculations will be undertaken in the near future.

The phonon peak positions as a function of $\vec{q}$ for $Mo_{0.85}Re_{0.15}$ are shown in Fig. 9 in comparison to the dispersion curves of pure Mo. The crosses near the zone boundaries represent the interpolated host lattice frequencies used in the second set of mass defect CPA calculations; at other $\vec{q}$ values, the measured peak positions were used. Compared to pure Mo, there is an overall softening in most of the branches, particularly along the zone boundary from P to H. The triply degenerate zone boundary frequency at the point H has decreased dramatically. Preliminary measurements of an alloy with 25 at. % Re indicate even lower frequencies in these regions.

McMillan [55], in the derivation of his well-known formula for the transition temperature of strong-coupled superconductors found that λ, the electron-phonon coupling constant, was equal to $N(0) <I^2>/M<\omega^2>$, where $N(0)$ is the electronic density of states at the Fermi energy and $<I^2>$ is the electron-phonon matrix element averaged over the Fermi surface, M is the ionic mass and $<\omega^2> = \int F(\omega)\omega d\omega/\int F(\omega)\omega^{-1}d\omega$. He proposed that $N(0)<I^2>$ is approximately constant within a given class of materials, and that within each class variations in λ are primarily determined by the phonon spectra. In a recent re-examination of the McMillan equation, Allen and Dynes [56] concluded that λ is not determined solely by the phonon frequencies, for $N(0)<I^2>$ is not a constant and is probably more responsible for variations in T_c than $M<\omega^2>$. Weber [57], in his analyses of the phonon sepctra of the TM carbides and the Nb-Mo alloy system, also questioned the validity of the constant $N(0)<I^2>$ assumption for these materials. A preliminary analysis of the phonon spectra for the $Mo_{0.85}Re_{0.15}$ alloy discussed above reveals that, even though the phonon softening is large at the zone boundary points H and N, the decrease in $<\omega^2>$ is partly compensated by the increase in the average mass of the alloy. This seems to indicate that the variation in λ may, in large part, be due to variations in $N(0)<I^2>$. Systematic studies of the phonon spectra in this and other alloy systems should prove useful in determining the importance of the various contributions to λ.

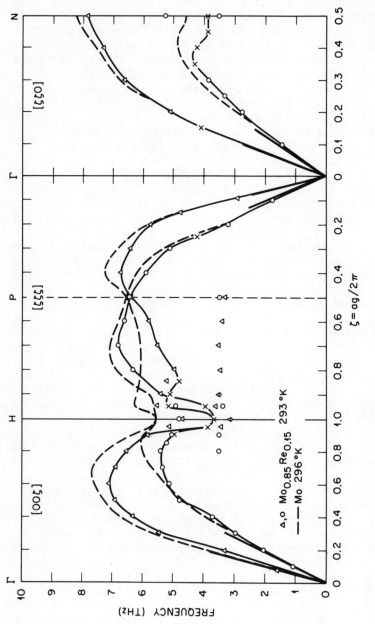

Figure 9: The phonon "dispersion curves" for $Mo_{0.85}Re_{0.15}$. The crosses near the zone boundaries represent the interpolated host lattice frequencies; at other $\vec{q}$ values, the measured peak positions were used. The solid curves are merely to guide the eye but the dashed curves are the measured curves for pure Mo (Ref. 4).

PHONON ANOMALIES

HCP Materials

Finally, we present a limited amount of data for some hcp materials. The hcp structure is not generally considered to be a favorable structure for superconductivity, but <u>technetium</u> is an exception with $T_c \approx 8°K$. While it may not be a twin brother to Nb perhaps it is a half-brother, and in fact, addition of 3% of Nb raises T_c to 12.5°K. The dispersion curves in the c-direction of some 4-d hcp metals [13] are shown in Fig. 10. We see a tremendous difference for Tc compared to Y [58] and Zr [59]; the Y curves are quite simple, but Zr, a low T_c superconductor, shows some softening in the LO mode. These results are similar to those for Sc [60] and Ti [61], but one must be careful for Co [62] also shows a slight softening for some modes on top of an overall hardening. Figure 11 compares technetium with neighboring ruthenium, a low T_c material;

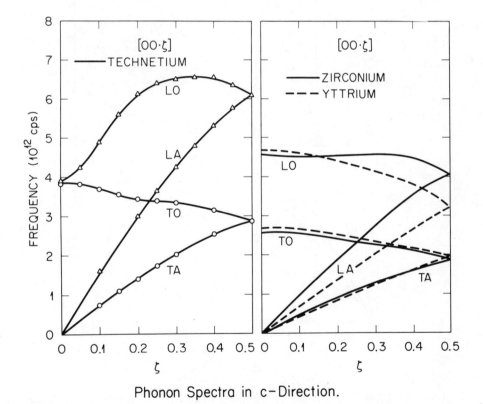

Phonon Spectra in c-Direction.

Figure 10: Phonon spectra for Tc, Zr, and Y in the c-direction at room temperature (Ref. 13).

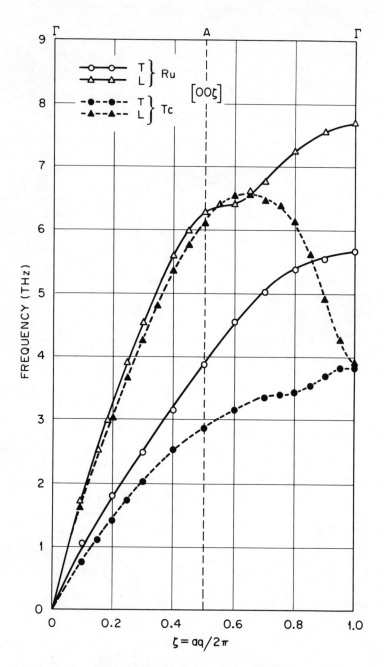

Figure 11: Comparison of phonon spectra in the c-direction for Tc (closed symbols) and Ru (open symbols) at room temperature. The circles (triangles) represent transverse (longitudinal) phonons.

note that the extended zone scheme has been used in the figure. The small anomaly in the Ru LO mode can perhaps be correlated with the shape of the Fermi surface, but the tremendous dip observed for Tc cannot [63], at least not by inspection. It is possible that the small anomaly in the TO mode for Tc is a Kohn anomaly, but the large dip in the LO mode is thought to be of other origin. However, preliminary measurements of phonons in Re along the c-direction show considerable softening of the LO mode even though its T_c (1.7^0K) is relatively low compared to that for Tc. This may indicate that there are exceptions to the apparent correlation of large dips in the phonon spectra with high T_c's, but on the other hand, a T_c of 1.7^0K is relatively high for elemental hcp metals.

An interplanar force constant analysis for the longitudinal modes of Tc in the c-direction requires interactions out to at least five planes, but a similar analysis based on a shell model formalism requires only two planes with five parameters. However, in both cases interactions with second neighbor planes are large, and this is thought to reflect significant metal-metal bonding along the c-direction.

A study of some dilute alloys of Tc [64] with Nb and Fe should prove informative, since 3 at. % Nb in Tc raises T_c to $\approx 12.5^0$K, whereas 3.7 at. % Fe in Tc lowers T_c to 4.9^0K. Single crystals of Tc and its alloys are difficult to prepare because of its radioactivity, but inelastic neutron scattering measurements on the polycrystalline alloys, similar to those described by Reichardt at this conference on the polycrystalline A-15 and Chevrel compounds, should be helpful in furthering understanding of the relationship of lattice vibrations to high superconducting transition temperatures.

CONCLUDING REMARKS

There seems to be little doubt that a relationship exists between the observed anomalies in phonon dispersion curves and the superconducting transition temperature in superconductors. The anomalies very likely reflect lattice instabilities in high T_c superconductors and are related somehow to the detailed electronic properties of the metal, alloy, or compound, particularly the properties relating to chemical bonding.

It is encouraging that _first principles_ approaches to a theoretical understanding of the lattice dynamics of superconducting materials are being made. This should encourage further neutron studies on other superconductors and additional, but more detailed, measurements on those materials already studied. However, the theoretical approaches should not be limited to the complex d- and f-band materials, for there is a wealth of experimental information on other superconducting materials available, for example, the extensive tunneling and neutron studies of the Pb-Tℓ-Bi alloys already cited, the In-Tℓ [65,66] system, the elements Sn [67,68], and Zn [69,70,71], and the recently reported α-Hg neutron studies [72], to name a few.

ACKNOWLEDGEMENTS

The authors wish to acknowledge many helpful discussions with R.F. Wood. The expert technical assistance of J.L. Sellers is gratefully appreciated.

REFERENCES

1. H.G. Smith and W. Gläser, Phys. Rev. Lett. 25, 1611 (1970).
2. H.G. Smith in Superconductivity in d- and f-Band Metals, ed. D.H. Douglass (AIP, Vol. 4, New York, 1972).
3. Y. Nakagawa and A.D.B. Woods, Phys. Rev. Lett. 11, 271 (1963).
4. R.I. Sharp, J. Phys. C 2, 421 (1969).
5. A.D.B. Woods and S.H. Chen, Solid State Comm. 2, 233 (1964).
6. C.B. Walker and P.A. Egelstaff, Phys. Rev. 177, 1111 (1969).
7. R. Collella and B.W. Batterman, Phys. Rev. B 1, 3913 (1970).
8. W.M. Shaw and L.D. Muhlestein, Phys. Rev. B 4, 969 (1971).
9. A.D.B. Woods, Phys. Rev. 136, A781 (1964).
10. S.H. Chen and B.N. Brockhouse, Solid State Comm. 2, 73 (1964).
11. B.N. Brockhouse, T. Arase, G. Caglioti, K.R. Rao and A.D.B. Woods, Phys. Rev. 128, 1099 (1962).
12. S.C. NG and B.N. Brockhouse in Neutron Inelastic Scattering, (IAEA, Vol. 1, Vienna) (1968).
13. H.G. Smith, N. Wakabayashi, R.M. Nichlow, S. Mihailovich in Proc. Low Temperature Physics-LT13, Vol. 3, ed. K.D. Timmerhaus, W.J. O'Sullivan and E.F. Hammel (Plenum Publishing Co., New York, 1974).
14. A.O.E. Animalu, Phys. Rev. B 8, 3555 (1973).
15. S.T. Chui, Phys. Rev. B 9, 3300 (1974).
16. M.I. Korsunskii, Ya. E. Genkin, V.I. Markovnin, V.G. Zarodinskii, and V.V. Satsuk, Soviet Phys.-Solid State 13, 1793 (1972).
17. B.N. Ganguly and R.F. Wood, Phys. Rev. Lett. 28, 681 (1972).
18. W. Weber, Thesis (Technische, Universität, München, 1972)(unpublished).
19. S.K. Sinha and B.W. Harmon, Phys. Rev. Lett. 35, 1515 (1975).
20. W. Weber, Phys. Rev. B 8, 5082 (1973).
21. W. Weber, H. Bilz and U. Schroeder, Phys. Rev. Lett. 28, 600 (1972).
22. M. Gupta and A. Freeman, Bull. Amer. Phys. Soc. 21, 258 (1976).
23. M.P. Verma and B.R.K. Gupta, Phys. Rev. B 12, 1314 (1975).

24. J.F. Cooke, H.L. Davis and Mark Mostoller, Phys. Rev. B $\underline{9}$, 2485 (1974).
25. A.D.B. Woods, W. Cochran and B.N. Brockhouse, Phys. Rev. $\underline{119}$, 980 (1960).
26. R.A. Cowley, W. Cochran and B.N. Brockhouse and A.D.B. Woods, Phys. Rev. $\underline{131}$, 1030 (1963).
27. W. Cochran, R.A. Cowley, G. Dolling, and M.M. Elcombe, Proc. Roy. Soc. (London) $\underline{A293}$, 433 (1966).
28. G.S. Pawley, W. Cochran, R.A. Cowley and G. Dolling, Phys. Rev. Lett. $\underline{17}$, 753 (1966).
29. Mark Mostoller, Phys. Rev. B $\underline{5}$, 1260 (1972).
30. J.M. Rowe, J.J. Rush, H.G. Smith, Mark Mostoller, and H.E. Flotow, Phys. Rev. Lett. $\underline{33}$, 1297 (1974).
31. A.P. Miller and B.N. Brockhouse, Can. J. Phys. $\underline{49}$, 704 (1971).
32. B.N. Ganguly, Z. Phys. $\underline{265}$, 433 (1973); Z. Phys. B$\underline{22}$, 127 (1975).
33. D.A. Papaconstantopoulos and B.M. Klein, Phys. Rev. Lett, $\underline{35}$, 110 (1975).
34. H. Rietschel, Z. Phys. B $\underline{22}$, 133 (1975).
35. R.C. Dynes and J.P. Garno, Bull. Amer. Phys. Soc. $\underline{20}$, 422 (1975).
36. A. Eichler, H. Wühl and B. Stritzker, Solid State Comm. $\underline{17}$, 213 (1975).
37. P.J. Silverman and C.V. Briscoe, Phys. Lett. $\underline{53A}$, 221 (1975).
38. N. Wakabayashi, H.G. Smith, and R. Shanks, Phys. Lett. $\underline{50A}$, 367 (1974).
39. W. Kohn, Phys. Rev. Lett. $\underline{2}$, 393 (1959).
40. N. Wakabayashi, H.G. Smith and R.M. Nicklow, Phys. Rev. B $\underline{12}$, 659 (1975).
41. D.E. Moncton, J.D. Axe and F.J. DiSalvo, Phys. Rev. Lett. $\underline{34}$, 734 (1975).
42. B.M. Powell, P. Martel and A.D.B. Woods, Phys. Rev. $\underline{171}$, 727 (1968).
43. N. Wakabayashi, Annual Report ORNL-5135 (1975).
44. J.G. Traylor, N. Wakabayashi and S.K. Sinha, Bull. Amer. Phys. Soc. $\underline{20}$, 300 (1975).
45. C.C. Koch, J.O. Scarbrough and D.M. Kroeger, Phys. Rev. B $\underline{9}$, 15 (1974).
46. H.G. Smith, Annual Report ORNL-4952 (1973).
47. J.M. Rowe, N. Vagelatos and J.J. Rush, Phys. Rev. B $\underline{12}$, 2959 (1975).
48. N. Wakabayashi, Annual Report ORNL-5135 (1975).
49. H.G. Smith and N. Wakabayashi, Bull. Amer. Phys. Socl $\underline{21}$, 410 (1976).
50. D.W. Taylor, Dynamical Properties of Solids, Chap. 5, Vol. 2, eds. G.K. Horton and A.A. Maradudin (North Holland Publishing Co., New York)(1975).
51. R.J. Elliott and P.L. Leath, Dynamical Properties of Solids, Chap. 6, Vol. 2, eds. G.K. Horton and A.A. Maradudin (North Holland Publishing Co., New York)(1975).
52. D.W. Taylor, Phys. Rev. $\underline{156}$, 1017 (1967).
53. T. Kaplan and Mark Mostoller, Phys. Rev. B $\underline{9}$, 353 (1974).

54. T. Kaplan and Mark Mostoller, Phys. Rev. B $\underline{9}$, 1783 (1974); Phys. Rev. B $\underline{10}$, 3610 (1974).
55. W.L. McMillan, Phys. Rev. $\underline{167}$, 331 (1968).
56. P.B. Allen and R.C. Dynes, Phys. Rev. B $\underline{12}$, 905 (1975).
57. W. Weber, Phys. Rev. B $\underline{8}$, 5093 (1973).
58. S.K. Sinha, T.O. Brun, L.D. Muhlestein and J. Sakuri, Phys. Rev. B $\underline{1}$, 2430 (1970).
59. H.F. Bezdek, R.E. Schmunk and L. Finegold, Phys. Stat. Solidi $\underline{42}$, 275 (1970).
60. N. Wakabayashi, S.K. Sinha and F.H. Spedding, Phys. Rev. B $\underline{4}$, 2398 (1971).
61. N. Wakabayashi, Annual Report ORNL-4779 (1971).
62. R.H. Scherm, N. Wakabayashi and H.A. Mook, Annual Report ORNL-4779 (1971).
63. J.S. Faulkner, Bull. Amer. Phys. Soc. $\underline{21}$, 308 (1976).
64. C.C. Koch, W.E. Gardner and M.J. Mortimer, Proc. Low Temperature Physics-LT13, Vol. 2, eds. K.D. Timmerhaus, W.J. O'Sullivan and E.F. Hammel (Plenum Publishing Co., New York)(1974).
65. H.G. Smith and W. Reichardt, Bull. Amer. Phys. Soc. $\underline{14}$, 378 (1969).
66. N. Wakabayashi, Annual Report ORNL-5135 (1975).
67. J.M. Rowe, Phys. Rev. $\underline{163}$, 547 (1967).
68. D.L. Price, Proc. Roy. Soc. (London) $\underline{A300}$, 25 (1967).
69. G. Borgonovi, G. Caglioti, and J.J. Antal, Phys. Rev. $\underline{132}$, 683 (1963).
70. P.K. Iyengar, G. Venkataraman, Y.H. Gameel and K.R. Rao in Neutron Inelastic Scattering (Inter. Atomic Energy Agency, Vienna) $\underline{1}$, 153 (1968).
71. L. Almqvist and R. Stedman, J. Phys. F. (Metal Phys.) $\underline{1}$, 785 (1971).
72. W.A. Kamitakahara and H.G. Smith, Bull. Amer. Phys. Soc. $\underline{20}$, 299 (1975).

QUESTIONS AND COMMENTS

B. Ganguly: The only thing that I wanted to say was when you do the shell model phonon calculations for the hydrides you have to bear in mind that the crystal is anharmonic.

H. Smith: This is certainly true. The transverse optic modes in $PdD_{0.63}$ are very sharp at low temperatures, but broaden considerably as the temperature is raised. Recent measurements by Rowe et al. under conditions of better resolution show structure in the longitudinal optic phonons even at low temperature. The crystal is very, very anharmonic.

J. Rowell: I did not quite follow one of your points. What was your estimate of the change in the average phonon frequencies of niobium when you add oxygen and when you add zirconium? Do they increase or decrease?

H. Smith: With the addition of 13 at.% Zr to Nb the phonon frequencies decrease about 3 to 5%. The greatest decrease ($\sim$ 10%) occurs in the vicinity of the point P in the Brillouin zone. The anomalous behavior of the TA[001] branch did not change very much. With the addition of 2.6 at.% oxygen to Nb the phonon frequencies increased, on the average, a few percent. The largest percentage changes occurred in the low energy portion of the TA[001] branch consistent with the specific heat studies.

H. Teichler: When discussing the resonance structure in Mo-Re you showed how to find the position of the correct phonon. Now my question: What is the meaning of the correct phonon, because, as far as I see it, only the resonance structure makes physical sense?

H. Smith: That is correct. However, we are trying to reproduce the observed intensity curves by model calculations. The frequencies of the Mo host lattice have been perturbed by the extra valence electron of the Re atoms. This is evidenced by the change in phonon frequencies far from the resonant frequency of the heavy Re atoms. In order to calculate the response function for the alloy (e/a = 6.15) a new $g(\nu)$ had to be determined and this required a knowledge of the phonon frequencies of the lattice without the effects of the mass defects. The procedure I described was a first attempt in this determination. A close approximation would probably be the phonon spectrum of a $Mo_{0.85}Tc_{0.15}$ lattice, which is unknown at present. The fact that there are force constant changes between the defects and the host lattice complicates matters further.

THEORY OF THE LATTICE DYNAMICS OF STRONG COUPLING SYSTEMS IN THE

RIGID MUFFIN TIN APPROXIMATION

W.E. Pickett and B.L. Gyorffy

University of Bristol, H.H. Wills Physics Laboratory
Royal Fort, Tyndall Avenue
Bristol BS8 1TL, United Kingdom

I. INTRODUCTION

At about the time of the first conference in this series it began to be clear that high temperature superconductivity in d- and f-band metals was closely related to anomalies in the phonon spectra, and that these phenomena often occurred near (in some sense) lattice instabilities. The interrelations of these phenomena are undoubtedly obscurred by a lack of a theoretical understanding of the phonon spectra. Although a rigorous theory of lattice dynamics is available, the complications inherent in the formulation make computation impractical, and progress toward a first principles understanding of phonon spectra has been slow indeed. This observation, together with the potential rewards to be expected, provide ample excuse for a reformulation of the theory of lattice dynamics. In this paper we discuss a formulation which includes among its virtues the feasibility of numerical evaluation.

The usual formulation of lattice dynamics is given in terms of the full (interacting) density reponse function χ. The force constant $\Phi_{\alpha\beta}(\ell,\ell')$ which gives the force in the direction α on the ℓ-th ion due to a displacement in the direction β of the ℓ'-th ion is given by the sum of a bare ion-ion contribution and an effective interaction mediated by electrons given by

$$\Phi_{\alpha\beta}^{el-el}(\ell,\ell') = \int dr \int dr' \frac{\partial v(r)}{\partial R_{\ell,\alpha}} \chi(r,r') \frac{\partial v(r')}{\partial R_{\ell';\beta}} , \quad (1)$$

where $v(r)$ denotes the bare electron-ion interaction. The calculation of χ involves the inversion of the dielectric function, and in a transition metal, for which the electron density is far from homogeneous, this presents a formidable problem. An important ingredient in the physics of transition metals (TM's) is a resonant d-wave phase shift which gives rise to localized d-electron states, and in order to use this localization to advantage the dielectric function inversion problem has recently been reformulated [1] in terms of a set of localized functions. Although approximate calculations for fcc Pd and Ni have been carried out [2] within this formulation with some success, the computational work necessary to do a full calculation for bcc TM's and TM compounds which exhibit high temperature superconductivity is too great to be useful at present. An additional objection to this scheme is that very little is known about the functions and matrix elements in the localized representation, and this complicates the process of finding suitable approximations.

The force constants can in principal be derived from the real space (cluster) calculations, and it might be argued that recent total energy calculations indicate that the theory of electronic systems is highly enough developed that this method would give useful results. Experimental data indicate, however, that important forces often extend to tenth neighbors [3] and beyond, and the cluster sizes necessary are so large that this method cannot be seriously considered.

In section II we summarize a reformulation of lattice dynamics recently proposed by Gomersall and Gyorffy [4]. In this theory the force constants are separated into a contribution from the band structure energy, which gives effective forces extending to large distances and which gives rise to interesting peculiarities in phonon spectra, and the remainder which is interpretable as an interaction between neutral objects and which should be appreciable for only a few neighbors. The main results of this paper are contained in sections III and IV, in which the band structure contribution to the force constants is examined in some detail. By using approximations which have been highly successful in electron band theory, and by utilizing the techniques of scattering theory, this part is expressed in a novel form which can be interpreted in terms of contributions from metallically bonding electrons, and a few idealizations which help in the understanding of strong scattering systems are discussed. A summary is given in section V.

II. A REFORMULATION OF LATTICE DYNAMICS

The density functional formalism of the total electronic energy given by Hohenberg, Kohn, and Sham [5] allows the effective potential energy of the ions to be written

$$U(\{R_\ell\}) = \tfrac{1}{2}\sum_{\ell\ell'}{}' \frac{Z^2 e^2}{|R_\ell - R_{\ell'}|} + \tfrac{1}{2}\int dr \int dr' \frac{n(r)n(r')}{|r-r'|} + \qquad (2)$$

$$\int dr\, v(r) n(r) + T[n] + E_{xc}[n]$$

given only the adiabatic approximation, in which it is assumed that for each configuration of the ionic positions $\{R_\ell\}$ the density $n(r) = n(r, \{R_\ell\})$ is the ground state density. Here Ze is the ionic charge, $v(r)$ is as in equation (1), and T and E_{xc} are kinetic and exchange-correlation energies defined by Hohenberg and Kohn [5]. Kohn and Sham have shown that a one-electron equation

$$\{-\nabla^2 + v_{sc}(r)\}\psi_i(r) = \varepsilon_i \psi_i(r) , \qquad (3)$$

where $v_{sc}(r)$ is a self-consistent potential, arises from the ground state property, and Gomersall and Gyorffy [4] have used this equation to make a separation of U as

$$U(\{R_\ell\}) = U^0(\{R_\ell\}) + U^{el}(\{R_\ell\}) , \qquad (4)$$

where U^{el} is the sum of the eigenvalues $\sum_{i=1}^{N} \varepsilon_i(\{R_\ell\})$ of the N electrons. The "neutral object" part U^0 is given by

$$U^0(\{R_\ell\}) = \tfrac{1}{2}\sum_{\ell\ell'}{}' \frac{Z^2 e^2}{|R_\ell - R_{\ell'}|} - \tfrac{1}{2}\int dr \int dr' \frac{n(r)n(r')}{|r-r'|} + U^{xc} , \qquad (5)$$

with U^{xc} given by

$$U^{xc} = E_{xc}[n] - \int \frac{\delta E_{xc}[n]}{\delta n(r)} n(r)\, dr$$
$$\simeq \int \left.\frac{dE_{xc}(n)}{dn}\right|_{n=n(r)} n^2(r)\, dr , \qquad (6)$$

where the approximate equality holds if a local density approximation

$$E_{xc}[n] \simeq \int dr\, \varepsilon_{xc}(n(r)) n(r) \qquad (7)$$

is sufficient for the exchange and correlation energy.

Expanding the potential U in powers of the lattice displacements from their equilibrium positions, the harmonic force constants

Φ are given by

$$\Phi_{\alpha\beta}(\ell,\ell') \equiv \frac{\partial^2 U}{\partial R_{\ell,\alpha} \partial R_{\ell';\beta}}\bigg|_{\{R_\ell^0\}} = \Phi^0_{\alpha\beta}(\ell,\ell') + \Phi^{e\ell}_{\alpha\beta}(\ell,\ell') , \qquad (8)$$

with Φ^0 and $\Phi^{e\ell}$ arising from the corresponding potentials in equation (4). At this stage this is a rigorous formulation of lattice dynamics, and in principle not even the one-electron approximation has been made, as the one-electron-like equation arising in the density functional formalism merely provides a self-consistent method of determining the density $n(r)$.

This formulation suggests the following approach to the difficulties of lattice dynamical calculations. The one-electron equation (3) is essentially that used in energy band calculations, which give highly reliable electron energies in TM's and many of their compounds, and the approximations of band theory will be utilized in treating the "band structure" contribution $\Phi^{e\ell}$. As the band structure term is known to give long range effective interactions between ions (for example, Kohn anomalies), these force constants must be transformed into k-space for computation. In sections III and IV the band structure part is investigated further.

The neutral object force constants, on the other hand, should be short range and should be attacked with real space models and approximations. In the limit of very tight binding the electron density from different ions do not overlap and move rigidly with the ions, so the Coulomb contributions cancel exactly. For transition metal systems the disturbance due to displacing an ion is expected to be localized, and the neutral object force constants will correspondingly be short range. Gomersall [4] has studied the simplest approximation for d-band metals, in which the atomic density is assumed to move rigidly with an ion. This may in fact give a reasonable approximation to the large force constants between first and second neighbors, but the exponential decrease for distant neighbors in this model is artificial.

An important aspect of the separation of force constants in this manner is that the neutral objects give a convenient definition of "bare" frequencies. The large Coulomb energies which roughly cancel give bare frequencies with three desirable features: (i) they are significantly smaller than the ionic plasma frequencies which result from the jellium model, (ii) the acoustic modes have the correct $q \to 0$ behavior, and (iii) they can reasonably be expected to be similar for similar metals, for the same lattice constant. Finally we note that in this definition of the bare frequencies we have swept aside the problem of dealing explicitly with the electron-electron interaction [1] [6], in much the same manner as has been

THEORY OF THE LATTICE DYNAMICS

so successful in band theory. This is not an essential approximation since the inclusion of the Coulomb interaction can be carried out (in principle) in the scattering formalism in parallel with the evaluation of the band structure term, but the simplification that results is substantial. For the remainder of this paper we concentrate on the understanding and calculation of the band structure contributions.

III. THE RIGID MUFFIN TIN APPROXIMATION

We now introduce a realistic approximation which will allow a calculation of the band structure contribution to the force constants and discuss the resulting expression. Assuming that the self-consistent potential in the one-electron equation (3) can be written as the sum of contributions from each ion

$$v_{sc}(r) = \sum_{\ell} v_{sc}^{\ell}(r) , \qquad (9)$$

it is easily shown that

$$\Phi_{\alpha\beta}^{e\ell}(\ell,\ell') = \int dr \int dr' \nabla_{\ell,\alpha} v_{sc}^{\ell}(r) \chi_o(r,r') \nabla_{\ell';\beta} v_{sc}^{\ell'}(r') \qquad (10)$$

for $\ell \neq \ell'$, where χ_o gives the density response to a change in the self-consistent potential. As a further approximation we assume that $v_{sc}^{\ell}(r)$ can be taken to be spherically symmetric inside non-overlapping "muffin tins" (MT's) and constant outside, and that it is sufficient in finding the scattering of the electrons to shift this potential rigidly: this is the rigid muffin tin approximation (RMTA).

The MT approximation is widely used in situations where the ions can be taken to be stationary. The obvious example is in the band theory of metals, for which it is well established that essentially the one-electron Schrödinger equation (3) gives the single-particle-like excitations of metallic systems in a region of energy near the Fermi energy ε_f. The response function χ_o needs to be calculated only for stationary ions and for this the MT approximation should be adequate. The further approximation of rigidly shifting the muffin tins in order to calculate the electron-ion cross section is one which has never been sufficiently tested in this context, although recent calculations [7] [8] of the electron-ion parameter η indicate that this is a reasonable approach. In the RMTA there will be no scattering from regions outside the MT, but since in TM's both the potential and charge density is much larger inside the muffin tin region than outside this should not be a severe approximation. We will show below that the results depend on the MT potential only through the phase shifts, and to the extent that these are not very sensitive to the uncertainties in the potential the scattering from inside the MT region should be treated correctly within the RMTA.

The great advantage of making the MT approximation is that the techniques of scattering theory become available. In the manipulations below only the essential elements of the theory are presented. The reader is referred to the literature [9] [10] for details.

In terms of the electron Green's function G for the crystal equation (10) can be rewritten

$$\Phi_{\alpha\beta}^{e\ell}(\ell,\ell') = \int d\varepsilon \int d\varepsilon' \frac{f(\varepsilon)-f(\varepsilon')}{\varepsilon-\varepsilon'} \int dr \int dr' [\nabla_\alpha v_{sc}(r-R_\ell) \cdot \text{Im}G(r,r';\varepsilon) \text{Im}G(r',r;\varepsilon) \nabla_\beta v_{sc}(r'-R_{\ell'})] , \quad (11)$$

where f is the Fermi function and ImA denotes the imaginary part of A. With the expansion in terms of Bloch states ψ_k

$$G(r,r';\varepsilon) = \sum_k \frac{\psi_k(r,\varepsilon_k)\psi_k^*(r',\varepsilon_k)}{\varepsilon-\varepsilon_k+i\delta} \quad (12)$$

The Fourier transform of $\Phi^{e\ell}$ becomes

$$\Phi_{\alpha\beta}^{e\ell}(q) = \sum_k M^\alpha(k,k+q) M^\beta(k+q,k) \frac{f(\varepsilon_k)-f(\varepsilon_{k+q})}{\varepsilon_k-\varepsilon_{k+q}} \quad (13)$$

and further evaluation is difficult because no simple form for the matrix element

$$M^\alpha(k,k+q) \equiv \int_{cell} d^3r \, \psi_k^*(r,\varepsilon_k) \nabla_\alpha v_{sc}(r) \psi_{k+q}(r,\varepsilon_{k+q}) \quad (14)$$

between states of different energy $\varepsilon_k \neq \varepsilon_{k+q}$ is known. Jewsbury [11] noted that the analyticity properties of G allow equation (11) to be written

$$\Phi_{\alpha\beta}^{e\ell}(\ell,\ell') = -\frac{2}{\pi} \text{Im} \int d\varepsilon f(\varepsilon) \int dr \int dr' [\nabla_\alpha v_{sc}(r-R_\ell) \cdot G(r,r';\varepsilon) G(r',r;\varepsilon) \nabla_\beta v_{sc}(r'-R_{\ell'})] \quad (15)$$

and that scattering theory allows an expansion of the Green's function in a set of functions <u>at energy ε</u>, given by

$$G(r,r';\varepsilon) = \sum_{LL'} \Delta_L(r-R_\ell)\tau_{LL'}^{\ell\ell'}(\varepsilon)\Delta_{L'}(r'-R_{\ell'}) \,, \tag{16}$$

for $r(r')$ inside the muffin tin at position R_ℓ ($R_{\ell'}$). The angular momentum index L is an abbreviation for (ℓ,m) and the functions $\Delta_L(r)$ are given in terms of the radial solutions $R_L(r)$ and the real spherical harmonics [12] $Y_L(\hat{r})$ by

$$\Delta_L(\underset{\sim}{r}) = -\frac{\sqrt{\varepsilon}}{\sin\delta_L(\varepsilon)} R_L(r,\varepsilon)Y_L(\hat{r}) \,, \tag{17}$$

where δ_L is the scattering phase shift. The scattering path operator τ introduced by Gyorffy and Stott [10] is the total t-matrix of the crystal resolved into site and angular momentum components.

The $\underset{\sim}{r}$ and $\underset{\sim}{r}'$ integrals in equation (15) can now be done [13], with the result

$$\Phi_{\alpha\beta}^{e\ell}(\ell,\ell') = -\frac{2}{\pi} \operatorname{Im}\int d\varepsilon f(\varepsilon)\operatorname{Tr}_L W^\alpha(\varepsilon)\tau^{\ell\ell'}(\varepsilon)W^\beta(\varepsilon)\tau^{\ell'\ell}(\varepsilon) \tag{18a}$$

for $\ell \neq \ell'$, or, in terms of lattice Fourier transforms,

$$\Phi_{\alpha\beta}^{e\ell}(q) = -\frac{2}{\pi} \operatorname{Im}\int d\varepsilon f(\varepsilon)\sum_k \operatorname{Tr}\{W^\alpha(\varepsilon)\cdot \tag{18b}$$
$$\tau(k,\varepsilon)W^\beta(\varepsilon)\tau(k+q,\varepsilon)-(q\to 0)\} \,,$$

where Tr_L denotes a trace over angular momentum indices and $(q\to 0)$ indicates the first term in the limit $q\to 0$. The matrix element W can be evaluated as

$$W_{LL'}^\alpha(\varepsilon) = \int d^3 r \Delta_L(r)\nabla_\alpha v_{sc}(r)\Delta_{L'}(r)$$
$$= A_{LL'}^\alpha \varepsilon \frac{\sin(\delta_{L>}-\delta_{L<})}{\sin\delta_L \sin\delta_{L'}} \,, \tag{19}$$

where all phase shifts are evaluated at energy ε, L> (L<) denotes the greater (lesser) of ℓ and ℓ', and A is given by

$$A_{LL'}^\alpha \equiv \int d\hat{r} Y_L(\hat{r})\hat{r}_\alpha Y_{L'}(\hat{r}) \tag{20}$$

and vanishes unless $|\ell-\ell'| = 1$.

At this point a comparison of equation (13) and equation (18) reveals that this approach has exchanged a complicated matrix element for the complicated scattering path operator τ. In the k-space representation $\tau(k,\varepsilon)$ is actually the inverse of the KKR matrix [14] whose vanishing determinant gives the band energies, and by using the spectral representation of $\tau(k,\varepsilon)$ we can show that equation (18b) can be evaluated. Invoking the Mittag-Leffler theorem we can write

$$\tau(k,\varepsilon) = \sum_n \frac{Z(kn)}{\varepsilon-\varepsilon_{kn}+i\delta} \quad , \tag{21}$$

where $Z(kn)$ is the residue matrix at $\varepsilon = \varepsilon_{kn}$, and equation (18b) becomes

$$\phi_{\alpha\beta}^{e\ell}(q) = \int d\varepsilon f(\varepsilon)\{ \phi_{\alpha\beta}(q,\varepsilon) - \lim_{q\to 0} \phi_{\alpha\beta}(q,\varepsilon)\} \quad , \tag{22}$$

with

$$\phi_{\alpha\beta}(q,\varepsilon) = \sum_{\substack{k \\ nn'}} \mathrm{Tr}_L \{ W^{\alpha}(\varepsilon) Z(kn) W^{\beta}(\varepsilon) \cdot Z(k+qn') + (\alpha\leftrightarrow\beta)\} \frac{\delta(\varepsilon-\varepsilon_{kn})}{\varepsilon_{kn}-\varepsilon_{k+qn'}} \quad , \tag{23}$$

and $(\alpha\leftrightarrow\beta)$ denotes the first term with α and β interchanged. This form is especially intriguing because $\phi(q,\varepsilon)$ can be interpreted as the contribution to the phonon frequency renormalization from electrons at energy ε: $W^{\alpha}(\varepsilon)$ describes the elastic scattering of an electron from an ion with coupling factors A^{α} appropriate to longitudinal or transverse processes, the Z matrices describe the wave functions of the electron between sites, and the susceptibility due to an electron at energy ε is given by

$$\bar{\chi}(q,\varepsilon) = \frac{1}{n(\varepsilon)} \sum_{\substack{k \\ nn'}} \frac{\delta(\varepsilon-\varepsilon_{kn})}{\varepsilon_{kn}-\varepsilon_{k+qn'}} \quad , \tag{24}$$

where $n(\varepsilon) = \sum_{kn} \delta(\varepsilon-\varepsilon_{kn})$ is the density of states. Equation (23) explicitly represents the attractive force between two ions due to an electron circulating between them.

By using the relation [7]

THEORY OF THE LATTICE DYNAMICS

$$Z_{LL'}(kn) = C_L(kn)C_{L'}(kn), \quad (25)$$

where $C_L(kn)$ are the Bloch wave expansion coefficients inside the muffin tin, equation (22) can be rewritten

$$\phi_{\alpha\beta}^{e\ell}(q) = \sum_k \left[\frac{R^\alpha(k,k+q;\varepsilon_k)R^\beta(k+q,k;\varepsilon_k)f(\varepsilon_k)}{\varepsilon_k - \varepsilon_{k+q}} - \frac{R^\alpha(k,k+q;\varepsilon_{k+q})R^\beta(k+q,k;\varepsilon_{k+q})f(\varepsilon_{k+q})}{\varepsilon_k - \varepsilon_{k+q}} - (q\to 0) \right], \quad (26)$$

where the matrix element is given by

$$R^\alpha(k,k+q;\varepsilon) \equiv \sum_{LL'} C_L(k) W_{LL'}^\alpha(\varepsilon) C_{L'}(k+q) \quad (27)$$

and band indices have been suppressed. Equation (26) has an essentially different form from that of equation (13), in which $\varepsilon_k > \varepsilon_f > \varepsilon_{k+q}$ (or $\varepsilon_{k+q} > \varepsilon_f > \varepsilon_k$) gives an interpretation in terms of virtually excited electron-hole pairs. In equation (26) there is an additional region of summation ε_k, $\varepsilon_{k+q} < \varepsilon_f$ which makes it clear that the interpretation is different. The new picture is of a real electron elastically scattering from ions rather than of virtual electron-hole pairs being inelastically excited by phonons. It is likely that this new picture will be the more useful, and in an attempt to understand it a few simple situations are discussed in the next section.

IV. DISCUSSION

In order to more fully understand the physics contained in the present formulation of lattice dynamics we need to study the behavior of the matrix elements, wave functions, and the energy dependent susceptibility as defined in equation (24). To this end a few simple examples are presented below, and a viable calculational scheme is presented.

A. Weak Scatterers

Jewsbury [11] has shown that, in the limit of small phase shifts, the force constant for two scattering centers a distance R apart is given by

$$\Phi^{e\ell}_{\alpha\beta}(R) = \frac{18\pi Z_v^2}{k_f^3} |v_p(2k_f)|^2 \frac{\cos 2k_f R}{R^3}, \qquad (28)$$

where Z_v is the valency, k_f is the Fermi momentum, and v_p is the pseudopotential. This is in precise agreement with results from pseudopotential theory.

B. Two Strong Scatterers

The problem of two strong scatterers in an electron gas corresponds to the neglect of band structure effects on the electron propagating between the scattering centers. In order to make this simplification directly we first rewrite equation (18a). The scattering path operator τ satisfies the equation

$$\tau = t + tG^0\tau, \qquad (29)$$

where t denotes the t-matrix for a single MT potential and G^0 is the free Greens function. The real space force constants can be written as a contribution from each energy

$$\Phi^{e\ell}_{\alpha\beta}(R) = \Phi^{e\ell}_{\alpha\beta}(\ell,\ell')|_{R_\ell - R_{\ell'}=R} = \int d\varepsilon f(\varepsilon) \phi_{\alpha\beta}(R,\varepsilon), \qquad (30)$$

with ϕ given by

$$\phi_{\alpha\beta}(R,\varepsilon) = -\frac{2}{\pi} \mathrm{Im Tr}_L \sum_{\ell_1 \ell_2} W^\alpha(\varepsilon) t(\varepsilon) G^0(\ell,\ell_1) \tau^{\ell_1 \ell'}(\varepsilon)$$
$$\times W^\beta(\varepsilon) t(\varepsilon) G^0(\ell;\ell_2) \tau^{\ell_2 \ell}(\varepsilon)|_{R_\ell - R_{\ell'}=\hat{R}} \qquad (31)$$

For ions separated by a large distance $R \gg 1/n$, where $\kappa^2 = \varepsilon - V_{mtz}$ and V_{mtz} is the constant potential outside the muffin tin region, we can use

$$G^0_{LL'}(\ell,\ell';\varepsilon)|_{R_\ell - R_{\ell'}=R} \sim -4\pi i^{L-L'} \frac{e^{i\kappa R}}{R} Y_L(\hat{R}) Y_{L'}(\hat{R}). \qquad (32)$$

The dominant contribution is retained if we make the replacement $\tau^{\ell\ell'} \to t\delta_{\ell,\ell'}$ in equation (31) and for a single resonant phase shift δ we obtain

$$\phi(R,\varepsilon) \sim \sin^2\delta(\varepsilon) \frac{\sin(2\kappa R + 2\delta(\varepsilon))}{R^2} . \tag{31}$$

The matrix elements give a strong energy dependence which is strongly peaked at the resonance energy. Integrating up to ε_f gives the expected Friedel oscillations in the force constants

$$\phi_{\alpha\beta}^{e\ell}(R) \sim \sin^2\delta(\varepsilon_f) \frac{\cos(2\kappa_f R+Q)}{R^3} . \tag{32}$$

C. The Energy Dependent Susceptibility

In order to understand the behavior of the energy dependent susceptibility $\bar{\chi}(q,\varepsilon)$ defined in equation (24) we refer to the density of states of Nb, which can be decomposed [15] into a peak arising from each of the five d-bands, with a bonding-antibonding "gap" separating the third and fourth bands. An idealization of this, typical of any bcc TM, is shown in Figure 1a. In the interband ($n \neq n'$) contributions to $\bar{\chi}$, denoted by $\bar{\chi}^{er}$, the energy denominator never gets very small, and for the purpose of exposing the energy dependence we can disregard the dependence on q. The behavior of $\bar{\chi}^{er}(\varepsilon)$ for our simple model of the density of states $n(\varepsilon)$ is shown (smoothed) in Figure 1b, and it is evident that although $\bar{\chi}^{er}(\varepsilon)$ is in general negative at the bottom of the bands and positive near the top, its value can vary sharply for closely spaced bands such as those found in TM's.

The contributions of $\bar{\chi}^{er}$ to the force constants is given by

$$\bar{\phi}^{er}(\varepsilon_f) = \int d\varepsilon f(\varepsilon) n(\varepsilon) \bar{\chi}^{er}(\varepsilon) \tag{33}$$

when wave function and matrix elements effects are ignored. From the plot of this in Figure 1c for our simple model it can be seen that $\bar{\phi}^{er}(\varepsilon_f)$ can vary significantly with ε_f, and the general behavior of a peak (in magnitude) occurring for Fermi energies just below the gap can be expected to persist in more realistic calculations. In as much as the bcc transition metals and alloys (with 5-6 electrons/atom) have Fermi energies which lie in the region of the third band, it is possible that some of the variation of the force constants (which has usually been correlated to $n(\varepsilon_f)$ [16]) may be due to this effect.

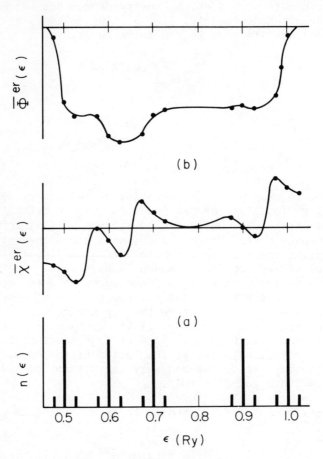

Figure 1: A model density of states $n(\varepsilon)$ for a bcc transition metal is shown in (A). The energy-dependent susceptibility $\bar{\chi}^{er}(\varepsilon)$ and the contribution to the force constants $\bar{\Phi}^{er}(\varepsilon)$ are pictured in (B) and (C) respectively. The graphs have been smoothed and the units on all ordinates are arbitrary.

The intraband contribution to the susceptibility must be considered separately, as for this part the energy denominators can be vanishingly small. The susceptibility $\bar{\chi}^{ra}(q,\varepsilon)$ is presumably very q-dependent and oscillates widely with ε. The energy integration in equation (22) will result in a cancellation of the contributions except for those near ε_f, which can give rise to Kohn anomalies and to the type of anomalies discussed by Chui [17]. The effects will be proportional to $n(\varepsilon_f)$ and will in general be sensitive to changes in both q and ε_f.

D. The Andersen Approximation

A simplification which compromises none of the important physics of strong scattering systems is the Andersen approximation. Andersen [18] has noted that the energy eigenvalues for TM's within the KKR band structure scheme are only weakly dependent on the constant energy V_{mtz} between muffin tins and has proposed that, for applications which do not require exact bands, this quantity be varied to advantage. The approximation of always taking V_{mtz} to be the energy ε, i.e., $\varepsilon - V_{mtz} = \kappa \to 0$, considerably simplifies the scheme. This approximation can be used to speed numerical calculation of ϕ^{el} and may even give good quantitative results [19].

In the $\kappa \to 0$ limit the expression in equation (26) can be written [20]

$$\phi^{el}_{\alpha\beta}(q) = \frac{2}{S^2} \sum_{\ell_1 \ell'_1 \ell_2 \ell'_2} \int d\varepsilon \int d\varepsilon' \left[\frac{P_{\ell_1 \ell'_1}(\varepsilon) P_{\ell_2 \ell'_2}(\varepsilon) f(\varepsilon)}{\varepsilon - \varepsilon'} \right. \\ \left. - \frac{P_{\ell_1 \ell'_1}(\varepsilon') P_{\ell_2 \ell'_2}(\varepsilon') f(\varepsilon')}{\varepsilon - \varepsilon'} \right] B^{\alpha\beta}_{\ell_1 \ell'_1 \ell_2 \ell'_2}(q,\varepsilon,\varepsilon') , \quad (34)$$

where we have used $\int d\varepsilon \delta(\varepsilon - \varepsilon_{kn}) = 1$. S denotes the Wigner-Seitz radius, and $P_{\ell-1,\ell}(\varepsilon) = (2\ell-1)!!(2\ell+1)!!(L_\ell(\varepsilon)+\ell+L)/(L_\ell(\varepsilon)-\ell)$ in terms of the logarithmic derivative $L_\ell(\varepsilon)$ [18]. The function B contains all of the wave vector and polarization dependence and is given by

$$B^{\alpha\beta}_{\ell_1 \ell_2 \ell_3 \ell_4}(q,\varepsilon,\varepsilon') = \sum_{\substack{k \\ nn'}} \{ \mathrm{Tr}_m A^{\alpha}_{L_1 L_2} Z_{L_2 L_3}(kn) \quad (35)$$

$$\times A^{\beta}_{L_3 L_4} Z_{L_4 L_1}(k+qn') \delta(\varepsilon - \varepsilon_{kn}) \delta(\varepsilon' - \varepsilon_{k+qn'}) - (q \to 0) \} ,$$

where $Z(kn)$ and ε_{kn} are the $\kappa \to 0$ quantities. The energy dependence of B is equivalent to that of similar functions studied by Kane [21] and by Pickett and Allen [22].

Using equation (34) $\Phi^{e\ell}$ can be evaluated numerically. The Brillouin zone integration in equation (35) must be done once for each wave vector q and can be accomplished by a Gilat-Raubenheimer [23] (or equivalent) method, the band energies, velocities, and Z-matrices having been computed (once) on a grid in reciprocal space and stored. Previous experience [15,22] with similar calculations indicate that $\sim 10^3$ cubes in 1/48 of the Brillouin zone will be sufficient. The double integral over energy is then carried out, and it is especially useful to note that within a rigid band approximation for alloys the result can be obtained for a variety of electron/atom ratios (by varying ε_f) with virtually no increase in computer time.

The $\ell_1 = 1 = \ell_2$, $\ell_1' = 2 = \ell_2'$ term (p-d scattering in each matrix element) has been coded, and preliminary results have been obtained for logarithmic derivatives appropriate to Nb. The additional terms involving d-f scattering have not yet been computed and they are not expected to be negligible. The magnitude of $\Phi^{e\ell}$ from p-d scattering alone are 3-4 eV/Å^2 compared to experimental values [3] of $(M\omega_q^2)_{max} \simeq 16$ eV/Å^2 for Nb. Depending somewhat on the energy grid chosen, the required computer (CPU) time on the CDC 7600 is ~ 15 seconds per q-point. Since the band quantities are calculated once only, calculations utilizing the full KKR and/or higher angular momentum values will be possible. Full details and results will be published later.

V. SUMMARY

In this paper it has been shown that within the rigid muffin tin approximation for electron-ion scattering processes, the band structure contribution to the dynamical matrix in strong scattering systems can be cast into a form suitable for numerical evaluation. This "calculability" is of considerable practical importance, as exact formulations which cannot at present be evaluated have not greatly increased our understanding of lattice dynamics in d- and f-band metals. Computations are currently under way and the successes and shortcomings of the approximations will soon be known.

Finally we note that this formulation in terms of scattering theory leads naturally to a picture of metallically bonding electrons scattering elastically from ions, resulting in an attractive interaction, rather than the usual picture of a "bare" phonon inelastically exciting electron-hole pairs. Using this new framework, simple models have been presented to indicate the behavior of the matrix elements and to study the role of the band structure in determining lattice dynamics.

ACKNOWLEDGEMENTS

The authors wish to thank W. Temmerman and G.M. Stocks for discussions and assistance with computer calculations. We also acknowledge illuminating communications with S.J. Gale and D. Pettifor. [See note added in proof at end of COMMENTS.]

REFERENCES

1. W.R. Hanke, Phys. Rev. $\underline{B8}$, 4585 (1973); L.J. Sham, Phys. Rev. $\underline{B6}$, 3584 (1972); S.K. Sinha, R.P. Gupta, and D.L. Price, Phys. Rev. $\underline{B9}$, 2564 (1974).
2. W.R. Hanke, Phys. Rev. $\underline{B8}$, 4591 (1973).
3. B.M. Powell, P. Martel and A.D.B. Woods, Phys. Rev. $\underline{171}$, 727 (1968).
4. I.R. Gomersall, Ph.D. thesis, University of Bristol (1975); I.R. Gomersall and B.L. Gyorffy, to be published.
5. P. Hohenberg and W. Kohn, Phys. Rev. $\underline{136}$, B864 (1964); W. Kohn and L.J. Sham, Phys. Rev. $\underline{140}$, A1133 (1965); L.J. Sham and W. Kohn, Phys. Rev. $\underline{145}$, 561 (1966).
6. See, for example, S.K. Sinha and B.N. Harmon, Phys. Rev. Lett. $\underline{35}$, 1515 (1975).
7. W.H. Butler, J.J. Olson, J.S. Faulkner, and B.L. Gyorffy, to be published.
8. B.M. Klein and D.A. Papaconstantopoulos, Phys. Rev. Lett. $\underline{32}$, 1193 (1974); D.A. Papaconstantopoulos and B.M. Klein, Phys. Rev. Lett. $\underline{35}$, 110 (1975).
9. P. Lloyd and P.V. Smith, Adv. Phys. $\underline{21}$, 69 (1972) and references therein.
10. B.L. Gyorffy and M.J. Stott, Band Structure Spectroscopy, ed. D.J. Fabian (Academic Press, 1971).
11. P. Jewsbury, Ph.D. thesis, University of Bristol (1973).
12. H.L. Davis, in Computation Methods in Band Theory, ed. P.M. Marcus, J.F. Janak, and A.R. Williams (Plenum Press, New York, 1971).
13. G.D. Gaspari and B.L. Gyorffy, Phys. Rev. Lett. $\underline{28}$, 801 (1972).
14. For the Korringa-Kohn-Rostoker method see B. Segall and F.S. Ham, Methods in Computational Physics (Academic Press, New York, 1968), Vol. 8, Ch. 7.
15. W.E. Pickett, Ph.D. thesis, State University of New York at Stony Brook (1975).
16. See, for example, I.R. Gomersall and B.L. Gyorffy, Phys. Rev. Lett. $\underline{33}$, 1286 (1974).
17. S.T Chui, Phys. Rev. $\underline{B9}$, 3300 (1974).

18. O.K. Andersen, Solid State Comm. 13, 133 (1973); O.K. Andersen, Phys. Rev. B12, 3060 (1975) and references therein.
19. W. Temmerman (private communication) and N.E. Christensen (private communication) have independently found that for bcc metals the $n \to 0$ limit can give substantially larger errors than have been found in fcc and hcp metals.
20. W.E. Pickett and B.L. Gyorffy, to be published.
21. E.O. Kane, Phys. Rev. 175, 1039 (1968).
22. W.E. Pickett and P.B. Allen, Phys. Rev. B11, 3599 (1975).
23. See, for example, J.F. Janak, in Computational Methods in Band Theory, ed. P.M. Marcus, J.F. Janak, and A.R. Williams (Plenum Press, New York, 1971), p. 323.

QUESTIONS AND COMMENTS

F. Mueller: I'm a little bit confused. I wonder if you can clarify for me the relationship between the formalism you have presented here based on the KKR secular equation (in terms of phase shifts) with a work which appears to me to be very similar (that done by Sinha in 1969) based on the APW method. Is there a relationship?

W. Pickett: Can you be a little more explicit about this work by Sinha?

F. Mueller: I thought it was very much the same sort of thing as you have done. He had also calculated electron-phonon interactions using Bloch functions, interactions, overlaps, etc. Part of what you said during your talk was that you wanted to make a model so that you could, in fact, integrate radial, secular equations based on wavefunctions at two different energies, for example. I thought that this was in fact contained in Sinha's formalism already.

B. Gyorffy: Well, some part of it is exactly the same. There are two part of the theory. One is setting up the dynamical matrix consisting of two parts. The point here is that only the non-interacting response function enters into the second part. It does not mean we are neglecting the screening in the problem; that is included in what has been called the neutral object problem. The second is that we are evaluating all the matrix elements exactly and the question is whether you have a rapidly converging calculation for the phonon spectrum.

F. Mueller: I have a question. Is there a significant difference in the KKR phase shift formalism you present here and Sinha's?

THEORY OF THE LATTICE DYNAMICS

B. Gyorffy: The formulations are formally equivalent. The difference is in the approach to the problem of screening. We have proposed a scheme to get around the very difficult numerical problem of including screening effects exactly.

F. Mueller: Are the radial matrix elements in your expressions the same as those evaluated by Sinha?

W. Pickett: Sinha has actually found expressions for the electron-ion matrix element for electron states at two different energies. In the formulation given here only matrix elements between states of the same energy are necessary, for which the Gaspari-Gyorffy form can be used. This is one of the useful simplifications of this approach.

<u>Note added in proof</u>. The expression for the energy dependence $P^{(\epsilon)}_{\ell-1,\ell}$ of the electron-ion matrix element, given below eq. (34) should be corrected to read

$$P^{(\epsilon)}_{\ell-1,\ell} = - \frac{(2\ell-1)!!(2\ell+1)!!}{(L_\ell(\epsilon)-\ell)(L_{\ell-1}(\epsilon)-\ell+1)} [(L_\ell(\epsilon)+\ell+1)(L_{\ell-1}(\epsilon)-\ell+1) + (\epsilon-V_{mtz})S^2].$$

The second term, proportional to $\epsilon-V_{mtz}$, must be retained to ensure that the electron crosses the discontinuity at the ionic sphere radius at <u>energy</u> ϵ, before allowed to propagate at energy $\kappa^2 \to 0$. Although the discontinuity in the muffin tin potential is unphysical, the inclusion of this discontinuity in the electron-ion matrix element will account roughly for scattering in the interstitial region of the real crystal which is missing in the rigid muffin tin approximation. We are indebted to D. Pettifor for communication on this topic. - W.E. Pickett.

PHONON ANOMALIES IN d-BAND METALS AND THEIR RELATIONSHIP TO SUPERCONDUCTIVITY[†]

S.K. Sinha

Solid State Science Division, Argonne National Laboratory

Argonne, Illinois 60439

and

B.N. Harmon

Ames Laboratory-ERDA and Department of Physics
Iowa State University

Ames, Iowa 50011

ABSTRACT

Many high T_c d- and f-band superconductors exhibit actual or incipient lattice instabilities as evidenced by structural transitions or anomalous dips in their phonon dispersion curves. We present a theory demonstrating that the dips and high T_c's in these materials are likely manifestations of the characteristic response (screening) by d-electrons near the Fermi level. In materials with a relatively large density of localized states at the Fermi level there is a tendency towards a collective electronic instability characterized by a periodic charge redistribution. However a charge density wave does not form because these charge fluctuations are strongly coupled to the lattice giving rise to anomalies in the phonon spectrum and in some cases eventually driving the lattice

---

[†]Work performed for the U.S. Energy Research and Development Administration under Contract No. W-7405-eng-82.

into a structural transformation. By including relaxation of the fluctuations the central peak observed by neutron diffraction in many of these materials can also be understood. In regions of $\vec{q}$-space where such incipient electronic instabilities are present there is lessened screening or relative enhancement of the electron-phonon matrix elements - giving rise to larger λ's and thus greater T_c's. The details of the theory are sketched and the physical ideas explained. A simplified application of the theory to Nb and NbC shows that the positions of the dips are readily accounted for and that the size or depth of these anomalies is very sensitive to the density of states at E_F. The relationship to other work is discussed.

INTRODUCTION

It has become increasingly clear in recent years that many d-band materials with a high density of states at the Fermi level tend to be unstable against periodic lattice distortions. The wavevector $\vec{q}$ characterizing these distortions depends on the type of material, and is sometimes incommensurate with the lattice periodicity. Even in materials of this type for which lattice instabilities do not actually occur, they manifest their incipient nature by the presence of anomalous softenings of the phonons in selected regions of $\vec{q}$-space. In many cases, materials which have such anomalies in their phonon spectra tend also to be rather good superconductors. Examples of such systems are the metals Nb and Ta and their alloys with Ti and Zr [1-4], the group VB transition metal carbides and nitrides [5-7], the layered transition metal chalcogenides [8,9], the high-T_c A-15 type compounds [10,11] and many others. In this paper, we develop in detail the idea that their instabilities are driven by the electrons in the d-bands at the Fermi level, and show how a consideration of the electron-phonon interaction and electronic screening in these materials leads to a successful correlation with the occurrence of such instabilities, their characteristic wavevectors, and also throws light on the relationship to high T_c. A brief report of this work has been published earlier [12]. Very similar ideas have been presented recently also by Hanke, Hafner, and Bilz [13].

As is by now very well established [12-17], in any microscopic theory for phonons in d-band materials it is crucial to take into account the effect of the off-diagonal elements of the dielectric matrix or the so-called "local field corrections" arising from the quasi-localized nature of the d-orbitals. In fact as we shall see, in certain cases, this will be essential in promoting such instabilities. One must therefore regard theories which involve only the diagonal elements of the dielectric matrix as being somewhat incomplete and occasionally misleading. In addition, the non-availability of the pseudopotential concept (at least in its simple local form) for such materials requires a satisfactory alternative treatment of the electron-phonon interaction. A well-known alternative to the pseudopotential method has been the so-called "Schrödinger Method" which utilizes the Schrödinger equation to transform the electron phonon matrix element (EPME). This method dates back to Bardeens's classic paper on the subject [18], and has recently been elegantly formulated by Gyorffy and co-workers [19-21] in terms of phase shifts. These formulations however are not valid when the states involved in the EPME are not of equal energy, as is the case for the virtual transitions involved in calculating the renormalized self-energy of the phonons. For such transitions one may use the version of the Schrödinger method developed by Sinha [14], and recently formulated by Sinha, Gupta and Harmon [22] in terms of a canonical transformation technique. The main physical ideas and results are discussed below. We do not claim here to provide a detailed numerical computation of phonon spectra using the rigorous expressions developed. Instead, we shall be guided by the structure of the rigorous microscopic expressions obtained to develop somewhat simplified models which manifest what we believe to be the basic physics involved.

The basic physical ideas are in fact quite simple. If one has partially occupied d-orbitals at the Fermi level, there is always the possibility of a periodic rearrangement of electrons on the atoms possessing d-orbitals, i.e. a so-called "charge density wave". We show that for certain wavevectors, the energy to produce such a charge fluctuation becomes small owing to the favorable nature of the Coulomb interactions between the d-electrons and the high density of states at E_F. These charge fluctuations however do not occur spontaneously, but rather as a resonant response to the lattice displacements. As this response grows stronger, they couple more strongly to the lattice displacements, first producing dips in the phonon spectrum, then critical scattering and eventually actual phase transitions. In materials where there exists such a tendency towards incipient charge density waves, it follows that in certain regions of $\vec{q}$-space there is a net decrease in the screening of the electron phonon interaction, i.e. a <u>relative</u> enhancement of the EPME's, which helps to raise T_c.

CALCULATION OF THE PHONON SELF-ENERGIES

For explanatory purposes, we shall restrict ourselves in this discussion to the case of monatomic crystals. For clarity of exposition, we shall restrict ourselves here to presenting the central result for the phonon frequencies renormalized by the electron-phonon coupling, and the consequences which follow from it. The derivation of this result is quite lengthy and complicated, and full details will be published elsewhere [22]. (A brief sketch of the derivation is given in the Appendix to this paper.) We obtain

$$\omega_{\vec{q}j}^2 = \omega_{\vec{q}j}^{(0)2} - \frac{1}{M} \sum_{\alpha\beta} \sum_{\mu_1\mu_2} e_\alpha^*(\vec{q}j) I_{\mu_1}^\alpha(-\vec{q}) \chi_{\mu_1\mu_2}(\vec{q}) I_{\mu_2}^\beta(\vec{q}) e_\beta(\vec{q}) \quad (2.1)$$

where $\omega_{\vec{q}j}^{(0)}$ is the partially renormalized phonon frequency due to the interactions between "rigid pseudoatoms" inside the muffin-tin spheres, and the second term represents the further renormalization due to polarization effects; $e_\alpha(qj)$ is the α-component of the phonon eigenvector for mode (qj), M is the atomic mass, μ_1 stands for the pair of orbital indices (ξ_1, ξ_1') corresponding to a charge fluctuation involving the <u>virtual</u> transition $\xi_1 \to \xi_1'$ (where ξ_1 is an index for an orbital in terms of which the Bloch functions are expanded inside each unit cell - see Appendix), and the quantities $I_{\mu_1}^\alpha$, $\chi_{\mu_1\mu_2}$ are defined below, and discussed more fully in the Appendix.

$I_{\mu_1}^\alpha$ is the α-component of a transformed electron-phonon matrix element for the transition $\xi_1 \to \xi_1'$ as is given in Eq. (A12). It consists of a local part which looks like the usual form obtained using pseudopotentials, and a non-local part which obeys an angular momentum selection rule $\Delta\ell = \pm 1$ for the transition $\xi_1 \to \xi_1'$; $\chi_{\mu_1\mu_2}(\vec{q})$ is the electron density response matrix, given by

$$\chi_{\mu_1\mu_2}(\vec{q}) = [\underset{\sim}{N}(\underset{\sim}{1} + \underset{\sim}{\tilde{V}}\underset{\sim}{N})^{-1}]_{\mu_1\mu_2} \quad . \quad (2.2)$$

This expression involves the inverse of a matrix in the space of all possible transitions μ_1, μ_2. The matrices $\underset{\sim}{N}$, $\underset{\sim}{V}$ are defined as follows:

$$N_{\mu_1\mu_2}(\vec{q}) = - \sum_{\substack{\vec{k}\sigma \\ \lambda_1\lambda_2}} \frac{n_{\vec{k}\lambda_1\sigma} - n_{\vec{k}+\vec{q}\lambda_2\sigma}}{E_{\vec{k}\lambda_1\sigma} - E_{\vec{k}+\vec{q}\lambda_2\sigma}} A_{\vec{k}\lambda_1}^{*\xi_1} \cdot \quad (2.3)$$

$$A_{\vec{k}+\vec{q}\lambda_2}^{*\xi_2} A_{\vec{k}+\vec{q}\lambda_2}^{\xi_1'} A_{\vec{k}\lambda_1}^{\xi_2'}$$

where $\lambda_1\lambda_2$ are band indices, σ a spin index, and the $A_{\vec{k}\lambda}^\xi$ are the expansion coefficients of the Bloch state $(\vec{k}\lambda)$ in terms of the orbitals ξ.

$$\tilde{V}_{\mu_1\mu_2}(\vec{q}) = \sum_j V_{\xi_1 i,\xi_2 j;\xi_1' i,\xi_2' j}\, e^{i\vec{q}(\vec{r}_j-\vec{r}_i)} \qquad (2.4)$$

$$-\tfrac{1}{2} V_{\xi_1 i,\xi_2 i;\xi_2' i,\xi_1' i}$$

where

$$V_{\xi_1 i,\xi_2 j;\xi_1' i,\xi_2' j} = \iint d\vec{r}d\vec{r}'\, \phi^*_{\xi_1}(\vec{r}-\vec{r}_i)\phi_{\xi_1'}(\vec{r}-\vec{r}_i) \cdot$$
$$[e^2/|\vec{r}-\vec{r}'|]\phi^*_{\xi_2}(\vec{r}'-\vec{r}_j)\phi_{\xi_2'}(\vec{r}'-\vec{r}_j) \qquad (2.5)$$

where $\phi_\xi(\vec{r}-\vec{r}_i)$ is the ξth orbital in the cell associated with lattice site $\vec{r}_i$. The second term in Eq. (2.4) subtracts off the leading exchange correction to the Coulomb coupling coefficient between charge fluctuations μ_1 and μ_2. If the ϕ_ξ are re-expressed in terms of overlapping atomic orbitals $\tilde{V}_{\mu_1\mu_2}(\vec{q})$ will be seen to contain all the conventional "exchange overlap" Coulomb integrals. $N_{\mu_1\mu_2}(\vec{q})$ is an "unenhanced charge fluctuation susceptibility" and $\chi_{\mu_1\mu_2}(\vec{q})$ is the "Coulomb enhanced charge fluctuation susceptibility". The expression for the latter is formally identical to the results of Hanke and Sham [24], except of course for the different choice of basis functions referred to in the Appendix. Eq. (2.1) is the central result used in this paper. An equivalent but physically more transparent way of expressing this result is to write it in terms of equations of motion

$$M\omega^2 e_\alpha = \sum_\beta D^{(0)}_{\alpha\beta} e_\beta + \sum_{\mu'} I^\alpha_{\mu'}(-\vec{q}) w_{\mu'} \qquad (2.6a)$$

$$0 = \sum_\beta I^\beta_\mu(\vec{q}) e_\beta + \sum_{\mu'} (\tilde{V} + \tilde{N}^{-1})_{\mu\mu'} w_{\mu'} \qquad (2.6b)$$

where $D^{(0)}_{\alpha\beta}$ is the dynamical matrix arising from the rigid pseudoatoms alone, and the $w_{\mu'}$ are the amplitudes of the driven charge fluctuations μ' in response to the lattice motion. Eq. (2.6b) is the "self-consistency" equation determining such amplitudes,

$I_\mu^\alpha(\vec{q})$ measures the coupling of unit lattice displacements along x_α to the charge fluctuation μ of unit amplitude.

PHONON ANOMALIES

Although we believe that Eq. (2.1) is computationally feasible once a detailed set of energy bands and wavefunctions is available, we shall not attempt here to present a rigorous calculation of the full phonon spectrum. Rather let us consider a material where the band complex at the Fermi level is dominated by $\ell = 2$ orbitals which we shall loosely call "d-orbitals", and that the only large elements of $N_{\mu_1\mu_2}(\vec{q})$ are where all orbitals are $\ell = 2$ orbitals. Let us consider the contributions to $\omega_{\vec{q}j}^2$ from the terms in Eq. (2.1) where μ_1, μ_2 are both d-d transitions. By the angular momentum selection rule in Eq. (A13), the non-local contributions to I^α vanish. Then by Eq. (A12), $I_{\mu_1}^\alpha$, $I_{\mu_2}^\beta$ can be expressed purely in terms of local contribution from the last term. Since we wish to concentrate on the effect of the d-orbitals, it is tempting to regard $(1 + \underset{\sim}{V}\underset{\sim}{N})_{\mu_1\mu_2}$ in Eq. (2.2) as a small matrix where μ_1, μ_2 are restricted to d-d transitions only. It may be shown that one may in fact do this _formally_ by replacing $\underset{\sim}{V}$ with a Coulomb coupling coefficient which is _screened_ by all other (non d-d) transitions, and similarly screening $W(\vec{q}+\vec{G})$, the local part of the electron-ion interaction. To do this rigorously is equivalent to inverting the full matrix $(1 + \underset{\sim}{V}\underset{\sim}{N})$ in the space of _all_ transitions. Nevertheless, an approximate way of screening (e.g. with a Lindhard dielectric function for the s-band) should be sufficient to reveal the main effect of intra-d-band transitions on $\omega_{\vec{q}j}^2$.

Before discussing the calculations, it is worth discussing the simplified version of Eq. (2.1) which would result if we had only a _single_ d-orbital involved instead of a degenerate set of such orbitals, in order to see clearly the origin and locations in $\vec{q}$-space of the anomalies. The results are in fact qualitatively similar to those obtained from the more detailed calculations. Eqs. (2.1) - (2.4) and Eq. (A12) then yield

$$\omega_{\vec{q}j}^2 = -\frac{1}{M} \sum_{\alpha\beta} e_\alpha^*(qj) I^\alpha(\vec{q}) [V(\vec{q}) + 1/\chi_o(\vec{q})]^{-1} I^\beta(\vec{q}) e_\beta(\vec{q}j) \quad (3.1)$$

where

$$I^\alpha(\vec{q}) = \sum_{\vec{G}} (\vec{q}+\vec{G})_\alpha [W(\vec{q}+\vec{G})/\varepsilon_s(\vec{q}+\vec{G})] f_d(\vec{q}+\vec{G}) \quad (3.2)$$

PHONON ANOMALIES IN d-BAND METALS

$f_d(Q)$ being the form factor of the d-orbital density, and $\varepsilon_s(\vec{Q})$ the screening function of the s-band, $V(q)$ is given by Eq. (2.4) with all the orbitals set equal to the single d-orbital, and $e^2/|r-r'|$ replaced by the screened Coulomb interaction;

$$\chi_o(\vec{q}) = -\sum_{\vec{k}} \frac{n_{\vec{k}d\sigma} - n_{\vec{k}+\vec{q}d\sigma}}{E_{\vec{k}d\sigma} - E_{\vec{k}+\vec{q}d\sigma}} \qquad (3.3)$$

implying a sum over the "d-band" only. We note that in Eq. (2.5) the exchange term subtracts off half the intra-atomic Coulomb integral for $j = i$. We expect that correlation and screening effects reduce this integral further so that we may replace the $i = j$ term by U, which is typically of the order of only 2-3 eV for a d-shell atom. $V(\vec{q})$ in fact measures the increase in potential energy due to mutual Coulomb interactions of a charge fluctuation of unit amplitude with periodicity $\vec{q}$ on the d-shells (see Fig. 1). $V(\vec{q})$ may be seen to decrease steadily from $\vec{q} = 0$ to negative values at $\vec{q} =$ a zone-boundary wavevector corresponding to an "ionic-like" distribution of charge. $1/\chi_o(\vec{q})$ measures the remaining energy cost for such a charge redistribution. If $\chi_o(\vec{q})$ is large, particularly

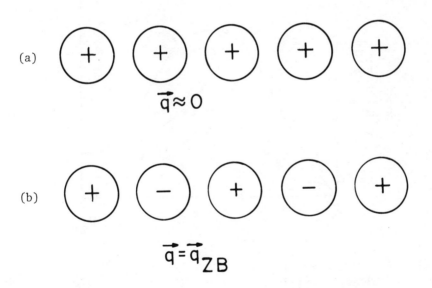

Figure 1: Schematic one-dimensional illustration of charge fluctuations on atomic d-shells for (a) $\vec{q} = 0$ (positive electrostatic interaction energy) and (b) $\vec{q} = q$(zone boundary) (negative electrostatic interaction energy).

in regions of $\vec{q}$ where $V(\vec{q})$ is most negative (e.g. at the zone boundaries) the system tends to become unstable with regard to such charge fluctuations. The scale of $\chi_0(\vec{q})$ is set by the density of states at the Fermi level so that in materials where this is large one may expect tendencies towards such instabilities. One would also expect them from what has been said above to occur for $\vec{q}$ vectors near the zone boundaries. However, we must also consider the coupling to the lattice. The dips in $\Delta\omega^2$ are also controlled by the q-dependence of $I^\alpha(\vec{q})$ which is illustrated in Fig. 2. $I^\alpha(q)$ measures the coupling between a unit charge fluctuation on a d-shell and unit amplitude ionic displacements along x_α. At both $\vec{q} = 0$ and at the zone boundary the neighboring ions move equally in the same direction about a given ion, and thus their couplings to the charge fluctuation on that ion <u>cancel</u> in pairs. Note that the denominator in Eq. (3.1) can never actually vanish for the <u>lattice</u> will have become unstable long before the electrons can themselves

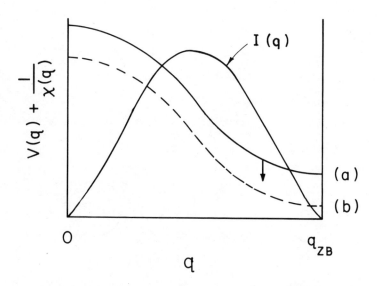

Figure 2: Schematic illustration of the function $[V(q)+(1/\chi_0(q))]$ for (a) a low $n(E_F)$ material (b) a higher $n(E_F)$ material. No structure in $\chi_0(q)$ has been assumed, although for some materials nesting features of the Fermi surface can produce peaks in $\chi_0(q)$ which modify this simple picture.

PHONON ANOMALIES IN d-BAND METALS

spontaneously form a charge density wave. Note also that if the intra-atomic U is particularly large, even for large n(E$_F$), the tendency towards a charge density wave instability and lattice mode softening is reduced. Instead, as is well known, there is a tendency towards spin density wave instabilities.

In our calculations for Nb, which are meant to be only semi-quantitative, we considered the d-bands near the Fermi level to be made up of equal mixtures of three t_{2g}-type orbitals. Thus $A^{\xi 1}_{\vec{k}+\vec{q}\lambda} \simeq 1/\sqrt{3}$. In addition we neglected the variation of the radial wavefunction. Because of the rapid variation of the actual $A^{\xi 1}_{\vec{k}+\vec{q}\lambda}$ across the bands, the 9x9 matrix $N_{\mu_1\mu_2}$ was taken as

$$N_{\mu_1\mu_2} \simeq \delta_{\mu_1\mu_2} \cdot \frac{1}{9} \cdot \chi_0(\vec{q}) \qquad (3.4)$$

where

$$\chi_0(\vec{q}) = - \sum_{\vec{k}\sigma} \sum_{\lambda_1\lambda_2} (n_{\vec{k}\lambda_1\sigma} - n_{\vec{k}+\vec{q}\lambda_2\sigma})/(E_{\vec{k}\lambda_1\sigma} - E_{\vec{k}+\vec{q}\lambda_2\sigma}) \qquad (3.5)$$

where the sum over bands is restricted to the bands which cross the Fermi level. $\chi_0(\vec{q})$ was evaluated for $\vec{q}$ along the three principal symmetry directions from a APW energy band calculation. The results are almost identical to those of Cooke, Davis, and Mostoller [29] for Nb, and show little structure along [100] and [110], and two slight peaks at (.23,.23,.23) $2\pi/a$ and (.73,.73,.73) $2\pi/a$ along [111]. The 9x9 $\underset{\sim}{\tilde{V}}$ matrix was evaluated using U = 2 eV and a Coulomb potential screened by $\varepsilon_s(\vec{Q})$ but cut off <u>inside</u> a radius of 1.3 Å so as not to include the self-interaction already represented by U. (For this purpose, the sum in Eq. (2.4) was converted into a sum in reciprocal space.) $I^\alpha_{\mu_1}(\vec{q})$ was obtained by an expression similar to Eq. (3.2) but with $f_d(q+G)$ replaced by $\langle \xi_1 | e^{-i(\vec{q}+\vec{G})\cdot\vec{r}} | \xi_1' \rangle$ where μ_1 stands for the virtual transition $\xi_1 \to \xi_1'$. $\varepsilon_s(\vec{Q})$ was taken to be a Lindhard function with $k_f = 0.2(2\pi/a)$. Although this corresponds to a rather small number of effective s-electrons per atom, it should be noted that the effect of s-d hybridization will be to decrease the ability of the s-electrons to screen. Using Eq. (2.1) the dips in $\Delta\omega^2$ were calculated for the longitudinal branches along the three principal symmetry directions. The results are shown in Fig. 3. Of particular interest are the peaks at (0.7, 0, 0) $2\pi/a$ and (0.8, 0.8, 0.8) $2\pi/a$. Also indicated on the figure are the effects of scaling $\chi_0(\vec{q})$ as obtained from Eq. (3.5) slightly up and down, simulating the effects of increasing or decreasing the density of states at E_F (or alternatively varying the intra-atomic Coulomb interaction U, since it combines additively with $9/\chi_0(\vec{q})$

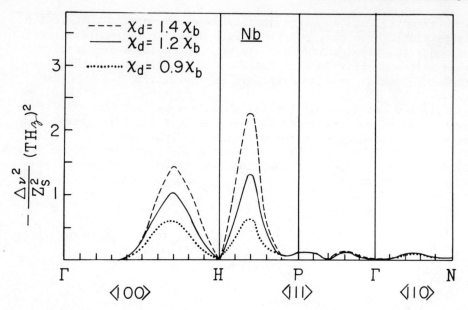

Figure 3: The decrease in ν^2 for the longitudinal branches of Nb (in units of Z_s^2) due to the intra-d-band contributions to the dielectric function shown for different values of χ_d.

from Eqs. (3.4) and (3.5)). We note that the magnitudes of the dips in the phonon spectrum are extremely sensitive to this variation. However, the <u>positions</u> of the dips turned out to be extremely <u>insensitive</u> to the values of any of the parameters used in the calculations. The dip along [100] correlates well in position and shape with that observed experimentally in Nb. This dip is also observed to decrease on alloying with Mo. The latter presumably causes $\chi_0(\vec{q})$ to decrease as E_F moves down the density of states curve. The anomaly along [111] is probably masked by the dip in the longitudinal brance which exists between P and H in most bcc metals due to the direct ion-ion interaction. It is interesting to note, however, that in pure Nb, a central peak is observed by neutron scattering at q = (0.75, 0.75, 0.75) $2\pi/a$. [4,30] This peak increases strongly upon alloying with Zr. It is also well-known that upon greatly increasing the Zr content, an instability to the so-called "ω-phase" occurs [3] characterized by lattice displacements with $\vec{q}$ = (2/3, 2/3, 2/3) $2\pi/a$. The actual value of 2/3 may be stabilized by third-order anharmonic terms in the free energy. (In the next section, we discuss how an extension of the theory given above to allow for electronic relaxation converts phonon frequency renormalization effects to a central peak in the phonon Green's function.)

We now turn to the calculations on the transition metal carbides. The dispersion curves of the group IVB transition metal carbides (with NaCl structure) show normal behavior, but the VB carbides exhibit interesting dips [5] along all three symmetry directions in the acoustic modes. For the calculation on NbC, we used Eqs. (2.6)(generalized to the case of two atoms per cell). The <u>interband</u> charge fluctuations were taken to be of the dipolar type, so that Eqs. (2.6) look like a set of conventional shell-model equations of motion [31,32] coupled to the d-band charge fluctuations. For the shell-model part only the Nb atoms were assumed to be polarizible. Again the d-orbitals on the Nb sites were taken to be of the t_{2g} type. In addition the p-orbitals on the C sites (which are known [33,34] to hybridize with the Nb d-orbitals) were approximated by a single spherically symmetric orbital. The overlap between orbitals on the Nb and C sites were neglected for this calculation, since we were interested mainly in qualitative effects. The $\tilde{V}$ and $\tilde{N}$ matrices now become 10x10 matrices (from the 9 possible d-d transitions on the Nb site and one p→p on the C site). $\varepsilon_s(\vec{Q})$ was taken to be a Lindhard function arising from the few s-electrons present with $k_f = (0.1)2\pi/a$. $\chi_0(\vec{q})$ in Eq. (3.5) is now replaced by $\chi_0^{Nb}(\vec{q})$ associated with the Nb sites and $\chi_0^C(\vec{q})$ associated with the C sites. For our calculations we have neglected the $\vec{q}$-dependence of these functions (see discussion in the next section) and taken them to be roughly equal to the partial densities of states at E_F for the two sites respectively. The coupling coefficients between the Nb and C site charge fluctuations and the ion displacements were taken to be of the same form as in the Nb calculations (with extra phase factors to allow for the different atoms in the cell) and the coupling between the charge fluctuations and the dipole moments on the Nb sites was taken to be of similar form. The adjustable parameters in the calculation were the shell-model parameters and

$$\left.\begin{aligned} A_{Nb} &= U_{Nb} + 9/n^{Nb}(E_F) \\ A_C &= U_C + 1/n^C(E_F) \end{aligned}\right\} \quad (3.6)$$

(see Eq. (3.4)), which were adjusted to give a fit to the experimental points on the longitudinal branches. Fig. 4 shows the fitted curves which are in surprisingly good agreement with experiment considering the crudeness of the model. In particular, it is to be noted that the dips in the longitudinal acoustic branches are predicted at precisely the $\vec{q}$-values at which they are observed. These <u>positions</u> in fact turned out to be extremely independent of the fitting parameters and again seem to be <u>intrinsic</u> to our model. However, the magnitude of the dips turned out to be extremely

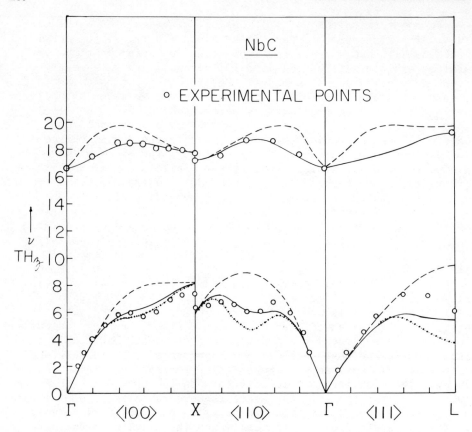

Figure 4: Dispersion curves for longitudinal branches of NbC compared with measurements of Smith, et al. [5]. Smooth curve corresponds to best fit values within model. Dotted curve corresponds to 10% decrease in A_{Nb}, A_C. Dashed curve to $A_{Nb}, A_C \to \infty$. For the optic modes the smooth and dotted curves coincide.

sensitive to the values of A_{Nb}, A_C so that negative ω^2 resulted by decreasing these by only 25% or so. The curves calculated by setting A_{Nb}, A_C to infinity (equivalent to removing the d-band charge fluctuation effects completely) are shown as the dashed curves in Fig. 4. One sees that they are quite normal and reminiscent of the group IVB carbide dispersion curves [5]. For these latter materials, it is interesting to note that the Fermi level falls right in the minimum between the p- and d-band peaks in the density of states [33-35]. The values of the parameters used are listed in Table I. From these and Eq. (3.6) it is possible to estimate order of magnitudes for $n^{Nb}(E_F)$ and $n^C(E_F)$ if we assume reasonable values for U_{Nb} and U_C. Taking these to be 2 eV and 1 eV respectively, we

Table I. Parameters for NbC

a) Shell-Model Parameters (Notation here identical to that of Weber (Ref. 6)). (All force constants are in units of e^2/Ω.)

$Z_C = -0.6 = -Z_{Nb}$ $B(12) = 1.7$

$Y_{Nb} = -0.8$ $A_1(22) = 19.2$

$k_{Nb} = 114.0$ $B_1(22) = -11.8$

$A(12) = 23.69$

b) Microscopic Parameters:

$A_{Nb} = 3.57$ Ryd

$A_C = 0.7$ Ryd

---

obtain $n_{Nb}(E_F) \simeq 2.5$ states/rydberg atom spin, $n_C(E_F) \simeq 1.6$ states/rydberg atom spin. It is to be noted that the values obtained in a band structure calculation are even larger. Thus it does not appear to be necessary to have a very large density of states at E_F to produce dips in the phonon spectra of these rocksalt structure metalloids. Our low values may be due to the crudeness of the simple shell-model used to fit the residual "smooth" part of the dispersion curves. In fact, the actual dips are somewhat deeper than the fitted curves, as seen in Fig. 4. An interesting point arising from the calculation is the fact that if A_C is set to infinity, and A_{Nb} adjusted, the dips appear more in the optic rather than the acoustic modes, indicating that some partial density of states at E_F associated with the carbon sites is essential for explaining the observed anomalies.

DISCUSSION

In the first part of this paper and in the Appendix, we have presented a fairly rigorous formulation of the problem of calculating the phonon spectrum of a solid, using a canonical transformation to eliminate the troublesome strong part of the bare ion-electron interaction, and also treating local field corrections in a manner essentially equivalent to the treatment of Hanke and Sham [24]. By examining the structure of the phonon self-energies due to the electron-phonon interaction, we are led to propose a physically simple microscopically justified model to account for the dips in

the phonon spectra of many d-band materials. This model is based
on the tendency, for materials where U is not too large but $\chi_0(q)$
is appreciable, for resonant behavior of the electron density-
response function, i.e. a tendency towards a charge fluctuation
instability on the d-shells, brought about by the attractive nature
of the Coulomb interaction when charge fluctuations on neighboring
shells get out of phase with each other. Strong coupling of such
virtual charge fluctuations to the lattice displacements brings
about the dips in the phonon spectra and eventually may drive an
actual lattice instability. It is obvious that an additional
effect will be <u>relative</u> enhancement (i.e. diminished screening) of
the electron-phonon interaction in regions of $\vec{q}$-space where the
density response function becomes large, which may explain why
such materials are usually good superconductors. This will be
made more quantitative below. The microscopic expressions for $\Delta\omega^2$
have been somewhat oversimplified and cast in parameterized form
to bring out the qualitative features of this model, however the
results are encouraging in that they appear to correctly predict
the occurrence and positions of the anomalies in $\vec{q}$-space. In
addition, these positions appear to be quite insensitive to the
<u>parameters</u> of the model, but rather intrinsic to the physics
underlying the model itself. Further applications of this type of
model to other compounds for which electronically driven instabilities appear have also been made with encouraging results. Details
will appear in a later publication [36]. In some materials appreciable phonon softening does not occur, but instead a central peak
appears at certain points in $\vec{q}$-space heralding an instability. The
present theory can be extended to account for this effect by introducing electron relaxation processes which in effect make $\chi_o(\vec{q})$
a complex entity. The one phonon Green's function can then be
shown to exhibit a central peak at those regions in $\vec{q}$-space where
the undamped version of the theory would predict a large dip in
the dispersion curve [36]. The simple models described here
neglect p-d hybridization effects. It may be shown that introducing
such effects would yield phonon anomalies at the zone boundary,
i.e. at the point H in a bcc metal. This is becuase $I_\mu^\alpha(\vec{q})$ is no
longer zero at the zone boundary as it would be if μ is simply a
d→d transition, while the electron density response matrix $\chi_{\mu_1\mu_2}$
will have as before, elements which peak at the zone boundary.
This seems consistent with the occurrence of dips in the phonon
spectrum at H in Mo, Cr and increasingly so in Mo-Re alloys as
reported in this conference [37]. For such materials it is known
that one has appreciable p-d hybridization around the Fermi level.
We do not claim that the model explains all features of the phonon
spectrum of d-based metals. Various additional anomalies remain
to be explained in some of the transverse branches, for which our
simple d→d model yields no dips. Some of these may be accounted
for by introducing p-d hybridization. Very similar ideas about the
phonon anomalies in these materials have recently also been put

forward by Hanke et al.[13] who also emphasize the importance of the local field corrections. Returning now to the rigorous theory, let us examine the form of the bare electron-phonon matrix element as given by Eq. (A11). It may be shown [22] that the <u>screened</u> matrix element may be obtained by replacing $I^\alpha_{\mu_1}(\vec{q})$ in that expression by

$$\sum_{\mu_2} (1 + \tilde{V}N)^{-1}_{\mu_1\mu_2} I^\alpha_{\mu_2}(\vec{q}).$$

Inserting this expression for the screened electron-phonon coupling matrix elements in the formula for the parameters λ given by McMillan [38], we obtain:

$$\lambda = \frac{1}{(2\pi)^3 \Omega n(E_F) M} \sum_{\lambda\lambda'} \sum_{\xi_1 \xi_2} \sum_{\xi'_1 \xi'_2} \int \frac{dS_{\vec{k}}}{|\nabla E_{\vec{k}\lambda}|} A^{*\xi_1}_{\vec{k}\lambda} A^{\xi'_1}_{\vec{k}\lambda}$$

$$\int \frac{dS_{\vec{k}'}}{|\nabla E_{\vec{k}'\lambda'}|} A^{*\xi_2}_{\vec{k}'\lambda'} A^{\xi'_2}_{\vec{k}'\lambda'} \sum_{\alpha\beta j} e_\alpha(\vec{q}j) \sum_{\mu_2} (1+\tilde{V}N)^{-1}_{\mu_1\mu_2}$$

$$I^\alpha_{\mu_2}(\vec{q}) \sum_{\mu_4} (1+\tilde{V}N)^{-1}_{\mu_3\mu_4} I^{*\beta}_{\mu_4}(\vec{q}) e^*_\beta(\vec{q}j) \cdot \frac{1}{\omega^2_{\vec{q}j}} \quad (4.1)$$

where $\vec{q} = \vec{k}' - \vec{k}$, μ_1 stands for $\xi_1 \to \xi_2$, and μ_3 for $\xi'_2 \to \xi'_1$. A similar expression has been given by Bar-Sagi and Hanke[39].

If we neglect screening, i.e. set $(1 + \tilde{V}N) \simeq 1$, and neglect all but the first term on the right in Eq. (A12), we obtain the so-called "rigid muffin-tin approximation". Then using Eq. (A13) to evaluate the I^α_μ, and certain identities and averaging procedures, we recover the well-known Gaspari-Gyorffy expression for λ [19] which is rather simply and elegantly expressed in terms of phase shifts. (Although we believe that an extra term must be applied to their formula as discussed more fully in our other paper in these proceedings [27].) In Ref. 27, we show that a full-fledged APW calculation for λ in this approximation yields values which are too large, thus indicating the effect of the $(1+\tilde{V}N)^{-1}_{\mu_1\mu_2}$ screening factors. However in certain regions of $\vec{q}$-space (near the zone boundaries) as we have shown many elements of this matrix will grow appreciably, giving rise to <u>relative</u> enhancement of the

$$I^{\alpha}_{\xi_1\xi_2}(q) = \tfrac{1}{2}(Q^{\alpha}_{\xi_1\xi_2} + Q^{*\alpha}_{\xi_2\xi_1}) - W^{ex,\alpha}_{\xi_1\xi_2}$$

$$- \frac{i}{\Omega} \sum_{\vec{G}} (\vec{q}+\vec{G})_{\alpha} W(\vec{q}+\vec{G})(\xi_1|e^{-i(\vec{q}+\vec{G})\cdot\vec{r}}|\xi_2) \qquad (A12)$$

where Ω is the unit cell volume, $\vec{G}$ is a reciprocal lattice vector, $W(\vec{k})$ is the Fourier transform of $W(\vec{r})$ given in Eq. (A8), $(\xi_1|\cdot|\xi_2)$ denotes an integral over the unit cell, $W^{ex,\alpha}_{\xi_1\xi_2}$ is the approximate electron-pseudo-atom exchange contribution which is $\vec{q}$-independent, and $Q^{\alpha}_{\xi_1\xi_2}$ is also q- independent and arises from the $M^{\ell,\alpha}_{\vec{k}+\vec{q}\lambda,\vec{k}\lambda'}$ terms defined in Eq. (A2). It is given by [22,27]

$$Q^{\alpha}_{\xi_1\xi_2} = \frac{4\pi r_s^2}{2m}[\delta_{\ell_1,\ell_2-1}\left(\frac{\ell_2}{2\ell_2+1}\right)^{1/2} C^{\ell_1\ 1\ \ell_2}_{m_1\ m_2-m_1} \{ \frac{\ell_2+1}{r_s^2} R_{\ell_1 s_1} R_{\ell_2 s_2}$$

$$+ \frac{\ell_2+1}{r_s}(R'_{\ell_1 s_1} R_{\ell_2 s_2} - R_{\ell_1 s_1} R'_{\ell_2 s_2}) + (R'_{\ell_1 s_1} R'_{\ell_2 s_2} - R_{\ell_1 s_1} R''_{\ell_2 s_2})\}$$

$$+ \delta_{\ell_1,\ell_2+1}\left(\frac{\ell_2+1}{2\ell_2+1}\right)^{1/2} C^{\ell_1\ 1\ \ell_2}_{m_1\ m_2-m_1} \{\frac{\ell_2}{r_s^2} R_{\ell_1 s_1} R_{\ell_2 s_2}$$

$$+ \frac{\ell_2}{r_s}(R'_{\ell_1 s_1} R_{\ell_2 s_2} - R_{\ell_1 s_1} R'_{\ell_2 s_2}) - R'_{\ell_1 s_1} R'_{\ell_2 s_2} - R_{\ell_1 s_1} R''_{\ell_2 s_2})\}]$$

$$\cdot A_{m_2-m_1,\alpha} \qquad (A13)$$

where $\ell_1 m_1 s_1$ refer to ξ_1, etc., r_s is the muffin-tin radius, the C's are the Clebsch-Gordon coefficients, the radial functions and derivatives are understood to be for $r = r_s$, and $A_{m,\alpha}$ is the matrix which transforms vectors from a Cartesian to spherical basis.

The Green's function for the phonons may then be obtained from the equation of motion technique, using the extended random phase approximation [23] to decouple the four-particle electron operators, making certain approximations for the exchange terms, and decoupling also the electron and phonon operators in the resulting Green's functions. The final result for the poles of the one-phonon Green's function is given in Eq. (2.1).

PHONON ANOMALIES IN d-BAND METALS

$$\psi_{\vec{k}\lambda} = \frac{1}{\sqrt{N}} e^{i\vec{k}\cdot\vec{r}_\ell} \sum_\xi A^\xi_{\vec{k}\lambda} \phi_\xi(\vec{r}-\vec{r}_\ell) \tag{A9}$$

where ξ indexes the type of orbitals involved. The types of orbitals suggested in Ref. 24 are the Wannier functions of the band complex. However, we propose here a representation in terms of angular momentum orbitals which appear to us to have definite advantages for d-band materials. They do not involve multicenter integrals, the size of the eventual matrix which must be inverted is in general smaller, all the quantities involved may be obtained directly from an APW or KKR-type band calculation, and finally strict selection rules may be obtained for the non-local part of the EPME. Accordingly, we write

$$\phi_\xi(\vec{r}) = Y_{\ell m}(\Theta,\psi) R_\ell(r,E) \qquad (\xi \to \ell,m,s) \tag{A10}$$

where $R_\ell(r,E)$ is a radial function truncated inside the atomic cell, which we may approximate by a Wigner-Seitz sphere. For $\ell = 0,1$ the ordinary solutions to the radial Schrödinger Equation do not vary appreciably across the d-band complex and the argument E may be dropped. For $\ell = 2$, the energy dependence of the radial function may be represented as a linear combination of two energy-independent radial functions, as shown by Pettifor [25]. In this case, we write $R_\ell(r,E) = \alpha_1 R_{\ell,1}(r) + \alpha_2 R_{\ell,2}(r)$, where the coefficients are energy-dependent. Substituting in Eqs. (A10) and (A9), we may absorb these coefficients in the $A^\xi_{\vec{k}\lambda}$ provided ξ now stands for a composite index ℓ,m and (in the case of $\ell = 2$), s, where $s = 1,2$. For the present we ignore higher ℓ components of the wavefunction. Note that the deviation in the crystal of the $\phi_\xi(\vec{r})$ from <u>free atomic</u> orbitals indicates that "overlap" effects are in principle included in the wavefunction given by Eq. (AS), only the expansion is about the <u>sites</u> themselves rather than in terms of overlapping orbitals from neighboring sites. Thus for the "bonding" d-states, the effective radial function is much larger in the outer regions of the cell than for the "antibonding" d-states [26].

With such a representation, it may be shown that Eq. (A7) reduces to

$$P_{\vec{k}+\vec{q}\lambda,\vec{k}\lambda'}(qj) = \left(\frac{1}{2NM\omega^{(0)}_{qj}}\right)^{\frac{1}{2}} \sum_\alpha e_\alpha(\vec{q}j) \sum_{\xi_1\xi_2} A^{*\xi_1}_{\vec{k}+\vec{q}\lambda} A^{\xi_2}_{\vec{k}\lambda'} I^\alpha_{\xi_1\xi_2}(\vec{q}) \tag{A11}$$

where

electron-phonon coupling. There is thus a direct correlation between such enhancement and the occurrence of dips in the phonon spectrum.

We conclude with a discussion of the relationship of the above theory to other explanations of the phonon anomalies in these materials. In an important paper, Weber [8] pointed out that the anomalies in the Group VB carbides could be explained in terms of a "double-shell" model if one chose parameters such that the coupling coefficient between the "super-shells" became very small in certain regions of $\vec{q}$-space. If we compare Eqs. (2.6) arising from the microscopic theory with his equations, they look formally very similar. In fact, in our theory the effects arise from the smallness of $(\tilde{V}+\underset{\sim}{N}^{-1})$. This is of course the coupling coefficient between the d-shell charge fluctuations. In Weber's model, however, the "supershell" degrees of freedom were dipolar fluctuations. Because of the intrinsically different $\vec{q}$-space behavior he had to postulate attractive forces between the electrons in his supershells and also that the second-neighbor interaction of these entities dominated the nearest neighbor interactions. Nevertheless, the resonance idea has turned our to be very useful and indicates the insight obtainable from model fittings of this type. In a more microscopically based version of the dipolar fluctuation idea, Wakabayashi [40] had also earlier noted that the anomalies in the longitudinal branches of Nb could be explained by removing the self-interaction of electrons on an atom, thereby allowing a net attractive interaction in certain regions of $\vec{q}$-space. Ganguly and Wood [41] proposed an explanation for the anomaly in Nb at (0.7, 0, 0) $2\pi/a$ in terms of interaction of the LA phonon branch with an acoustic plasmon associated with d-shell charge fluctuations. The idea has some similarities to ours, except that in our model, the d-shell charge fluctuations do not spontaneously occur as plasmons but are rather driven by the phonons in an adiabatic manner. It is in fact doubtful if the d-band velocities are low enough to allow acoustic plasmons to exist in Nb.

Finally, we discuss the relevance to this work of the "peaks in $\chi(q)$". There appears to be some confusion in the current literature regarding the exact way this function affects the phonon dispersion curves. The actual non-interacting $\chi_0(\vec{Q},\vec{Q}')$ function is given by

$$\chi_0(\vec{Q},\vec{Q}') = - \sum_{\vec{k}\lambda_1\lambda_2\sigma} \frac{n^0_{\vec{k}\lambda_1\sigma} - n^0_{\vec{k}+\vec{q}\lambda_2\sigma}}{E_{\vec{k}\lambda_1\sigma} - E_{\vec{k}+\vec{q}\lambda_2\sigma}} \langle k\lambda_1 | e^{-i\vec{Q}\cdot\vec{r}} | \vec{k}+\vec{q}\lambda_2 \rangle$$

$$\times \langle \vec{k}+\vec{q}\lambda_2 | e^{i\vec{Q}'\cdot\vec{r}} | \vec{k}\lambda_1 \rangle \qquad (4.2)$$

where $\vec{q}$ is $\vec{Q}$ or $\vec{Q}'$ reduced to the first zone. This function is not however directly applicable to calculation of the phonon spectrum as one needs $\varepsilon^{-1}(\vec{Q},\vec{Q}')$ [14,15,42]. For d-band metals, it is in fact impossible to treat $\varepsilon(\vec{Q},\vec{Q}')$ as diagonal, since for $\vec{q}$ near the zone boundary, some of the off-diagonal elements of $\chi_0(\vec{Q},\vec{Q}')$ become of equal size to the diagonal elements. Thus it is misleading to say that purely diagonal elements of $\chi_0(\vec{Q},\vec{Q}')$ dominate the phonon spectrum. However, it is common nowadays in calculations from actual energy bands to calculate the contribution to Eq. (4.2) by ignoring the matrix elements and keeping only the contributions from the few bands around E_F. This function is usually called $\chi_0(q)$. If there are nesting features on the Fermi surface $\chi_0(q)$ often has considerable structure. However, in the few calculations which have included the matrix elements in Eq. (4.2) the effect of the matrix elements has been to wash out the structure - even to the extent of removing very large peaks in $\chi_0(q)$ [29,43]. Nevertheless, peaks in $\chi_0(q)$ can be important as is most easily seen from our simplified model for monatomic crystals where $\chi_0(q)$ appeared. $\chi_0(q)$ approximates our $N_{\mu_1\mu_2}(\vec{q})$ in the more rigorous formulation (see Eq. 2.3). Thus it is now possible to see from Eq. (3.1), for instance, that pronounced peaks in $\chi_0(q)$ could yield dips in the dispersion curves (which we may call generalized Kohn anomalies) in addition to those present even in the absence of such structure. Further, the values of $\tilde{V}(\vec{q})$ and $I_\mu^\alpha(\vec{q})$ in those regions will control the extent to which such structure will actually be reflected in the dispersion curves. In the case of Nb there is no structure in $\chi_0(\vec{q})$ along the [100] direction so that the dip in the longitudinal branch is due to the q-dependence of $\tilde{V}(\vec{q})$ along [111] as mentioned, in fact the larger one at $(.23, .23, .23)2\pi/a$ does not manifest itself at all and calculations based on our model show that the dip at $(0.75, 0.75, 0.75)2\pi/a$ would be present even with a flat $\chi_0(\vec{q})$. In the case of NbC it appears [34] that $\chi_0(q)$ does indeed have pronounced peaks at the positions of the observed anomalies, and would thus appear to reinforce the mechanism described here. At present this appears to be a remarkable coincidence. It would be interesting to see if NbN (a still higher density of states material) also had pronounced anomalies at the same $\vec{q}$-values as NbC in order to determine whether the mechanism described here or Fermi surface effects were most important in determining the position of these anomalies.

The authors wish to acknowledge helpful conversations with Dr. S.G. Das.

REFERENCES

1. B.M. Powell, P. Martel, and A.D.B. Woods, Phys. Rev. <u>171</u>, 727 (1968); also Y. Nakagawa and A.D.B. Woods, Phys. Rev. Lett. <u>11</u>, 271 (1963) and also in <u>Lattice Dynamics</u> (Proc. of Int. Conference,

Copenhagen) p. 39, 1963; also R. Sharp, J. Phys. C $\underline{2}$, 421 (1969).
2. A.D.B. Woods, Phys. Rev. $\underline{136}$, A 781 (1964).
3. J.C. Williams, D. de Fontaine, and N.E. Paton, Metall. Trans. $\underline{4}$, 2701 (1973); S.C. Moss, D.T. Keatings, and J.D. Axe, Solid State Commun. $\underline{13}$, 1465 (1973).
4. N. Wakabayashi and J.G. Traylor, to be published.
5. H.G. Smith, in Superconductivity in d- and f-band metals, ed. by D.H. Douglass (AIP, New York, 1972); H.G. Smith and W. Glesser, in Proceedings of the International Conference on Phonons, Rennes, France, 1971, ed. by M.A. Musimovici (Flammarion, Paris, 1972) p. 145.
6. W. Weber, Phys. Rev. B $\underline{8}$, 5082 (1973); W. Weber, H. Bilz, and U. Schröder, Phys. Rev. Lett. $\underline{28}$, 600 (1972).
7. W. Weber, Phys. Rev. B $\underline{8}$, 5093 (1973).
8. J.A. Wilson, J. DiSalvo, and S. Mahajan, Phys. Rev. Lett. $\underline{32}$, 882 (1974); Adv. in Physics $\underline{24}$, 117 (1975).
9. D.C. Moncton, J.D. Axe, and F.J. DiSalvo, Phys. Rev. Lett. $\underline{34}$, 734 (1975).
10. G. Shirane and J.D. Axe, Phys. Rev. Lett. $\underline{27}$, 1803 (1971); G. Shirane, J.D. Axe, and R.J. Birgeneau, Solid State Commun. $\underline{9}$, 397 (1971).
11. C.W. Chu, C.Y. Huang, P. Schmidt, and K. Sugawara, these proceedings.
12. S.K. Sinha and B.N. Harmon, Phys. Rev. Lett. $\underline{35}$, 1515 (1975).
13. W. Hanke, J. Hafner, and H. Bilz, Proceedings of Conference on Low-lying Lattice modes and their relationship to Superconductivity and Ferroelectricity, Univ. of Puerto Rico, 1976.
14. S.K. Sinha, Phys. Rev. $\underline{169}$, 477 (1968).
15. N. Singh, J. Singh, and S. Prakoash, Proc. Nucl. Phys. and Sol. State Phys. Symp., Bangalore, India (New Delhi, India, Dept. of Atomic Energy, 1973) p. 49.
16. W. Hanke, Phys. Rev. B $\underline{8}$, 4585 (1973).
17. L.J. Sham, in Dynamical Properties of Solids, edited by G.K. Harton and A.A. Maradudin (North-Holland, Amsterdam, 1974), Vol. 1, Ch. 5.
18. J. Bardeen, Phys. Rev. $\underline{52}$, 688 (1937).
19. G.D. Gaspari and B.L. Gyorffy, Phys. Rev. Lett. $\underline{28}$, 801 (1972).
20. R. Evans, G.D. Gaspari, and B.L. Gyorffy, J. Phys. F $\underline{3}$, 39 (1973).
21. B.M. Klein, D. Papaconstantopoulos, Phys. Rev. Lett. $\underline{32}$, 1193 (1974); D. Papaconstantopoulos and B.M. Klein, Phys. Rev. Lett. $\underline{35}$, 110 (1975).
22. S.K. Sinha, R.P. Gupta, and B.N. Harmon, to be published.
23. H. Yamada and M. Shimizu, J. Phys. Soc. Japan $\underline{22}$, 1404 (1967).
24. W. Hanke and L.J. Sham, Phys. Rev. B $\underline{12}$, 4501 (1975).
25. D.G. Pettifor, J. Phys. C $\underline{5}$, 97 (1972).
26. K.H. Oh, B.N. Harmon, S.H. Liu, and S.K. Sinha, Phys. Rev. (in press).
27. B.N. Harmon and S.K. Sinha, these proceedings.
28. C. Herring, in Magnetism, ed. by G.T. Rado and H. Suhl (Academic

Press, New York, 1966) Vol. 4.
29. J.F. Cooke, H.L. Davis, and M. Mostoller, Phys. Rev. B $\underline{9}$, 1485 (1974).
30. S. Shapiro and R. Pynn, private communication.
31. W. Cochran, CRC Critical Rev. Solid State Sci. $\underline{2}$, 1 (1971) and also H. Bilz, in Dynamical Properties of Solids, edited by G.K. Harton and A.A. Maradudin (North-Holland, Amsterdam, 1974) Vol. 1.
32. S.K. Sinha, CRC Critical Rev. Sol. State Sci. $\underline{3}$, 273 (1973).
33. D.J. Chadi and M.L. Cohen, Phys. Rev. B $\underline{10}$, 496 (1974).
34. M. Gupta and A.J. Freeman, these proceedings.
35. B. Klein, D. Papaconstantopoulos, and L. Boyer, these proceedings.
36. S.G. Das and S.K. Sinha, to be published. See also Y. Yamada, H. Takatera, and D.L. Huber, J. Phys. Soc. Jpn. $\underline{36}$, 641 (1974).
37. H.G. Smith, N. Wakabayashi, and M. Mostoller, these proceedings.
38. W.L. McMillan, Phys. Rev. $\underline{167}$, 331 (1968).
39. J. Bar-Sagi and W. Hanke, Proceedings of Conference on Low-lying Lattice Modes and their relationship to Superconductivity and Ferroelectricity, Univ. of Puerto Rico, 1976.
40. N. Wakabayashi, private communication.
41. B.N. Ganguly and R.F. Wood, Phys. Rev. Lett. $\underline{28}$, 681 (1972).
42. L.J. Sham, Proc. Roy. Soc. (London) A $\underline{283}$, 33 (1965).
43. R.P. Gupta and A.J. Freeman, Phys. Rev. B $\underline{13}$, (in press).

APPENDIX

We sketch here the ideas leading to the results given in Eq. (2.1). Full details will be published elsewhere [22]. As is well-known, the usual electron-phonon Hamiltonian possesses certain unsatisfactory features due to the large values of ∇V_b (where V_b is the bare electron-ion potential) in the vicinity of the ion cores. This leads to poor convergence in perturbation sums over excited states. On the other hand, the Schrödinger Equation yields the identity [14]

$$-\langle \psi_{\vec{k}\lambda}|\nabla_\alpha U(\vec{r}-\vec{r}_\ell)|\psi_{\vec{k}'\lambda'}\rangle = \tfrac{1}{2}(M^{\ell,\alpha}_{\vec{k}\lambda,\vec{k}'\lambda'} + M^{*\ell,\alpha}_{\vec{k}'\lambda',\vec{k}\lambda})$$

$$+ (E_{\vec{k}\lambda} - E_{\vec{k}'\lambda'})\langle \psi_{\vec{k}\lambda}|A|\psi_{\vec{k}'\lambda'}\rangle$$

(A1)

where $U(\vec{r}-\vec{r}_\ell)$ is the total crystal potential cut off inside the muffin-tin volume V_0 in the ℓth cell, $\psi_{\vec{k}\lambda}$, $\psi_{\vec{k}'\lambda'}$ are Bloch states with Bloch wavevectors, $\vec{k},\vec{k}'$ and band indices λ,λ', energies $E_{\vec{k}\lambda}$, $E_{\vec{k}'\lambda'}$ respectively, α signifies a Cartesian component,

$$M^{\ell,\alpha}_{\vec{k}\lambda,\vec{k}'\lambda'} = \frac{1}{2m} \int_{S_o(\ell)} dS \left\{ \left(\frac{\partial}{\partial n} \psi^*_{\vec{k}\lambda}\right)\left(\frac{\partial}{\partial x_\alpha} \psi_{\vec{k}'\lambda'}\right) \right.$$

$$\left. - \psi^*_{\vec{k}\lambda} \left(\frac{\partial^2}{\partial n \partial x_\alpha} \psi_{\vec{k}'\lambda'}\right) \right\} \qquad (\hbar = 1) \qquad (A2)$$

where the integration is over the surface of the ℓth muffin-tin sphere, and $\partial/\partial n$ denotes differentiation normal to the surface, and

$$<\psi_{\vec{k}\lambda}|A|\psi_{\vec{k}'\lambda'}> = \tfrac{1}{2} \int_{V_o(\ell)} d\vec{r} \left\{ \psi^*_{\vec{k}\lambda}\left(\frac{\partial}{\partial x_\alpha}\psi_{\vec{k}'\lambda'}\right) \right. \qquad (A3)$$

$$\left. - \left(\frac{\partial}{\partial x_\alpha}\psi^*_{\vec{k}\lambda}\right)\psi_{\vec{k}'\lambda'} \right\}.$$

We now transform to a representation where the unperturbed basis set corresponds to wavefunctions which are carried rigidly with the ions inside each muffin-tin, with due care paid to the effects of the discontinuity at the muffin-tin surface. This is accomplished by means of the following canonical transformation

$$\left. \begin{array}{l} \tilde{C}^+_{\vec{k}\lambda} = \sum\limits_{\vec{k}'\lambda'} <\psi_{\vec{k}'\lambda'}|e^A|\psi_{\vec{k}\lambda}> C^+_{\vec{k}'\lambda'} \\[2ex] \tilde{C}_{\vec{k}\lambda} = \sum\limits_{\vec{k}'\lambda'} <\psi_{\vec{k}'\lambda'}|e^A|\psi_{\vec{k}\lambda}> C_{\vec{k}'\lambda'} \end{array} \right\} \qquad (A4)$$

where the operators on the right hand side are the new electron creation and annihilation operators. The transformed Hamiltonian contains "bare" phonon frequencies which are already partly renormalized in the sense that the ion-cores have been replaced by the "pseudo-atoms" inside the muffin-tin spheres $V_o(\ell)$. The bare electron-phonon matrix element is replaced by a part which comes from surface integrals of the type $M^{\ell,\alpha}_{\vec{k}\lambda,\vec{k}'\lambda'}$, and a part which comes from the tails of the potentials due to the pseudo-atoms in the regions outside their own muffin-tin spheres, i.e. weak local potentials. We obtain [22]

$$H = \sum_{\vec{q}j} (\omega^{(0)}_{qj} + \tfrac{1}{2}) b^+_{\vec{q}j} b_{\vec{q}j} + \sum_{\vec{k}\lambda\sigma} E_{\vec{k}\lambda\sigma} C^+_{\vec{k}\lambda\sigma} C_{\vec{k}\lambda\sigma}$$

$$+ \tfrac{1}{2} \sum_{\substack{\vec{k}_1\vec{k}_2\vec{k}_3\vec{k}_4 \\ \lambda_1\lambda_2\lambda_3\lambda_4}} \sum_{\sigma\sigma'} W_{\vec{k}_1\lambda_1\sigma,\vec{k}_2\lambda_2\sigma',\vec{k}_4\lambda_4\sigma',\vec{k}_3\lambda_3\sigma} C^+_{\vec{k}_1\lambda_1\sigma} C^+_{\vec{k}_2\lambda_2\sigma'}$$

$$C_{\vec{k}_3\lambda_3\sigma'} C_{\vec{k}_4\lambda_4\sigma} - \sum_{\substack{k_1 k_2 \\ \lambda_1\lambda_2\lambda_3}} \sum_{\sigma\sigma'} W_{\vec{k}_1\lambda_1\sigma',\vec{k}_2\lambda_2\sigma,\vec{k}_1\lambda_3\sigma',\vec{k}_2\lambda_2\sigma} <n_{\vec{k}_2\lambda_2\sigma}>$$

$$C^+_{\vec{k}_1\lambda_1\sigma'} C_{\vec{k}_1\lambda_3\sigma'} + \sum_{\substack{k_1 k_2 \\ \lambda_1\lambda_2\lambda_3}} \sum_{\sigma} W_{\vec{k}_1\lambda_1\sigma,\vec{k}_2\lambda_2\sigma,\vec{k}_2\lambda_2\sigma,\vec{k}_1\lambda_3\sigma} <n_{\vec{k}_2\lambda_2\sigma}>$$

$$C^+_{\vec{k}_1\lambda_1\sigma} C_{\vec{k}_1\lambda_3\sigma} + \sum_{\vec{q}j} \sum_{\vec{k}_o} \sum_{\sigma} P_{\vec{k}_o+\vec{q}\lambda_1,\vec{k}_o\lambda_2}(\vec{q}j)$$

$$C^+_{\vec{k}+\vec{q}\lambda_1\sigma} C_{\vec{k}\lambda_2\sigma} (b_{\vec{q}j} + b^+_{-\vec{q}j}) \tag{A5}$$

where $\omega^{(0)}_{qj}$ are the ion-plasma frequencies associated with the bare pseudo-atoms discussed above, $\vec{q}j$ denote phonon wavevector and branch index respectively, σ is an electron spin index, $n_{\vec{k}\lambda\sigma}$ is the occupation number for state $(\vec{k}\lambda\sigma)$,

$$W_{\vec{k}_1\lambda_1\sigma_1,\vec{k}_2\lambda_2\sigma_2;\vec{k}_4\lambda_4\sigma_4,\vec{k}_3\lambda_3\sigma_3} = \delta_{\sigma_1\sigma_4}\delta_{\sigma_2\sigma_3} \iint d\vec{r}\, d\vec{r}'$$

$$\psi^*_{\vec{k}_1\lambda_1}(\vec{r})\psi^*_{\vec{k}_2\lambda_2}(\vec{r}')(e^2/|\vec{r}-\vec{r}'|)\,\psi_{\vec{k}_3\lambda_3}(\vec{r}')\psi_{\vec{k}_4\lambda_4}(\vec{r}) \tag{A6}$$

and

$$P_{\vec{k}+\vec{q}\lambda,\vec{k}\lambda'}(\vec{q}j) = \left(\frac{1}{2NM\omega^{(0)}_{\vec{q}j}}\right)^{\frac{1}{2}} \sum_\alpha e_\alpha(\vec{q}j)\, \sum_\ell \{\tfrac{1}{2}(M^{\ell,\alpha}_{\vec{k}+\vec{q}\lambda,\vec{k}\lambda'}$$

$$+ M^{*\ell,\alpha}_{\vec{k}\lambda',\vec{k}+\vec{q}\lambda}) - \langle \psi_{\vec{k}+\vec{q}\lambda} | \frac{\partial}{\partial x_\alpha} W(\vec{r}-\vec{r}_\ell) | \psi_{\vec{k}\lambda'}\rangle$$

$$- \langle \psi_{\vec{k}+\vec{q}\lambda} | W^\ell_{ex} | \psi_{\vec{k}\lambda'}\rangle\}\, e^{i\vec{q}\cdot\vec{r}_\ell} \tag{A7}$$

where N is the total number of atoms, M is the atomic mass, and $W(\vec{r}-\vec{r}_\ell)$ is defined by

$$W(\vec{r}-\vec{r}_\ell) = -Z_S\, e^2/|\vec{r}-\vec{r}_\ell| \quad (\vec{r}\text{ outside } V_0(\ell))$$

$$= \sum_{\ell'\neq\ell} Z_S\, e^2/|\vec{r}-\vec{r}_{\ell'}| + V_{cH}(r_s) \quad (\vec{r}\text{ inside } V_0(\ell)) \tag{A8}$$

where Z_S is the <u>total</u> charge inside the muffin-tin sphere, and $V_{cH}(r_s)$ is a constant which ensures the continuity of $W(\vec{r}-\vec{r}_\ell)$ across the surface of $V_0(\ell)$. W^ℓ_{ex} is the non-local exchange correction to $W(\vec{r}-\vec{r}_\ell)$ and we will not discuss it in detail here.

Eq. (A5) has been written down in a way analogous to the Hamiltonian of Yamada and Shimizu [23] with respect to the electron-electron interactions. It forms the starting point for our calculation of the phonon frequencies.

In order to treat correctly the local field corrections, we follow Hanke and Sham [24] in writing the Bloch wavefunctions in a localized representation of the form

QUESTIONS AND COMMENTS

B. Klein: In the discussion of the phonon anomalies, I suggest that when a theory confirms the existence of an anomaly at a high symmetry point where it occurs experimentally, it is a very weak piece of support for the theory.

S. Sinha: P is not a high symmetry point. It is a very low symmetry point.

B. Klein: Well, it is certainly a point of symmetry. It has symmetry other than the identity.

S. Sinha: It was not a high symmetry point in niobium, or in niobium carbide, for which we got the dips.

B. Klein: Just answer that one point.

S. Sinha: Without relevance to our calculations, okay? There is always an extremum there so you always obtain a maximum or a minimum. As long as it was a maximum you would be okay. But it turns out that is irrelevant for these calculations because, in fact, none of them are at high symmetry points.

B. Gyorffy: We evaluated the integral $\int_0^{R_{MT}^+} dr\, R_\ell (dv/dr) R_{\ell+1}$ to R_{MT}^+ being infinitesimally larger than R_{MT}. This means we included the jump in the potential at R_{MT}. This jump mimics the force an electron would have felt in the interstitial region if we did not flatten out the potential there. Therefore we think it should be included. I also have a question. If I calculated the Hubbard U using Hartree-Fock-Slater wave functions I get 20 eV which is a factor of ten too big. How can you expect that the same wave functions give a more reliable value for the interatomic Coulomb integrals which determines your U(g)?

S. Sinha: I will answer the question first. In the case of the intra-atomic Coulomb integral, it is well known because of the strong correlation effects in the same atom and strong screening effects that you really over-estimate U. Herring discussed this in detail in his book. You completely over-estimate U by a factor of 5 or an order or magnitude. If you just take Hartree-Fock you decrease it already by a factor of 2. Now, admitting our ignorance of doing the many-body theory properly on the same atom, we stick with a value of U about 2 eV. Now the inter-atomic integral can be calculated much more accurately. You can put in the Coulomb interaction and subtract the exchange part and it turns out that because the intra-atomic U has beem reduced so much, the interatomic ones, eight of them, compared to one intra-atomic will tend to overwhelm it as $\vec{q}$ goes towards the zone boundary. Now I would like to respond to the first part because that is an important point. Very briefly, when you say that this is a technical matter and that this discontinuity in the potential is artificial I quite agree with you. However, it is in fact the inside part of the second derivative that is involved in the expression. In your formalism, you have essentially used the outside part as the one to take and that is an artifice of the potential because is is cut off and made flat. You should really take the inside part and that is what we have corrected for.

F. Mueller: I have a question which I hope will shed light on this subject and not generate more heat and in fact be useful. The question focuses on the point of where in transition metals (in r-space) does the electron-phonon interaction take place? Now Suny, in your theory you have said that for the ℓ equal to zero and one components of the radial wavefunction inside the muffin-tin sphere --- one has relatively the same wave function (which does not change). We know that these are parts of Bloch states, which are connected to the interstitial parts and of course which vary in a systematic way. So the impression that one might get that these are constant is not correct because the Bloch states will change in the interstitial region. It is this interstitial region which I would like to focus on. The question is: 'Is this interstitial region and its interaction important for d-band metal electron-phonon interaction and does it make a contribution specifically toward the d-d bonding (and anti-bonding) orbitals in the interstitial region?' Now since this conference is filled with calculations and experiments which are not being reported, I will also similarly report some work which Harold Myron and myself have not done. This suggests that the change of the bonding of the d electrons in the interstitial region is what is important for λ. Where in your theory is such an effect?

S. Sinha: Okay, it is there and it is very true. You recall that the mixing coefficients $A_{\ell,m}$ are functions of $\underset{\sim}{k}$. And they mix in all the possible angular momentum states. You do not necessarily have to have a radial function that varies with energy to get the kind of distribution in the outer part of the cell that you are talking about. You can do that by suitable combinations of the various ℓ,m states. Now the second part is: should one completely neglect the interstitial part of the potential?

F. Mueller: Of course not! Charging effects --- especially in compounds are focused in the interstitial region because you have charge-transferred them (in making the compounds).

S. Sinha: I agree. Especially in compounds...

M. Cohen: For NbO_2, David Bullett in my group has done some preliminary tight-binding calculations; we have not yet done pseudopotential calculations. In fact, it appears that there is a peaked region in the d-density of states and that the Fermi level can lie in this peak. On the other hand, if you start going into the "perverted" structure, not the rutile, you still get a metal because correlation effects are not included. One of the exciting things about NbO_2 is the fast switching times. Do you have any idea what this instability does to the switching? Can you get the metallic behavior to semiconducting behavior through this instability which is not a soft mode but is some kind of a hopping conductivity change? Have you though about that?

S. Sinha: I have not thought about it. But it is an interesting question to consider. I will have to think about it. Actually NbO_2 in fact is probably a material which if it had not had this transition would probably have gone on to be a good superconductor because of the high density of states. Presumably that is the reason why Professor Foner mentioned $LiTiO_2$ may be trying to do the same thing but not as drastically as to make a phase transition. I see Dr. Matthias disagrees.

B. Matthias: That is completely wrong because it [NbO_2] will never be a high superconductor since is is only tetragonal.

S. Sinha: You mean $LiTiO_2$ is not tetragonal?

B. Matthias: Dear God, no.

S. Sinha: Okay, that is what I was wondering. I assumed it [$LiTiO_2$] was the same structure.

B. Matthias: Spinel, which is radically different.

S. Sinha: Okay, I take all that back.

ELECTRON AND PHONON PROPERTIES OF A-15 COMPOUNDS AND
CHEVREL PHASES*

F.Y. Fradin, G.S. Knapp, S.D. Bader, G. Cinader

Argonne National Laboratory

Argonne, Illinois 60439

and

C.W. Kimball[†]

Northern Illinois University

DeKalb, Illinois 60115

ABSTRACT

A review of the information on the electron and phonon spectra in the A-15 compounds and Chevrel phase superconductors obtained from nuclear magnetic resonance, heat capacity, and Mössbauer effect experiments is presented. Relationships of the Fermi energy electrons and the soft phonons to the strength of the electron-phonon interaction are discussed.

---

*Work supported by the U.S. Energy Research and Development Administration.

[†]Work supported by the National Science Foundation.

I. INTRODUCTION

High critical temperature (and critical field) superconductivity is often associated with anomalous electron and phonon properties [1,2]. This is due to the fact that the strength of the electron-electron coupling mediated by the phonons is dependent upon the details of the normal state properties. In this paper we will focus on the unusual electron and phonon properties of two important classes of superconductors; the A-15 compounds that have the highest T_c values and the Chevrel phases that have the highest H_{c2} values. In Section II of this paper we review a number of experimental results on the A-15 compounds that yield information about the symmetry of the dominant electron character near the Fermi level and the nature of the average phonon behavior. We will show that anharmonic effects as well as λ are related to the electronic density of states at E_F. In Section III, we present data on the recently discovered Chevrel phases that illustrates the extremely unusual phonon behavior both from a microscopic and macroscopic point of view. We will indicate the relationship of the unusual phonon spectra of these compounds to the crystal structure and indicate the importance of the particular features of the spectra to λ. We will also show that there is a strong correlation between electronic properties and the magnitudes of T_c for these compounds as well. For the sake of brevity the extensive literature will not be reviewed; details of the analyses can be found in the references.

II. A-15 COMPOUNDS

Although there have been a number of band structure calculations of the A-15 compounds [1,3,4,5], they have, to this point, lacked the accuracy and resolution necessary to explain the electronic properties. Therefore, we have had to rely on various probes of the electronic structure to obtain information about the Fermi surface electrons. Except for the very low T_c, low density of states $N(0)$, compound [6] Nb_3Sb, no deHaas van Alphen work exists on the A-15 compounds. Similarly, the optical experiments have been unable to determine features of $N(\varepsilon)$ for ε near ε_F. However, the electronic heat capacity coefficient γ, the spin susceptibility χ_s, the electrical resistivity ρ, and the nuclear spin-lattice relaxation rate $1/T_1$ yield information about $N(0)$. In addition, the temperature dependences of these quantities yield information [7] about the shape of $N(\varepsilon)$ near ε_F, which is quite sharp for V_3Ga, V_3Si, and Nb_3Sn. We will focus on the results for $\gamma_0 = \gamma/(1+\lambda)$, where γ is obtained from the low temperature heat capacity [8] using the entropy constraint at T_c, and on the ^{51}V or ^{93}Nb values [9,10] of $1/T_1$ to construct the subband density of states at the Fermi level.

The A-15 (Cr_3O) structure in which the A_3X (A = V or Nb and X = non-transition element) compounds crystallize is a cubic lattice

with the X atoms on a body-centered-cubic sublattice. We focus
attention on the A sites. These are the sites for which a LCAO
description of the $\ell = 2$ component of the wave functions will yield
some insights into the dominant character of the Fermi surface
electrons. The A sites have tetragonal symmetry; the point group
is $D_{2d}(\bar{4}2m)$, and the A atoms lie on chains arranged in three orthogonal families. In Table I we list the irreducible representation
of the d wave functions ($\ell = 2$) for an A atom on a chain with axis
parallel to the $\bar{z}$ direction.

TABLE I. Irreducible representation of $D_{2d}(\ell = 2)$.

Bases	Assignments	Atomic functions	Spherical[a] harmonics
A_1	σ_z	$(5/4\pi)^{1/2}(3z^2 - r^2)/2r^2$	Y_2^0
E	π_{1z}	$(15/4\pi)^{1/2}\, zx/r^2$	Y_2^1,c
E	π_{2z}	$(15/4\pi)^{1/2}\, zy/r^2$	Y_2^1,s
B_1	δ_{1z}	$(15/16\pi)^{1/2}(x^2 - y^2)/r^2$	Y_2^2,c
B_2	δ_{2z}	$(15/4\pi)^{1/2}\, xy/r^2$	Y_2^2,s

[a] $Y_\ell^{m,c} \equiv (Y_\ell^m + Y_\ell^{-m})/\sqrt{2};\ Y_\ell^{m,s} = -i(Y_\ell^m - Y_\ell^{-m})/\sqrt{2};\ Y_\ell^m(\theta,\delta)$ are normalized spherical harmonics. The polar angle θ is measured with
respect to a z axis parallel to the tetragonal or chain axis.

In a tight-binding picture of the V(Nb) d-states in the A-15
compounds, the nuclear spin-lattice relaxation rate $1/T_1$ of ^{51}V
(^{93}Nb) is given in terms of bilinear products of the subband
density of states $N^m(\varepsilon)$ evaluated at the Fermi level [9], where

$$N^m(0) = \sum_\mu \sum_k |C_{\mu m k}|^2\, \delta(E_{\mu k} - E_F).$$

Here the $C_{\mu m k}$ are the fractional admixture coefficients in the LCAO
prescription for the wave function for band index μ and wave-vector
k, and m represents the appropriate tight binding basis function,
i.e., σ, π, δ_1, or δ_2. The results for the subband densities of
states derived from the experimental values [9,10] of $1/T_1$ are
listed in Table II.

TABLE II. Subband densities of states
(states/Ry-V(Nb)-spin)

Compound	$N^\sigma(0)$	$N^\pi(0)$	$N^\delta(0)$	$N_{tot}(0)$	$N_{\gamma_0}(0)$
V_3Ga	5	27	5	39	44
$V_3Ga_{0.8}Si_{0.2}$	5	19	6	32	-
$V_3Ga_{0.6}Si_{0.4}$	5	14	9	30	-
$V_3Ga_{0.4}Si_{0.6}$	5	15	7	29	-
$V_3Ga_{0.2}Si_{0.8}$	5	18	6	31	-
V_3Si	5	22	6	35	35
$V_3Ga_{0.9}Sn_{0.1}$	5	20	7	34	-
$V_3Ga_{0.7}Sn_{0.3}$	5	15	8	29	-
$V_3Ga_{0.5}Sn_{0.5}$	4	12	8	26	25
$V_3Ga_{0.1}Sn_{0.9}$	4	11	4	21	-
V_3Sn	4	8	3	17	18
Nb_3Sn	11	~ 0	10	22	22
Nb_3Al	3	~ 0	11	15	15
Nb_3Sb	2	~ 1	~ 1	~ 4	4

a. $N_s(0) \approx 2$ for V_3X compounds, $N_s(0) \lesssim 1$ for Nb_3X compounds.

b. For Nb_3X compounds, $N_{tot}(0)$ fixed at value of $N_{\gamma_0}(0)$.

c. $N^\delta(0) = N^{\delta 1}(0) + N^{\delta 2}(0)$.

In general for the V_3X compounds, we find a correlation of $1/T_1$ with $[N_{\gamma_0}(0)]^2$ principally due to the dominance of $N^\pi(0)$. No correlation between $1/T_1$ and $[N_{\gamma_0}(0)]^2$ is found in the Nb_3X compounds; also, $N^\pi(0)$ is quite small for the Nb_3X compounds. The large value of $N_{\gamma_0}(0)$ for Nb_3Sn relative to Nb_3Al, is due to a large value of $N^\sigma(0)$ for Nb_3Sn. The value of $N_{\gamma_0}(0)$ for the very low T_c compound Nb_3Sb is very small; the Fermi-level appears to lie in a valley in $N(\varepsilon)$.

The high temperature lattice entropy data on the A-15 compounds can be analyzed [8] to determine the geometric mean frequency (zero-th moment of the phonon spectrum)

$$\omega_g(T) = \left[\prod_{i=1}^{3N} \omega_i(T) \right]^{1/3N}$$

and its temperature dependence. The results are illustrated in Fig. 1 as an effective Debye temperature $\theta(T) = e^{1/3}\hbar\omega_g(T)/k_B$. [We note that the values of $\theta(T)$ for Nb_3Sn, V_3Si and V_3Ga are within 5 to 10% of the values obtained from inelastic neutron scattering experiments [11] on polycrystals.] Nb_3Sb behaves normally indicating phonon softening with increasing temperature due to lattice expansion. On the other hand, the high-T_c compounds show considerable softening of $\theta(T)$ with decreasing temperature. A measure of the amount of anharmonicity is given by the logarithmic derivative of $\omega_g(T)$, i.e., $(1/\omega_g)(\partial\omega_g/\partial T)$. We find an interesting correlation between the anharmonicity and $N(E_F)$, which is illustrated in Fig. 2.

The correlation of $(1/\omega_g)(\partial\omega_g/\partial T)$ with $N(E_F)$ could be a manifestation of strong selective electronic screening. The high $N(E_F)$ can cause the frequencies of certain phonon modes to decrease significantly. These frequencies can be temperature dependent for two reasons. First, electronic screening could cause the effective second-order term in the lattice potential energy to be reduced relative to the third and fourth-order terms. This reduction enhances the anharmonicity. Second, electronic screening, by near-Fermi energy-electronic states, will be quite temperature dependent because of sharp structure [7] in $N(E)$ near E_F.

Some insight into the microscopic aspects of the anharmonicity of the A-15 compounds has been obtained from a ^{119}Sn Mössbauer investigation [12] of the mean-square displacement $<X^2>_{Sn}$ and the mean-square velocity $<V^2>_{Sn}$ obtained from the recoil-free fraction and the thermal shifts in the $V_3Ga_{1-x}Sn_x$ compounds, respectively. In the Ga-rich high T_c compound, the thermal shift results are found to be consistent with a softening of the Sn modes at low

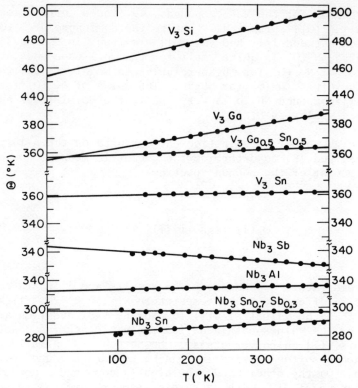

Figure 1: The temperature dependence of the effective Debye temperature associated with the geometric mean phonon-mode frequencies for the indicated A-15 compounds.

temperature. Comparison with the anharmonicity derived from the heat capacity (discussed above) indicates that the V vibrational motion also softens with comparable magnitude at low temperature. Further, it is found that the Sn vibrational motion is softer in the low T_c compounds than in the high T_c compounds; hence the V site motion is inferred to be somewhat softer in the high T_c compounds than in the low T_c compounds. (Note that in Fig. 1 the macroscopic phonon moment is nearly constant for the $V_3Ga_{1-x}Sn_x$ compounds.)

It is clear from the results in the foregoing sections that the systematic correlation of T_c or λ with the total bare electronic density of states at the Fermi level, which exists for the V_3X compounds, is broken by Nb_3Al. Also, there clearly is not a correlation of λ with $[M\omega(0)^2]^{-1}$ (see Table III), although we can not rule out a correlation of λ with $[M_V\omega(0)^2_V]^{-1}$. In Figure 3 we show the dependence of λ on the subband density of states for the Nb_3X and some of the V_3X compounds (some of the V_3X data is not shown for clarity).

TABLE III. Dependence of λ on $M\omega(0)^2$ and $N^\pi(0) + N^\delta(0)$.

Compound	λ	$\lambda M\omega(0)^2 \left(\frac{eV}{A^2}\right)$	$\lambda/[N^\pi(0)+N^\delta(0)]$ [Ry-V(Nb)-spin]
V_3Ga	0.91	5.9	.028
V_3Si	0.86	7.4	.031
$V_3Ga_{0.4}Si_{0.6}$	0.72	-	.032
$V_3Ga_{0.5}Sn_{0.5}$	0.62	4.5	.031
V_3Sn	0.56	4.6	.051
Nb_3Sn	1.17	8.4	.12
Nb_3Al	1.07	8.4	.10
Nb_3Sb	0.3	3.3	.15

a. An unrestricted multivariate linear least squares fit to all of the V_3X data in Table II yields λ = 0.11 (4) $N^s(0)$ + 0.05 (3) $N^\sigma(0)$ + 0.016 (1) $N^\pi(0)$ + 0.006 (5) $N^\delta(0)$ with a value of $\sigma = [(1/(11-4)) \sum_i (\lambda_i - \lambda_{i_{cal}})^2]^{\frac{1}{2}} = 0.014$.

b. For Nb_3X, λ = 0.024 $N^\sigma(0)$ + 0.16 $N^\pi(0)$ + 0.091 $N^\delta(0)$.

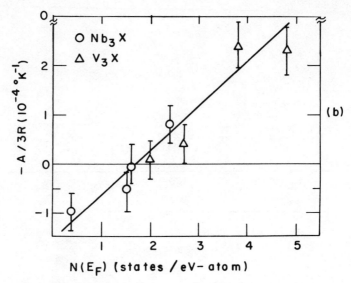

Figure 2: The phonon frequency shift parameter, $(-A/3R) \equiv (1/\omega_g) \cdot (\partial \omega_g/\partial T)$, as a function of the bare $N(E_F)$ for the A-15 compounds.

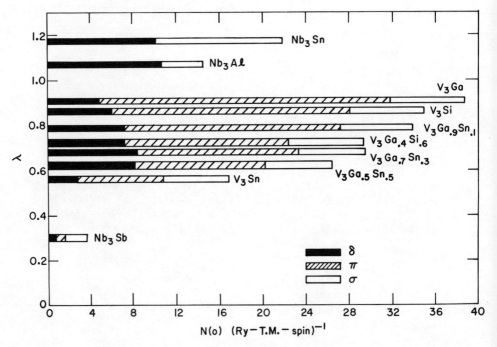

Figure 3: Dependence of the strength of the electron-phonon interaction on the bare subband density of states for the A-15 compounds. $N^\delta(0) = N^{\delta 1}(0) + N^{\delta 2}(0)$.

If the σ-subband density $N^\sigma(0)$ is ignored, there is a rough correlation of λ with $N^\delta(0) + N^\pi(0)$. We note that the ℓ = 2 wave functions of σ-symmetry (see Table I) are expected to have the largest intra-chain overlap. It would be very interesting to see if this result could be understood in terms of the different magnitudes of the two-center overlap integrals (intra-chain and inter-chain) that characterize the d-states of σ-symmetry from those of π or δ-symmetry [13].

III. CHEVREL PHASES

The Chevrel phases show even more unusual lattice properties than the A-15 compounds. We will show that these materials have large numbers of low-lying phonon modes, and that certain of these modes are very anharmonic. From the heat capacity measurements we are able to estimate the bare electronic density of states at the Fermi level of these metals, and we find the density of states N(0) per molybdenum atom is comparable to the density of states per V atom, observed in the V_3X compounds. We also find a strong correlation between the densities of states and the λ values, which is similar to the V_3X compounds.

In Figure 4 we show the results of our low temperature heat capacity measurements. The magnitude of these heat capacities are very large when compared with ordinary transition metals, an indication of the soft modes present in these materials. Recently, Bader et al. [14] proposed a simple model which is capable of explaining the unusual lattice properties of these materials. They assumed that these metals could be viewed as "molecular crystals" consisting of X atoms (where X is Sn or Pb) and quasi-rigid Mo_6S_8 (or Mo_6Se_8) units. The lattice dynamics would then simplify since the 45 normal modes of the unit cell of the ternaries, derived from the 15 atoms of the formula unit, would group into 9 soft "external" modes and 36 hard internal modes. The 9 soft external modes consist of 3 acoustic, 3 optic, and 3 torsional modes. In the case of the binary Mo_6Se_8, where there are only 42 normal modes, there would be 6 soft external modes (3 acoustic and 3 torsional) and 36 hard internal modes. At low temperatures, the heat capacity of Mo_6Se_8 (see Figure 4) is considerably smaller than for the other compounds, an indication that there are less soft modes for this material.

In Figure 5, we show the effective Debye temperatures θ^C and θ^S, obtained from the lattice heat capacities and entropies, respectively, from 2-400 K obtained from a sample of $PbMo_{5.1}S_6$ [15]. Note the increase in the effective Debye temperatures by a factor of ~ 2. This variation is approximately an order of magnitude larger than found for simple metals, but is one that is similar to that found for molecular crystals.

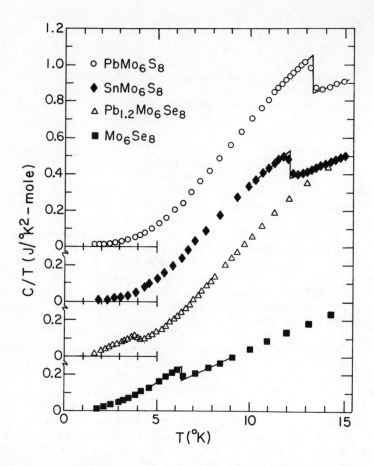

Figure 4: Low temperature heat capacities of the Chevrel phase.

Mössbauer measurements on ^{119}Sn in SnMo$_6$S$_8$, performed by Kimball et al. [16], shed light on the local character of the lattice vibrations of the tin atoms. The Sn atoms are in sites with axial symmetry; it is possible to determine the anisotropy of the Debye Waller factors from the recoil-free fraction in directions parallel and perpendicular to the axis of symmetry. By assuming that the tin atoms vibrate at a single frequency in each direction (Einstein model) it is possible to determine the magnitude and temperature dependences of the tin normal modes. The results of this analysis are shown in Figure 6. Note that the frequencies, both parallel and perpendicular to the axis of symmetry, are very low and are very temperature dependent; the latter behavior indicates that the tin atoms are undergoing very anharmonic motions. Since the Sn

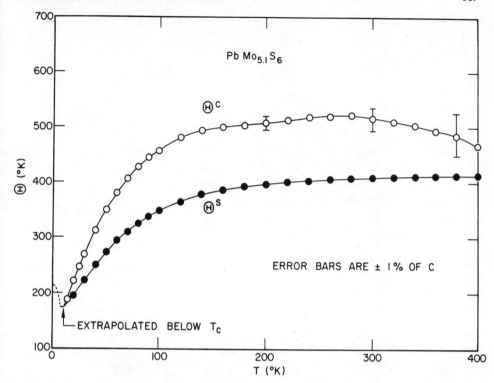

Figure 5: Effective Debye temperatures θ^C and θ^S obtained from the lattice heat capacity and entropy, respectively, of $PbMo_{5.1}S_6$.

vibrational motion should hybridize with the external modes, we would expect anharmonicity of these modes, as well. Recent inelastic neutron scattering determinations of phonon spectra [11,17] in polycrystalline binary and ternary Chevrel phase compounds support the molecular crystal model and also indicate that the external modes soften on cooling.

The large magnitudes of the lattice heat capacities above T_c of the Chevrel phases make it difficult to determine the electronic contribution to the heat capacity. At T_c the lattice heat capacities are well outside the T^3 region, even for the lowest T_c material, $Pb_{1.2}Mo_6Se_8$ (T_c = 3.8 K). Methods that use the electronic entropy constraint to make the C/T vs. T^2 extrapolation from above T_c less arbitrary may be unreliable because the electronic entropy is only a small fraction of the total above T_c. The magnitude of the jump in the heat capacity ΔC at T_c can be utilized to estimate the enhanced electronic heat-capacity coefficient γ, if the strong-coupling corrections to the BCS jump $\Delta C_{BCS} = 1.43 \gamma T_c$ are known. Kresin, et al. [18] have estimated the correction and find

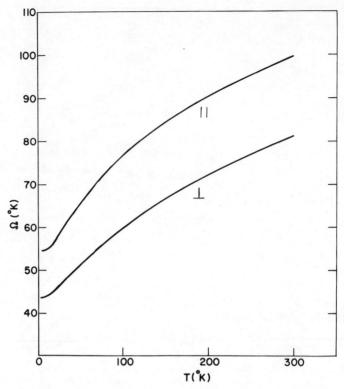

Figure 6: Temperature dependence of the Einstein frequency describing the Debye-Waller factor of ^{119}Sn in SnMo$_6$S$_8$ for motion parallel and perpendicular to the rhombohedral axis.

$$\gamma = \frac{\Delta C}{T_c} \left[1.4 + 2.52 \left[\ln\left(\frac{\tilde{\omega}}{T_c}\right) + 0.5 \right] \left(\frac{\pi T_c}{\tilde{\omega}}\right)^2 \right]^{-1},$$

where $\tilde{\omega}$ is an estimated characteristic phonon frequency. For the Chevrel phases we choose $\tilde{\omega} = 12$ meV, a value that roughly characterizes the external phonon modes of the compounds studied in the heat capacity and the neutron scattering experiments. The maximum strong-coupling correction to γ is 45% for PbMo$_6$S$_8$, and the minimum is 5% of Pb$_{1.2}$Mo$_6$Se$_8$. We estimate that the absolute errors in the γ values thus determined are ±30% for the higher-T_c materials and and ~ ±10% for the lower-T_c materials. These errors will not, of course, affect the trends we find. To obtain unenhanced densities of states, we proceed by calculating λ from the McMillan equation

[19], assuming a characteristic phonon energy of 12 meV, again, for the prefactor, and then use the expression $\gamma = (2\pi/3)k_B^2 N(E_F)[1+\lambda]$. The results of the analysis are given in Table IV, and λ versus $N(E_F)$ is plotted in Figure 7. A strong dependence of λ on $N(E_F)$ is observed, similar to that for the V_3X, A-15, compounds [9] and for $4d$ and $5d$ transition metals [13]. Note that Fischer [20] found a correlation between the magnitude of the spin susceptibility and T_c, which corroborates the correlation of Figure 7.

The mechanism by which the high $N(E_F)$ enhances T_c is not clear at this time. Whether the numerator of λ, which carries the important electronic information, is directly enhanced, or the phonon properties in the denominator of the λ expression are renormalized due to electronic screening effects, for instance, remains an open question. It is clear, however, that the crystal structure with its weak bonding between Mo_6S_8 (Mo_6Se_8) units and between these units and Sn (or Pb) is clearly favorable for low lying phonon modes, which are certainly of importance for high transition temperatures.

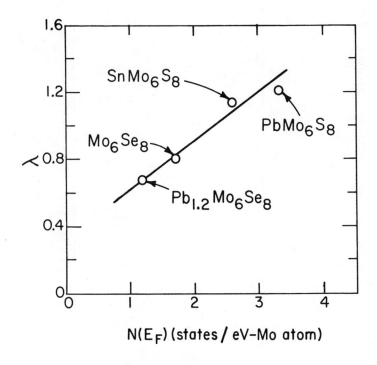

Figure 7: Dependence of the mass enhancement on the bare density of states at the Fermi level for the Chevrel phases.

TABLE IV. Electronic parameters for the Chevrel phases.

	T_c (°K)	$\frac{\Delta C}{T_c}$ (mJ/°K²-mole)	$\gamma \left(\frac{mJ}{°K^2\text{-mole}} \right)$	λ	$N(E_F) \left(\frac{\text{states}}{\text{eV-Mo-atom}} \right)$
PbMo$_6$S$_8$	13.3	215	105	1.20	3.4
SnMo$_6$S$_8$	12.2	155	79	1.13	2.6
Mo$_6$Se$_8$	6.3	70	44	0.80	1.7
Pb$_{1.2}$Mo$_6$Se$_8$	3.9	42	28	0.67	1.2

IV. CONCLUSION

A number of experimental results, which shed light on the macroscopic and microscopic aspects of the Fermi energy electrons and the lattice vibrational modes in the A-15 compounds and the Chevrel phases, have been reviewed. In the A-15 compounds, it was found that large densities of states at the Fermi level in certain d-subbands are highly favorable for large λ, and coincidentally important for low temperature anharmonicity. In the Chevrel phases, a high density of low energy (and anharmonic) phonon modes associated with weak sulfur bonds in "external" modes was found. These modes along with a high electronic density of states at the Fermi level in the Chevrel phases are, undoubtedly, favorable for superconductivity.

ACKNOWLEDGMENTS

The authors would like to thank Dr. R.N. Shelton and Dr. Z. Fisk for supplying some of the samples used in this study. We would also like to thank J.W. Downey, T.E. Klippert, and C.L. Wiley for technical assistance.

REFERENCES

1. M. Weger and I.B. Goldberg, in *Solid State Physics*, edited by H. Ehrenreich, F. Seitz, and D. Turnbull (Academic Press, New York 1973), Vol. 28, p. 1.
2. L.R. Testardi, in *Physical Acoustics*, edited by W.P. Macon and R.N. Thurston (Academic Press, New York, 1973), Vol. 10, p. 193.
3. L.F. Mattheiss, Phys. Rev. <u>138</u>, A112 (1965); L.F. Mattheiss, Phys. Rev. <u>B12</u>, 2161 (1975).
4. G. Barak, I.B. Goldberg, and M. Weger, J. Phys. Chem. Solids (1975).
5. B.M. Klein and D.A. Papaconstantopoulos, *Int'l. Conf. on Low Lying Lattice Vibrational Modes and Their Relationship to Superconductivity and Ferroelectricity*, to be published.
6. A.J. Arko, to be published.
7. S.D. Bader and F.Y. Fradin, *2nd Rochester Conference on Superconductivity in d- and f-Band Metals*; F.Y. Fradin, *Int'l. Conf. on Low Lying Lattice Vibrational Modes and Their Relationship to Superconductivity and Ferroelectricity*, to be published.
8. G.S. Knapp, S.D. Bader, and Z. Fisk, Phys. Rev. B (in press).
9. F.Y. Fradin and D. Zamir, Phys. Rev. B <u>7</u>, 4861 (1973); F.Y. Fradin and J.D. Williamson, Phys. Rev. B <u>10</u>, 2803 (1973).
10. F.Y. Fradin and G. Cinader, to be published.
11. W. Reichardt, P. Schweiss, E. Schneider, and B. Renker, *2nd*

Rochester Conference on Superconductivity in d- and f-Band Metals.
12. C.W. Kimball, L.W. Weber, and F.Y. Fradin, Phys. Rev. B (in press).
13. R.C. Dynes and C.M. Varma, 2nd Rochester Conference on Superconductivity in d- and f-Band Metals.
14. S.D. Bader, G.S. Knapp, S.K. Sinha, P. Schweiss, and B. Renker, to be published.
15. S.D. Bader, G.S. Knapp, and A.T. Aldred, Int'l. Conf. on Low Lying Lattice Vibrational Modes and Their Relationship to Superconductivity and Ferroelectricity, to be published.
16. C.W. Kimball, L. Weber, G. Van Landuyt, F.Y. Fradin, B.D. Dunlap, and G.K. Shenoy, Phys. Rev. Letters $\underline{36}$, 412 (1976).
17. S.D. Bader, S.K. Sinha, and R.N. Shelton, 2nd Rochester Conference on Superconductivity in d- and f-Band Metals.
18. V.Z. Kresin and V.P. Parkhomenko, Sov. Phys. Solid State $\underline{16}$, 2180 (1975).
19. W.L. McMillan, Phys. Rev. $\underline{167}$, 331 (1968).
20. Ø. Fischer, 2nd Rochester Conference on Superconductivity in d- and f-Band Metals.

DIRECT CORRELATION OF OBSERVED PHONON ANOMALIES AND MAXIMA IN THE GENERALIZED SUSCEPTIBILITIES OF TRANSITION METAL CARBIDES*

Michèle Gupta [†]

Physics Department and Materials Research Center

Northwestern University, Evanston, Illinois 60201

and

A.J. Freeman

Physics Department, Northwestern University

Evanston, Illinois 60201 and Argonne National

Laboratory, Argonne, Illinois 60439

The generalized susceptibility, $\chi(\vec{q})$, of both NbC and TaC determined from APW energy band calculations show large maxima to occur at precisely those $\vec{q}_{max}$ values at which soft phonon modes were observed by Smith. Maxima in $\chi(\vec{q})$ are predicted for other directions. The locus of these $\vec{q}_{max}$ values can be represented by a warped cube of dimension $\sim 1.2(2\pi/a)$ in momentum space - in striking agreement with the soft mode surface proposed phenomenologically by Weber. In sharp contrast, the $\chi(\vec{q})$ calculated for both ZrC

---

*Supported by the Air Force Office of Scientific Research (Grant No. 76-2948), the National Science Foundation (Grant Nos. DMR-72-03101-A02 and DMR-72-03019) and the Energy Research and Development Administration.

[†]On leave from the University of Paris-Sud, 91405 Orsay, France and Le Centre de Mécanique Ondulatoire Appliquée du Centre National de la Recherche Scientifique, 23 rue du Maroc, 75019 Paris.

and HfC - for which no phonon anomalies have been observed - fall off in all symmetry directions away from the zone center. We thus interpret the phonon anomalies in the transition metal carbides as due to an "overscreening" effect resulting from an anomalous increase of the response function of the conduction electrons.

I. INTRODUCTION

The phonon dispersion curves of the transition metal carbides (TMC) of the Metal IV and V series have been studied extensively by Smith and Gläser[1], and Smith[2]. These coherent inelastic neutron scattering experiments showed the important correlation that the high critical temperature (T_c) superconducting 9 valence electron compounds like NbC (T_c= 11.1 K) and TaC (T_c= 10.35 K) possess anomalies in their phonon spectrum which are not present in the low T_c ($T_c \leq$ 0.05 K) 8 valence electron compounds like ZrC and HfC. Such a correlation between phonon anomalies and high values of T_c has been also observed for a large number of d electron superconducting compounds or alloys, e.g., the well known Nb-Mo system[3] for which Powell et al. showed that the low T_c alloy $Nb_{0.25}Mo_{0.75}$ ($T_c \simeq$ 0.04 K) no longer possesses any phonon anomaly. The A-15 superconductors like V_3Si and Nb_3Sn also display a correlation between high values of T_c and the occurrence of a low temperature structural phase transformation[4] (from the cubic to tetragonal phase), or the existence of anomalous mode softening for the non-transforming compounds. However, even though instability or anomalous phonon softening in many systems appear to be correlated with the occurrence of a high T_c, there also exist some counter examples[5]. Quantitative relations between these experimental facts are more difficult to draw from first principles calculations. However, in the existing model theories, such as the strong coupling theory[6], the occurrence of soft phonon modes reflected in the variation of the second moment of the phonon frequency, $<\omega^2>$ [defined in Ref. 6], does not appear to be sufficient to explain the variation of the electron phonon coupling constant λ which essentially determines T_c. From the compilation of experimental data and using his calculated phonon density of states, Weber[7] showed that a large variation of the electron-phonon matrix elements $<I^2>$ [defined in Ref. 6] is required to explain the variation of λ in the 8 and 9 valence electron TMC. The same conclusions were later reached by Klein and Papaconstantopoulos[8] using $<\omega^2>$ given by Weber, and the Gaspari-Gyorffy[9] expression of the electron-phonon interaction. More recently, after analyzing experimental data, Varma and Dynes[10] also concluded that for transition metals "soft phonon are not responsible for high T_c".

Since the two physical phenomena, phonon mode softening and high T_c occur concomittantly in the TMC, it is very important to understand the origin of the phonon anomalies in these materials. Several theoretical approaches to the lattice dynamics of the TMC are already available in the literature. Weber, Bilz and Schröder [11] and Weber[12] developed a very successful phenomenological model, reproduced the phonon dispersion curves of the 8 and 9 valence electron TMC and predicted anomalies in some symmetry directions (e.g., the [110] in TaC), which were later confirmed experimentally[13]. Weber[12] showed that the phonon dispersion curves of the 8 valence electron compounds are well reproduced by a simple shell model with free-electron dielectric screening. However, for the 9 valence electron compounds, the anomalies cannot be reproduced without introducing a new degree of freedom to describe the short range d-d metal ion interaction. In this so-called double shell model (DSM), the anomalies were explained[12] as being due to a "resonance-like increase of the electonic polarizability" when the metal ions vibrate in phase opposition. The evidence of a certain amount of d metallic charge density of d_{xy} symmetry between the metal ions has been confirmed by the charge density analysis of Chadi and Cohen[14] resulting from their empirical pseudopotential band structure calculation. Additional evidence that the screening in the 9 valence electron TMC is inadequately described by a free-electron Lindhard function was given by Mostoller[15] who showed, using a pseudopotential to describe the electron-ion interaction and a free electron dielectric screening, that he could reproduce the shapes of the phonon dispersion curves but not the anomalies. Recently, a phenomenological three body force shell model (TSM) was applied successfully to HfC and TaC by Verma and Gupta[16]. Whereas the parameters in this model cannot be interpreted in an obvious way as "resonance" effects due to the instability of the metallic d charge density, these authors pointed out that the TSM describes the long range ion-ion interaction which was represented in Weber's DSM by a short range screened ion-electron-ion interaction term. It thus appears that the two models describe the same physical phenomenon.

A different approach for describing the longitudinal acoustic branches of NbC is given by Sinha and Harmon[17] in detail elsewhere in this proceedings. The electron-ion interaction is described in terms of a local pseudopotential and using an ansatz, the dielectric screening matrix is factorized and inverted analytically. The parameters entering into this calculation are determined by a fit to the experimental curves. The transverse branches, which were not obtained because the d orbital overlap was neglected as a simplifying assumption, could in principle be included in the model. We have been informed that a calculation by Hanke[18] et al. on NbC, is essentially similar to the work of Sinha and Harmon.

From the previously reported calculations and the physical interpretation of the anomalies as discussed by Weber[12], the role of the dielectric screening appears to be the essential factor in the occurrence of phonon anomalies; this is also in agreement with the idea of "overscreening" developed by Phillips[19]. Due to the complexity of the problem, no ab-initio calculation of the dielectric screening matrix for the TMC has been performed to date.

We have examined the physical information contained in the generalized susceptibility function $\chi(\vec{q})$, which in linear response theory is the response of the conduction electrons to an external pertubation (here, the change in the crystal potential due to the lattice vibrations).[20] Let us assume that the lattice is driven into a vibrational mode defined by the wave vector $\vec{q}$. If, for this particular mode, the response function of the conduction electrons of the system shows a divergence, the screening of this vibrational mode will be rather strong and will lead to softening of the mode. This trend can be seen directly from the mathematical expressions. The phonon frequency ω corresponding to a mode $\vec{q}$ is obtained by solving the well-known secular equation:

$$|| D^{\alpha\beta}_{KK'}(\vec{q}) - M_K \omega^2(\vec{q}) \delta_{KK'} \delta_{\alpha\beta} || = 0 , \qquad (1)$$

where D is the dynamical matrix for the wave vector $\vec{q}$. The indices α, β refer to the coordinates $1 \leq \beta \leq 3$; K, K' are the indices of the atoms, M_K their mass. The dynamical matrix can be written as the sum of two terms, the core-core interaction $\underset{\sim}{C}$ and the electronic contribution $\underset{\sim}{E}$

$$\underset{\sim}{D} = \underset{\sim}{C} + \underset{\sim}{E} \qquad (2)$$

If, for simplicity we represent the electron-ion interaction by a pseudopotential, the electronic contribution to the dynamical matrix can be written as

$$E^{\alpha\beta}_{KK'}(\vec{q}) = \sum_{\vec{G},\vec{G}'} \frac{(\vec{q}+\vec{G})_\alpha (\vec{q}+\vec{G}')_\beta}{V(\vec{q}+\vec{G})} W_K(\vec{q}+\vec{G}) W_{K'}(\vec{q}+\vec{G}') [\varepsilon^{-1}(\vec{q}+\vec{G},\vec{q}+\vec{G}') - \delta_{\vec{G},\vec{G}'}]$$

$$\times e^{i\vec{G}\cdot\vec{r}_K} \cdot e^{-i\vec{G}'\cdot\vec{r}_{K'}} \qquad (3)$$

The summation is carried over the reciprocal lattice vectors $\vec{G}$ and $\vec{G}'$, and $W_K(\vec{q}+\vec{G})$ is the Fourier transform of the electron-ion pseudopotential; $\vec{r}_K$ is the position vector of atom K in the unit cell and the dielectric screening matrix ε can in turn be expressed in terms of the generalized susceptibility matrix χ:

$$\varepsilon(\vec{q}+\vec{G},\vec{q}+\vec{G}') = \delta_{\vec{G},\vec{G}'} + V(\vec{q}+\vec{G})\chi(\vec{q}+\vec{G}, \vec{q}+\vec{G}') \tag{4}$$

$V(\vec{q}+\vec{G})$ is the Fourier transform of the electron-electron interaction (including exchange and correlation).

From these expressions and using simplifying asssumptions, it can be seen that if χ becomes large, the corresponding phonon frequency is expected to be small, as ω is related to χ through the inverse of the dielectric screening matrix.

If $E_{\vec{k}}^n$ and $\psi_{\vec{k}}^n$ are respectively the energy and wavefunction of state $\vec{k}$, band index n and $f_{\vec{k}}^n$ is the corresponding Fermi distribution function, the elements of the generalized susceptibility matrix is given by:

$$\chi(\vec{q}+\vec{G},\vec{q}+\vec{G}') = \sum_{\substack{\vec{k},\vec{k}'\\n,n'}} \frac{f_{\vec{k}}^n - f_{\vec{k}}^{n'}}{E_{\vec{k}'}^{n'} - E_{\vec{k}}^n} \cdot \langle\psi_{\vec{k}}^n | e^{-i(\vec{q}+\vec{r})\cdot\vec{r}} | \psi_{\vec{k}'}^{n'} \rangle$$

$$\langle \psi_{\vec{k}}^{n'} | e^{i(\vec{q}+\vec{G})\cdot\vec{r}} | \psi_{\vec{k}}^n \rangle \tag{5}$$

We have performed an accurate ab-initio determination of the diagonal matrix element $\chi(\vec{q}) = \chi(\vec{q},\vec{q})$ for several of the 9 and 8 valence electrons TMC [NbC, TaC, ZrC, HfC] and have found a striking correlation between the existence of large maxima in the band 4 contribution to $\chi(\vec{q})$ for the 9 valence electron compounds at exactly the $\vec{q}$ values where the phonon anomalies are found experimentally. The generalized susceptibility function of the 8 valence electron compounds is drastically different, does not show any strong maxima, but rather decreases in all symmetry directions from its value at the Γ, the zone center. This result correlates very well with the fact that no anomalies are observed for these compounds and emphasizes the role played by the conduction electron response function in the occurrence of phonon anomalies.

II. ELECTRONIC BAND STRUCTURE, DENSITY OF STATES AND FERMI SURFACES

In order to calculate the energy denominator dependence of Eq.(5), we performed a band structure calculation for NbC and TaC (which crystallize in the rocksalt structure) using the Augmented Plane Wave (APW) method. The crystal charge density was obtained by superposition of the neutral atom charge densities obtained from the Herman-Skillman self-consistent atomic program and including 14 shells of neighbors. The original Slater local exchange potential[21] was used in the determination of the atomic charge densities as well as for the band structure calculation. Expanding the APW wave function in a basis set of 68 plane waves, we obtained a convergence of the order of 1 mRy in the energy eigenvalues. The ab-initio calculation of the valence and conduction band energy eigenvalues was performed at 89 points in the 1/48th irreducible Brillouin zone (BZ). The calculations for both compounds were carried out in an exactly similar fashion and the resulting bands for NbC are plotted in Fig. 1 along some high symmetry directions; similar results were obtained for TaC.[20] The two compounds show a great similarity in their band structure as we can expect from the position of the metal atoms in the same column of the periodic table. In Fig. 1, the zero of the energy refers to the constant value of the potential outside the muffin-tin spheres.

The lowest band shown in Fig. 1 is the C-2s band (with some slight admixture of metal states) and is separated from the next group of levels by a direct gap at L. The next three bands above the C-2s band are essentially C-2p states with an important bonding-metal d contribution (and a slight admixture of higher metal states). In both compounds, the C-2p and metal d bands overlap as can be seen in the Γ to X (Δ) direction. This result is in qualitative agreement with the X-ray emission spectra by Ramqvist et al.[22] and Källne and Pessa[23]. The essential features of the band structure found in this calculation agree with previously reported results by Chadi and Cohen[14], and by Simpson[24] using the semi-self-consistent APW method in the muffin-tin approximation. The main difference between our results and Simpson's is that the latter found a greater overlap between the C-2p and Nb-4d bands. Although a fully self-consistent result would be desirable, it is well known that SCF calculations within the approximation of the muffin-tin potential have to be taken with caution, especially in covalently bonded compounds like NbC as, at each iteration, the charge density is spherically averaged inside the muffin-tin spheres leading to a loss of the directional properties of the bonds at each cycle. Furthermore, the fact that a limited number of points in the BZ were used by Simpson to calculate the charge density at each cycle, can also lead to a substantial loss of accuracy.

A preliminary calculation[25] using a molecular cluster model of the form MC_6 was used to understand some features of the bonding

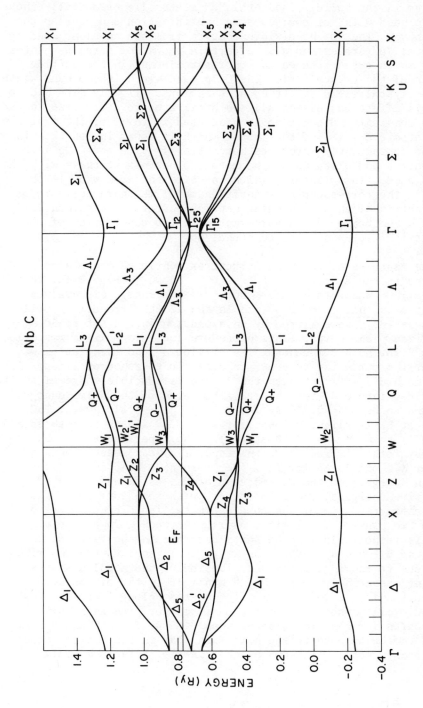

Figure 1: The energy bands of NbC (a=8.4472 au) along several symmetry directions. The zero of energy corresponds to the averaged potential in the interstitial region.

mechanism in the solid. The metal "d" states lie essentially above the C-2p band with, of course, some hybridization and overlap as noted above. The "d" band width results from the covalent bonding C-2p-metal "d" states and also a certain amount of metal d-d interaction, which appears to be an important element in the lattice dynamics of the crystal[12]. The C-2p band width, the d band width and the crystal field splitting at Γ increase slightly from NbC to TaC; this effect can essentially be attributed to a decrease in the lattice constant and can also be predicted in terms of the position of the atomic d level of the metal (which is lower in TaC than in NbC) and consequently interacts more strongly with the C-2p level. The higher s and p states of the metal do not mix strongly with the metal d band. This feature, which is different from what is observed in the corresponding transition metal is characteristic of the formation of an anti-bonding metal s-p states and carbon 2s-2p states in the compound, in agreement with the observation of Mattheiss[26] for a similar compound, NbN.

The features of the band structures mentioned above are reflected in the density of states (DOS) shown in Fig. 2 for TaC; a similar DOS was obtained for NbC.[20] The energy eigenvalues, calculated <u>ab initio</u> at 89 points in the 1/48th irreducible Brillouin zone (BZ), were fitted by an expansion into a set of 50 symmetrized plane waves. This procedure led to a maximum rms error of ~ 2mRy for both compounds. The Brillouin zone integration was then carried out using an accurate analytic tetrahedron linear energy method[27,28] in which the 1/48th irreducible wedge of the BZ was divided into 6144 tetrahedra. The value of the Fermi energy was found to be E_F = 0.770 Ry for NbC and E_F = 0.810 Ry for TaC. The DOS at E_F is slightly lower for TaC (8.67 states of both spin/Ry-unit cell) than for NbC(9.08 states). The experimental values reported by Toth[29] from electronic specific heat measurements are 9.79 for NbC and 10.88 for TaC. The carbon 2s band is well separated in energy from the C_{2p}-metal d bands, as shown in Fig. 2. At positive energies, the C-2p states, with some admixture of d states characteristic of the covalent metalligand bonding, give rise to two peaks. The limited overlap between C-2p and metal d states is responsible for the sharp decrease in the value of the DOS around E ~ 0.6 Ry. Then, as we go up in energy, the DOS rises again, due to the increasing contribution of the metal d states belonging essentially to the "t_{2g}" manifold (with, of course, some contribution from C-2p states). The states belonging to the "e_g" manifold overlap those of the t_{2g} manifold but give an important contribution for energies greater than 0.85 Ry for NbC and 0.9 Ry for TaC and contribute to the next main peak in the DOS of height ~ 35 states/Ry-cell. The DOS plotted in Fig. 2 includes only

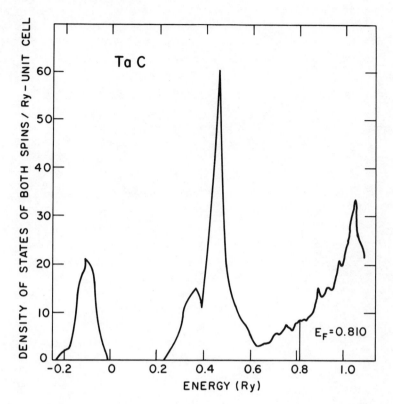

Figure 2: The density of states of TaC for 9 bands (including the C-2s band).

9-valence and conduction bands. For both compounds, the Fermi energy (E_F) falls in the lower part of the metal 4d (5d) bands essentially in the "t_{2g}" band complex. This is in agreement with the results obtained from different band structure calculations[14,24] and confirms that these compounds are essentially d type superconductors, in contradiction with a model proposed by Geballe et al.[30], where the s and p states were assigned the major role at E_F.

It is seen from Fig. 1 that three bands cross the Fermi energy; they belong essentially to the 4d band complex of "t_{2g}" symmetry. If we omit the C-2s band, which is filled and well-separated in energy, the 3 bands crossing the Fermi energy can be

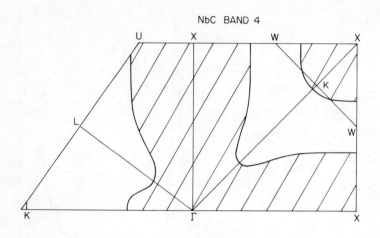

Figure 3(a)

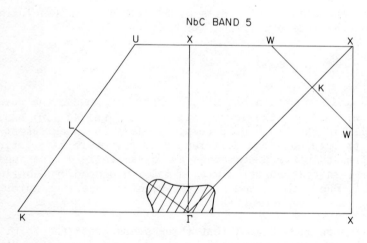

Figure 3(b)

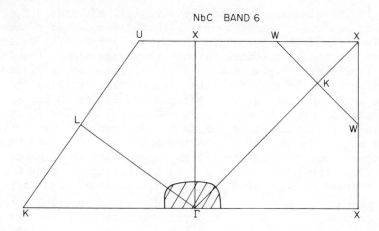

Figure 3(c)

Figure 3: Fermi surface cross sections of NbC.
a) band 4; b) band 5; c) band 6
The bands are labeled omitting the C-2s band. The hatched portions show the occupied regions.

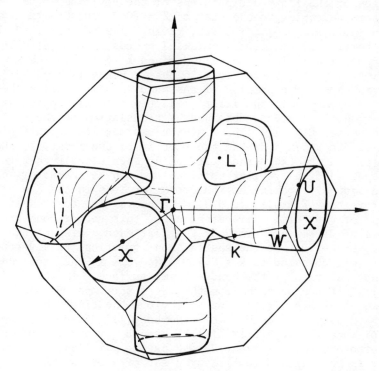

Figure 4: The band 4 Fermi surface of NbC is the first Brillouin zone.

labeled (in increasing values of the energies) as bands 4,5,6. An analysis of the DOS of these bands up to the Fermi energy shows that band 4 plays the dominant role in the Fermi surface (FS) related properties, as it accommodates in both compounds almost the full totality (~96%) of the remaining one "d" electron to be filled. The cross sections of the FS for the three bands crossing the Fermi energy in NbC are drawn in two high symmetry planes in Fig. 3. Similar cross-sections have been obtained for TaC. The FS of bands 5 and 6 can be described as small electron surfaces centered around Γ; band 5 Fermi surface has the shape of a deformed "cube" with stretched and rounded corners as can be seen from the cross sections along the Γ to K and the Γ to L directions. Band 6 gives rise to a "cube" with rounded corners of dimension ~0.32 in $\frac{2\pi}{a}$ units for both NbC and TaC. Since band 4 is of particular interest for both compounds, we sketch its corresponding Fermi surface in Fig. 4 for NbC in the first BZ; it has a similar shape, with slightly larger volume, for TaC. This FS is a multiply connected "jungle-gym" electron surface which can be roughly characterized, away from Γ, as formed by 6 flattened cylinders with axes ΓX. In fact, as is clearly seen in Fig. 3(a), the dimension of the cross sections of each arm of the "cylinder" perpendicular to the ΓX axes decreases from the point X in the BZ to points closer to Γ. The maximum dimension of the flattened portions of the "cylinders" are spanned by vectors belonging to the star of $\vec{k} = \{0.7,0.0, 0.0\}\frac{2\pi}{a}$ for NbC, and $\vec{k} = \{0.72,0.0,0.0\}\frac{2\pi}{a}$ for TaC, giving rise to "nesting" features of the Fermi surface (nearly parallel pieces which "nest" into each other upon translation by a well-defined $\vec{k}$ vector). The minimum dimension of the cross sections in the xy plane is k ~0.55$(\frac{2\pi}{a})$ for both compounds. The portion of the band 4 Fermi surface sketched in Fig. 4 around Γ in the first Brillouin zone is supplemented by another identical piece, obtained by translation by the reciprocal lattice vector $\vec{G} = (1,1,1)\frac{2\pi}{a}$. The cross section of this piece of Fermi surface appears around the K point of the first BZ, shown in Fig. 3(a), in the ΓXW plane. It can be seen clearly that this cross section displays flat portions and can be characterized as a rounded square. Looking at the cross sections in the ΓXW plane, another vector $\vec{k} \sim \{0.3,0.0,0.0\}\frac{2\pi}{a}$ appears also to connect flat portions of the Fermi surface; however, this vector connects portions of <u>different</u> "arms" of the flattened cylinders and consequently, if we look at the spatial configuration, this vector plays a minor role compared to $\vec{k} \sim \{0.7,0.0,0.0\}\frac{2\pi}{a}$ in the occurrence of "nesting" features of the band 4 Fermi surface. In the [110] direction, it can be seen from the cross sections of band 4 in the ΓKL plane [see Fig. 3(a)] that an obvious "nesting" vector is $\vec{k} \sim \{0.55,0.55,0.0\}\frac{2\pi}{a}$. However, in the [111] direction, a <u>continuous range of vectors</u> $0.7|\Gamma L|<|\vec{k}|<|\Gamma L|$ span a given "arm" of the jungle gym; moreover, two different "arms" having parallel axes can be spanned in the [111] direction by a set of vectors such as $0.6|\Gamma L|<|\vec{k}|<|\Gamma L|$.

The general characteristics of the band 4 Fermi surface obtained in this work for NbC agree with the cross sections available in the literature like the ΓXW plane cross sections shown by Chadi and Cohen[14]; the band 4 FS cross sections of Simpson[24] also display similar features to those obtained in the present calculation. This fact shows that the Fermi surface obtained for band 4 is not very sensitive to the atomic configuration used to construct the crystal potential. Recently, similar conclusions were drawn for the Fermi surface of Nb[31] on the basis of a comparison between a non-self-consistent and self-consistent APW band structure calculation. For TaC, no relativistic corrections were included; however, considering that essentially only the band 4 determines the FS and this band crosses the Fermi energy at steep angles, these effects are expected to be small[32].

III. GENERALIZED SUSCEPTIBILITIES

Using the results of the APW band structure calculations, we calculated the generalized susceptibility function $\chi(\vec{q})$ in the constant matrix elements approximation, defined by,

$$\chi(\vec{q}) = -\sum_{\substack{\vec{k} \\ n,n'}} \frac{f_{\vec{k}}^n - f_{\vec{k}+\vec{q}}^{n'}}{E_{\vec{k}}^n - E_{\vec{k}+\vec{q}}^{n'}} , \qquad (6)$$

where all symbols have been previously defined in Eq. (5). As can be seen from Eq. (6), large contributions to $\chi(\vec{q})$ are expected at those values of the wave vector $\vec{q}$ connecting states having nearly the same energy and different occupancies, hence maxima in $\chi(\vec{q})$ are expected when large portions of the FS are spanned by a given vector $\vec{q}$ ("nesting" effect) or when parallel bands near E_F are spanned by a $\vec{q}$ vector (so-called "volume" effect).

The necessity of using a very accurate procedure to calculate $\chi(\vec{q})$ was emphasized by Evenson et al.[33] To carry out the BZ integration, these authors divided the irreducible wedge of the BZ into microcubes and assumed the energy to be constant inside each microzone; they showed that a coarse mesh could give rise to spurious structures (peaks) when computing the well-known analytic Lindhard function for a free electron gas. In order to avoid such inaccuracies, we adapted to the rocksalt structure, the analytic tetrahedron linear energy method derived by Rath and Freeman[34] as an extension of the tetrahedron scheme devised by Jepsen and Andersen[28] and Lehmann et al.[27] for computing density of states. In this method, the irreducible BZ is divided into microzones chosen as tetrahedra. If one assumes, following Gilat-Raubenheimer[35], that the energy can be linearly expanded inside each tetrahedron (in which case the surface of constant energy is a plane), then the contribution to $\chi(\vec{q})$ from each microzone can be obtained from simple <u>analytic</u> expressions involving only the volume of the tetrahedron and the energies at its corners. The accuracy of this method has been discussed at length by Rath and Freeman[34] where model calculations, such as that of the Lindhard function, were found to be in very good agreement with the well-known analytic result. The susceptibility of a "box" energy surface proposed by Fehlner and Loly[36] was also checked successfully in order to test whether the method can also reproduce the effect of very strong "nesting" features. The average deviation of the entire curve was found to be around 0.75%.

In our calculations for the carbides, we also used the analytic tetrahedron integration method[34], with 2048 tetrahedra in the 1/48th BZ. As can be seen from Eq. (6), the intraband contribution (n=n') to the generalized susceptibility function is only non-zero for the 3 bands crossing the Fermi energy. Figs. 5 and 6 display the individual intraband contribution to $\chi(\vec{q})$, from bands 4, 5, and 6 along several symmetry and off-symmetry directions, as well as the total interband contribution in both NbC and TaC. Here $\chi(\vec{q})$ was calculated at 20 equally spaced $\vec{q}$ points for each of the directions plotted in Figs. 5 and 6. In addition, the mesh around each of the maxima was further decreased in order to better define the peaks shown.

The remarkable result of these calculations is the existence of strong structure in the contribution from band 4 to $\chi(\vec{q})$. More specifically, the maxima in $\chi(\vec{q})$, which occur at $\vec{q} = (0.6,0.0,0.0)$; $\vec{q} = (0.55,0.55,0.0)$, $\vec{q} = (0.5,0.5,0.5)$ in $\frac{2\pi}{a}$ units for NbC match exactly with the positions of the dips in the longitudinal acoustic branches of the dispersion curves measured by Smith[2] in the [100], [110] and [111] symmetry directions. For TaC, the position of the maxima in the calculated $\chi(\vec{q})$ band 4 contribution, found at $\vec{q} = (0.63,0.0,0.0)$, $\vec{q} = (0.55,0.55,0.0)$, and $\vec{q}=(0.5,0.5,0.5)$, are also in excellent agreement with experiment for the [110] and [111] directions. Even more remarkable, in the [100] direction, the small shift in the q value at which the anomaly occurs in going from NbC to TaC (q_{th} = 0 60 vs. 0.63) reproduces the trend in the shift observed experimentally (q_{exp} = 0.60 vs. 0.65).

We must also emphasize the large magnitude of the calculated maxima in $\chi(\vec{q})$. For band 4, the peaks in [100], [110] and [111] directions show an increase of 33.3%, 55.2%, 91.7% for NbC and an increase of 32%, 53%, 90% for TaC from the value of the function at q=0 (which is the contribution from band 4 to the DOS at E_F). A striking feature of these results is that the ratios of these maxima also match closely to the magnitude of the depth of the mode softening - as can be seen by calculating $\omega^2(ZrC) - \omega^2(NbC)$ or

ω^2(HfC) - ω^2(TaC) at those $\vec{q}$ values where anomalies occur for NbC or TaC (note that ZrC and HfC do not possess any phonon anomalies). The calculated peaks are broad, as are the dips observed in the dispersion curves.

A possible correlation between "nesting" vectors and phonon anomalies in the [100] and [110] directions has been proposed previously by Chadi and Cohen.[14] However, as mentioned earlier, both "nesting" (FS) and band structure (or "volume") effects contribute to the position of the maxima singled out in the $\chi(\vec{q})$ calculation. From a visual inspection of the geometrical properties of the FS, some features of the $\chi_4(\vec{q})$ contribution can be interpreted: in the [100] direction, the shoulder in $\chi_4(\vec{q})$ observed around $\vec{q} \simeq (0.3,0.0,0.0)$ $\frac{2\pi}{a}$ can be associated with the limited "nesting" properties pointed out previously, between different arms of the flattened cylinders; more significant is the sharp decrease of $\chi_4(\vec{q})$ for $\vec{q} > (0.7,0.0,0.0)$ $\frac{2\pi}{a}$ which is the maximum dimension of the "arms" of the jungle-gym. The exact location of the peak cannot be predicted by simple visual inspection of the FS; as mentioned previously, the dimension of the rounded squares cross sections varies between q = 0.55 to q = 0.7 for NbC and to q = 0.73 for TaC. Only an accurate calculation can remove the uncertainty of a qualitative estimate based on FS properties of a predicted structure in $\chi(\vec{q})$. Along the [110] direction, the peak in $\chi_4(\vec{q})$ at q = (0.55,0.55,0.0) and the drop in $\chi(\vec{q})$ for larger $\vec{q}$ values, are related to the previously mentioned nesting vector q = 0.78 $\simeq 0.55\sqrt{2}$ [see Fig. 3]. The response function of the system contains not only information about FS states ("nesting") but also about band structure effects for states below and above E_F ("volume effects"); it is also obvious that the relative height of the peaks cannot be given at all by FS nesting features even when the "nesting" properties are the dominant factor, as each nesting vector which contributes to a peak must be "weighted" by the portion of surface spanned. For example, in the [111] direction, the sharp increase in $\chi(\vec{q})$ for $0.7|\Gamma L| < q < |\Gamma L|$ can be related to the previously mentioned $\vec{q}$ vectors spanning the Fermi surface in this direction, as well as to band structure contributions, but only the $\chi(\vec{q})$ calculation yields the correct position of the peak.

The intraband contributions to $\chi(\vec{q})$ from bands 5 and 6, plotted separately in Figs. 5 and 6, are small in magnitude and do not show any structure. The corresponding curves can be described as "step" functions, the dimension of the "step" being commensurate with the dimension (q~0.3) of the small electron surfaces discussed previously. In all directions, the value of $\chi_n(\vec{q})$ for (n=5,6) decreases from its value at the zone center Γ. The total interband contribution to $\chi(\vec{q})$ from bands 4, 5, and 6, shown at the top of Figs. 5 and 6, is remarkably structureless in the total (as well as in the individual) interband contributions. It is important in the constant

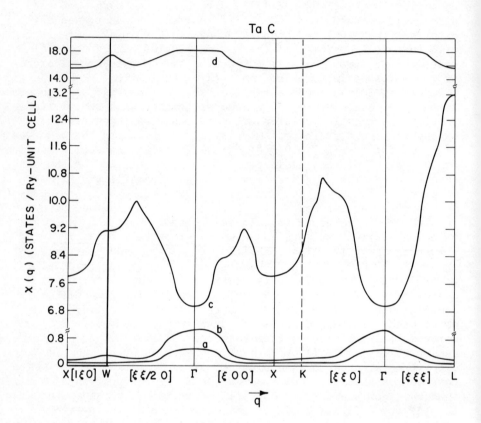

Figure 5: The generalized susceptibility $\chi(\vec{q})$ of NbC along several symmetry and off-symmetry directions:
a) the intraband contribution from band 6;
b) the intraband contribution from band 5;
c) the intraband contribution from band 4 and
d) the total intraband contribution from bands 4, 5, 6.

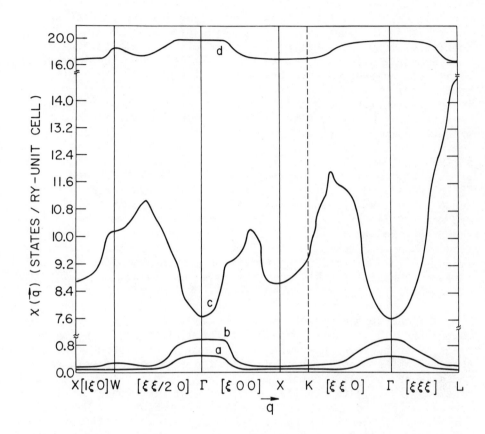

Figure 6: The generalized susceptibility $\chi(\vec{q})$ of TaC along several symmetry and off-symmetry directions:
a) the intraband contribution from band 6;
b) the intraband contribution from band 5;
c) the intraband contribution from band 4; and
d) the total interband contribution from bands 4, 5, 6.

matrix elements (ME) approximation to plot and discuss separately the intra- and inter-band contributions, as we will explain below.

As mentioned in the Introduction, Weber interpreted the origin of the soft phonon modes as a "resonance-like increase of the polarizability"; from the $\vec{q}$ dependence of the dynamical matrix, the $\vec{q}$ values at which this resonance occur are expected to span the surface of a cube centered at Γ and of dimension $\sim \frac{2\pi}{a}$. Slight departures from this soft mode surface with a shift towards the zone boundary were expected in Weber's model, from the introduction of nearest neighbors force constants. This model of a "soft" surface has proven to be credible, as predictions of the soft modes in some directions were followed by experimental confirmation; consequently, we thought to be of interest to check whether the predictions made by Weber could be present in a microscopic model in which the polarizability of the conduction electrons is described in terms of the generalized susceptibility function. We studied an arbitrary off-symmetry direction Γ to W defined by $[\xi,\xi/2,0]$ with $0 \leq \xi \leq 2$. Our results, also presented in Figs. 5 and 6, show that the $\chi_4(\vec{q})$ function has a maximum at $\xi \sim 0.65$ for NbC and TaC in the Γ W direction. Taken together, our calculations show that the maxima of the $\chi_4(\vec{q})$ function lie on the surface of a warped cube in $\vec{q}$ space, centered at Γ and of approximate dimension $1.2(\frac{2\pi}{a})$. This result is in striking agreement with the soft mode surface proposed by Weber starting from an entirely different (an phenomenological) formulation.

The one to one correspondence between the large maxima in the calculated $\chi(\vec{q})$ function and the position of the soft phonon modes leads us to believe that (for the compounds under study) the essential physical phenomenon which is driving the lattice into soft phonon modes is essentially contained in the electronic response function of the system. Although the possibility of divergences in $\chi(\vec{q})$ leading to phonon anomalies was mentioned by Sinha and Harmon [17] for NbC, they made the simplifying assumption that $\chi(\vec{q})$ was constant and equal to the DOS at E_F. Within the limitation of this approximation and assuming that the ansatz used to factorize ϵ is valid, these authors found that the "bell" shape of the Fourier transform of the assumed electron-ion pseudopotential and the magnitude of the DOS at E_F were the essential elements in reproducing the soft phonon modes. We note that our results are not in contradiction With Sinha and Harmon's work, as their calculation could be reparametrized to take into account the structure of $\chi(\vec{q})$.

As a further test of the role played by structure in the response function in the occurrence of phonon anomalies we calculated $\chi(\vec{q})$ for the 8 valence electrons compounds ZrC and HfC, by applying the rigid band model, respectively, to the NbC and TaC energy band results. The Fermi energy of ZrC and HfC falls in a minimum region of the density of states (DOS), $E_F = 0.595$ Ry. and $E_F = 0.630$ Ry. respectively for ZrC and HfC. For the 8 valence electrons compounds, four bands cross E_F (bands 1, 2, 3, 4 omitting the C-2s band); the first three bands have essentially a C-2p character. The corresponding intraband and interband contributions to $\chi(\vec{q})$ are plotted in Figs. 7 and 8 along symmetry directions. Unlike the case of NbC and TaC, we find for ZrC and HfC that both the individual and the total intraband contributions to $\chi(\vec{q})$ decrease (in all directions) from the value at $\chi(\vec{q})=0$. This results in a maximum at Γ and, consequently, in an overscreening at the zone center. These results correlate very well with the experimental results of Smith[2] where the optic modes at Γ in ZrC and HfC have a significantly lower value than in NbC and TaC. The decrease at Γ is from $\sim$ 17 THz to $\sim$ 13 THz from NbC to ZrC. The general decrease of $\chi(\vec{q})$ away from Γ is also consistent with the fact that no anomalies are present in this low T_c compound. We thus conclude from this study of the 9- and 8-valence electrons TMC that the phonom anomalies of these compounds can be explained essentially as being due to an anomalous increase in the response function of the conduction electrons, resulting in a strong screening of the corresponding phonon modes.

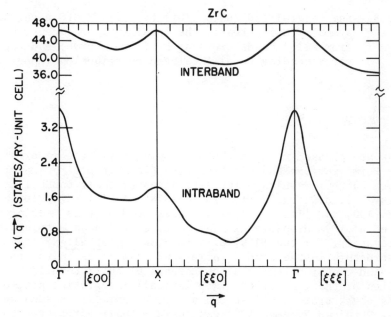

Figure 7: The generalized susceptibility $\chi(\vec{q})$ of ZrC along some symmetry directions: the total intraband contribution from all 4 bands crossing E_F (lower curve) and the corresponding total interband contribution (upper curve).

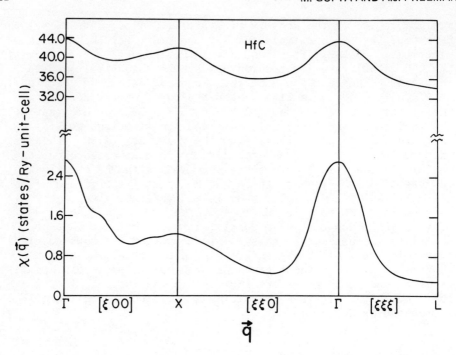

Figure 8: The generalized susceptibility $\chi(\vec{q})$ of HfC along some symmetry directions: the total intraband contribution from all 4 bands crossing E_F (lower curve) and the corresponding total interband contribution (upper curve).

We are pleased to note that following our first reports[20,37] of our Fermi surface and $\chi(\vec{q})$ results, Klein et al.[38] undertook to check out our results using extensions of their unpublished band structure calculations performed in 1974. A $\chi(\vec{q})$ curve for band 4 in the [100] and [111] directions, presented at this meeting by Klein et al.[38] displays similar features to those obtained by us. Unfortunately, for the [110] direction, Klein et al.[38] obtained more structure in the band 4 intra-band as well as in the total inter-band contribution. In the absence of their detailed description of their numerical (statistical) integration procedure and its check against well-known analytic expressions such as were used by Rath and Freeman,[34] we consider their extra structure to be a further demonstration that a lack of numerical accuracy can lead to spurious peaks, as is well known in the literature concerning $\chi(\vec{q})$ calculations.[33,39-41]

IV. ROLE OF THE MATRIX ELEMENTS; THE WANNIER FUNCTION REPRESENTATION

In order to assess the variation of the oscillator strength matrix elements which in Eq. (5) multiply the energy dependent term, let us expand the Bloch function $\psi_{\vec{k}}^n$ in a set of Wannier function a_n

$$\psi_{\vec{k},n}(\vec{r}) = \frac{1}{\sqrt{N}} \sum_{\ell} e^{i\vec{k}\cdot\vec{R}_\ell} \cdot a_n(\vec{r}-\vec{R}_\ell), \qquad (7)$$

where $\vec{R}_\ell$ is the coordinate of site ℓ and N the number of unit cells in the crystal. Conversely, the Wannier function can be written as,

$$a_n(\vec{r}-\vec{R}_\ell) = \frac{1}{\sqrt{N}} \sum_{\vec{k}} e^{-i\vec{k}\cdot\vec{R}_\ell} \cdot \psi_{\vec{k},n}(\vec{r}). \qquad (8)$$

Using the expansion of Eq. (7) in Eq. (5) and performing the lattice summation, Eq. (5) becomes rigorously:

$$\chi(\vec{q}+\vec{G},\vec{q}+\vec{G}') = \sum_{\substack{\vec{k},\vec{k}'\\n,n'}} \frac{f_{\vec{k}}^n - f_{\vec{k}'}^{n'}}{E_{\vec{k}'}^{n'} - E_{\vec{k}}^n} \sum_{\ell\ell'} e^{i\vec{k}\cdot\vec{R}_\ell} \cdot e^{i\vec{k}\cdot\vec{R}_{\ell'}} \cdot$$

$$\langle a_n(\vec{r}-\vec{R}_\ell)|e^{-i(\vec{q}+\vec{G})\cdot\vec{r}}|a_{n'}(\vec{r})\rangle \langle a_{n'}(\vec{r})|e^{i(\vec{q}+\vec{G}')\cdot\vec{r}}|a_n(\vec{r}-\vec{R}_{\ell'})\rangle, \qquad (9)$$

with the condition $\vec{k}' = \vec{k}+\vec{q}+\vec{H}$ (10), $\vec{H}$ being a reciprocal lattice vector which brings $\vec{k}+\vec{q}$ to the 1st BZ. If we now make the approximation of neglecting the overlap between sites ℓ and ℓ' Eq. (9) can be written in a simpler form:

$$\chi(\vec{q}+\vec{G},\vec{q}+\vec{G}') = \sum_{\substack{\vec{k}\\n,n'}} \frac{f_{\vec{k}}^n - f_{\vec{k}+\vec{q}}^{n'}}{E_{\vec{k}+\vec{q}}^{n'} - E_{\vec{k}}^n} \langle a_n(\vec{r})|e^{-i(\vec{q}+\vec{G})\cdot\vec{r}}|a_{n'}(\vec{r})\rangle \cdot$$

$$\langle a_{n'}(\vec{r})| e^{i(\vec{q}+\vec{G}')\cdot\vec{r}}|a_n(\vec{r})\rangle. \qquad (11)$$

In Eq. (11) unlike in Eq. (5), the matrix elements no longer depend upon $\vec{k}$ and $\vec{k}'$, and it is easier to make predictions about their variation, particularly in the $\vec{G}=\vec{G}'=0$ limit (normal processes). We can see that for the element $\chi(\vec{q},\vec{q})$, the energy dependent term will be multiplied by the square of the form factor, which is a slowly varying function of $\vec{q}$. Keeping in mind the orthogonality relations satisfied by the Wannier functions, one can see from Eq. (11) that only the intraband (n=n') part contributes to $\chi(\vec{q},\vec{q})$ in the $\vec{q} \to 0$ limit. Generally speaking, the dominant contribution to $\chi(\vec{q},\vec{q})$ will be derived from the intraband part and the effect of the matrix elements will be to very much suppress the interband part since for these elements $\vec{q}$ is rather small (restricted to the first BZ). The interband part can be expected to play an important role only if there is a very strong peak at some wave vector[39], but this is not the case here. Since normal processes (i.e., terms containing $\vec{G}=\vec{G}'=0$) play a major role in determining the longitudinal branches, we believe that the anomalies observed in the longitudinal acoustic branches can be explained in terms of the behavior of the diagonal elements $\chi(\vec{q},\vec{q})$ of the susceptibility matrix $\chi(\vec{q}+\vec{G}, \vec{q}+\vec{G}')$. Some contribution will, of course, come also from the Umklapp processes as well as from local field corrections ($\vec{G},\vec{G}'\ne 0$).

The possibly important role played by the matrix elements (ME) for several transition metals has been stressed by several authors:[39-41] Gupta and Sinha[39] (for Cr), and Gupta and Freeman[41] (for Sc) make an explicit use of the APW Bloch functions while Cooke et al.[40] (for Nb) calculate the matrix elements in the tight binding approximation. In the case of NbC and TaC, we can see from Figs. 5 and 6 that the interband value (without ME) provides a large background for all $\vec{q}$ values, ranging from ∼16 to 20 states/Ry unit cell; this is about 2 times larger than the DOS at E_F. It is well known that the large value of the interband contribution at $\vec{q}=0$ is zero when ME are included. Moreover, from all previously reported results with non-constant ME,[39-41] it is seen that the background is largely suppressed for all $\vec{q}$ values. [For example, whereas $\chi(\vec{q})$ without ME in Cr ranges from 132 to 140 states/Ry. atom, it never exceeds 9 states/Ry. atom when the ME are included (due to a drastic reduction of the interband value) and so the variation of $\chi(\vec{q})$ (about 8 states/Ry. atom) remains unaffected by the inclusion of ME. Similarly, $\chi(\vec{q})$ without ME for Nb varies between ∼60 to ∼68 states/atom-Ry-spin in the ΓH direction while the result with ME varies between ∼11 (at q=0) to 3.8 states/atom-Ry-spin at H.]

For this reason, and because the interband contribution without ME is remarkably structureless, we do not expect the interband part with ME to give rise to sharp structure. On the other hand, the inclusion of ME into the intraband contribution would modify the results shown in Figs. 5 and 6 by only a slowly varying function (essentially the square of the form factor), as discussed previously. But if we recall the large heights of the peaks emphasized earlier, it seems very improbable that the ME could smear out the strong structure observed without ME, and so we do not expect our general conclusions to be severely affected.

V. CONCLUSION

Using detailed ab-initio APW band structure calculations, the FS of the 9 valence electron TMC: NbC and TaC, were fully investigated and revealed remarkable nesting properties[14,20], which contribute, together with band structure effects, to a strongly $\vec{q}$ dependent generalized susceptibility function of the conduction electrons. The large peaks in the response function correlate exactly with the position of the phonon anomalies observed experimentally and the locus of the maxima [a warped cube of dimension $\sim 1.2(\frac{2\pi}{a})$] in $\chi(\vec{q})$ is in striking agreement with the soft mode surface proposed by Weber.[12] It appears from these results that the phonon anomalies in the TMC can be essentially interpreted as due to an "overscreening effect"[19] originating from a resonance-like increase of the response function of the system. The important role of the diagonal elements $\chi(\vec{q},\vec{q})$ of the susceptibility matrix (which are the only elements calculated here) in the onset of phonon softening was further checked successfully for the 8 valence electron compounds which do not show peaks in $\chi(\vec{q})$ and do not possess any anomalies.

REFERENCES

1. H.G. Smith and W. Gläser, Phys. Rev. Lett. 25, 1611 (1970).
2. H.G. Smith in Superconductivity in d and f Band Metals, ed. by D.H. Douglass (AIP, New York, 1972).
3. B.M. Powell, P. Martel and A.D.B. Woods, Phys. Rev. 171, 727 (1968).
4. B.W. Batterman and C.S. Barret, Phys. Rev. Lett. 13, 390 (1964); L.R. Testardi in Physical Acoustics, Vol. X, 193 (W.P. Mason and R.N. Thurston, Eds., Academic Press, New York, 1973); M.Weger and I.B. Goldberg in Solid State Physics, Vol. 28, 1 (F. Seitz and D. Turnbull, Eds. Academic Press, New York, 1973).
5. L.R. Testardi, Comments Sol. State Phys. 6, 131 (1975).
6. W.L. McMillan, Phys. Rev. 167, 331 (1968).
7. W. Weber, Phys. Rev. B 8, 5093 (1973).
8. B.M. Klein, D. Papaconstantopoulos, Phys. Rev. Lett. 32, 1193 (1974).
9. G.D. Gaspari and B.L. Gyorffy, Phys. Rev. Lett. 28, 801 (1972).
10. C.M. Varma and R.C. Dynes, Bull. Am. Phys. Soc., 259 (March 1976).
11. W. Weber, H. Bilz and U. Schröder, Phys. Rev. Lett. 28, 600 (1972).
12. W. Weber, Phys. Rev. B 8, 5082 (1973).
13. H.G. Smith, Phys. Rev. Lett. 29, 353 (1972).
14. D.J. Chadi and M.L. Cohen, Phys. Rev. B 10, 496 (1974).
15. M. Mostoller, Phys. Rev. B 5, 1260 (1972).
16. M.P. Verma and B.R.K. Gupta, Phys. Rev. B 12, 1314 (1975).

17. S.K. Sinha and B.H. Harmon, Phys. Rev. Lett. 35, 1515 (1975).
18. W. Hanke, J. Hafner and H. Bilz International Conference of Low Lying Vibrational Modes and their Relationship to Superconductivity and Ferroelectricity, San Juan, Puerto Rico 1975 (to be published).
19. J.C. Phillips, Phys. Rev. Lett. 26, 543 (1971).
20. Michèle Gupta and A.J. Freeman, Bull. Am. Phys. Soc., 21, 258 (1976); Phys. Rev. Lett. 1976 submitted; Phys. Lett. 1976 (in press).
21. J.C. Slater, Phys. Rev. 81, 385 (1951).
22. L. Ramqvist, B. Ekstig, E. Källne, E. Noreland and R. Manne, J. Chem. Phys. Solids 32, 149 (1971).
23. E. Källne and M. Pessa, J. Phys. C 8, 1985 (1975).
24. R.W. Simpson, Ms. Sc. Thesis, University of Florida, Gainesville (1969).
25. Michèle Gupta, V.A. Gubanov and D.E. Ellis (to be published).
26. L.F. Mattheiss, Phys. Rev. B 5, 315 (1972).
27. G. Lehmann, P. Rennert, M. Taut and H. Wonn, Phys. Status Solidi 37, K27 (1970); G. Lehmann and M. Taut, ibid. 54, 469 (1972).
28. O. Jepsen and O.K. Andersen, Solid State Comm. 9, 1763 (1971).
29. L.E. Toth, Transition Metal Carbides and Nitrides (Academic, New York, 1971).
30. T.H. Geballe, B.T. Matthias, J.P. Remeika, A.M. Clogston, V.H. Compton, J.P. Maita and H.J. Williams, Phys. 2, 293 (1966).
31. N. Elyashar and D.D. Koelling, Bull. Am. Phys. Soc., 21, 309 (1976).
32. L.F. Mattheiss, Phys. Rev. B 1, 373 (1970).
33. W.E. Evenson, G.S. Fleming and S.H. Liu, Phys. Rev. 178, 930 (1969).
34. J. Rath and A.J. Freeman, Phys. Rev. B 11, 2109 (1975).
35. G. Gilat and L.J. Raubenheimer, Phys. Rev. 144, 390 (1966).
36. W.R. Fehlner and P.D. Lolly, Sol. State Comm. 14, 653 (1974).
37. S.K. Sinha, Naval Research Laboratory Colloquium, January 18 (1976).
38. B. Klein, D. Papaconstantopoulos and L. Boyer (in this Proceedings).
39. R.P. Gupta and S.K. Sinha, Phys. Rev. B 3, 2401 (1971).
40. J.F. Cooke, H.L. Davis and Mark Mostoller, Phys. Rev. B 9, 2485 (1974).
41. R.P. Gupta and A.J. Freeman, Bull. Am. Phys. Soc. 20, 364 (1975) and Phys. Rev. B May 15 (1976).

QUESTIONS AND COMMENTS

C. Varma: For pure niobium, has anybody divided the susceptibility into interband-intraband parts?

M. Gupta: No, there exists a calculation that I mentioned by Cooke, Davis, and Mostoller of the susceptibility of Nb without matrix elements and with matrix elements, but they do not separate the intraband and interband contribution in either of the cases. In their calculation, they find only a small structure in the [100] direction, but which is not at all as important as the large peaks which we find for the transition metal carbides. We are presently recalculating $\chi(\vec{q})$ for Nb using the self-consistent RAPW energy bands of Elyasher and Koelling.
I would like to make comment on what I mentioned earlier. The remarkable features of the Fermi surface of NbC were pointed out by Chadi and Cohen, who tried to establish a correlation between the "nesting" vectors observed in some cross-sections of the Fermi surface and the position of the phonon anomalies. What is new, I believe, in our calculation, is that $\chi(\vec{q})$ contains far more information that a simple inspection of the geometrical properties of the Fermi surface can provide.

CALCULATIONS OF THE SUPERCONDUCTING PROPERTIES OF COMPOUNDS

B.M. Klein, D.A. Papaconstantopoulos* and L.L. Boyer

Naval Research Laboratory

Washington, D.C. 20375

ABSTRACT

Using the results of self-consistent relativistic APW band structure calculations, and the measured phonon spectra, we have calculated the electron-phonon interaction and λ for the compounds: PdD, NbC, TaC, HfC and V_3Si, based on an extension of the theoretical formulation of Gaspari and Gyorffy. We discuss and contrast the nature of superconductivity in these different materials, and give an extended explanation of the relationship between superconductivity, phonon anomalies and chemical bonding (as mirrored in the Fermi surface) of the refractory carbides.

---

* Permanent address: George Mason University, Fairfax, Virginia 22030

I. INTRODUCTION

Since the first Rochester Conference in 1971 there has been considerable theoretical progress in gaining a fundamental understanding of superconductivity in real d- and f-band materials. This has been largely due to the formulation and implementation of methods of doing explicit numerical computations of the parameters that enter into the strong coupling version of the BCS theory.[1-4]

The theory of the electron-phonon (e-p) interaction as formulated by the Bristol group[4] has the important virtue of being in a form in which the results of a band structure calculation are the key ingredients. Although there may be problems with non-muffin-tin corrections, potential discontinuities, and perhaps some subtle many body effects, it seems clear at this stage that both a good qualitative and quantitative theoretical understanding of superconductivity in rather complicated materials is available. It is also important to keep in mind that superconductivity can be correlated with crystal bonding via the calculated band structure.

In this paper we will summarize the calculations of the group at the Naval Research Laboratory on three interesting types of materials. While some of these results have been reported previously we hope that the reader will find it useful and interesting to contrast our results for these different compounds. All of our results are based on relativistic self-consistent augmented plane wave (APW) calculations. For most materials self-consistency is an important part of achieving an accurate representation of the energy bands, and hence of the superconducting properties. This is especially true for compounds where the possibility of charge transfer makes guessing at the actual crystal charge density, as is done in non-self-consistent calculations, extremely unreliable in general.

The paper is organized into six sections. In Section II we describe the theory that we use to do our superconductivity calculations. We then discuss results for the palladium hydrogen system (Section III); the refractory carbides (Section IV), including a somewhat extensive discussion of superconductivity and e-p scattering with regard to the unusual Fermiology of these materials; and V_3Si (Section V). We then summarize our results in Section VI.

II. FORMALISM

Gaspari and Gyorffy (GG)[4] have derived a formalism for calculating $\eta = n(E_F)<I^2>$ (quantities defined below), in the rigid-ion, muffin-tin approximation, which involves scattering phase shifts and angular momentum decomposed densities of states. Klein and Papaconstantopoulos

SUPERCONDUCTING PROPERTIES

have recently proposed an extension of the GG equations for compounds which takes the form of the following set of equations:

$$\lambda = \sum_{s=1}^{N_s} \lambda_s \qquad \lambda_s = 2 \int_0^\infty (\alpha^2 F(\omega)/\omega) \, d\omega \qquad (1a)$$

$$\lambda_s = \frac{n \langle I_s^2 \rangle}{M_s \overline{\omega_s^2}} \qquad (1b)$$

$$\overline{\omega_s^2} = \langle \omega \rangle_s / \langle \tfrac{1}{\omega} \rangle_s \qquad (1c)$$

$$\langle \alpha_s^2 \rangle = \frac{n \langle I_s^2 \rangle}{2 M_s \langle \omega \rangle_s} \qquad (1d)$$

$$\langle \omega^m \rangle_s = \int_0^\infty \omega^m F_s(\omega) \, d\omega \qquad (1e)$$

$$n \langle I_s^2 \rangle = \frac{E_F}{\pi^2 n(E_F)} \sum_{\ell=0}^{2} \frac{2(\ell+1) \sin^2(\delta_{\ell+1,s} - \delta_{\ell,s}) n_{\ell,s} n_{\ell+1,s}}{n_{\ell,s}^{(1)} n_{\ell+1,s}^{(1)}} \qquad (1f)$$

$$F_s(\omega) = \frac{1}{3N} \sum_{\vec{q}\nu} | \vec{e}(\vec{q}|\nu s) |^2 \delta(\omega) - \omega_{\vec{q},\nu}) \qquad (1g)$$

Notation:

$n(E_F)$: density of states (DOS) at the Fermi energy, E_F.
s, N_S: s is one of the N_S different sites in the unit cell.
$F_S(\omega)$: site decomposed phonon DOS
$\vec{e}(\vec{q}|\nu s)$: phonon polarization vector (eigenvector of the dynamical matrix) a wavevector $\vec{q}$, branch ν, site s.
$n_{\ell,s}$: angular momentum and site decomposed DOS at E_F.
$\delta_{\ell,s}$: scattering phase shifts at E_F.
$n_{\ell,s}^{(1)}$: "free scatterer" DOS (see Ref. 4)
M_s: atomic mass
$\alpha^2(\omega) F(\omega)$: tunneling DOS
N: number of unit cells

This set of equations is a particular generalization for compounds of the "constant α^2" approximation.[3,4] It is shown in Ref. 5 that for binary material ($N_S = 2$) for which $M_1 \gg M_2$, $\overline{\omega_1^2}$ and $\overline{\omega_2^2}$ are equivalent to averages over the acoustic and optic modes respectively to order $(M_2/M_1)^{1/2}$. In this sense one may speak of atom 1 as giving an acoustic mode and atom 2 an optic mode contribution to λ.

For compounds in general, we feel that Eqs. (1) represent a good starting point for discussing the superconducting properties. However, Reitschel[6] has shown that <u>intersite</u> contributions are formally non-negligible; and also that λ depends implicitly on the atomic masses only through the inverse force constant matrix. This latter point is consistent with our formulation since $\{M_s \omega_s^2\}$ are combinations of such force constants. The decomposition of λ into separate <u>intrasite</u> contributions is very convenient and revealing for analyzing the physics of superconductivity for compounds, as we discuss below.

We note that "off-diagonal" correction terms to Eq. (1f) discussed by Butler, et al.[7] have not been included in our calculations. For the monatomic materials that they studied, these corrections are quite small, and we assume that this prevails for compounds. Both the effects of non-muffin-tin corrections which have been discussed by us previously for the pure transition metals,[8] and the corrections due to the discontinuity at the muffin-tin radius of the potential, discussed by Harmon and Sinha at this Conference have not been incorporated in our calculations for compounds. At this stage in the development of the theory, it is not clear to us whether or not these corrections are more properly included or left out of the calculations.

Making use of our calculated values of λ and $\langle\alpha^2\rangle$, and the measured phonon spectra, $F(\omega)$, T_c can be calculated using one of the McMillan type equations[4,9] if we assume a value of μ^*. Since for most superconductors, μ^* is not known to high precision, we take the following approach: we calculate the value of μ^* which, when used with our λ calculation gives the measured T_c, and call this value μ^* (implied). A value of $\mu^* \sim 0.1$ is generally considered reasonable, and that is what we find. Until a first-principles theory of μ^* is implemented, this is as far as one can go.

III. PALLADIUM HYDROGEN SYSTEM

The palladium hydrogen systems, PdH_x and PdD_x, form rocksalt structure crystals and are superconducting[10] for $x \gtrsim 0.8$. The maximum T_c's occur when $x = 1.0$, and are approximately $9°$ K and $11°$ K

TABLE I

Calculated results for Pd and PdD.

	Pd	PdD Pd	PdD D
$n(E_F)$ (states/Ry/spin/unit cell)	15.1	3.3	
$n<I^2>$ (eV/Å^2)	3.59	0.865	0.392
$<\alpha^2>$ (°K)	39.7	12.0	109.8
$M \overline{\omega^2}$ (eV/Å^2)	7.55[a]	6.01[b]	0.865[b]
λ	0.47	0.144	0.453
		0.597	
μ^* (implied)	$\gtrsim$0.3 for no superconductivity	~ 0.08	

Phonon moments from measurements of:

[a] A.P. Miller and B.N. Brockhouse, Can. J. Phys. 49, 704 (1971).
[b] Ref. 16

for PdH and PdD respectively.[11] These materials are especially interesting due to the fact that the host metal Pd, is itself not a superconductor, and also because of the possibility of even higher T_c's in other hydride systems. In fact, Stritzker and Buckel have already found T_c's in excess of 15°K in several palladium-noble metal-hydrogen systems,[12] and Satterthwaite and Toepke[13] have reported superconductivity in Th_4H_{15} and Th_4D_{15}. The inverse isotope effect in the palladium hydrogen system has not been quantitatively explained, although there is the general opinion that it is related to the large anharmonicity associated with the hydrogen or deuterium sites.[14,15]

Here we would like to report on calculations of the superconducting properties of PdD, based on our self-consistent relativistic APW results, and the phonon spectrum of $PdD_{0.63}$ measured by Rowe, et al.[16] Since $(M_D/M_{Pd}) \sim 0.05$, one can consider λ_{Pd} and λ_D as acoustic and optic mode contributions as we have discussed in Section II. Table I shows some of the results of our calculations for PdD and Pd metal for comparison. The values of $n \langle I^2 \rangle$ and λ for PdD are based on an improved DOS calculation, and are smaller by approximately 30% from our previously reported results.[17] The new DOS results were obtained by fitting the bands to 89 first-principles APW $\vec{k}$-points in the 1/48 irreducible zone using the QUAD scheme,[18] versus fitting only 20 $\vec{k}$-points in our previous results. The rms fitting error has decreased from ~ 25 mRy to ~ 7 mRy near E_F, and our new $n(E_F)$ and $n_\ell^{Pd,D}(E_F)$ are uniformly ~ 20% smaller. We are gratified that our more accurate values of $n(E_F)$ and λ are now in good agreement with the measured specific heat coefficient γ.[19] From Table I we calculate the value of $\gamma \propto n(E_F)(1+\lambda)$ to be 1.83, while the measurements yield 1.72 ± 0.2 (mJ/mole-K^2),[19] based on a linear extrapolation of the results from $PdH_{0.88}$.[19] We also note that our previous qualitative conclusions regarding the physics of the situation remain mainly unaltered.[17] Switendick has also done similar calculations which support our conclusions.[20]

Table I shows that $n(E_F)$ and $n \langle I^2 \rangle$ for PdD are much smaller than in Pd metal. Deuterium makes a large contribution to λ in PdD primarily due to the small value of $M_D \overline{\omega_D^2}$ - the optic mode "forces" are soft by about a factor of 7 compared to those of Pd in PdD. This means that the rms displacements of deuterium (or hydrogen) are very large ($\gtrsim 10$%) in the PdD lattice. This anharmonicity is in large measure responsible for superconductivity in PdH(D),[13,14,17] as first suggested by Ganguly,[14] and is probably related to the inverse isotope effect. We also find that since $\lambda = 0.47$ for Pd metal, there must be a large value of $\mu^* \gtrsim 0.30$ which suppresses superconductivity.[17,21]

IV. REFRACTORY CARBIDES

The refractory carbides and nitrides are notable for their extreme hardness, high melting points and the high T_c's of several of them.[22] We denote a particular rocksalt structure carbide as MC, where M is a metal atom from group IV (e.g. Zr of Hf), or group V (e.g. Nb or Ta). There has been considerable theoretical progress made on these materials since Phillips'[23] discussion at the first Rochester Meeting. Here we mainly concentrate on discussing our explicit calculations for NbC, TaC and HfC.

The most striking feature of the carbides relevant to their low-temperature properties is the observation that those with M from group V are relatively high T_c superconductors, and have anomalies in their phonon spectra (dips in phonon dispersion curves).[24] Both of these properties are absent from the group IV carbides. To study these phenomena, we have done self-consistent relativistic APW calculations for the three carbides, and used the results in calculations of the e-p interactions, Fermi surfaces and generalized susceptibilities.

Figs. 1 and 2 show the electronic DOS's of TaC and HfC respectively. The DOS of NbC, which is not shown, is very similar to TaC. Table II shows the results of the calculations of the superconducting parameters of these materials. It should be noted that the results in Table II for NbC and TaC differ by ~ 10% from our previously reported values of $n<I^2>$ and λ.[25] As in the case of PdD, this is due to an improvement in our DOS calculations (89 $\vec{k}$-point mesh versus the earlier 20 $\vec{k}$-point mesh); and we also note that our earlier results for λ were calculated using the decomposition of λ into site contributions proposed by Phillips.[23] Our present results also include relativistic corrections for NbC. Schwarz and Weinberger[26] have recently reported e-p interaction calculations for several carbides and nitrides using an approximate version of the GG^4 theory.

From Fig. 1 we see that at E_F, the DOS contains significant contributions from both the Ta (Nb) d-states and the carbon 2p-states (in ration ~ 2:1). This fact, along with the calculated charge transfer of a few tenths of an electron from the M atom to carbon shows that ionic-like bonding is very important in these materials. From Fig. 2 and Table II we see that HFC is not a superconductor because of the low value of $n(E_F)$, which leads to a small value of $n<I^2>$. This is likely a general property of all of the group IV carbides.[25]

We note that compared to the pure transition metals, the carbides are relatively low DOS materials. Both NbC and TaC have $n(E_F)$ values of about one-half that of the pure M elements, yet T_c is considerably enhanced over the pure metals, especially for TaC. The values of $<I^2>$ are also substantially larger than their pure metal counterparts.

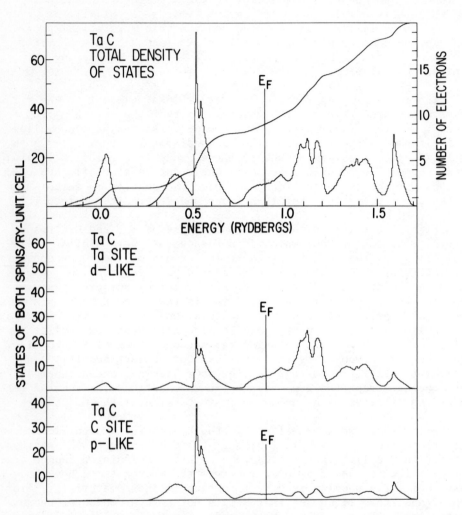

Figure 1: Electronic density of states results for TaC. The bottom two panels are the two largest site-angular momentum-decomposed densities of states near E_F.

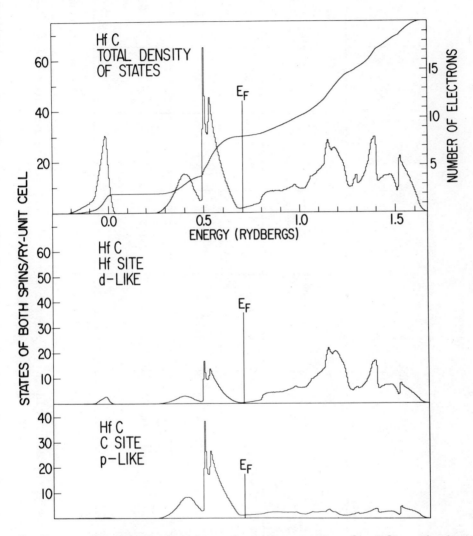

Figure 2: Electronic density of states results for HfC. The bottom two panels are the two largest site-angular-momentum-decomposed densities of states near E_F.

TABLE II

Calculated results for NbC, TaC and HfC.

	NbC		TaC		HfC	
	Nb	C	Ta	C	Hf	C
$n(E_F)$ (states/Ry/spin/unit cell)	4.76		4.33		0.92	
$n\langle I^2\rangle$ (eV/Å²)	4.85	2.99	4.82	2.92	0.19	0.33
$\langle\alpha^2\rangle$ (°K)	50.5	83.4	35.9	76.9	1.3	10.0
$M\overline{\omega^2}$ [a] (eV/Å)	12.87	14.38	12.92	16.22	15.66	12.45
λ	0.38	0.21	0.37	0.18	0.01	0.03
	0.59		0.55		0.04	
μ^* (implied)	0.09		0.06		not superconducting	

[a] Phonon moments computed by J.L. Feldman (unpublished for his fit to the measures spectra (Ref. 24)

We find for both NbC and TaC that $(\lambda_M/\lambda_C) \sim 2$. This is the same as the ratio of metal-d to carbon-p charge components contained in $n(E_F)$ (that is n_d^M/n_p^C at E_F), and indicated that the strong hybridization in the carbides results in a significant contribution to λ from the carbon atom, which enhances T_c. However, the metal atom contribution does dominate, in distinction to the situation in the palladium hydrogen system. We draw a further distinction between these isostructural systems by noticing that for the carbides, $[M\omega^2]_M \approx [M\omega^2]_C$ while for PdD, $[M\omega^2]_{Pd} = 7[M\omega^2]_D$.

To gain some insight into the origins of the large e-p interaction and the occurrence of the phonon anomalies in NbC and TaC, we have done accurate Fermi surface studies of these compounds.[27] Fig. 3 shows the 5th band Fermi surface of NbC or TaC as calculated from first principles APW points (band 1 is the carbon 2s band which, as seen in Fig. 1, is ≈ 10 eV below E_F). This "jungle gym" (JG) surface enclosing slightly less than one electron, has amazingly flat parallel surfaces, with the deviations from parallelism outside the dashed lines shown on Fig. 3 being remarkably small, except at the rounded corners. These "nesting" features of the Fermi surfaces of NbC and TaC result in there being a large amplitude of e-p scattering from one parallel sheet to another, as exemplified by the prototype $\vec{q}$-vector shown on Fig. 3. These $\vec{q}$-vectors are close to the location of the observed phonon anomalies[24] in the (100), (110), and (111) directions.[27] We believe that the sharp Fermi surfaces of NbC and TaC are the fundamental cause of the large e-p interaction, high T_c and phonon anomalies in these materials. In HfC, where both T_c and phonon anomalies are absent, our calculated Fermi surfaces have none of the nesting features shown in Fig. 3.[27] We note that Chadi and Cohen[28] have done nonlocal-pseudopotential calculations of energy bands and Fermi surface of NbC (non-self-consistent and non-relativistic), and they discussed the relationship between the dimensions of their calculated Fermi surface and the phonon anomalies. However, there are serious disagreements between their band structure and ours, with the cross-sections of their JG surfaces having considerable curvature.

It is interesting to see how the sharp JG features are manifest in the generalized susceptibility function which enters into the phonon theory.[29] Although the ab initio phonon theories are rather complicated functions of the electronic structure, it is expected that a strong wavevector dependence of the susceptibility indicates the presences of phonon anomalies. Neglecting for the moment the wavefunction matrix elements, we write the susceptibility as,

$$\chi_{nn'}(\vec{q}) = -\sum_{\vec{k}} \frac{f_{\vec{k},n} - f_{\vec{k}+\vec{q},n'}}{E_{\vec{k},n} - E_{\vec{k}+\vec{q},n'}}, \qquad (2)$$

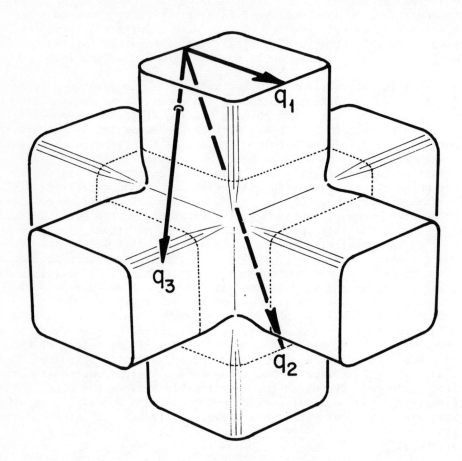

Figure 3: The fifth band Fermi surface of NbC or TaC showing the uniform planar arms pointing in the (100) directions, and containing occupied electron states. $\vec{q}_1$, $\vec{q}_2$ and $\vec{q}_3$ correspond to electron-phonon scattering which are related to the phonon anomalies in these materials, as described in the text. The dashed lines indicate the limits of the very flat surfaces.

where (n,n') are band indices, $f_{\vec{k},n}$ is a Fermi-Dirac function, $E_{\vec{k},n}$ is a band energy, and $\vec{q}$ is a phonon wavevector. From Fig. 3 we anticipate that especially interesting structure in $\chi_{nn'}(\vec{q})$ is expected to show up for intraband scattering for which $n = n' = 5$. We have done two different susceptibility calculations: 1) using an interpolation scheme fit (QUAD[18]) to our self-consistent APW energy bands, which, as expected, to some extent smears out part of the sharp Fermi surface structure; and 2) using a model analytic band structure which exactly reproduces the nesting features of the band 5 constant energy surfaces shown in Fig. 3. Gupta and Freeman[30] have also done susceptibility calculations based on interpolation fits to their energy bands. Although their APW calculations are non-self-consistent and non-relativistic, their Fermi surfaces of NbC and TaC have the same general shapes as do ours, and so do their calculated susceptibilities. Our susceptibility calculations were performed by evaluating the sum in Eq. (2) directly, with the $\vec{k}$-vectors selected randomly in the first Brillouin zone. We chose a sufficient number of $\vec{k}$-points ($\sim 50{,}000$) to ensure convergence to $\sim 1\%$.

In the bottom part of Fig. 4, the peaks at $\vec{q} = (0.3, 0, 0)$ and $\vec{q} = (0.3, 0.3, 0)$ are due to scattering between the two nested JGs, while the peaks at $(0.7, 0, 0)$ and $(0.7, 0.7, 0)$ are due to intra JG scattering ($\vec{q}_1$ and $\vec{q}_2$ of Fig. 3). For a JG width of 0.5 Γ-X (top of Fig. 4) the inter and intra JG nesting dimensions are the same ($q^i = 0.5$), and the sets of two peaks coalesce to one. In the (111) direction there is always only one nesting peak in the first Brillouin zone due to the intersection of the Brillouin zone boundary at $\vec{q} = (0.5, 0.5, 0.5)$, the L point. The model calculations verify that sharp nesting features will be manifest in sharp structure in the intraband $\chi(\vec{q})$; and in addition, show that the geometry of the Fermi surface JG arms alone indicates where the anomalies in $\chi(\vec{q})$ due to nesting will occur, as we have discussed previously.[27]

Results of calculations of $\chi(\vec{q})$ for TaC are shown in Fig. 5. Here $E_{\vec{k},n}$ was determined using a QUAD[18] fit to 89 APW energies per band in the 1/48 irreducible Brillouin zone. The upper curve, labeled interband, represents the total interband 5, 6, 7 (all partially occupied) contributions. Guided by our model calculations, or as we have shown, equivalently by the geometry of the Fermi surface of TaC, we interpret the peaks at $\sim (0.3, 0, 0)$ and $(0.7, 0, 0)$ and the shoulders at $(0.3, 0.3, 0)$, $(0.7, 0.7, 0)$ and $(0.3, 0.3, 0.3)$ in $\chi_{55}(\vec{q})$ in Fig. 5 as due to the JG nesting. A more accurate fit to the APW band structure is likely to make this structure more pronounced. It is also worthwhile to note that our $\chi(\vec{q})$ results for HfC (not shown) are essentially structureless.

It is difficult at this stage to draw any further detailed conclusions from the $\chi(\vec{q})$ results. The interband contributions are certainly large and to some extent anti-correlate with the intraband

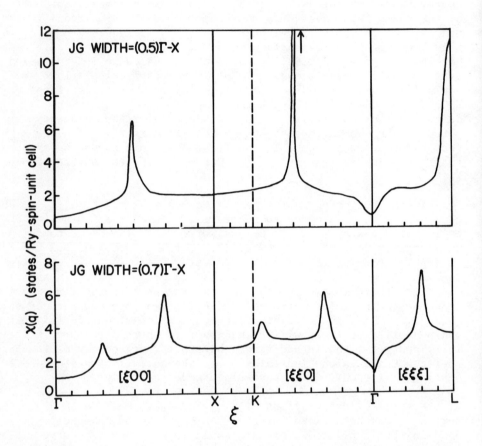

Figure 4: Results of model $\chi(\vec{q})$ calculations which shows peaks corresponding to scatterings between nested portions of a Fermi surface such as the one shown in Figure 3.

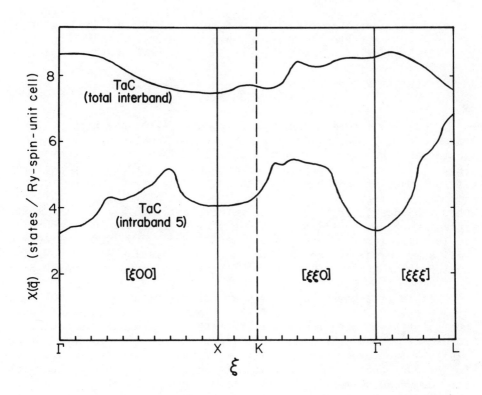

Figure 5: Results for $\chi_{nn'}(\vec{q})$ for TaC calculated from an interpolation fit to the APW energy bands. The curve labeled interband corresponds to the total interband contribution from scatterings between the three partially occupied bands. Matrix elements have not been included in these calculations.

contribution. Furthermore it is important to study in detail the $\vec{q}$ dependence of the matrix elements which have so far been left out of the $\chi(\vec{q})$ calculations for the carbides by both ourselves and Gupta and Freeman.[30] This may be a serious omission since, for instance, the interband contribution rigorously must vanish at $\vec{q} = 0$ with the inclusion of matrix elements. What is really needed is the implementation of a full phonon spectrum calculation for transition metals and transition-metal compounds using band theory results. Undoubtedly though, the sharp Fermi surfaces of NbC and TaC, and the resultant strong e-p scattering amplitudes, are an important factor in the unusual properties of these materials.

V. A15 MATERIALS: V_3Si

The A-15 materials include compounds having the highest T_c's now known (e.g. Nb_3Ge, Nb_3Sn and V_3Si). Although this fact alone would make them a very important class of materials to study theoretically, the occurrence of martensitic structural phase transitions in V_3Si and Nb_3Sn, and the attendant anomalous low-temperature electrical, magnetic and elastic properties of these materials further enhances their theoretical interest. Since all but the most recent work on the A15 materials has been discussed in three recent review articles,[31-33] and in a recent paper by two of us,[34] here we will only summarize the results of our calculations on the e-p interaction. The APW band structure of V_3Si is discussed in Refs. 34-36.

Table III shows the results of our calculations for V_3Si which are described in detail in Ref. 34. The most striking, and perhaps not surprising thing to note is that the Si sites have essentially zero $n<I^2>$, and make almost no direct contribution to superconductivity. The large value of $n<I^2>$ for the V sites is primarily due to the fact that $n(E_F)$ per V atom is approximately twice the value for bcc vanadium metal. Perhaps the most enlightening way to view the A15 structure is to picture it as enhancing the pure transition metal properties, at least as far as superconductivity is concerned. We note that the phonon DOS has only been measured[37] for the acoustic mode region, and the phonon moments used to calculate λ and T_c were calculated using this data.[38] We therefore believe that our calculated λ is uncertain by $\sim \pm 25\%$.

VI. CONCLUSIONS

In this paper we have elucidated the superconducting properties of three different sets of compounds, having rather different theoretical explanations. The hydride systems behave as "anharmonic" or optic mode superconductors; while in the isostructural refractory carbide

TABLE III

Calculated results for V_3Si.

	V_3Si	
	V	Si
$n(E_F)$ (states/Ry/spin/unit cell)	120.1	
$n \langle I_o^2 \rangle$ (eV/Å²)	9.5	0.04
$\overline{M \langle \omega_o^2 \rangle}$ ᵃ (eV/Å²)	7.4	-
λ	1.28	∼0.0
μ^* (implied)	∼ 0.16	

[a] Phonon moments computed by J. Hui[38] from measurements of P. Schweiss.[37]

superconductivity is mainly associated with the heavy metal site, or acoustic phonon mode coupling. Hybridization of the metal-d with the carbon-p states is, however, important in these latter materials, with the carbon site making a smaller, but significant contribution to λ. As far as superconductivity is concerned in the A15 structure (if V_3Si can be considered typical), the vanadium atom (or A site) electron-phonon scattering is totally dominant. These materials may be called "enhanced transition metal superconductors".

Now that it appears that band theory, or one-electron theory based calculations of superconductivity in compounds can give good quantitative estimates and physical explanations of the relevant physical parameters, what is the next step? With the development of reliable ab initio phonon spectra calculations for transition metals over the next several years, perhaps it is reasonable to expect some theoretical predictions with regard to higher T_c materials at the 3rd Rochester Conference.

ACKNOWLEDGMENTS

We wish to acknowledge helpful discussions with E.M. Forgan, B.L. Gyorffy, L.F. Mattheiss and A.C. Switendick. We also thank L.S. Birks and D.J. Nagel for a critical reading of the manuscript. Work on PdH at George Mason University was supported by NASA through Subcontract No. 953918 from the Jet Propulsion Laboratory.

REFERENCES

1. J. Bardeen, L.N. Cooper and J.R. Schrieffer, Phys. Rev. 106, 162 (1957); 108 1175 (1957).
2. G.M. Eliashberg, Zh. Eksperim.; Teor. Fiz. 39, 966 (1960); 39, 1437 (1960); English trans. Sov. Phys. JETP 11, 696 (1960); 12, 1000 (1961).
3. W.L. McMillan, Phys. Rev. 167, 331 (1968).
4. G.D. Gaspari and B.L. Gyorffy, Phys. Rev. Lett. 29, 801 (1972); R. Evans, G.D. Gaspari and B.L. Gyorffy, J. Phys. F: Metal Phys. 3, 39 (1973).
5. B.M. Klein and D.A. Papaconstantopoulos, J. Phys. F: Metal Phys. (in press, 1976).
6. H. Rietschel, Z. Physik B22, 133 (1975).
7. W.H. Butler, J.J. Olson, J.S. Faulkner and B.L. Gyorffy (to be published).
8. L.L. Boyer, B.M. Klein and D.A. Papaconstantopoulos (to be published in the Proc. of the Conference on Low Lying Vibrational Modes and Their Relationship to Superconductivity and Ferroelectricity, San Juan, Puerto Rico, 1975).
9. R.C. Dynes, Solid State Commun. 10, 615 (1972); P.B. Allen and R.C. Dynes, Phys. Rev. B12, 905 (1975).

10. T. Skoskiewicz, Phys. Stat. Sol. (a) 11, K123 (1972).
11. B. Stritzker and W. Buckel, Z. Physik 257, 1 (1972).
12. W. Buckel and B. Stritzker, Phys. Lett. 43A 403 (1073); B. Stritzker Z. Physik 268, 261 (1974).
13. C.B. Satterthwaite and I. Toepke, Phys. Rev. Lett. 25, 741 (1970).
14. B.N. Ganguly, Z. Physik 265, 433 (1973); B 22, 127 (1975).
15. R.J. Miller and C.B. Satterthwaite, Phys. Rev. Lett. 34, 144 (1975).
16. J.M. Rowe, J.J. Rush, H.G. Smith, M. Mostoller and H.E. Flotow, Phys. Rev. Lett. 33, 1297 (1974).
17. D.A. Papaconstantopoulos and B.M. Klein, Phys. Rev. Lett. 35, 110 (1975); B.M. Klein and D.A. Papaconstantopoulos, Proceedings of the 14th International Conference on Low Temperature Physics, M. Krusios and M. Vuorio, eds. (North Holland Publishing Co., Amsterdam, 1975), Vol. 2, pp. 399-402.
18. F.M. Mueller, J.W. Garland, M.H. Cohen and K.H. Bennemann, Ann. Phys. (New York) 67, 379 (1976).
19. C.A. Mackliet, D.J. Gillespie and A.I. Schindler, J. Phys. Chem. Solids 37, 379 (1976).
20. A.C. Switendick, Bull. Am. Phys. Soc. 20, 420 (1975); and private communication.
21. N.F. Beck and J.R. Schrieffer, Phys. Rev. Lett. 17, 433 (1966); S. Doniach and S. Englesherg, Phys. Rev. Lett. 18, 554 (1967).
22. L.E. Toth, Transition Metal Carbides and Nitrides, (Academic Pr Press, New York, 1971).
23. J.C. Phillips, in Superconductivity in d- and f-Band Metals, AIP Conference Proceedings No. 4, edited by D.H. Douglass, Jr. (American Institute of Physics, New York, 1972).
24. H.G. Smith and W. Glaser, Phys. Rev. Lett. 25, 611 (1970), and Phonons, Proc. of the International Conference on Phonons, Rennes, France, edited by M. Nusimovici (Flammarion, Paris, 1972); H.G. Smith, in Superconductivity in d- and f-band Metals, ibid.
25. B.M. Klein and D.A. Papaconstantopoulos, Phys. Rev. Lett. 32, 1193 (1974).
26. K. Schwarz and P. Weinberger, J/ Phys. C: Solid State Phys. 8, L573 (1975).
27. B.M. Klein, L.L. Boyer and D.A. Papaconstantopoulos (unpubl.)
28. P.J. Caadi and M.L. Cohen, Phys. Rev. B. 10, 496 (1974).
29. See for instance: S.K. Sinha, Phys. Rev. 169, 477 (1968); W. Hanke, Phys. Rev. B8, 4585 (1973).
30. M. Gupta and A.J. Freeman (unpublished).
31. M. Weger and I.B. Goldberg, in Solid State Physics, edited by H. Ehrenreich, F. Seitz and D. Turnbull (Academic Press, NY, 1973), Vol 28.
32. L.R. Testadri, in Physical Acoustics, edited by W.P. Mason and R.N. Thurston (Academic Press, NY 1973), Vol X.

33. Yu. A. Izyumov and Z.Z. Kurmaev, Usp. Fiz. Nauk, 113, 193 (1974); English Trans: Sov. Phys. Usp. 17, 356 (1974).
34. B.M. Klein and D.A. Papaconstantopoulos (to be published in the Proceedings of the Conference on Low Lying Vibrational Modes and Tehir Relationship to Superconductivity and Ferroelectricity, San Juan, Puerto Rico, 1975).
35. L.F. Mattheiss, Phys. Rev. 138, A112 (1965); B.12, 2161 (1975).
36. B.M. Klein, L. F. Mattheiss and D.A. Papaconstantopoulos (unpublished).
37. P. Schweiss (unpublished)
38. J. Hui (private communication).

QUESTIONS AND COMMENTS

B. Matthias: Now that μ^* is negative, how does this change your results?

P. Vashishta: All these calculations where T_c is calculated do not mean a damn thing because μ^* is a completely free parameter and the comparison of T_c is completely meaningless.

B. Klein: Would you prefer we did not calculate T_c's anymore? Since $\mu^* \approx 0.1$, we can certainly ascertain whether our calculated λ's give a reasonable T_c.

B. Ganguly: So far as Palladium hydride and deuteride are concerned, I think it is a very important system. I never understood myself why in fact in the case of Palladium hydride the hydrogen mode or the deuterium mode is so soft when compared to other hydrides. Now do you have any ideas on that?

B. Klein: Well, the only thing I would guess at is that it has something to do with the rock-salt structure. Now the other hydride does not form in the rock-salt structure with the octahedral coordination, and perhaps that is favorable for very loose bonding of the hydrogen atom. I know that is not a particularly strong statement but it is the best I can do.

B. Ganguly: In Palladium you have the extremely sharp density of states at the Fermi surface and as soon as you add either hydrogen or deuterium the Fermi surface shrinks and you have a tremendous fall-off of the density of states.

B. Klein: So does niobium. Niobium has the same kind of density of states structure.

B. Ganguly: I think your palladium is almost paramagnetic, and its density of states is, I think, 4 or 5 times higher than niobium.

B. Klein: I can't say off hand. If I used the rigid band model and added one electron to the niobium density of states, I am not sure it would show such a lowering of the density of states.

PRESSURE DEPENDENCE OF T_c FOR CARBIDES OF THORIUM, YTTRIUM AND SCANDIUM*

A.L. Giorgi, H.H. Hill, E.G. Szklarz and R.W. White

Los Alamos Scientific Laboratory

Los Alamos, New Mexico 87545

ABSTRACT

The effect of pressure on the T_c of Y_2C_3, Th_2C_3, and two ternary $(Th_xY_{1-x})_2C_3$ alloys (all b.c.c. Pu_2C_3-type structure) was determined. Measurements were also made on the new superconducting germanium stabilized scandium carbide and on a germanium stabilized thorium-yttrium-carbide. The pure binary compounds exhibited negative values for the dT_c/dP. Relatively strong positive pressure dependence was observed for the ternary compositions.

INTRODUCTION

Using high-temperature, high-pressure techniques, Krupta et al [1,2] recently succeeded in preparing new yttrium carbide and thorium carbide phases which crystallized in the b.c.c. (Pu_2C_3-type) structure. These materials were found to be superconducting with T_c values of 4.1 K for the thorium sesquicarbide and a maximum of 11.5 K for the yttrium sesquicarbide. Ternary compositions with the general formula $(Th_xY_{1-x})_2C_3$ were also prepared [3] and found to be superconducting over the entire range of compositions with the T_c going through a maximum of 17.0 K. A suggestion [4] that the high T_c values for these phases were related to the instability of the lattice at normal pressures had led to the expectation that an increase

---

*Work done under the auspices of the Energy Research and Development Administration.

in T_c should result with applied pressure. The effect of applied pressure on the T_c of the binary carbides as well as on two ternary compositions was determined in the present investigation. Similar pressure dependence measurements were also made on the new germanium stabilized b.c.c. scandium carbide [5] discovered in an earlier study. During the course of this work it was discovered that the addition of small quantities of germanium to certain compositions of yttrium-carbon stabilized a superconducting phase in the arc furnace preparations at normal pressures. Further when thorium was substituted for part of the yttrium metal, an increase in the observed T_c values resulted. The pressure dependence of T_c for this germanium stabilized yttrium-thorium-carbide was evaluated and compared to the ternary sesquicarbide prepared by the high-temperature, high-pressure technique.

EXPERIMENTAL

All of the samples examined in this study were prepared from the elemental materials: spectrographic quality graphite, crystal bar thorium, germanium (99.99%), yttrium (99+%), and scandium (99.9%). The following general procedure was used in all preparations. Weighed quantities of the elemental starting materials were arc-melted on a water-cooled copper hearth in a zirconium "gettered" helium atmosphere. To insure homogeneity, the samples were melted a total of four times, with the button being turned over between each melting. For the preparation of the yttrium sesquicarbide, thorium sesquicarbide, and the ternary thorium-yttrium-sesquicarbide samples, the arc-melt material was subjected to the high-temperature, high-pressure treatment described in an earlier publication [1].

A clamp device similar to the one described by T.F. Smith [6] was used for the pressure dependence measurements. A 1:1 mixture of n-pentane and isoamyl alcohol was used as the hydrostatic pressure medium in the pressure cylinder with 99.9999%-tin bead added for determination of the applied pressure at temperature. The pressure was applied to the cell at room temperature, the pressure was then "locked in", and the cell transferred to the cryostat for the T_c measurement. The transition was detected by the resonant arc bridge method of Lindsay, White, and Fowler [7]. A calibrated germanium resistance thermometer was used for the temperature measurements.

RESULTS AND DISCUSSION

The narrow ranges of composition in the yttrium-germanium-carbon system where superconductivity was found to occur in the arc-melt preparations at ambient pressure are shown in the diagram in Fig. 1.

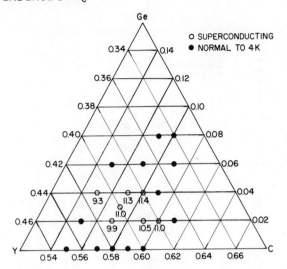

Figure 1: Superconducting compositions in the Y-Ge-C ternary system.

The T_c values for the superconducting compositions varied from 9.3 K to 11.4 K and are quite similar to the values obtained in the high temperature-high pressure study of the yttrium-carbon system [1]. X-ray diffraction patterns for the various yttrium-germanium-carbon samples showed that the major phase present was the non-superconducting α-Y_2C_3 normally formed in the yttrium-carbon system. The crystal structure of this phase is as yet unknown [8]. For the compositions with superconducting transitions additional diffraction lines corresponding to the strong diffraction lines of the b.c.c. Pu_2C_3-type structure also occurred suggesting that the superconductivity in the samples was associated with the stabilization of this phase in the arc-melt preparations. Based on the relative strength of the diffraction lines for the two phases, the concentration of the superconducting phase is estimated to be between 5 and 10%. Addition of thorium to one of the samples resulted in an arc-melt composition $Y_{.85}Th_{.15}Ge_{.1}C_{1.35}$ with a T_c = 12.4 K. The effect of applied pressure on the T_c of this sample and on the ternary sesquicarbide $Y_{.70}Th_{.30}C_{1.55}$ (T_c = 17.0 K) prepared by the high temperature-high pressure technique are illustrated in Fig. 2. Both samples had positive values of dT_c/dP supporting the belief that the superconducting phase in the arc-melt preparation has a b.c.c. Pu_2C_3-type structure. A second sample of the ternary sesquicarbide ($Y_{.65}Th_{.35}C_{1.35}$ T_c = 16.67 K) was also examined and the applied pressure was again observed. These results are shown in Fig. 3.

A change in slope (dT_c/dP) for the data appears to occur at approximately 12 kbar and the data can be fitted by two straight lines. Although this type of behavior is normally indicative of a

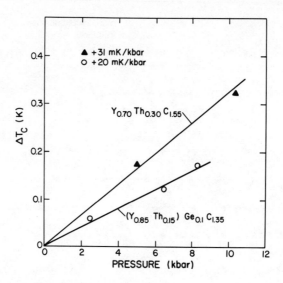

Figure 2: dT_c/dP curves for samples containing thorium.

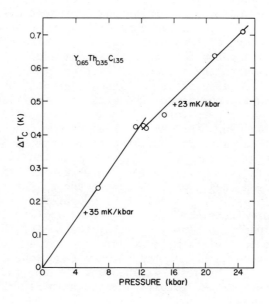

Figure 3: dT_c/dP for applied pressures up to 24 kbar.

phase change in the superconducting material, this is not believed to apply in this case. A broadening of the superconducting transition for the tin manometer was also observed at applied pressures of 12 kbar and above showing that a deviation from true hydrostatic conditions had occurred. Thus, the decrease in dT_c/dP for applied pressures above 12 kbar can be explained as due to a slight distortion in the crystal lattice of the superconducting phase by the quasi-hydrostatic conditions. When measurements were made on the binary sesquicarbides of yttrium ($YC_{1.25}$ T_c = 8.0 K) and thorium ($ThC_{1.35}$ T_c = 4.1 K) as well as on the germanium stabilized scandium carbide ($Sc_4Ge_{.1}C_3$ T_c = 8.0 K), a completely opposite effect was observed. Relatively large negative values of dT_c/dP resulted for all three samples. These data are shown in Fig. 4 which is a plot of the change in T_c against applied pressures up to 9 kbar. This change from a positive pressure effect for the ternary sesquicarbide to a negative pressure effect for the constituent binary sesquicarbides is surprising. In an earlier study [2] it was shown that thorium sesquicarbide and yttrium sesquicarbide are isostructural phases that are readily soluble in each other to form the ternary sesquicarbide. Further, the change in lattice parameter with composition of the ternary phase follows a Vegard relation indicating that the solution of one phase into the other occurs without any significant distortion of the crystalline lattice. The various properties of such materials normally change monotonically as the composition is varied from one binary through the ternary field to the other binary phase. This is obviously not true for the superconducting properties

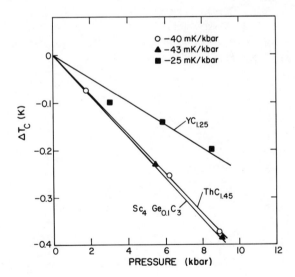

Figure 4: dT_c/dP for binary sesquicarbides and the germanium stabilized scandium carbide.

of these sesquicarbides. Not only does dT_c/dP change sign but the T_c values go through a maximum at an intermediate composition of the ternary. To properly interpret these results requires a knowledge of the pressure dependence of the various parameters controlling T_c. The three parameters through which pressure may effect T_c, according to the BCS theory, are the Debye temperature θ_D, the density of electron states at the Fermi surface $N(0)$ and the electron interaction parameter V. Two of these parameters, namely θ_D and $N(0)$, can be determined directly by experiment. Unfortunately measurements of this type have not been made as yet so an explanation of the unique behavior of this sesquicarbide system must await further investigations. As a result of the application of 24 kbar pressure to the ternary composition, the maximum T_c for the yttrium-thorium sesquicarbide has been increased to 17.4 K.

REFERENCES

1. M.C. Krupka, A.L. Giorgi, N.H. Krikorian and E.G. Szklarz, J. Less-Common Metals 17, 91 (1969).
2. M.C. Krupka, J. Less-Common Metals 20, 135 (1970).
3. M.C. Krupka, A.L. Giorgi, N.H. Krikorian and E.G. Szklarz, J. Less-Common Metals 19, 113 (1969).
4. B.T. Matthias, Superconductivity in d- and f-Band Metals, D.H. Douglass, Ed. (American Institute of Physics, New York, 1972) p. 367.
5. N.H. Krikorian, A.L. Giorgi, E.G. Szklarz, M.C. Krupka and B.T. Matthias, J. Less-Common Metals 19, 253 (1969).
6. T.F. Smith, J. Low Temp. Phys. 6, 171 (1972).
7. J.D.G. Lindsay, R.W. White and R.D. Fowler, Cryogen. 6, 213 (1966).
8. E.K. Storms, High Temp. Sci. 3, 99 (1971).

SUPERCONDUCTIVITY IN NIOBIUM: IMPLICATIONS FOR STRONG COUPLING

SUPERCONDUCTIVITY THEORY*

J. Bostock, K.H. Lo[+], W.N. Cheung, V. Diadiuk, and
M.L.A. MacVicar

Center for Materials Science and Engineering
Massachusetts Institute of Technology

Cambridge, Massachusetts 02139

ABSTRACT

The microscopic parameters λ and μ^* for high purity single crystal and polycrystalline Nb have been determined from superconductive tunneling junctions. These parameters are not in agreement with the strong coupling theory of superconductivity. This result is due to the fact that the associated electron-coupled phonon spectrum, $\alpha^2 F$, agrees with the phonon density of states, $F(\omega)$, up to the longitudinal phonon peak but differs from $F(\omega)$ at higher energies. Manipulating the value of the energy gap of Nb to lower than the measured bulk value of 3.93 $k_B T_c$ changes the result of the deconvolution significantly. Computer modeling studies of possible $\alpha^2 F$ spectra for Nb give μ^*, λ values that agree with theoretical predictions only when $\alpha^2 F$ has all the features of $F(\omega)$; otherwise negative values of μ^* are necessary to explain the experimentally observed superconductivity of Nb.

Although the strong coupling theory of superconductivity based on the Eliashberg equations [1] is quite successful in describing the deviations of various sp-band materials from the BCS behavior, there is great difficulty in correlating superconductive tunneling information from transition metal compounds with this theory [2].

Specifically, only tunneling data [3] on Ta fits this theory as modified by McMillan [4].

We recently reported the results of tunneling studies on both single crystal and polycrystalline Nb in which the effective Coulomb pseudopotential is negative [5]. As is well known, a negative μ^* has no real meaning within the context of the strong coupling theory; μ^* is supposed to represent the net repulsive interaction of electrons far from the Fermi surface. A statement that μ^* is negative means that structure manifest in superconductive tunneling data is far too small to account for the measured gap value. In fact, only if Δ_{Nb} used in the deconvolution is reduced from the measured 1.56 meV to ~ 1.3 meV do we obtain a μ^* and λ (the electron phonon coupling constant) in relatively good agreement with theory. It is our belief, however, that such an arbitrary decrease in the value of Δ_{Nb} cannot be assumed without a rigorous justification in terms of strong coupling theory. The implication, without such a justification, is that bulk-like Nb is not described by the currently accepted theory of strong coupling superconductivity.

Two different types of tunnel junction substrates were used in this investigation [5]: single crystal Nb cylinders grown in uhv by the floating-zone electron-beam method [6] and polycrystalline Nb rods made from thoroughly out-gassed and annealed high purity commercial stock. The resistivity ratios of the final single crystals ranged between 485 and 185 at 4.2 K in a magnetic field of 3 kG, while those for the polycrystalline rods were ~ 60.

For the single crystals, after a cooling period of two hours or less [7], a sample was thermally oxidized in situ, sectioned and masked; then, using conventional techniques, Au, In, or Pb counterelectrode thin films were deposited to complete a tunnel junction [8]. The resulting junction resistances were 5-100 Ω. In the polycrystalline Nb junctions, the substrates were acid-etched and acid-oxidized. Junctions were completed by an In counterelectrode deposition. Again, junction resistances were from 5-100 Ω.

Typical junction characteristics for a single-crystal-Nb/Ox/In-film junction are shown in Figures 1 and 2. The I-V curve (Fig. 1) has a ratio of currents above and below the current jump of ~ 40 and an excess current slope at zero bias of ~ 1 part in 10^3. The derivative (Fig. 2) shows the very sharp sum and difference peaks of the junction. The gap values [6] of the two junction materials are $2\Delta_{Nb}$ = 3.12 ± .05 meV and $2\Delta_{In}$ = 1.10 ± .05 meV.

Conventional modulation techniques using an ac bridge circuit [9] were used to measure the junction conductance in the superconducting and normal state at biases beyond the sum peak. Measurements from a polycrystalline junction Nb/Ox/In junction are shown in Figure 3. These data are typical of those obtained from all

SUPERCONDUCTIVITY IN NIOBIUM

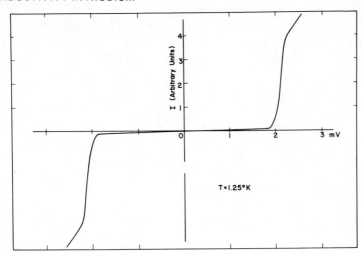

Figure 1: The I-V characteristic of a single-crystal-Nb/Ox/In junction at 1.25 K.

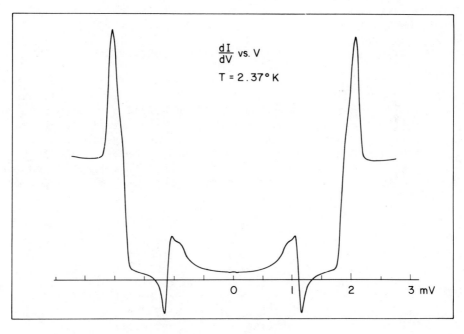

Figure 2: Derivative characteristic of a single-crystal-Nb/Ox/In junction at 2.37 K.

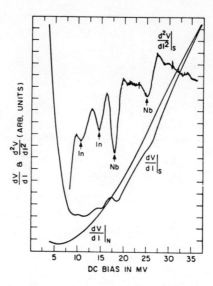

Figure 3: Tunneling characteristics from an annealed polycrystalline-Nb:In junction.

junctions, both single crystal and polycrystalline [10]. The phonon spectra for both In and Nb are clearly seen in the second derivative curve shown for the superconducting state. The normal state second derivative curve (not shown) has the same general shape, without phonon peaks, as does its superconducting counterpart [11].

The normalized conductance data from a junction, along with the measured superconducting energy gap value of the electrodes, was then used as input to a modified McMillan-Rowell $\alpha^2 F$ deconvolution program [12]. First the counterelectrode behavior (In or Pb) was extracted from the data [13] and a reduced density of states for Nb determined; then $\alpha^2 F$ for Nb was calculated. The essential feature of any self-consistent deconvolution is that the $\alpha^2 F$ generated for the material in question must reproduce the slope of the experimentally determined reduced density of states. The results obtained from a typical Nb/Ox/In junction are shown in Figure 4. The lower curve is the data extracted from a Nb/Ox/In data set; the upper curve is the calculated reduced density of states. Two aspects of this figure are remarkable. The first is that the deconvolution program, in fact, reproduces the slope exactly. However, the calculated curve is displaced from the experimental curve. To our knowledge, this kind of behavior has never been observed on any material. It is not an artifact of one Nb/Ox/In junction; it occurs for all junctions, irrespective of counterelectrode [14]. The second aspect is the fact that at lower energies the density of states for Nb seem to lie below the BCS density of states [15]. This, also, has not been reported previously [16].

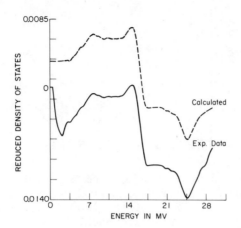

Figure 4: The reduced density of states of single crystal Nb extracted from the experimental data, compared to that calculated using the appropriate $\alpha^2 F$.

An electron-coupled phonon spectrum typical of the Nb junctions investigated in this study is plotted in Figure 5 along with the phonon density of states, $F(\omega)$, of Nb extracted from neutron scattering data [17], [21]. As is usual with tunneling studies, the shape of $\alpha^2 F$ agrees well with $F(\omega)$ up to the longitudinal peak; and, in fact, for Nb there is a one-to-one correspondence of the two curves up to this energy. Comparison of individual junction $\alpha^2 F$ spectra shows only minor deviations in the shoulder structure preceding the longitudinal peak and only slight changes in the height of that peak. The optimum values [18] of λ and μ^* obtained from these electron-coupled phonon spectra are $\mu^* \sim -0.05$ and $\lambda \sim 0.58$. These values are in disagreement with theoretical expectations of +.13 and .82 respectively.

The implications of a negative μ^* can be understood by considering briefly the parameterization of the strong coupling theory of superconductivity, in which λ represents the attractive pairing potential of electrons in the superconducting state while μ^* represents the effective mutual repulsion of electrons. The effect of the Coulomb term in the strong coupling equations is to reduce the ability of electrons to interact attractively and, thus, to reduce the contribution of the attractive interaction to the electron self-energy. The measured gap edge Δ_o, which is the value of the real part of the complex energy gap $\Delta(\Delta_o) = \Delta_o$, is a direct measure of the net attractive interaction that a pair feels and thus is essentially a sum of λ and $(-\mu^*)$.

To determine a value for λ and μ^*, the spectral weight function

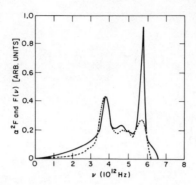

Figure 5: Comparision of $\alpha^2(\omega)F(\omega)$ from a polycrystalline-Nb:In junction (----) with $F(\omega)$ from neutron scattering (——) [21]. (The heights of the two spectra at the transverse peak are arbitrarily set equal.)

$\alpha^2 F$ must be known. Then

$$\lambda = 2 \int_0^\infty \frac{\alpha^2 F(\omega)}{\omega} d\omega \quad ,$$

and the value of μ* is obtained by balancing the attractive interaction represented by λ, against the mutual repulsion represented by μ* in such a way that the zero-frequency gap function $\Delta(\Delta_o)$ equals the experimentally determined gap value Δ_o. Clearly, if the structure in $\alpha^2 F(\omega)$ is weak (i.e., the area under the curve is small), λ will have a small value. As long as the measured gap value is small too, μ* will be positive, as predicted theoretically. But <u>if</u> the measured gap value is large, a small λ will not suffice to explain Δ_o. An additional attractive mechanism must then exist to explain the simultaneous weak $\alpha^2 F$ structure of high measured gap value. This latter situation corresponds to that of a negative μ*, which supplements the small λ to reproduce the high measured energy gap value.

To show graphically how λ and μ* change for a given shape of $\alpha^2 F$ when Δ_o is varied, a computer modeling study was carried out using the measured density of states for Nb in the usual way [12] to evaluate an $\alpha^2 F$ appropriate to a specified gap value. The results [19] are shown in Figures 6 and 7, and the numerics are presented in Table I. Observe that as the gap is decreased, the value of λ grows and μ* becomes positive. In fact, a value of $\Delta_o \sim 1.3$ meV gives microscopic parameters for Nb consistent with those predicted by theory. However, associated with those reduced parameters is a calculated T_c far higher that the experimentally measured one.

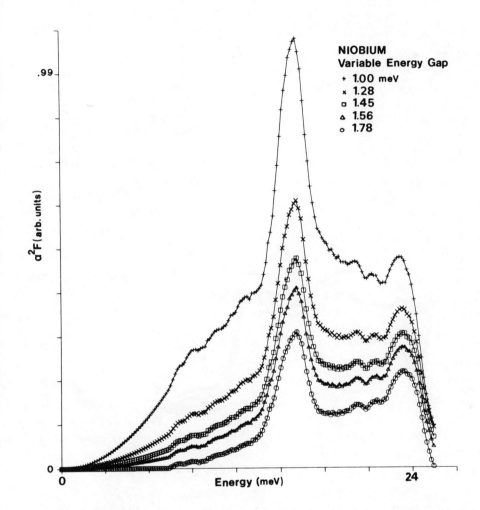

Figure 6: Comparison of $\alpha^2 F(\omega)$ for single crystal niobium ($\Delta_{experimental}$ = 1.56 meV) when the energy gap value used in the calculation is varied. (See Table I.)

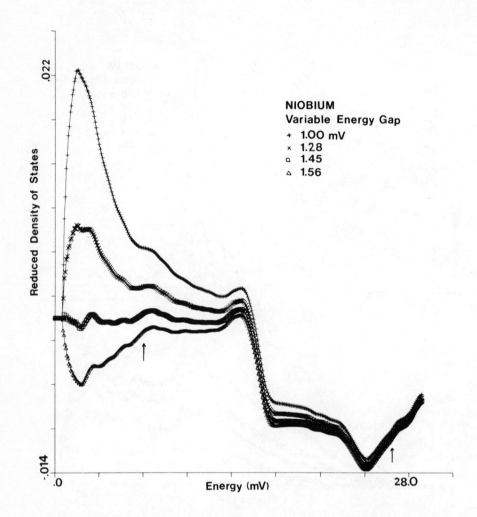

Figure 7: The reduced density of states calculated for single crystal niobium ($\Delta_{experimental}$ = 1.56 meV) when the gap value is varied. The arrow at 7 meV marks the end of the region wherein $\alpha^2 F$ is taken to vary as ω^2 in Figure 6. The arrow at ~ 27 meV marks the end of the spectrum.

Table I: Single Crystal Nb

Δ meV	μ*	λ	$<\omega>$ meV	A2	CALC. T_c (°K)
1.00	.178	1.350	13.93	9.40	13.0
1.28	.016	.764	14.97	5.72	12.1
1.45	-.049	.550	15.60	4.29	10.9
1.56	-.084	.430	16.23	3.49	9.7
1.78	-.140	.266	17.60	2.34	8.0
2.00	-.187	.135	20.44	1.38	6.3

The effects on $\alpha^2 F$ of varying the gap (Fig. 6) are mainly small changes in the relative sizes of structures in the spectra along with an overall amplification of the curve with decreasing Δ_0. More interesting, perhaps, are the changes in the calculated reduced density of states shown in Fig. 7 where the major effect of lowering the gap from its experimentally measured value is to change the amplitude and slope of the calculated curve in the very low energy region of the data. Since the shape of $\alpha^2 F$ is determined primarily by the slope of the reduced density of states in the moderate and higher energy regions ($\gtrsim 14$ meV), it is clear that the $\alpha^2 F$ calculation tends to be insensitive to lowering the gap value. It might also explain why in some earlier studies [2] it has been possible to manipulate the gap value to obtain more theoretically acceptable values of λ and, particularly, μ^* without significant effect on $\alpha^2 F$.

To determine if the suppression of longitudinal structure in $\alpha^2 F$ was a result of barrier or interface contamination, various amplification techniques [2] were applied to the experimentally obtained data sets. The results of the more interesting cases are given in Table II. In each, as μ^* increases toward theoretically acceptable values, λ grows very large and the associated T_c values become unacceptably larger than the measured value [20]. The inability of amplification to prove the existence of surface layer effects is not unexpected because $\alpha^2 F$ is in one-to-one correspondence with the phonon spectrum right up to the longitudinal peak; a contamination layer would have resulted in a progressive degradation of tunneling structures as a function of increasing energy. Experimentally, SEM

Table II: Game Playing (Nb-Au)

Description	μ*	λ	<ω>	A2	T_c(K)
Nb-Au (usual)	-.07	.53	13.65	3.56	9.9
Tomasch Attenuation Factor (Experimental Data)	.01	.91	14.14	6.45	15
Gaussian Enhancement (of $\alpha^2 F$)	.00	.87	14.23	6.00	14.9
	.12	2.00	14.23	14.23	23
Structure Amplification (of $\alpha^2 F$)	.12	2.76	13.38	18.45	26.5

analysis showed junction areas to be uniform and smooth on the scale of a coherence length, and the acid etch barrier technique gives assurance that no contamination effects originated at the barrier interface during thermal oxidation.

In order to determine the constraints on λ, μ^* and $\alpha^2 F$ if these parameters are to describe the experimentally observed superconductivity of Nb (Δ_0 = 1.56 meV, T_c = 9.2 K), computer modeling studies were carried out for three different types of $\alpha^2 F$ spectra. The first was a Nakagawa-Woods spectra [21] consisting of two Gaussians with the longitudinal peak height twice the height of the transverse peak; the second, a double Gaussian of equal peak heights; and the third, a spectrum typical of the experimentally observed $\alpha^2 F$: a double Gaussian such that the transverse peak is twice the height of the longitudinal peak. The experimental gap value was used as a given; a value was <u>chosen</u> for μ^*; and λ and T_c were then calculated [22]. The results of these calculations are summarized in Table III and the prototype of the experimental $\alpha^2 F$ is presented in Figure 8. It is seen that: (1) for any of the spectral shapes, as μ^* is varied from +.15 to -.11, T_c goes through a maximum; (2) for a given spectra shape and for positive μ^*, T_c is lowered as μ^* grows; (3) for all of the spectral shapes, calculated T_c's for negative μ^* are far closer to the actual experimental value for Nb than are the calculated T_c's for positive μ^*; and finally, (4) as one might expect, only the Nakagawa-Woods spectrum reproduces McMillan's original calculation [4],[23]. Even the best tunneling data available on the sp-band materials [24] has never resulted in a Nakagawa-Woods shaped $\alpha^2 F$; the shapes that come closest so far look like the

Table III: Model: 2 Gaussians

SHAPE α²F (Δ₀=1.56 meV)	μ^*	λ	$<\omega>$ meV	$\bar{\omega}$ meV	T_c °K
	.15	.95	20.25	21.21	9.23
	.11	.87			10.67
	0	.61			12.57
	-.11	.29			8.77
	.15	1.14	18.96	19.92	11.94
	.11	1.02			13.41
	0	0.69			14.36
	-.11	0.31			8.88
	.15	1.34	17.85	18.71	15.08
	.11	1.23			16.63
	0	0.78			16.08
	-.11	0.32			8.81

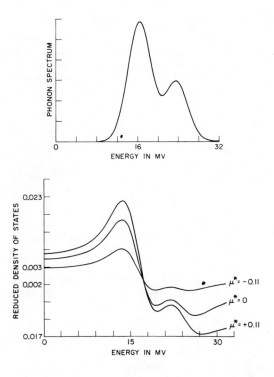

Figure 8: The Gaussian replica of experimental data set and the size of phonon structure in the reduced density of states as a function of μ^* value (Δ_0 = 1.56 meV).

equal-peak-height spectrum used in this computer study. As seen from Table III, such a shape for Nb would require either a negative μ* or a positive one of <u>extreme</u> magnitude.

The reduced density of states shown in Fig. 8 illustrates, once again, that positive μ* requires large structure in the low energy region, which translates into a large amplitude for $\alpha^2 F$. Thus, in this figure it is clear that negative μ* corresponds to very suppressed tunneling structures, such as are observed in the experimental data set.

The results of these model calculations on possible Nb-like electron-coupled phonon spectra are convincing evidence of the validity of the previously reported experimental tunneling results on pure crystalline bulk Nb.

REFERENCES

* This work supported by the U.S. Army Research Office and the U.S. Advanced Research Projects Agency.

+ Submitted in partial fulfillment of the Ph.D. thesis requirement, Department of Materials Science and Engineering, M.I.T. (1975). Current address: Gordon McKay Physics Laboratory, Harvard University, Cambridge, Massachusetts 02138.

1. J.R. Schrieffer, D.J. Scalapino, and J.W. Wilkins, Phys. Rev. Letters 10, 335 (1963); D.J. Scalapino, J.R. Schrieffer, and J.W. Wilkins, Phys. Rev. 148, 263 (1966).
2. L.Y.L. Shen, <u>Superconductivity in d- and f-Band Metals</u>, ed. D. H. Douglass (AIP, New York; 1972) p. 31; L.F. Lou and W.J. Tomasch, Phys. Rev. Letters 29, 858 (1972); L.Y.L. Shen, Phys. Rev. Letters 29, 1082 (1972).
3. L.Y.L. Shen, Phys. Rev. Letters 24, 1104 (1970).
4. W.L. McMillan, Phys. Rev. 167, 331 (1968); P. Morel and P.W. Anderson, Phys. Rev. 125, 1263 (1962).
5. J. Bostock, V. Diadiuk, W.N. Cheung, K.H. Lo, R.M. Rose, and M.L.A. MacVicar, Phys. Rev. Lett. 36, 603 (1976); J. Bostock, K.H. Lo, W.N. Cheung, V. Diadiuk, M.L.A. MacVicar, to be published.
6. M.H. Frommer, J. Bostock, K. Agyeman, R.M. Rose, and M.L.A MacVicar, Sol. St. Comm. 13, 1357 (1973); J. Bostock, Kofi Agyeman, M.H. Frommer, and M.L.A MacVicar, J. Appl. Phys. 44, 5567 (1973).
7. To eliminate tunneling anomalies, this cooling time must be less than the monolayer time for oxygen contamination in uhv.
8. Some single-crystal substrates were also acid-etched and acid-oxidized for comparison. Data from these samples is indistinguishable from that of the thermally oxidized samples.

9. R.V. Coleman, R.C. Morris, and J.E. Christopher, <u>Methods of Experimental Physics</u>, Vol. 11, ed. R.V. Coleman (Academic Press, New York; 1974). The bridge accuracy is 1 part in 10^4; signal-to-noise ratio was less than 5 µV rms. For the first derivative measurement, ac modulation was 90 µV rms; for the second, 200 µV rms.
10. The steepness of the rise of the normal state conductance is a function of counterelectrode material. For Au counterelectrode junctions the curve is not steep as for In; for Pb counterelectrodes, the curve is steeper than for In.
11. We are continuing our investigation of the shape of the second derivative in order to determine if, in fact, there exist normal state emission processes which have not been considered explicitly in the theory.
12. W. Hubin, Department of Physics, University of Illinois at Urbana, Technical Report No. 182, 1970; W.L. McMillan and J.M. Rowell, <u>Superconductivity</u>, ed. R.D. Parks (Marcel Dekker, New York; 1969) p. 561.
13. In the case of gold (normal) counterelectrodes, this is unnecessary. The results from these junctions were identical to those from In and Pb junctions.
14. There were 15 junctions studied: 3 Nb/Ox/Au, 1 Nb/Ox/Pb, and 8 Nb/Ox/In on single crystal substrates, and 3 Nb/Ox/In on polycrystalline substrates.
15. This was true for all but one data set. The deconvolution results of this data, on the other hand, are identical to that of all other junctions tested.
16. It is, perhaps, because of this behavior that we have found it impossible to "fit" the endpoint of the $\alpha^2 F$ spectrum for Nb, as one can do for all other materials deconvoluted. For Nb, a cut-off point must be imposed, irrespective of gap value associated with the spectrum. On the other hand, this behavior might also merely follow from the unusual shape of both normal and superconducting state second derivative data commented on in Ref. 11.
17. R.I. Sharp, J. Phys. C. (Solid St. Phys.) $\underline{2}$, 421 (1969).
18. Data from the bias region just above the sum gap in the dI/dV curve is critical in the evaluation of the parameters λ and μ^*, as well as being extremely difficult to measure accurately.
19. This particular data set assumes $\alpha^2 F \sim \omega^2$ from 0 to 7 meV such that the calculated reduced density of states fits smoothly to the measured reduced density of states. This means that μ^* is slightly too negative and λ somewhat too small, resulting in a calculated value of T_c which is slightly high.
20. One of the reasons why this numerical procedure seems to work so poorly for Nb, in contrast to studies on the simple metals, might be because niobium's density of states falls beneath the appropriate BCS curve.
21. Y. Nakagawa and A.D.B. Woods, Phys. Rev. Letters $\underline{11}$, 271 (1963).

22. The equation usually used to calculate T_c is:

$$T_c = \frac{\langle\omega\rangle}{1.2} \exp \frac{-1.04(1+\lambda)}{\lambda - \mu^* - (\frac{\langle\omega\rangle}{\bar{\omega}})\lambda\mu^*}$$

where $\langle\omega\rangle$ is the first moment of the $\alpha^2 F$ spectrum, $\langle\omega^2\rangle$ is the second moment, and $\bar{\omega} = \langle\omega^2\rangle/\langle\omega\rangle$.

23. The latter observation serves as a check on the entire calculation since this is the shape spectrum McMillan actually used to parameterize the Eliashberg gap equations and obtain his T_c equation.

24. J.M. Rowell, W.L. McMillan, and R.C. Dynes, A Tabulation of the Electron Phonon Interaction in Superconducting Metals and Alloys: Part I, unpublished; R.C. Dynes and J.M. Rowell, Phys. Rev. B11, 1884 (1975).

QUESTIONS AND COMMENTS

Discussion of this paper appears after the paper of B. Harmon and S. K. Sinha.

TUNNELING STUDY OF NIOBIUM USING ALUMINUM-ALUMINUM

OXIDE-NIOBIUM JUCTIONS*

>Bennett Robinson
>
>Stanford University
>
>Stanford, California 94305
>
>>and
>
>T.H. Geballe and J.M. Rowell
>
>Department of Applied Physics, Stanford University
>
>Stanford, California 94305
>
>>and
>
>Bell Laboratories
>
>Murray Hill, New Jersey 07974

INTRODUCTION

Given the success of the tunneling techique in the determination of the important superconducting parameters, [$\Delta(\omega)$, $\alpha^2(\omega)F(\omega)$, λ, $\langle\omega\rangle$ etc.] in the s-p metals [1], it was natural to hope that similarly detailed studies could be made of the transition metals, where it would be particularly valuable to have measurements of these parameters for families of alloys traversing the well-known peaks

---

* Work at Stanford University supported by the National Science Foundation through the Center for Materials Research and Grant DMR 74-21982.

in T_c versus electron per atom ratio. As outlined by Shen [2] in the previous Rochester conference, only in the case of tantalum has an entirely satisfactory tunneling study of a transition metal been possible [3]. This is still true today, despite considerable effort in the intervening years. In the case of niobium, investigators have used cleaner and cleaner bulk samples and employed ultrahigh vacuum systems to outgas bulk samples for long periods of time at elevated temperatures before preparation of the tunnel junctions using, as the insulating barrier, the oxide layer grown on these Nb surfaces. Extensive studies of this kind have been carried out by MacVicar and coworkers [4] and by Gärtner [5]. In this paper we first present a speculation, based on existing studies of the oxidation of clean Nb, as to why limited progress has been made with tunnel junctions prepared on bulk Nb. We then describe an alternative approach which relies on evaporation of a Nb film over an Al counterelectrode, thus using aluminum oxide as the tunneling barrier. We will show that the results obtained illustrate the importance of ultrahigh vacuum in the preparation of the Nb film and that initial tunneling results are extremely encouraging. This system allows not only the study of annealed Nb films, but also those deposited at low temperature (1.5 K) and at all stages of their anneal.

PROBLEMS ASSOCIATED WITH THE OXIDATION OF BULK Nb

In the case of tantalum, Shen [2,3] was able to specify clearly the steps taken to obtain a tunnel junction on bulk material which was comparable to the more familiar junctions utilizing s-p metals. For Nb he found that similar steps were not sufficient to assure a junction of this quality. (Junction "quality" can be loosely defined by the magnitude of current flow at voltages less than $\Delta_1 + \Delta_2$ for a junction of superconductors 1 and 2. Ideally only thermally excited quasiparticles should give rise to current in this voltage range. For a M-I-S junction, the conductance at V = 0, versus that when S is in the normal state, can be similarly used as a rough measure of "quality".)

As the tunneling experiment probes material within a coherence length of the surface, and this coherence length decreases with increasing energy, the "phonon induced structure" in dI/dV versus V at high voltages is even more seriously affected by junction quality than the I-V characteristic near the energy gap. This has been illustrated recently in a dissertation by Gärtner [5], where he shows the phonon induced structure for both tantalum-silver and niobium-silver junctions with the transition metal surfaces being prepared by various techniques. However, in his cleanest junction, these "phonon induced structures", measured as the deviation of the density of states from the BCS density of states for a superconductor with the same gap, resulted in a value for λ of 0.62 and for μ^* of .01. Both these values are smaller than those expected theoretically.

It is tempting to speculate that the difficulties experienced by all those working with bulk Nb are inherent to the Nb surface itself. For example, it is well known that the superconducting transition temperature of Nb decreases by 1 K with 1 at.% of dissolved oxygen. As an oxide layer grown on the Nb is forming the tunnel junction, some oxygen diffusion will be difficult to avoid.

As an example of what may happen in the preparation of a Nb tunnel junction, we can draw on the study by Lindau and Spicer [6] of the oxidation stages of a clean Nb film. In Fig. 1 we use a schematic representation of this oxidation taken from their publication. They report that the first stage of oxidation is absorption of oxygen into the fresh surface of the Nb. Next, a few monolayers of NbO are formed on the surface. This oxide is a metal which superconducts at 1.38 K and this is now in proximity with the clean bulk Nb beneath. However, another oxide of Nb, NbO_2, is next formed at the expense of NbO, but this process does not completely remove NbO from the interface. Therefore, at the end of this oxidation stage the surface in covered with a mixture of NbO and NbO_2. Finally, above this mixed layer the stable insulator, Nb_2O_5, is formed. While there is no guarantee that these stages of oxidation are followed exactly under the conditions used by all who have prepared tunnel junctions on Nb, it is clear that this study does indicate the problems which might arise.

The logical alternative to preparation of the junctions on bulk Nb is to use a geometry such as Al-Al oxide-Nb, where Nb is now an evaporated film. This presents three problems, the first being that Nb made in a conventional evaporation system at room temperature may not even be superconducting. Secondly, evaporation of metals such as Nb over Al at room temperature generally results in shorted junctions, and third, even if a successful junction is prepared, it is very possible that the Nb film will effectively compete for the oxygen of the aluminum oxide, or will getter any excess oxygen from the oxide, thus again contaminating the Nb layer near the oxide tunnel barrier. However, it is clear that the relevant activation energies for these self-oxidation processes will be such that much could be gained by completing a tunnel junction of this geometry at low temperatures.

EXPERIMENTAL DETAIL

In our experiment we avoided the use of Nb oxides and fabricated our tunnel junctions in the Al-Al oxide-Nb configuration.

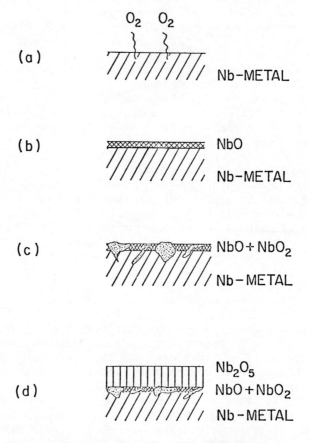

Figure 1: Stages of oxidation of a clean Nb surface, as discussed by Lindau and Spicer and copied with their permission.

The Al was evaporated onto a quartz substrate and allowed to oxidize during the time taken to mount it in a combined high vacuum and cryogenic station. After careful outgasing of the system, and particularly the Nb source, the station reached a pressure of 10^{-10} to 10^{-11} torr with the substrate held at 1.5 K. The Nb was evaporated onto this substrate and over the Al-Al oxide through suitable masks, the pressure remaining in the 10^{-10} torr scale during evaporation. Defining the "quality" of the tunnel junctions as described above, the junctions, as deposited, had a leakage conductance of approximately 1%. Upon annealing the junction to 150 K and 265 K, there was some evidence for deterioration of the Nb surface, or for a slight short developing through the oxide, in that the leakage conductance increased to 3 and 3.5% respectively.

RESULTS AND DISCUSSION

As deposited, the Nb films do not exhibit any sharp phonon structure in the derivative of the tunneling characteristic. The shape of $\alpha^2 F$ and annealing studies of these films will be reported elsewhere. After cycling to 150 K, it appears that the Nb film is fully annealed, as we do not observe any further sharpening of the phonon induced structure with further anneals. We therefore report results from the 150 K anneal only.

We first demonstrate in Fig. 2 the great importance of maintaining a vacuum of 10^{-10} torr or better during deposition. We show dynamic resistance plots for two junctions, one made in the best vacuum available, and one at 10^{-8} torr when a small leak was present in the system. It can be readily seen that the sharp increases in resistance near 18 and 25 mV, which are due to transverse and longitudinal phonons in the Nb, are drastically reduced in the 10^{-8} torr sample. It is interesting to note that not only the longitudinal structure is reduced, as might be expected from an energy dependent coherence length argument, but the transverse structure is also smaller. The energy gaps of these two samples were 1.21 and 1.48 meV.

Concentrating now on the cleaner junction of Fig. 2, we show in Fig. 3 (σ/σ_{BCS} - 1), where σ is the density of states, for the junction as annealed to 150 K, annealed briefly to 265 K and measured after a further three weeks' anneal at 290 K. To obtain σ_{BCS}, the energy gap Δ_o must be determined. As the aluminum gap does not appear in our I-V characteristics at our lowest operating temperature, we resort to a careful fit of σ, measured near the Nb gap, to the density of states for a BCS superconductor at finite temperature, as taken from the tables prepared by Bermon [7]. Such a fit gives Δ_o to ± .02 meV. After annealing the film to 150 K, the best fit to σ, as measured at 2 K, was obtained for

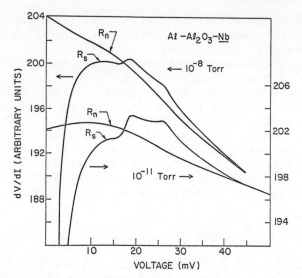

Figure 2: Dynamic resistance (dV/dI) versus voltage plots for two Al-Al oxide-Nb junctions, one made in a vacuum of 10^{-8} torr, the other in 10^{-10} torr. Both normal and superconducting states are shown.

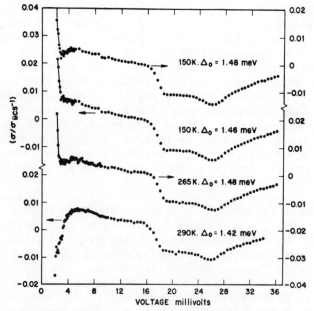

Figure 3: The reduced density of states $(\sigma/\sigma_{BCS}-1)$ versus voltage, for an Al-Al oxide-Nb junction after annealing briefly to 150 K (with σ_{BCS} generated for gap values of 1.46 and 1.48 meV), briefly to 265 K, and for three weeks at 290 K.

TUNNELING STUDY OF NIOBIUM

Δ_o = 1.46 meV. To show the sensitivity of (σ/σ_{BCS} - 1) to the value of gap, we have shown in Fig. 3 the results obtained by taking both 1.46 and 1.48 meV for the value of Δ_0 used in generating σ_{BCS}. The main effect of increasing the gap is to reduce the strength of the deviations from σ_{BCS} at low voltages. The gap value was unchanged after the brief anneal to 265 K, but after the long anneal at 290 K, Δ_0 had decreased to 1.42 meV. (The transition temperatures after these anneals were 8.8, 9.05 and 9.15 K.) To obtain σ at higher voltages, resistances were read every 0.5 mV directly from recorder plots, such as shown in Fig. 2, with an overall accuracy in resistance of about 5 in 10^4, except at low voltages where σ rises rapidly.

Despite the fact that the anneals take place in situ and in a vacuum ~ 10^{-11} torr, there is clear indication in Fig. 3 that the long anneal at room temperature deteriorates the junction. As the Nb film is 1700 Å thick, we believe it is most likely that oxidation of the Nb is taking place from the Nb/Al oxide interface. Note that even with the energy gap of 1.42 meV the reduced density of states (σ/σ_{BCS} - 1) obtained after the long anneal becomes negative at voltages just above the energy gap of Nb. This is very similar to results obtained on bulk Nb-Nb oxide-metal junctions.

We have not yet carried out enough annealing studies to say definitely that 150 K is the optimum temperature of anneal, and it is quite possible that measurements of both the tunneling characteristics and the resistance along the Nb film to study this annealing process in detail will result in slightly different results. However, taking the 150 K data of Fig. 3 results in an $\alpha^2 F$ as shown in Fig. 4 for Δ_o = 1.46 meV. The energies of transverse and longitudinal peaks agree well with those found by neutron scattering, but the respective heights of these peaks are grossly different from those in $F(\omega)$ obtained from single crystals [8], agreeing much better with the results determined more directly from polycrystalline samples [9]. A comparison of Fig. 4 with a calculated $\alpha^2(\omega)F(\omega)$ will be made by Harmon later in this conference.

The data of Fig. 3 for Δ_o = 1.46 meV gives the following values for some important parameters; λ = 0.81, μ^* = .051, $\bar{\omega}$ = 16.3 meV, $\langle\omega\rangle$ = 13.8 meV, $<\omega^2>^{\frac{1}{2}}$ =15.0 meV. When the energy gap is increased to 1.48 meV, these values become λ = 0.75, μ^* = .032, $\bar{\omega}$ = 16.4 meV, $\langle\omega\rangle$ = 14.1 meV, $<\omega^2>^{\frac{1}{2}}$ = 15.2 meV. This illustrates the sensitivity of the results to the determination of the gap value. Using these parameters, McMillan's T_c equation gives T_c = 10.3 K, well within the accuracies obtained for many materials [10]. However, the recent work of Allen and Dynes [11] suggests that the value of 0.81 for λ is still somewhat too low. Their plot of $T_c/\langle\omega\rangle$ versus λ, which is roughly linear, would suggest λ = .95 ± .01 for $T_c/\langle\omega\rangle$ = 0.66. This linear dependence of $T_c/\langle\omega\rangle$ on λ has suggested

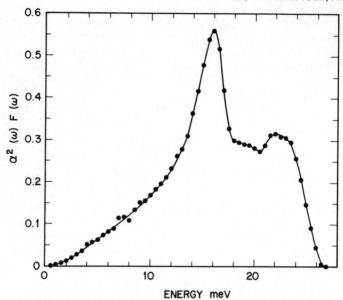

Figure 4: The phonon spectrum $F(\omega)$, weighted by the electron-phonon coupling parameter $\alpha^2(\omega)$, determined from the data of Fig. 3 after the 150 K anneal with $\Delta_0 = 1.46$ meV. From this $\alpha^2(\omega)F(\omega)$ we obtain $\lambda = 0.81$ and $\langle\omega\rangle = 13.8$ meV with $\mu^* = .051$.

a simple T_c equation [12] given by

$$T_c = \frac{\theta_D}{20}(\lambda - 0.25),$$

which, using $T_c = 9.4$ K and $\theta_D = 277$ K, indicates $\lambda = 0.93$ for Nb.

In Fig. 4, it is instructive to note that $\alpha^2 F$ is roughly linear from about 4 to 10 meV. This result is also found from neutron scattering [8,9]; namely, that the spectrum is far from Debye-like over this energy range. This behavior results from the soft transverse phonon modes in Nb which are analogous to those in Nb_3Sn. The presence of these modes must, of course, be reflected in the density of states of the tunnel junction in Fig. 3. Remembering that a peak in phonon density results in a sharp drop in the electron density of states, it is clear that the smooth decrease of $(\sigma/\sigma_{BCS} - 1)$ in Fig. 3 from 4 to 15 mV is the result of these soft modes.

In conclusion, we have demonstrated that tunneling into transition metals can be successfully carried out when the films are evaporated on Al oxide at 1.5 K and then annealed to higher temperatures. This technique not only allows the determination of $\alpha^2 F$ as a function of annealing temperature, but it should be possible to study other transition metals, such as vanadium, which have given particular difficulty in the past. For Nb, we have obtained $\lambda = .81$ and $\mu^* = .05$ with $\Delta_o = 1.46$ meV. Given the sensitivity of these results to the method of preparation and anneal of the junction, we suspect that even cleaner techniques will result in somewhat larger values for both λ and μ^*.

We would like to acknowledge valuable discussions with R.H. Hammond, who pointed out that aluminum-aluminum oxide transition metal tunnel junctions could be made using evaporation onto low temperature substrates.

REFERENCES

1. W.L. McMillan and J.M. Rowell, Superconductivity, ed. by R.D. Parks (Marcel Dekker, New York, 1969). For a list of the materials studied by tunneling, see R.C. Dynes, Sol. St. Comm. 10, 615 (1972) and P.B. Allen and R.C. Dynes, Phys. Rev. B12, 905 (1975).
2. L.Y.L. Shen, Superconductivity in d- and f-Band Metals, ed. by D.H. Douglass (AIP, New York, 1972).
3. L.Y.L. Shen, Phys. Rev. Lett. 24, 1104 (1970).
4. J. Bostock, V. Diadiuk, W.N. Cheung, K.H. Lo, R.M. Rose, and M.L.A. MacVicar, Phys. Rev. Lett. 36, 603 (1976) and references therein.
5. Klaus Gärtner, Dissertation, Abteilung fur Elektrotechnik an der Ruhr-Universitat Bochum.
6. I. Lindau and W.E. Spicer, J. Appl. Phys. 45, 3720 (1974).
7. S. Bermon, Technical Report No. 1, Dept. of Phys. Univ. of Illinois, Urbana, 1964.
8. Y. Nakagawa and A.D.B. Woods, Phys. Rev. Lett. 11, 271 (1963).
9. F. Gompf, H. Lau, W. Reichardt and J. Salgado, Neutron Inelastic Scattering 137, 1AEA, Vienna (1972), also W. Reichardt, this conference.
10. R.C. Dynes, Sol. St. Comm. 10, 615 (1972).
11. P.B. Allen and R.C. Dynes, Phys. Rev. 12, 905 (1975).
12. J.M. Rowell, Sol. St. Comm., to be published.

QUESTIONS AND COMMENTS

Discussion of this paper appears after the paper of Harmon and Sinha.

CALCULATION OF THE ELECTRON-PHONON SPECTRAL FUNCTION OF NIOBIUM[†]

B.N. Harmon

Ames Laboratory-ERDA and Department of Physics

Iowa State University, Ames, Iowa 50011

and

S.K. Sinha

Solid State Science Division, Argonne National

Laboratory, Argonne, Illinois 60439

ABSTRACT

The electron-phonon spectral distribution function, $\alpha^2(\omega)F(\omega)$, has been calculated for niobium. The energy bands and wavefunctions were obtained from a self-consistent APW muffin-tin potential, and the matrix elements were evaluated using the so called rigid ion approximation. The details of the calculation are presented and the results are compared with recent tunneling experiments, which confirm that the electron-phonon interaction is nearly three times larger for transverse phonons compared to the longitudinal phonons. It is shown that screening is of great significance and evidence is given which suggests that the same screening by d-electrons which causes the dips in the phonon dispersion curves of Nb is responsible for enhancing the electron phonon matrix elements and hence, T_c. An omission in the formulas of Gaspari and Gyorffy is pointed out.

[†]Work performed for the U.S. Energy Research and Development Administration under Contract No. W-7405-eng-82.

For d- and f-band materials the relationship between normal state and superconducting state properties is still a major problem. In their phenomenological work McMillan [1] and, more recently, Allen and Dynes [2] have shown the importance of and need for a better understanding of such normal state parameters as the electron-phonon interaction, the density of electron states at the Fermi level, and the phonon spectral distribution. These properties are related in a very complicated manner, and sorting out the coupling mechanism is a formidable task essential for determining the details governing superconductivity in high T_c materials.

We present a first-principles numerical calculation of $\alpha^2 F$ for niobium which provides new insight into the electron-phonon interaction. In the first section the details of the calculation are presented and compared with the approach of Gaspari and Gyorffy [3]. The second section gives the results of the calculations and presents a comparison with tunneling experiments made by two different groups [4,5]. (The experiments are reported in the proceedings of this conference.)

I. CALCULATIONAL CONSIDERATIONS

The electron-phonon (e-p) spectral distribution function is given by

$$\alpha^2(\omega) F(\omega) = \frac{1}{N(E_F)} \oint_{E_F} \frac{dS}{|\nabla_k E|} \oint_{E_F} \frac{dS'}{|\nabla_{k'} E|}$$

$$\times \sum_j \frac{|\langle k | \nabla V \cdot \hat{\varepsilon}_j(\vec{k}-\vec{k}') | \vec{k}' \rangle|^2}{2M \omega_{\vec{k}-\vec{k}',j}} \times \delta(\omega - \omega_{\vec{k}-\vec{k}',j})$$
(1)

where the integrals are over the Fermi surface, $\hat{\varepsilon}_j$ and ω_j denote the phonon eigenvectors and frequencies of the j-th branch, and $\nabla V \cdot \hat{\varepsilon}$ stands for the screened e-p interaction. The e-p mass enhancement is given by

$$\lambda = 2\int \alpha^2(\omega) F(\omega) \omega^{-1} d\omega$$
(2)

where $F(\omega)$ is the phonon density of states. It has been customary to approximate the expression by

$$\lambda = \eta/M\langle\omega^2\rangle$$
(3)

where

$$\eta = \frac{1}{N(E_F)} \oint_{E_F} \frac{dS}{|\nabla_k E|} \oint_{E_F} \frac{dS'}{|\nabla_{k'} E|} \langle k|\vec{\nabla}V|\vec{k}'\rangle \cdot \langle \vec{k}'|\vec{\nabla}V|\vec{k}\rangle \quad (4)$$

and the approximation is made by taking

$$\langle \omega^2 \rangle = \int F(\omega)\omega d\omega \, / \, \int F(\omega)\omega^{-1} \, d\omega. \quad (5)$$

A simple method of evaluating η has been devised by Gaspari and Gyorffy [3] (hereafter referred to as G-G). They used the so called "rigid ion" approximation in which ∇V is the gradient of the muffin-tin potential. This procedure ignores the tails of the atomic potentials in the interstitial regions as well as the electronic charge deformation and redistribution associated with the ionic displacements. G-G also averaged wavefunction coefficients over the Fermi surface, a procedure which John [6] has shown to be valid for cubic monatomic crystals. A drawback of the G-G method is that it precludes the consideration of individual phonons, hence the evaluation of Equation (1) is not feasible within the G-G framework. A wide application of the G-G method for calculating η has been made by Klein et al. [7].

The calculations presented in this paper used the APW energy bands and wavefunctions of a self-consistent muffin-tin potential [8]. The actual expression used to evaluate the e-p matrix elements is given by Sinha [9]. In testing the computer programs which evaluated Sinha's expression it was discovered that two terms had been overlooked. For completeness we include the correct expression in the Appendix to this paper. It can be shown that if $E_k = E_{k'} = E_F$ then $I_\alpha^{k'k}$ as given in the Appendix is equivalent to the matrix elements assumed by Gaspari and Gyorffy. Thus we are also neglecting screening effects as well as the tails of the atomic potentials in the interstitial region.

There may be some concern as to the appropriateness of a muffin-tin potential for these calculations [10]. Two separate questions naturally arise 1) How good are the wavefunctions and eigenvalues (Fermi surface) obtained from a muffin-tin potential and 2) How good is the electron-phonon matrix element evaluated within this approximation. Fortuitously the first question has been throughly investigated very recently by Naim Elyashar and Dale Koelling who performed a relativistic, self-consistent, $\alpha = 2/3$, APW calculation using a completely general potential [11,12,13]. They found that the energy eigenvalues were surprisingly accurate within the muffin-tin approximation (e.g., the Fermi energy changed by only 3 mRy in going from a muffin-tin self-consistent potential to a completely general self-consistent potential) [12]. The wavefunctions (and therefore the charge density) did show sensitivity to the non-spherical terms in the potential. There was an increased contribution to the L = 4 angular component of the density and non-uniform shifts of the order of 2% to 3% in the angular momentum character of the wavefunctions - generally in the direction of increased anisotropy [11,12]. Although these changes are significant for accurate X-ray analysis or for understanding the anisotropy in accurate induced form factor measurements by neutron diffraction, the changes are not as important for our calculations. In fact the greatest signigicance to us of the Elyashar and Koelling results is to give confidence in the wavefunctions we are using from the muffin-tin potential. Of particular concern (as emphasized in the G-G technique) is the ℓ-character of the wavefunctions at the Fermi surface. A comparison between the α = 2/3 muffin-tin and general self-consistent potential results indicated that there were less than 2% differences in the different ℓ components of the density of states at the Fermi level [13, 14].

The problem of assessing the quality of muffin-tin e-p matrix elements has not received as much attention. By only considering the contribution to the matrix element inside the muffin-tin spheres we are neglecting the contribution in the interstitial region. This is probably not too serious an approximation for the following reasons: 1) 85% of the d electron density is contained within the muffin-tin sphere for d-states at the Fermi energy, and p-d and d-f coupling is by far the dominant contribution to η; 2) the potential in the interstitial region is slowly varying. In future work we intend to investigate the size of the interstitial contribution, however we believe the more serious approximation which deserves immediate attention is the neglect of screening for these matrix elements. A procedure for including screening is outlined in our other paper in these proceedings, and further details will be published elsewhere.

The integrations over the Fermi surface were accomplished by adopting the clever tetrahedron methods which have been developed for evaluating the density of states [15,16]. The irreducible 1/48th of the Brillouin zone is divided into many small tetrahedrons. Inside each tetrahedron the energy bands are linearly interpolated between the four corner energies. As the number of small tetrahedrons increases the linear approximation converges to the correct analytic result. Within a tetrahedron a constant energy surface is a plane whose area can easily be obtained, as can the gradient, $\nabla_k E$. In our calculations the Fermi surface cut through 231 tetrahedrons in the 1/48th of the Brillouin zone. Wavefunctions were evaluated at the center of mass of each of the 231 planes compromising the 1/48th section of the Fermi surface. The summation over $\vec{k}'$, in Eq. (1) requires $\vec{k}'$ to vary over the entire Fermi surface so that group theory is required to rotate those wavefunctions obtained in the 1/48th into the other sections of the complete zone. Our procedure then was to pick a $\vec{k}$ in the 1/48th for which the incremental area ΔS and the gradient were known and let $\vec{k}'$ vary over the other 231 x 48 independent wavefunctions. In this manner $\alpha^2(\omega)F(\omega)$ or $\lambda_{\vec{k}}$ can be obtained for each $\vec{k}$. In principle we should have summed over all 231 $\vec{k}$ in the 1/48th, but because of computational expense the summation was truncated. The 73 $\vec{k}$ which had the largest weight ($\Delta S/|\nabla_k E|$) were included. These account for 67% of the total density of states at the Fermi energy, and result in 816,651 total matrix elements. Because each $\lambda_{\vec{k}}$ was calculated the convergence could be watched and we believe our final results are accurate to within 3%. This is supported by the comparison with the calculation of η by the G-G method as described below. The phonon frequencies and eigenvectors were obtained for each of the 816,651 matrix elements using an 8th nearest neighbor force constant fit [17].

While evaluating $\alpha^2 F$ we also calculated η as given in Eq. (4). Expanding the wavefunctions in the angular momentum representation inside the muffin-tin spheres the s-p, p-d, and d-f contributions to η have been evaluated and are given in the first row of Table I. The row labeled G-G(1) results using the G-G method as published [3,6]. The discrepancy between the results was traced to the neglect by G-G of a term arising from the discontinuity in the second derivative of the radial wavefunctions caused by the muffin-tin potential discontinuity at the sphere radius. Understanding how and why this term arises is made clearer by considering appendix 1 in the paper by Evans, Gaspari, and Gyorffy [18]. The radial integral to be evaluated is

$$I = \int_0^{r_s} dr\, U_\ell \frac{dV}{dr} U_{\ell+1} \tag{6}$$

where in their notation U_ℓ is the radial solution to Schrödinger's

Table I: Comparison of the results of Eq. (4) using 816,651 matrix elements with the results using the method of ref. 3. G-G (2) is with the corrected formulas. Units are eV/Å^2.

η	s-p	p-d	d-f
Eq. (2)	0.80	4.91	4.18
G-G (1)	0.61	3.78	4.53
G-G (2)	0.82	4.83	4.11

equation multiplied by r. Reference 18 shows that the integral is equal to

$$I = [\tfrac{1}{2}(U''_{\ell+1}U_\ell + U''_\ell U_{\ell+1}) - U'_{\ell+1}U'_\ell + \frac{(\ell+1)}{r}(U'_{\ell+1}U_\ell - U'_\ell U_{\ell+1})]_{r=0}^{r=r_s} \quad (7)$$

To evaluate this expression the value of the radial wavefunction and its derivatives at the sphere radius are required. Evans, Gaspari, and Gyorffy substitute the values of the radial wavefunction and its derivatives that are valid just outside the muffin-tin sphere - which is equivalent to keeping the unphysical singularity at the muffin-tin sphere in Eq. (6). An appropriate form of the radial function in the interstitial region where the potential is zero is

$$\frac{U_\ell}{r} = R(r) = j_\ell(\sqrt{E}\, r)\cos\delta_\ell(E) - n_\ell(\sqrt{E}\, r)\sin\delta_\ell(E) \quad (8)$$

where $\delta_\ell(E)$ are the scattering phase shifts of the muffin-tin potential evaluated at the sphere radius. With the normalization of Eq. (8) and using atomic units Eq. (7) yields

$$I = \sin(\delta_\ell - \delta_{\ell+1}) + V(r_s)U_\ell U_{\ell+1} \quad (9)$$

where $V(r_s)$ is the value of the muffin-tin potential just inside the muffin-tin sphere. The integral in Eq. (6) can easily be evaluated numerically and shows that Eq. (9) is indeed correct. The second term in Eq. (9) has been neglected previously. The row labeled G-G(2) are the results using the corrected formula. The small remaining differences are probably due to the lack of convergence in our summations. Our calculation includes cross terms [6] (arising from p-d transitions in one matrix element in Eq. (4) but d-f transitions in the other) and yields a total value of η of 10.33 eV/Å^2.

II. RESULTS

The calculated $\alpha^2 F$ function, Eq. (1), is shown as a histogram in Figure 1 along with the phonon density of states $F(\nu)$. The value of λ obtained from Eq. (2) was 1.87 (equivalent to setting $\langle\omega^2\rangle^{\frac{1}{2}}$ to 183 K). The usual approximation of Eq. (5) yields 195 K for $\langle\omega^2\rangle^{\frac{1}{2}}$ and using Eq. (3) would give $\lambda = 1.64$, which shows that the

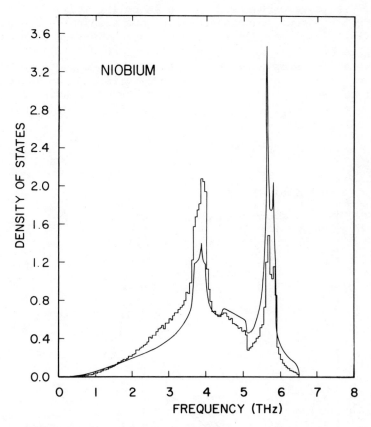

Figure 1: The electron-phonon spectral distribution function, $\alpha^2 F$ (histogram), compared with the phonon spectrum $F(\nu)$. Both curves normalized to 3.0.

so called "constant α^2" approximation is not very good. The value of 1.87 for λ is about a factor of 2 larger than the range of accepted values [1,2] and strongly suggests that screening effects are important. (We do not expect λ to change greatly with different potential prescriptions for calculating the band structure.)

Figure 1 shows that α^2 for the lower transverse peak is enhanced by nearly a factor of three over the α^2 for the longitudinal peak. This is also observed in the tunneling measurements [19]. The tunneling spectra appear to have some broadening probably caused by intrinsic phonon lifetime effects and/or surface effects [20]. Figure 2 shows the tunneling data of Bostock et al.[4,21] from a polycrystalline sample. Although there is some question over the values of μ^* and λ obtained in these experiments the shape of α^2F seems to be rather insensitive, and we therefore have some justification for comparing the shape of our calculated α^2F with the experimental results.

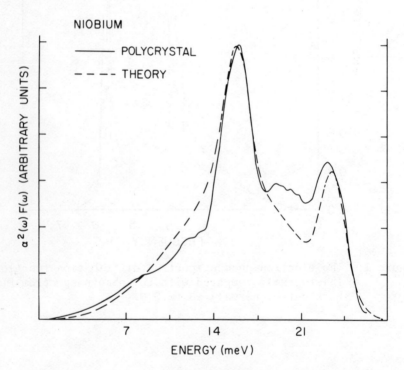

Figure 2: The polycrystal tunneling results for α^2F (ref. 4) compared with the energy dependent broadened theoretical α^2F of Fig. 1. The curves are normalized to have the same transverse peak height.

Surface contamination is a well known problem in tunneling experiments using d-band metals. Some arguments have been made that the high energy phonon induced structure is smeared more than the low energy structure because the high energy electrons have a shorter mean free path and thus are affected proportionately more by the surface crud [20]. In the absence of any rigorous way of taking into account the effect of the surface, we assumed the inverse of the penetration depth to be proportional to the width of a Gaussian function with which we broaden the calculated $\alpha^2 F$ histogram of Fig. 1. Results by Lo et al.[21] indicate that this broadening should go as ω^2. We thus broaden with a Gaussian of standard deviation $\sigma = b\omega^2$; where $b = 2.5 \times 10^{-3}$ meV^{-1}, and is adjusted roughly to fit the observed width of the transverse peak. The heights of the two curves have been normalized at the transverse peak as shown in Fig. 2. The agreement at high energy is not very good. Since the energy dependent broadening arguments seemed to be made in order to account for the small longitudinal peak, and since our calculation indicated this is mostly due to α^2; we tried broadening the histogram of Fig. 1 without any energy dependence. The theory curve in Fig. 3 is obtained using a Gaussian of standard deviation .82 meV - which was again picked to give approximately the experimental transverse peak width. Now the high energy region is in substantially better agreement. The preliminary results of Robinson, Geballe, and Rowell [22] are shown in Fig. 4. The theory curve in this figure is exactly the same as the theory curve of Fig. 3 except for the change of scale. With the theory curve as a guide the experimental curves of Figs. 3 and 4 are seen to be very similar, with only the low energy region being significantly different. The conclusions given below are qualitative and do not depend significantly on the assumed broadening function.

It is immediately clear that our calculations corroborate the experimental result that the longitudinal phonon peak is suppressed relative to the lower transverse peak, and that this is not an effect caused solely by poor surfaces. In addition the region between the two peaks in the calculated $\alpha^2 F$ is about 30% lower than the observed renormalized spectrum. This indicates screening affects this part of the spectrum less than it does the overall spectrum. In a previous paper [23] we showed that screening by d-electrons near the Fermi level could account for the phonon anomalies observed in many high T_c superconductors (in particular Nb and NbC). A consequence of the ideas developed there was that there should be lessened screening (i.e., relative enhancement) of the e-p interaction in the regions of the anomalies. Specifically the screening of the e-p matrix elements require the elements of the matrix $(1 + V\tilde{\chi})^{-1}$ where the elements involve pairs of virtual transitions between pairs of orbital states. $\tilde{\chi}$ is a generalized susceptibility matrix and $\tilde{V}$ the q-dependent Coulomb coupling coefficient between charge density waves corresponding to the transition. For Nb we

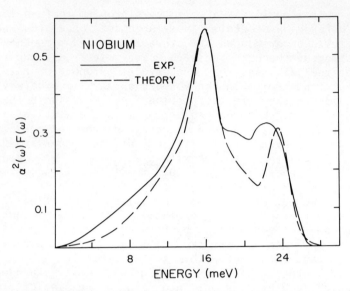

Figure 3: The same polycrystal tunneling results compared with the broadened theoretical $\alpha^2 F$. The broadening is energy independent and the curves are again normalized at the transverse peak.

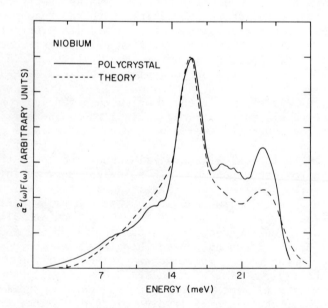

Figure 4: The same theoretical curve as shown in Fig. 3 but compared with the preliminary experimental $\alpha^2 F$ of ref. 5.

expect the transitions within the predominantly t_{2g}-like d-band complex at the Fermi surface to dominate. As discussed in Ref. 23 and in our other paper in these proceedings, for a sufficiently high density of states at E_F and a correspondingly large χ, the intra d-band elements of $(\underline{1} + \underline{V}\chi)$ can become quite small for q near the zone boundary, leading to both anomalies in the dispersion relations and relative enhancement of the e-p matrix elements in those regions. Unfortunately the $\alpha^2 F$ spectrum is ω- rather than q- selective, but it is significant that the differences between the two curves in Figs. 3 and 4 correspond exactly to the frequency range of the longitudinal phonons in the regions of the anomalies. In the low frequency region there also appears to be a relative enhancement of $\alpha^2 F$ over our calculated spectrum. This may be due to the softening of the low frequency transverse phonons in Nb at low temperatures as observed by Shapiro et al. [24], compared to the values we used which were based on room temperature data. Experiments are now in progress at the Ames Laboratory research reactor to obtain a more complete analysis of the low temperature phonon dispersion curve of Nb. We should also remark that Shapiro et al. observed stronger coupling to the transverse phonons (about twice as large as the average over all the phonons).

In conclusion, the results of this calculation together with the recent tunneling measurements clearly indicate the importance of screening and suggest that the same d-electron screening effects responsible for the dips in the Nb phonon spectrum also enhance the e-p interaction and hence T_c as we had speculated earlier [23]. Since this effect is rather sensitive to the density of states at the Fermi level, a corresponding comparison for Mo would be very worthwhile. We have also shown that the rigid ion matrix element calculation yields the correct qualitative features regarding how $\alpha^2 F$ differs from F, but that quantitatively it is not a good approximation and neither is the usual practice of separately averaging over the phonons.

The authors wish to acknowledge helpful discussions with D.K. Finnemore, W. Kamitakahara, D.D. Koelling, W. Butler, B.M. Klein, L. Boyer, and D. Papaconstantopoulos. They also wish to thank B. Robinson, T.H. Geballe, and J.M. Rowell as well as J.M. Bostock and M.L.A. MacVicar for sending details of their tunneling data. The authors are grateful to D. Papaconstantopoulos for the Nb self-consistent potential, and M. Mostoller for independently checking the phonon frequency calculations.

REFERENCES

1. W.L. McMillan, Phys. Rev. 167, 331 (1968).
2. P.B. Allen and R.C. Dynes, Phys. Rev. B12, 905 (1975).

3. G.D. Gaspari and B.L. Gyorffy, Phys. Rev. Lett. $\underline{28}$, 801 (1972).
4. J. Bostock, V. Diadiuk, W.N. Cheung, K.H. Lo, R.M. Rose, and M.L.A MacVicar, Phys. Rev. Lett. $\underline{36}$, 603 (1976).
5. B. Robinson, T.H. Geballe, and J.M. Rowell, Bull. Amer. Phys. Soc. $\underline{21}$ #3, 402 (1976).
6. W. John, J. Phys. F. $\underline{3}$, L231 (1973).
7. B.M. Klein, D. Papaconstantopoulos, Phys. Rev. Lett. $\underline{32}$, 1193 (1974); D. Papaconstantopoulos and B.M. Klein, Phys. Rev. Lett. $\underline{35}$, 110 (1975).
8. J.R. Anderson, D.A. Papaconstantopoulos, J.W. McCaffrey, and J. Schirber, Phys. Rev. $\underline{B7}$, 5115 (1973). A more accurate determination of $N(E_F)$ per spin gives 9.54 states/Ry-atom rather than the value quoted in this paper.
9. S.K. Sinha, Phys. Rev. $\underline{169}$, 477 (1968).
10. C. Varma in the proceedings of this conference expressed some concern that the muffin-tin approximation was too crude an approximation for his taste.
11. N. Elyashar, Ph.D. thesis, University of Illinois, Chicago (1975) unpublished.
12. N. Elyashar and D.D. Koelling, Phys. Rev. B (in press).
13. N. Elyashar and D.D. Koelling, Phys. Rev. B (to be published).
14. D. D. Koelling (private communication).
15. O. Jepson and O.K. Anderson, Solid State Commun. $\underline{9}$, 1763 (1971).
16. G. Lehman and M. Taut, Phys. Status Solidi $\underline{B54}$, 469 (1972).
17. Y. Nakagawa and A.D.B. Woods, Phys. Rev. Lett. $\underline{11}$, 271 (1963).
18. R. Evans, G.D. Gaspari, and B.L. Gyorffy, J. Phys. F $\underline{3}$, 39 (1973).
19. References 4,5, and the proceedings of this conference.
20. L.Y.L. Shen, Superconductivity in d- and f-Band Metals, edited by D.H. Douglass, A.I.P. Conference Proceedings No. 4 (A.I.P. New York 1972).
21. K.K. Lo, J.L. Bostock, M.L.A. MacVicar, and R.M. Rose, M.I.T. Technical Report No. 11, 1975 (unpublished).
22. Reference 5 and the proceedings of this conference.
23. S.K. Sinha and B.N. Harmon, Phys. Rev. Lett. $\underline{35}$, 1515 (1975).
24. S.M. Shapiro, G. Shirane, and J.D. Axe, Phys. Rev. $\underline{B12}$, 4899 (1975).

APPENDIX

A thorough discussion of the electron-phonon matrix element has already been given by Sinha [9]. In this appendix we restrict ourselves to correcting and rewriting the expression [Eq. (A6) of Ref. 9] for the electron-phonon matrix element. The interested reader is referred to Sinha's paper for details concerning notation.

For matrix elements taken within the muffin-tin sphere it is convenient to expand the Bloch functions in the angular momentum representation:

$$\psi_k(\vec{r}) = \sum_{\ell,m} B_{\ell,m}(k) R_\ell(r) Y_{\ell,m}(\hat{r}) \quad |\vec{r}| < r_s$$

where r_s is the muffin-tin radius. Then Sinha's Eq. (A6) can be correctly written as

$$I_\alpha^{k'k} = \frac{\hbar^2}{2m} r_s^2 \sum_\ell \left\{ \left[\frac{\ell(\ell+2)}{r_s^2} + \frac{2m}{\hbar^2}[V(r_s)-E_k] - (\frac{2}{r_s} + L_{\ell+1}) L_\ell + \frac{\ell}{r_s}(L_{\ell+1}-L_\ell) \right] \right.$$

$$\times (-i) \left(\frac{\ell+1}{2\ell+1}\right)^{1/2} \sum_{m,m'} B_{\ell,m}(k) B^*_{\ell+1,m'}(k') C^{\ell+1,1,\ell}_{m',m-m'} A_{m-m',\alpha}$$

$$+ \left[\frac{(\ell-1)(\ell+1)}{r_s^2} + \frac{2m}{\hbar^2}[V(r_s)-E_k] - (\frac{2}{r_s}+L_{\ell-1}) L_\ell + \frac{\ell+1}{r_s}(L_\ell - L_{\ell-1}) \right]$$

$$\left. \times (i) \left(\frac{\ell}{2\ell+1}\right)^{1/2} \sum_{m,m'} B_{\ell,m}(k) B^*_{\ell-1,m'}(k') C^{\ell-1,1,\ell}_{m',m-m'} A_{m-m',\alpha} \right\}$$

where L_ℓ represents the logarithmic derivative (R'_ℓ/R_ℓ) evaluated at the muffin-tin radius. L_ℓ is evaluated at E_k and $L_{\ell\pm1}$ at $E_{k'}$. For $E_k = E_{k'}$ the expressions inside the square brackets can be simplified using the scattering phase shifts similar to Eq. (9). In fact if $E_k = E_{k'} = E_F$, and if $I_\alpha^{k'k}$ is used as the matrix elements in Eq. (4) then one obtains exactly the Gaspari-Gyorffy formulation for η (including the correction term discussed with Eq. (9) and the off diagonal term discussed by John [6]).

QUESTIONS AND COMMENTS FOR THE PAPERS OF BOSTOCK ET AL.; ROBINSON, GEBALLE, AND ROWELL; AND HARMON AND SINHA.

P. Vashishta: To the arguements of the MIT and Stanford data I would like to add that, apart from the understanding or the lack of it on the part of the MIT group, the results are nearly the same.

J. Bostock: They are identical.

B. Harmon: Yes. By comparing figures 3 and 4 of my paper in which the theoretical curves are the same, one can see that the shape of the $\alpha^2 F$ obtained by both groups is very nearly identical except for the low energy region.

P. Vashishta: Actually the results will come out to be the same. The purpose of the inversion, which is rather artificial is to determine $\alpha^2 F$ from using experimental tunneling results. In principle it is possible to determine uniquely $\alpha^2 F$ and μ^*. In practice it is not. It will be a miracle if you could determine all uniquely--- you would have to use some kind of physical intuition. It is physical intuition that is being used by the MIT group to obtain a reasonable value of T_c. On the other hand, you know from physics that μ^* can not be negative. So one should take a reasonable value of μ^* and let the $\alpha^2 F$ adjust to that reasonable value.

P. Allen: There is an absolute measurement of α^2 by Shapiro et al. [Harmon and Sinha, Ref. 24] which could serve as a benchmark for testing theoretical calculations. Unfortunately, α^2 is only measured to a few specific phonons (by linewidth measurements). This means you would have to modify your computer codes to calculate it. Also, this might help the tunneling experimentalists to settle on an absolute value of λ.

S. Nam: I would like to know if the disagreement [between the MIT and Stanford groups] comes from the experimental raw data or the disagreement comes from analyzing data? It is my feeling that it comes from analyzing data and not from experiment---is this correct?

J. Rowell: I would like to show viewgraphs taken entirely from the last Rochester meeting on d- and f-band superconductors, in which Shen talked about problems of tunneling into transition metals. I think it is a pity that I have to show these but I want to stress the points he made again. First, there was no problem with the value of λ or μ^* when he tunneled into very clean tantalum. This is what the I-V characteristics looked like in Fig. 1 of his paper which I now show you. [See Fig. 1 next page.] I want you to note a few things. If you go to a temperature below T_c of tantalum in this tantalum-lead junction, the tantalum gap opens very cleanly with a very sharp rise at the sum of the gaps (1.9 mV) and there is no hint of any structure just above the sum of the gaps. The current within the gap is very small; if you expand the current scale you can see that you have to expand the scale 5000 times before

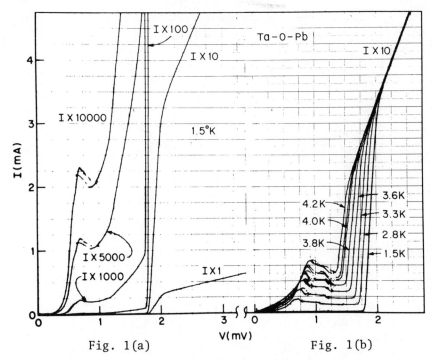

Figure 1: The I-V characteristics of a Ta-O-Pb junction. In Fig. 1a the 1.5 K I-V curves are plotted with different current gains. In Fig. 1b the I-V curves are shown at different temperatures.

you see a current comparable to the current that would flow with both Pb and Ta in the normal state. What does that mean? It means there are no shorts through the oxide and no states within the gap; there are no states to this limit of one part in 5000 in the tantalum gap. Now with this junction, or junctions made like this with a normal electrode, he obtained straightaway from the inversion program a very reasonable value for μ^* (0.11) and a very reasonable value of λ (0.69). But, in addition, what he showed was the effect on the derivative traces of the way the Ta-oxide-Ag junctions were made. The good data was taken from a tantalum-silver junction like the lower one of his Fig. 2 [see Fig. 2 next page], where the Ta surface was outgassed. But on a surface cleaned by sputtering he obtained derivatives as in the upper trace, this looks very similar to the one Bennett [Robinson] showed you for tunneling into a niobium film which was made at 10^{-8} Torr. As far as I was concerned Shen's data settled

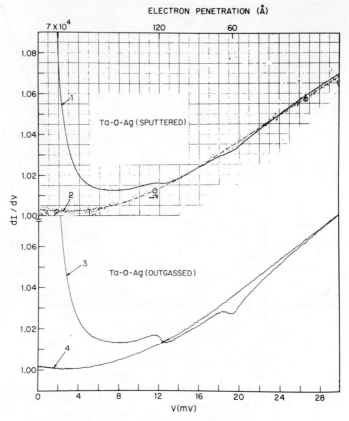

Figure 2: The dI/dV curves of Ta-O-Ag junctions. Curves 1 and 2 are superconducting and normal state conductances of a sputtered bulk tantalum junctions. Curves 3 and 4 are superconducting and normal state conductances of an outgassed tantalum junction.

the problem of whether the phonon interaction was all important in the superconductivity of at least one transition metal. When you go to niobium you immediately have experimental problems---Shen showed this at the last Rochester meeting in Fig. 4 [see Fig. on next page] of his paper. If you take a really dirty niobium surface, the gap is very substantially depressed on the surface. There is this S shape (sometimes called a knee) above the gap, which everybody now interprets in terms of a proximity effect between clean niobium and a contaminated surface. As Shen cleaned the surface the gap grew, and for his most carefully cleaned surface he reached a value of about 2.9 meV for the sum of the Nb and Pb gaps. To the eye there is now no problem just above the sum of the gaps, in that the knee seems to be absent, but a

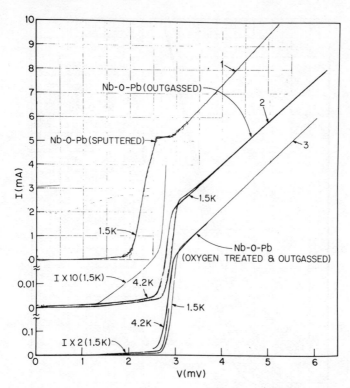

Figure 4: The I-V characteristics of Nb-O-Pb junctions. Curve 1 is from a sputtered Nb junction. The set of curves 2 contains the I-V curves at 4.2K and 1.5K as well as the increased-gain, 1.5K, I-V curve of an outgassed Nb junction. The set of curves 3 contains the I-V curves at 4.2K and 1.5K as well as the increased-gain, 1.5K, I-V curve of an oxygen-treated, outgassed Nb junction.

derivative trace will still show a dip in dI/dV in this region. In addition, there is still a leakage current within even this best junction of about a fiftieth of the normal state current and that is in my mind unacceptably large (the lowest trace of Fig. 4). There should be no current flowing below the gap in a clean tunnel junction as there are no states below the gap in either electrode, so you should be concerned over why this current is there at all. As far as I know, this junction of Fig. 4 is comparable to the best niobium junctions made by anyone, in that they all seem to have this leakage current of about 1 part in 50. As the current flow is much smaller at voltages less than the Pb gap, this leakage is not due to shorts through the oxide. When Shen made a niobium-gold junction, which is the one shown in his Fig. 5 [see next page] and is the one from which he took tunneling data, you can again

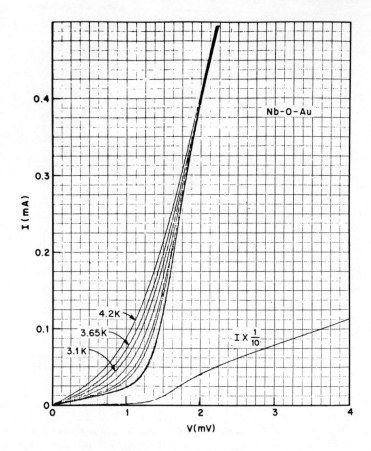

Figure 5: The I-V characteristics of a Nb-P-Au junction at different temperatures.

see an appreciable leakage current, which is larger than in the junction Bennett [Robinson] showed you. When Shen took the superconducting and normal derivatives from this junctions of Fig. 5 and tried to obtain $\alpha^2 F$, etc., μ^* turned out to be negative. This was because, when he evaluated the ratio of the density of states to the BCS density (for a superconductor with the same gap), the ratio went to less than one at voltages just above the gap, as Bennett pointed out and as the MIT data does. Now, if that ratio goes to less than one just above the gap then you are missing the weight of all those phonons down below the transverse peak.

J. Bostock: May I have the last rebuttal? The first thing that I think should be pointed out is---well, that's Shen's

ELECTRON-PHONON SPECTRAL FUNCTION

J. Rowell: data---we don't have knees, we don't have those excess currents, we don't have any of that kind of phenomena.

J. Rowell: The fact that the derivative goes negative---rather, the ratio of density of states to the BCS density goes to less than one, is a hint that some of the knee still remains, and I believe the leakage currents in your junctions are comparable to Shen's.

J. Bostock: I want to show a niobium-gold junction with a positive deviation for niobium that we obtained [see below].

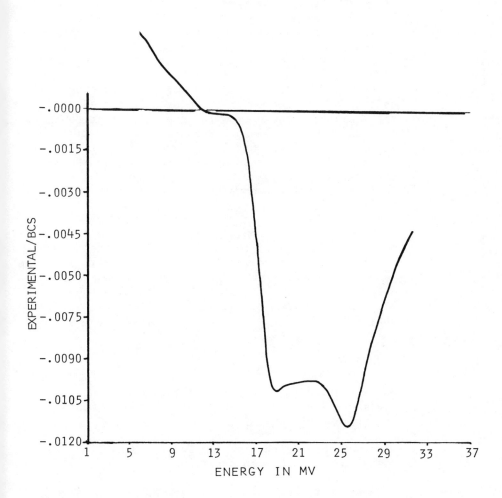

Tunneling curve for a Nb-Au junction.

You can see that it's positive, all right. The results on this junction are: λ is 0.57, μ^* is $-.05$. The point in fact is that deviation, per se, is not as crucial as you would make it out to be. It is true that most of the niobium junctions do not show it, but you can have junctions which do have positive deviations and in fact, still give you negative values of μ^*. I have a slide [Bostock et al., Fig. 7] which I think might be worthwhile showing. It shows the reduced density of states for the gap parameter. You see the reduced density of states as you play with the gap value. You can see that the amplification factor is such as to make that deviation positive. That, however, carries with it implications of changing the phonon structure at the low frequency end [Bostock et al., Fig. 6]. You have not chosen to talk to us beforehand so for example, I am not sure exactly how you did it, but the statement was made that values of λ and μ^* were chosen [see Fig. 8 and Table III of Bostock et al.]. All right, μ^* was fixed. It was put back into the program and different values of T_c, etc., were calculated. But, when you do a deconvolution you have to balance μ^* and λ against the measured structure so the gap value that you have determines both values of λ and μ^* and λ is $2\int \alpha^2 F(\omega)\frac{d\omega}{\omega}$; so, I do not really understand how you did that [fix μ^* and $\alpha^2 F$] and I would appreciate knowing. Perhaps Bennett [Robinson] would want to say. I really think that is important because it sounds like the game playing I did.

B. Robinson: I think if you want to do this in a very self-consistent way--- what you want to do is to take the measured parameters from the experiment, then put them into the gap inversion program. You get out of the program $\alpha^2 F(\omega)$, μ^*, and λ. If μ^* is positive then you can ask yourself, 'if you use the $\alpha^2 F(\omega)$ and then fix μ^* what does λ turn out to be?' That is a very reasonable thing to do. What you do not want to do, I feel, is, if you come up with a negative μ^* to play around with it and say well...

J. Bostock: But you are doing that---you are game playing. That is precisely what your slide shows! You can "measure" that parameter [Δ].

F. Fradin: I do not think either of the competing parties are going to convince each other.

J. Leslie: A question to Robinson. When you make your junction by

evaporating at this very low pressure you say that
you have a good oxide and obtain a Nb film that may
be disordered or what have you, but when you do the
inversion on such samples that you get a very small
μ*. Now let me just ask the question. If every
superconductor has to have a positive μ* of 0.1 why
didn't you get a μ* of 0.1 for the as-deposited Nb
film? You say only when you anneal did you get a
good value of μ*. What happened during the anneal?
If the oxide is just the same and the only thing
that your film has become a bit more ordered, why has
μ* changed?

B. Robinson: I should have made a comment before I presented the
data. What I presented to you was a preliminary
result. What I should have shown you was that in the
as-deposited case for the sample of niobium we did
get positive μ*. We actually saw a shift---I mean
the peaks were the same but we actually generated a
positive μ* and everything was self-consistent in that
respect. Okay, so what happens: For the as-deposited
Nb using a gap of 1.04 mv we got μ* essentially equal
to zero. The density of states drops below BCS at
values just below the gap. We do not fully understand
this behavior in the as-deposited Nb samples. We
annealed the as-deposited sample to the crystalline
state and then reduced the data again now using Δ_0=1.46
mv and then we got a positive μ*. Let me say one other
thing. Suppose you say that Δ_0 is proportional to
(λ-μ*) and then you measure a Δ_0 and you put that value
into the gap unfolding program and come out with a
negative μ*. Since Δ_0 is a net coupling strength and
λ is the strength of the effect in the first derivative,
then if μ* comes out to be negative it tells you that
the strength in the effect of the first derivative
doesn't support the measured gap value. It is very
straightforward, but in order to make a statement like
that you have to believe in the Eliashberg equations-
you can not say that is wrong and then try to use the
program to make adjustments.

N. Cheung: [To Robinson] May I know how the energy gap was determined because I noticed there was a difference between
the energy gap, the MIT value was 1.56 and yours is
1.48. Since you know the deviation from BCS is very
sensitive near there and μ* is only 0.04, so I think...

B. Robinson: Our tunnel junction were aluminum, aluminum-oxide,
superconductor and the aluminum was not superconducting.

We looked at the conductance near the energy gap and used the Bermon's tables---we fit the gap parameter that way.

N. Cheung: How thick was your film?

B. Robinson: I would say about 2,000 Å.

EXTENDED DEFECTS IN A-15 SUPERCONDUCTORS

J.C. Phillips

Bell Laboratories

Murray Hill, New Jersey 07974

ABSTRACT

A brief review is given of the nature of lattice instabilities in high-T_c superconductors. The Frohlich instability in A-15 compounds is indicated to be a microdomain which acts as an embryo for the Martensitic transformations in these compounds.

For several years I have studied [1] the structural systematics of high-T_c superconductors in an effort to develop a phenomenological description of lattice (de)stabilization in these materials. Although the formal results of this effort have been meagre, the physical insight that I have gained has been substantial and has been helpful in understanding related materials. Thus the present paper could be understood as part of a chapter entitled "Superconductivity in Soft Lattices" to be included in a book entitled "The Architecture of Exotic Crystals". Other chapters in the book might be titled "Scaffolded Superionic Conductors" [2] and "Bilinear Organic Addition Compounds" [3].

I should stress that the central reason for this work is the desire to avoid an overly microscopic (Hamiltonian many-body theory or generalized Landau-Gizburg statistical scaling model) approach to exotic materials. It is my feeling that while such approaches are appropriate to simple materials, for the materials listed above the microscopic models necessarily contain simplifications which are likely to be self-defeating, at least from the viewpoint of understanding trends in material properties.

High-T_c superconductors are a good illustration of the dangers inherent in the mechanical application of formulae derived from microscopic models. According to the BCS theory, electrons form Cooper pairs because of the attractive character of the electron-phonon interaction. The formula $T_c \sim \theta_D \exp(-N(0)V)^{-1}$ suggests that high values of T_c could be expected either by increasing θ_D (if $N(0)V$ is constant among related materials) or by increasing the electron-phonon interaction V (if θ_D and $N(0)$ can be kept nearly fixed), or by increasing the density of states at the Fermi energy $N(0)$ (if θ_D and V do not vary appreciably).

A number of papers can be found in the literature taking one or another of these viewpoints, and there is no doubt some value in each of them, especially for discussing trends in elemental and pseudoelemental low-T_c superconductors. However, if from among all of nature's thousands of binary and ternary compounds we select for study only those with the highest T_c's, little can be gained from such simplistic models, e.g., the $N(0)$ models [4] for A-15 compounds. Instead we must ask the obvious question: why are these materials (e.g., Nb_3Sn and more recently Nb_3Ge) the ones with the highest T_c's? In a less obvious way, that is, why are there no materials (e.g., Nb_3Si) with still higher T_c's?

At this point if we have been mechanically modeling microscopic formulae we will almost certainly reach an incorrect conclusion, e.g., we may conclude [5] that because of canceling electronic interactions $T_c(\lambda)$ has a maximum for $\lambda = N(0)V = \lambda_0$. Of course, this mistake would only be made by a theorist; experimentalists have long realized that the fundamental limitation to T_c is not electronic but is rather caused by lattice instabilities. In fairness to the theorists, we should note that the stronger theorists, beginning with Frohlich [6], have also recognized the determinative role played by lattice instabilities.

Are there some rules of thumb which can be gleaned from an architectural overview which will guide us in our choice of models, and will enable us to see whether electronic or lattice models are preferable? There are. If we consider a family of related materials, and through alloying or other methods (suggested by the periodic table) increase the electron-lattice coupling, there are two extreme cases to consider (actual crystals will fall closer to one extreme or the other). These are : (A) normal, and (B) exotic.

(A) The increase electron-lattice coupling applies rather uniformly to all normal modes. In this case the simplistic, mechanical approaches may have some value. Examples of this type are $T\ell$-Pb-Bi alloys and bcc transition metal alloys. Both cases are pseudoelemental and in both cases the highest value of T_c is limited by a first-order phase transition to an entirely different crystal structure. The highest known transition temperature in materials of this type is achieved in Nb_{1-x}-Zr_x alloys and is about 11 K. It is possible that electronic effects (especially N(O)) dominate the trends in T_c in all of these materials, and electronic effects may explain the discontinuity in $\partial^2 T_c/\partial p2x$ in these alloys [7]. On the other hand, Nb itself is rather unstable, as shown by neutron studies of the lattice vibrations near q = 0, and it has been suggested [8] that Nb-Zr alloys may contain microdomains similar to those predicted for the A-15 compounds. In any event, this case is marginal, and much more interest attaches to the A-15 compounds discussed below.

(B) The increased electron-lattice coupling is concentrated largely in one normal mode. Binary and ternary compounds appear to be particularly favorable cases for this kind of behavior. There are two binary families that I will discuss here, NbN (B-1 structure) and Nb_3Sn (A-15 structure). There are several interesting ternary families, such as $PbMo_6S_8$ [9] and $BaMO_3$ [10], which remain (at least for me) undeciphered at present, but their transition temperatures are anomalously high, so I would include them in this group.

The two families of exotic structures which are becoming increasingly well understood are the interstitial (B-1) carbides, nitrides and oxides, with a maximum value of T_c occurring near x = 0.1 in $NbN_{1-x}C_x$, and the A-15 compounds. For many years the primary structural defect that was emphasized in the B-1 compounds was nonstoichiometry. As I have pointed out [1], there are really two kinds of structural defects to consider, those arising because of the structure of the sample at high temperatures near the melting point T_m, and those arising at low temperatures because of incipient lattice instabilities (of an essentially Martensitic type, which we define more carefully below). This simple theoretical separation would not appear to be of much practical value, however, since "all" samples "must" be formed initially at T near T_m, "of course". But not "of course", because by sputtering on substrates of various temperatures in the presence of variable partial pressures of background inert gases, we may be able to grow films at temperatures T of order $T_m/2$. Then depending on the details of the phase diagram, and in particular on the extent to which retrograde solubility [11] occurs at lower temperatures, we may be able to suppress the high-temperature defects, raise T_c, and concentrate our attention on the low-temperature defects which provide the ultimate limitation on attainable values of T_c.

The low-temperature instability may well be called a Frohlich instability, to distinguish it from the high-temperature formation instabilities. To my knowledge the Frohlich configuration coordinate for a given gamily of structurally homologous materials was first identified [1] in the NbN family; it turns out simply to be the cubic lattice constant. That a hydrostatic lattice instability is

involved can be established by studying a(Z) for each sequence TM
(M = C, Z = 4, M = N, Z = 5, M = O, Z = 6). It is found that the
coefficient of the term Z corresponds to nonstoichiometry (formation
instability) while the term in Z^2 corresponds to the Frohlich instability [1]. The critical prediction of the model is the sign of
$\partial T_c/\partial p$, which is given correctly in NbC, NbN and ZrN.

The identification of the Frohlich instability in the A-15
compounds has proved much more difficult. As the title of this
paper suggests, it is now believed by several others as well as myself that the Frohlich instabiltiy in V_3Si, Nb_3Sn and Nb_3Ge is
probably an extended defect of overall tetragonal symmetry, which I
would call a microdomain. The microdomain is probably an ellipsoid
of revolution, with its principal axis along a (110) direction of
the cubic host crystal. When long-range order develops in the
coordinates of the centers of these microdomains, and when the microdomains are oriented in a staggered fashion, we obtain [12] the
Martensitic (100) deformed structure first identified in V_3Si.

We see, then, that microdomains are natural candidates for the
Frohlich instability in A-15 compounds. These instabilities characteristically involve macroscopic lengths, e.g., $a^2/\delta a$ in the B-1
compounds, where δa is the departure of a(Z) from Vegard's law
(linearity in Z). Here the microdomain dimensions can be estimated
from Axe and Shirane's classic neutron scattering data [13] on Nb_3Sn;
the domain has a pancake shape, and the shortest radius (thickness)
of the pancake along a (110) direction is about 10a. The situation
is similar to the one found in Nb-Zr alloys where ω-phase nuclei
(double hcp symmetry) have been found [14] in samples supposedly
in the bcc region of the phase diagram. In other words, because of
a lattice instability associated with electrons near the Fermi energy, we have an intrinsic coexistence of phases where conventional
theory would lead us to expect a single phase. The nuclei or
embryos of the second phase are "frozen in" by strain fields associated with grain boundaries. This point of view is most helpful
in attempting to understand why nucleation energies for Martensitic
transformations in, e.g., steel, are so low [15].

The importance of Frohlich lattice instabilities is considerable
to experimentalists because ferroelastic interactions can often
appear to make it impossible to obtain reproducible data. Recognition of the presence of Frohlich instabilities is also essential
to theorists. At present it is fashionable to discuss phase transitions either entirely in microscopic electronic terms (Fermi surface
models) or more macroscopically using free-energy density models of
the Landau-Ginzburg type. However, an important assumption to both
types of models, which is almost never stated explicitly, is that
the systems in question are single phase. In some compounds, such
as layer compounds which have charge-density-wave transitions,

these trivial models are quite appropriate because of the inherent lattice stability of the underlying material. (After all, with many layer compounds, the layers can be peeled off almost one-by-one with scotch tape.) On the other hand, whenever microdomains are present (e.g., A-15 materials, linear organic conductors), such trivial field-theoretic models obscure the basic physics of coexisting phases.

New superlattice patterns in A-15 compounds have been discovered by Schmidt, et al. [16]. The spots ascribed to CDW observed at room temperature disappear abruptly > 486 K. Superstructure simplification also occurs at elevated temperatures. Similar anomalies in other materials will be reported elsewhere [16]. The transitions can be explained as the condensation of charge density waves, probably on defects, to form microdomains. The growing intensity of the CDW corresponds to larger and larger atomic displacements in the microdomains. The similarity to ω-phase microdomains in bcc transition metal alloys [14] is striking. The microdomains finally order and macroscopic tetragonal symmetry is recovered below the Martensitic transformation temperature T_m, which are of order 20 - 40 K.

Between T_{cd} and T_m anomalous enhancement of the resistivity is observed [17], [18], which goes to zero (e.g., by comparing [17] Nb_3Sn with Nb_3Sb) very close to T_{cd}. The additional resistivity appears to be thermally activated with an excitation energy of order a cubic transverse acoustic zone boundary phonon [18]. While this decomposition is probably oversimplified, it is qualitatively suggestive of a cubic → hexagonal lattice instability, which can be described by such a phonon mode [19]. The "freezing in" of this distortion in Nb_3Sn accounts for its excess resistivity over Nb_3Sb. Very similar behavior has been observed [20] in several other structurally homologous metal pairs. This raises the interesting question of whether the resistivity bulges are always associated with the formation of microdomains, and whether the latter always result from the condensation of CDW's. (Evidently in the alloy case composition fluctuations broaden the transition, making it more difficult to identify the ω-phase as a true "strongly condensed" CDW.)

In another paper presented at this conference [21] it is suggested that "excitonic superconductivity" may exist at $Ge-NbGe_2$ interfaces, with T_c ~ 16 K. An alternative explanation for this phenomenon, more closely related to current praxis of Schottky barriers [22], is that an interfacial phase has been formed which is metastable in a quasi-two-dimensional geometry, although it does not appear in the bulk phase diagram. The reason for the occurrence of such a phase can be given picturesquely as a Gibbs oscillation in $x(z)$ for the system Nb_xGe_{1-x} where z is the coordinate normal to the interface, x = 0 to the left, x = 1/3 to the right, and x = 1/2 (Gibbs oscillation) in a narrow, metastable interfacial region. Presumably NbGe in the interfacial phase is quasi-cubic [23] with the NbN (T_c = 18 K) structure.

REFERENCES

1. J.C. Phillips, Phys. Rev. Lett. 26, 543 (1971); 29, 1551 (1972); 28, 1196 (1972); in Superconductivity in d- and f-Band Metals (ed. D. Douglass, AIP Conf. Proc. No. 4, 1972); Sol. State Comm. 18, 831 (1976).
2. J.C. Phillips, J. Electrochem. Soc. 123, 434 (1976).
3. J.C. Phillips (to be published).
4. J. Labbe, Phys. Rev. 172, 451 (1968).
5. W.L. McMillan, Phys. Rev. 167, 331 (1968). But see P.B. Allen and R.C. Dynes, Phys. Rev. B12, 905 (1975).
6. H. Frohlich, Proc. Roy. Soc. A215, 291 (1952).
7. T.F. Smith, in Superconductivity in d- and f-Band Metals (ed. D. Douglass, AIP Conf. Proc. No. 4, 1972), p. 306.
8. C.M. Varma, J.C. Phillips and S.T. Chui, Phys. Rev. Lett. 33, 1223 (1974); J. Noolandi and C.M. Varma, Phys. Rev. B11, 4743 (1975).
9. M. Marezio, et al, Mat. Res. Bull. 8, 657 (1973).
10. A.W. Sleight, et al, Sol. State Comm. 17, 27 (1975).
11. Retrograde solubility is an effect which is well known for impurities in semiconductors. Because the lattice instabilities which give rise to large electron-phonon coupling and high values of T_c have a covalent character, it is convenient to use this term to describe bowing of, e.g., the A_3B and A_5B_2 phase boundary in Nb_3Ga. This bowing was first noted by G.W. Webb in Superconductivity in d- and f-Band Metals (ed. D. Douglass, AIP Conf. Proc. No. 4, 1972), p. 142, where a correlation between the bowing, the formation temperature T_f, and the transition temperature T_c was shown.
12. B.W. Batterman and C.S. Barrett, Phys. Rev. 145, 296 (1966).
13. J.D. Axe and G. Shirane, Phys. Rev. B8, 1965 (1973).
14. J.D. Axe, D.T. Keating and S.C. Moss, Phys. Rev. Lett. 35, 530 (1975).
15. P.C. Clapp, Phys. Stat. Sol. (b) 57, 561 (1973).
16. P. Schmidt, et al (this conference), and private communication.
17. Z. Fisk and G.W. Webb, Phys. Rev. Lett. 36, 1084 (1976) and this conference.
18. S. Williamson and M. Milewits, this conference.
19. W. Bührer and P. Brüesch, Sol. State Comm. 16, 155 (1975).
20. Z. Fisk and A.C. Lawson, Sol. State Comm. 13, 277 (1973).
21. A.K. Ghosh and D.H. Douglass, this conference.
22. J.M. Andrews and J.C. Phillips, Phys. Rev. Lett. 35, 56 (1976).

See also a forthcoming review of interfacial phases by J.C. Phillips and J.A. VanVechten.
23. A.K. Sinha, et al, J. Appl. Phys. $\underline{43}$, 3637 (1972).

QUESTIONS AND COMMENTS

F. Mueller: Jim, in your introductory remarks where you were rejecting what I will call the standard explanation for anomalies in the phonon spectra of materials---based on electron-phonon interactions and calculations of Fermi surfaces, etc.---I wonder whether in your considerations, where you were talking about how Kohn anomalies developed and so on, whether it is possible that in fact the anomalies could be caused by a related mechanism based on the difference between normal coupling and what I will call "umklapp" coupling. Doesn't it depend sensitively on the way the umklapps enter and isn't it in fact true that if one considers (carefully) umklapp coupling that it is possble it could be the explanation? Now the standard material one looks at is Pb where one does get the "normal" Kohn effect---but I wonder whether in the exotic materials, such as Nb_3Sn that you were discussing, it is possible that the "standard explanation" works, too.

J. Phillips: My basic inclination is to say the experimental data at present in the compounds systems indicate that the lattice instability is the dominant mechanism and the Fermi surface interactions are too weak to be the dominant mechanism. I realize there are---I am aware there are Umklapp terms---but, for instance, in the niobium carbides which I did not show but which Harold Smith probably talked about. Some of these anomalies do not move with alloying, so if you want to say the Umklapp terms correspond to the off-Fermi energy interactions that is all right with me. I would not deny there are electronic explanations for the carbides but I would not call them Fermi surface explanations in the case of the A-15 compounds. The Axe and Shirane data, in my opinion, cannot be explained in Umklapp terms at all. I think that is ordering of defects and the martensitic transformation which takes place is connected with the ordering of the microdomains, the tetragonal microdomains of the kind Chui and I calculated. There is basically a order-disorder transformation going on there and that is not convenient to describe in electronic terms because the microdomains in question might be 15 or 20 Å in diameter. Those things, by the way, were known to metallurgists---even Zener put forth nearly 30 years ago an idea that one would be especially likely to find martensitic transformations in soft lattices. About 10 years later Kaufmann pointed out that the problem with the martensitic transformations at low temperatures would be nucleating them and the easiest things to assume would be that you had embryos or microdomains which nucleate these phase transformations. So we have a system here with these soft lattices where I think before we get to the electronic degrees of freedom we should pay close attention to the degrees of freedom associated with the atoms and the lattice itself and the mechanical instabilities of the lattice itself.

G. Webb: Could you give us a prediction for the number of defects in some representative materials which are needed to pin and/or order these microdomains?

J. Phillips: We think that in the A-15 compounds that the characteristic number for stabilizing the lattice is about 1%. (Point defects.) But around each point defect is a microdomain of local tetragonal distortion.

IMPURITY STABILIZATION OF Nb$_3$Ge

J.R. Gavaler

Westinghouse Research Laboratories

Pittsburgh, Pennsylvania 15235

ABSTRACT

Experimental data are reported which support the hypothesis that the high-T_c Nb-Ge phase is impurity stabilized. Oxygen, nitrogen, and silicon have thus far been used as impurities to deposit Nb-Ge films which have superconducting onset temperatures of $\gtrsim$ 20 K.

In 1965, Matthias, Geballe, and co-workers reported the preparation of bulk Nb-Ge samples which remained superconducting up to $\sim$ 17 K [1]. These samples, prepared by splat cooling, had the A15 crystal structure and were believed to be close to the stoichiometric Nb$_3$Ge composition. However, they were also crystallographically disordered with almost half of the germanium sites being occupied by niobium. These authors suggested that an ordered stoichiometric Nb$_3$Ge phase would have an even higher critical temperature. Due to the chemical instability of such an Nb$_3$Ge phase, efforts to test this hypothesis were unsuccessful until recently, when Nb-Ge films were prepared which had T_c's of over 22 K [2]. These films were deposited by a sputtering process which employed unusually low sputtering voltages ($\lesssim$ 1000 V), high argon pressures ($\gtrsim$ 0.2 Torr), and high substrate temperatures (700 to 1000°C). Sputtering, similar to splat cooling, can under certain conditions be a very fast quenching process. However, the use of low voltages

---

*Supported in part by AFOSR Contract No. F44620-74-C-0042.

and high argon pressures would minimize the quenching aspect of sputtering. Therefore, it was concluded that the main importance of this low energy sputtering method was that it provided a means by which the desired Nb-Ge phase could be formed at the temperatures at which it was stable, i.e. $\lesssim 1000°C$. Initial results, however, provided no explanation why this phase was in fact stable at these temperatures. More recent data which are summarized in this paper now suggest that the stabilization results from the incorporation of impurities, such as oxygen, into the film structure during the deposition process.

When very high T_c's were originally observed in sputtered Nb-Ge films, it was assumed that the impurity content of these films was very low. This was based on the fact that they were deposited in an ultra-high vacuum enclosure whose background impurity level was typically reduced to the 10^{-10} Torr range prior to sputtering. Compositional analyses showed, however, that the films in fact contained significant levels of oxygen [3,4]. For example, it was found that even in the highest-T_c films the oxygen content in the center portion of the film averaged about 0.1 atom per Nb_3Ge. These analytical results, raised the obvious question regarding what effect the oxygen was having on T_c. To help answer this question, a series of films were prepared which had even higher oxygen concentrations. This was done by sputtering Nb-Ge under otherwise optimum conditions but with various levels of oxygen mixed in with argon at partial pressures ranging from 10^{-9} to 10^{-4} Torr. It was found that films sputtered at O_2 pressures greater than $\sim 10^{-4}$ Torr were not superconducting down to 4.2 K. The films sputtered at lower O_2 pressures were superconducting and the T_c's of six of these films are shown in Fig. 1 where they are plotted as a function of oxygen content. The oxygen analysis in this case was made by Auger spectroscopy. The thickness of each of these films was about 5,000 Å and the data shown in the figure were obtained at depths of $\sim 2,000$ Å. Although, as can be seen, high T_c's persist in films, which have quite high oxygen concentrations, there is no indication that the presence of oxygen at levels greater than ~ 0.1 atom/Nb_3Ge is having a beneficial effect on T_c. To investigate the influence of oxygen at concentrations below this level a new growth method was employed to deposit Nb-Ge films. The method involved sputtering niobium in an argon and germane (GeH_4) atmosphere. Other conditions were similar to those used previously. It was felt that germane would provide a very strong gettering action for oxygen both as a gas and after it was decomposed into finely divided germanium and hydrogen. The amount of germanium introduced into the system, and relatedly the composition of the Nb-Ge films, was controlled by regulating the germane partial pressure. Using this method a series of films were prepared which had a wide range of Nb/Ge ratios. In these films, through x-ray analyses, the niobium, hexagonal Nb_5Ge_3, $NbGe_2$ and germanium phases were identified. The samples which had

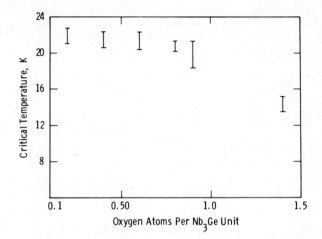

Figure 1: Data illustrating the effect of oxygen on T_c in Nb-Ge sputtered films. X-ray analyses show that most of the oxygen in these films is in the form of a Nb-O second phase.

a Nb/Ge ratio of ~ 3 showed only the presence of the niobium b.c.c. structure. The absence of any A15 structure material correlated with T_c measurements which indicated that none of these films were superconducting down to 4.2 K. Some additional films were then prepared using a two-piece niobium and vanadium sputtering target. The sputtering was done with a germane partial pressure which produced a 3/1 ratio of Nb and/or V to Ge in the films. The critical temperatures of these films, plotted as a function of substrate location, are shown in Figure 2. As can be noted, T_c's up to 8.2 K were obtained in the V-Ge samples. This is slightly higher than the best reported bulk value of 6.5 K. X-ray analysis of these samples showed they had the A15 structure with a lattice parameter of ~ 4.77 Å. The Nb-Ge samples, on the other hand, which were located beneath the Nb portion of the target showed no evidence of the presence of an A15 phase, either from x-ray or from T_c measurements. Analysis of Nb-Ge films, located at the extreme end of the substrate holder beneath the Nb portion of the target, showed that they had the b.c.c. niobium structure. However, the structure had a greatly contracted lattice with an a_o of $\lesssim 3.25$ Å. (The normal lattice parameter for niobium is 3.306 Å.) It is quite conceivable that an Nb-Ge solid solution containing 25 atomic % germanium would in fact have such a contracted lattice.

It was concluded that the apparent inability to prepare an A15 Nb-Ge phase, using the germane method under high purity conditions, supported the hypothesis that oxygen (or some other defect) is

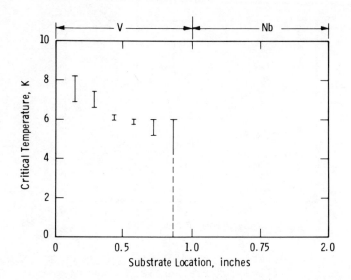

Figure 2: T_c data on films sputtered from a composite niobium and vanadium target in an argon-germane gas mixture. The partial pressure of germane was set to obtain a niobium and/or vanadium to germanium ratio of 3.

required to stabilize the Nb-Ge A15 phase. To try to obtain more direct evidence regarding the importance of impurities, an attempt was made to sputter Nb-Ge films with very low oxygen contents, using the originally reported method. It was felt that the most probable source of oxygen in the original experiments was from outgassing from hardware in the deposition chamber. To minimize this source of contamination, the sputtering chamber was completely disassembled and cleaned, primarily by etching in a HNO_3-HF acid solution. After reassembly, the niobium-germanium target was also sputter-cleaned for several hours using a high (2500 V) sputtering voltage. Following these operations the usual 0.5" x 0.125" x 0.020" polished sapphire substrates were positioned on the holder and Nb-Ge films were deposited using the low energy sputtering conditions previously described [2]. Contrary to the original results, none of these Nb-Ge films were found to be superconducting down to 4.2 K. A second test was then run but in this case oxygen at a pressure of 10^{-6} Torr was introduced into the system via a leak valve. In this experiment several of the Nb-Ge films were superconducting with onsets up to as high as 20.9 K. Repeating the test without oxygen again produced films all of which were non-superconducting down to 4.2 K. This series of experiments was then repeated using nitrogen and also silicon as the impurity element. [The silicon was introduced into the deposition chamber in the form of silane (SiH4).] The results were similar in that only in the tests where the impurity gas was added were any of the films found to be superconducting. The significant data from these experiments are summarized in Table 1 with the T_c's for the best film in each of the experiments shown.

Table 1. Effect of Impurities on T_c of Nb-Ge Films

Impurity Gas	Pressure (measured prior to sputtering)	T_c (onset)	Mid-Point	ΔT_c
O_2	10^{-6} Torr	20.9 K	17.0 K	5.2 K
N_2	6×10^{-7}	20.3	18.8	2.0
SiH_4	5×10^{-4}	18.8	15.7	5.7

Undue emphasis should not be attached to the differences in T_c values obtained with the different impurities. Due to the inability to monitor and regulate the impurity gas pressure during the actual sputtering operation, it is probable that the optimum impurity concentrations for obtaining the maximum T_c's were not achieved in any of these experiments. In essence, therefore, all of these results support the conclusion that the high-T_c A15 structure Nb-Ge is a low temperature phase which is impurity stabilized. There is an historical irony here worth noting. As is now well known, the original A15 structure material, the so-called β form of tungsten, is a low temperature phase of tungsten which is believed to be stabilized by a small amount of oxygen atoms [7].

Accepting as fact that high T_c Nb-Ge films are impurity stabilized allows one to offer some plausible hypotheses for previously inexplicable experimental results. For example, it is now probable that the occasional, i.e. non-reproducible, appearance of very high critical temperatures near ~ 23 K, that we and others have observed [5,6], is due to the fortuitous presence of an optimum impurity content in these films. Another not previously understood observation is the presence of high T_c's in Nb-Ge films which are far removed from ideal 3/1 stoichiometry [6,8]. In this case impurity stabilization may permit formation of an A15 structure containing a large number of vacancies in the 'B' (germanium) sites. Alternatively, the impurity atoms may themselves occupy 'B' sites. In either case, the 'A' (niobium) sites could be completely occupied by niobium atoms, and the 'B' sites by some combination of germanium atoms, impurity atoms and vacancies. Under these conditions the integrity of the three mutually orthogonal chains formed by the 'A' atoms in the A15 structure would be maintained. Also, since there would be no niobium atoms on the 'B' sites there would be no d-orbital bonding between 'A' and 'B' atoms. Although there is only a partial theoretical understanding of the unique properties found in A15 superconductors, the consensus seems to be that these are two of the requisites for obtaining high T_c's in such compounds [9].

Assuming the validity of the impurity stabilization hypothesis there are still several important and unanswered questions remaining such as by what mechanism does the stabilization occur, what other impurities can be used, and what concentrations of the impurities are required. These questions can, of course, be answered only after further study.

ACKNOWLEDGEMENTS

I thank Alex Braginski and Martin Ashkin for the benefits accrued from many discussions, Jan Schreurs for the Auger data of Figure 2, and A.L. Foley for valuable technical assistance.

REFERENCES

1. B.T. Matthias, T.H. Geballe, R.H. Willens, E. Corenzwit, and G.W. Hull, Jr., Phys. Rev. 139, A1501 (1965).
2. J.R. Gavaler, Appl. Phys. Letts. 23, 480 (1973).
3. J.R. Gavaler, J.W. Miller, B.R. Appleton, Appl. Phys. Letts. 28, 237 (1976).
4. A.T. Santhanam and J.R. Gavaler, J. Appl. Phys. 46, 3633 (1975).
5. J.R. Gavaler, M.A. Janocko, and C.K. Jones, J. Appl. Phys. 45, 3009 (1974).
6. L.R. Testardi, R.L. Meek, J.M. Poate, W.A. Royer, A.R. Storm, and J.H. Wernick, Phys. Rev. B11, 4304 (1975).
7. A.F. Wells, Structural Inorganic Chemistry 3rd Edition, Clarendon Press, Oxford (1962).
8. G.W. Roland and A.I. Braginski, Proceedings of the Int. Cryogenic Materials and Cryogenic Engineering Conf., Kingston, Ont., 1975 unpublished.
9. See for example, M. Weger and I.B. Goldberg, Solid State Phys. 28, 1 (1973) edited by F. Seitz and D. Turnbull, Academic Press, New York.

QUESTIONS AND COMMENTS

M. Cohen: Do you know if it is atomic or molecular oxygen?

J. Gavaler: The indication is that it should be atomic, or at least the effect is stronger when it is atomic.

M. Cohen: Even close to the surface?

J. Gavaler: I do not know. The basis on which I answered the question is not from my results particularly. I understand, and perhaps I can be corrected on this, that the Los Alamos group had some results which showed an exhanced T_c in their CVD films when they had oxygen contamination originating from the source material; it thus provided a source for atomic oxygen. They did not get the same enhancement when they introduced molecular oxygen.

SUPERCONDUCTIVE TUNNELING INTO NIOBIUM-TIN THIN FILMS [†]

D.F. Moore

Department of Applied Physics, Stanford University[*]

J.M. Rowell

Department of Applied Physics, Stanford University[*]
and
Bell Laboratories, Murray Hill, New Jersey 07974[**]

and

M.R. Beasley

Departments of Applied Physics and Electrical Engineering
Stanford University[*], Stanford, California 94305

ABSTRACT

Oxide layer tunnel junctions exhibiting both quasi-particle and Josephson tunneling were produced by overlaying lead on niobium-tin thin films. The superconducting transition temperature and average energy gap of the Nb-Sn both increased as the tin concentration was increased toward Nb_3Sn and reached maximum values of 17.2°K and 3.2 meV, respectively, or $2\Delta = 4.3$ kT_c.

[†] A full account of this work will be published in Solid State Communications.

[*] Work at Stanford supported by the National Science Foundation through the Center for Materials Research at Stanford University, by the Joint Services Electronics Program, and by the Air Force Office of Scientific Research, Air Force Systems Command, USAF, under Grant No. AFOSR 73-2435.

[**] Permanent address.

THE STRUCTURE OF SPUTTERED SUPERCONDUCTING Nb_3Ge FILMS REVEALED BY X-RAY DIFFRACTION, TRANSMISSION ELECTRON DIFFRACTION, AND BY TUNNELING

P.H. Schmidt, E.G. Spencer, D.C. Joy and J.M. Rowell

Bell Laboratories
Murray Hill, New Jersey 07974

ABSTRACT

Tunneling studies of sputtered Nb_3Ge films have revealed the presence of two energy gaps, one which has been identified with the high T_c A15 phase and the other ascribed to highly disordered or amorphous Nb_3Ge. In this paper we first give some details of the tunneling experiments which have not been discussed previously. The surprising tunneling results have led us to investigate the structure of these films using both x-ray diffraction at 300°K and transmission electron diffraction between 300 and 900°K. The complex electron diffraction patterns show both superstructure and diffuse scattering. This is interpreted as evidence for structural defects. A striking transition in the TEM patterns has been observed at 486°K.

INTRODUCTION

At present, the superconductor with the highest transition temperature, Nb_3Ge, is most readily prepared by sputtering,[1] co-deposition from electron beam sources,[2] or by chemical vapor deposition.[3] The reason why these preparation techniques result in films with transition temperatures as high as 23°K, whereas bulk material commonly has a T_c of 6 to 17 K, remains a mystery. It has been suggested by both Gavaler[4] and by Hallak et al.[5] that it is necessary to incorporate oxygen in the films to stabilize the A15 structure. Although we have, at present, no evidence for oxygen within our sputtered films, we have already demonstrated[6] that tunneling experiments clearly indicate the presence of highly disordered or amorphous material within even the highest transition temperature Nb_3Ge films. In this paper we will give some further

details of the tunneling results which led us to this surprising
conclusion. To investigate the structure of these films further,
we have also carried out x-ray diffraction and transmission
electron diffraction studies. As we will show below, x-ray diffraction indicates that the films are practically single phase
A15 Nb_3Ge. Only two unusual diffraction peaks are observed, along
with quite strong diffuse scattering beneath some of the A15 lines.
However, the transmission electron microscope reveals in much
greater detail the structure of these Nb_3Ge films. The resolution
of this instrument allows essentially single crystal regions of the
films to be examined, and the resulting complex diffraction patterns
contain structural information in the form of both superstructure
and diffuse scattering detail.

A. TUNNELING EXPERIMENTS

The details of the preparation of the superconducting films
and the manufacture of the tunnel junctions on the freshly sputtered
surface has been described previously[6] and will not be repeated here.
A current vs. voltage characteristic typical of junctions made on
many of the highest T_c films is shown in Fig. 1. The rise in current
near 5 mV is ascribed to the high T_c material and occurs at the sum
of Δ_{Pb} and Δ_H, where Δ_H is the energy gap of Nb_3Ge. However, a
sharp rise in current also occurs between 1.7 and 2 mV, as can be
easily seen when the current scale is expanded 10 or 100 times.
We have argued earlier that this rise in current must be interpreted
as indicating the presence of a second phase within the Nb_3Ge film,
which therefore has a second lower energy gap Δ_L. Further expansion
of the current measurement in Fig. 1 shows a structure at Δ_{Pb}, which
is not uncommon in junctions exhibiting Josephson effects (the
simplest interpretation is that the Josephson radiation of frequency
2 eV is absorbed by the Pb film of energy gap $2\Delta_{Pb}$), and a further
shoulder at about 1 mV, which can be seen to be $\Delta_{Pb} - \Delta_L$.

To demonstrate the range of behavior we have observed in the
current-voltage characteristics for a large number of Nb_3Ge-oxide-Pb
junctions, we have selected some extreme examples in Fig. 2. In
junction 21-76-B4, although the transition temperature of the film
was still reasonably high (see Table I), there is only weak evidence
in the I-V characteristic to indicate the present of the high gap
phase. In fact, if the tunneling characteristic can be interpreted
at all quantitatively, this trace would indicate that less that 10%
of the material is the high temperature phase. Junction 52-75-1-3
was made on a film which exhibited the highest transition temperature
onset (23 K), but the rise in current near 1.8 mV is still strong
and the magnitude of the larger energy gap, Δ_H, is not the highest
we have observed, despite the high T_c of the film. Junction 1-76-B5
is included because, of those made on films sputtered at 975°C, it

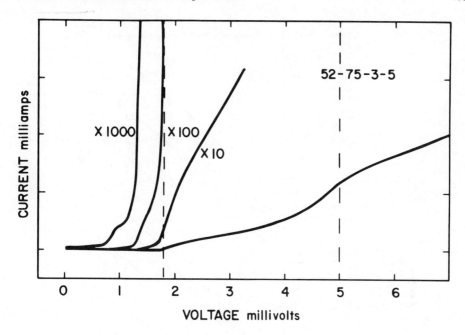

Figure 1: Current versus voltage traces for a Nb_3Ge-oxide-Pb junction at 1.3 K. The current scale has been expanded 10 times for each consecutive trace.

shows least of the low gap phase and the most obvious evidence of a strong rise in current from the high T_c material. However, the largest energy gap we have seen in any film is that from 8-76-B5. By subtracting Δ_{Pb} = 1.35 meV from the voltage where the current-voltage characteristic has maximum derivative, we obtain Δ_H = 3.9 meV. With the resistive superconducting transition having a midpoint of 21.55 K, this indicates that $2\Delta/kT_c$ = 4.2.

The result for junction 4-76-D4 is crucial to our argument that highly disordered material is responsible for the low gap, Δ_L, within these films. This particular junction was prepared on a film sputtered from the same target as all those discussed so far, but the table temperature was set at 200°C. At these temperatures the Nb/Ge ratio is maintained but the constituents of the film apparently have no chance to order into the A15 structure. The transition temperature for this film is 3.06°K (an identical value was obtained with the table set at 100°C), which is very close to that obtained when high T_c Nb_3Ge films are very heavily damaged,

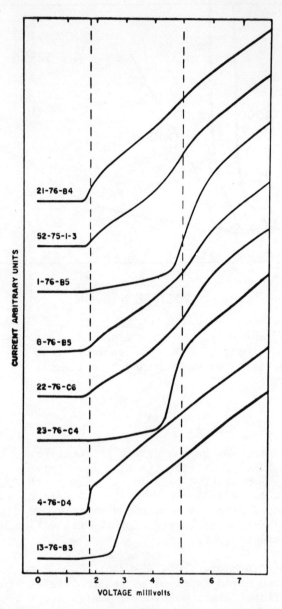

Figure 2: The variety of current-voltage characteristics that can be obtained from Nb_3Ge-oxide-Pb junctions, and from one $Nb_{80}Ge_{20}$-oxide-Pb junction. For details of the film preparation, see Table I and the text. Traces have been offset vertically for clarity.

Table 1

Deposition #	Table(1) Temperature	Pressure(2) (microns)	Resistance Ratio	Transition Temperature			ΔT_c	Target Composition
				Onset(3)	Midpoint(4)	End(5)		
52-75	975	600	2.35	23.0	22.2	21.6	1.4	75:25
1-76	975	700	2.22	21.9	21.35	20.95	0.95	75:25
4-76	200	600	.94	3.085	3.066	3.020	.065	75:25
7-76 Film on substrate	975	600	1.88	21.00	20.00	19.65	1.35	75:25
Free standing film				20.75	19.90	19.10	1.65	
8-76	975	800	2.00	21.95	21.55	21.0	0.95	75:25
10-76	775	600	1.12	1.98	1.80	1.70	0.28	62.5:37.5
13-76	975	600	1.26	18.95	15.25	11.60	7.35	80:20
21-76	975	600	2.09	21.95	21.60	21.2	0.75	75:25
22-76	925	600	2.12	21.25	20.90	20.45	0.80	75:25
23-76	1025	600	1.88	20.95	19.62	19.50	1.45	75:25

(1) The table temperature is read with a thermocouple directly attached to the platinum.
(2) Argon pressure is read with a capacitance manometer.
(3) Onset of transition is chosen as the first deviation (∼ 0.5%) from the normal state resistance.
(4) Midpoint is the temperature giving half the normal state resistance.
(5) The end of the transition is when less than ∼ 0.2% of the normal state resistance remains.

as reported by Poate et al.[7] It is interesting to note that we had previously found that Nb_3Sn films made in the same way had a T_c of 2.9°K. Only later was it demonstrated that the T_c of highly damaged Nb_3Sn films also saturates at this low transition temperature.[8] The nature of this highly disordered material will be discussed in more detail below.

To date most of the Nb_3Ge films have been made from a $Nb_{75}Ge_{25}$ sputtering target, and only a few attempts have been made to vary the film stoichiometry. The film 13-76-B was made from a $Nb_{80}Ge_{20}$ target, and shows only a single energy gap at 1.45 meV, which is close to the energy gap of pure Nb. However, tunneling studies at high voltages can be used to reveal the phonon spectrum of this material, clearly indicating that this film is not Nb. The resistive transition of this film was broad, extending from 11.6 to 19 K, but the energy gap indicates that the bulk of the film has a transition closer to 9°K.

All the films for which we have previously published data were made at a heater temperature of 975°C. The temperature of the substrate is probably 150°C lower than this, and it is known that substrate temperature is one of the important parameters in all the growth techniques that result in high T_c Nb_3Ge. In Fig. 2, film 22-76-C was made with the table temperature set 50° lower than usual, at 925°C. Clearly the current-voltage characteristic is not too dissimilar from those of many of the 975°C films, although T_c has dropped by about 1 K. The effect of increase in temperature is much more drastic, as can be seen in film 23-76-C, which was prepared at a table temperature of 1025°C.

In Fig. 3 we show expanded current measurements on this junction 23-76-C. There is clearly a rise in current at Δ_{Pb}, probably due to the interaction of the a.c. Josephson current with the Pb film. However, only one junction made on this film showed even weak indications of a rise in current near 1.8 mV, showing that this film is considerably more ordered than those made at 975°C. The magnitude of the gap, Δ_H, is reduced to 3.35 meV, and this is accompanied by a relatively low transition temperature of 19.6 K. This result is in agreement with the phase diagram of Nb_3Ge for films made by electron beam codeposition,[5] where an increase in growth temperature results in movement of the A15 phase boundary away from the 3:1 composition. This further illustrates the apparent delicate balance between order and stoichiometry which is achieved in all the preparation methods used for these high T_c films.

The presence of the disordered material in the highest T_c films leads to serious difficulties in quantitative analysis of "phonon induced structure" in the tunneling characteristic, which

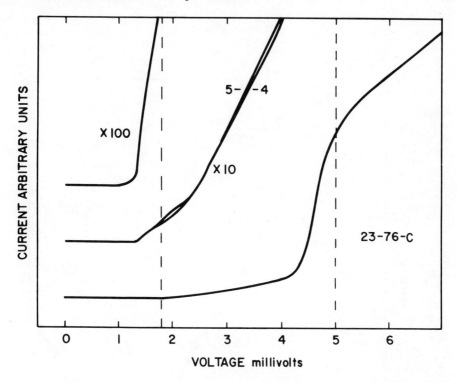

Figure 3: The effect of using a high table and substrate temperature (1025°C) is shown in the current-voltage characteristic of this Nb_3Ge-oxide-Pb junction. To show detail at low voltages, the current scale has been offset and expanded 10 and 100 times.

is measured at voltages up to 40 mV in these junctions. However, we show in Fig. 4 the reduced density of states for junction 1-76-B5, whose I-V characteristic has been shown in Fig. 2. As directly measured, the density of states at low voltages exceeds that of a BCS superconductor with the same gap, but dros between 11 and 24 mV and again near 30 mV. After subtracting the energy gap of 3.7 meV from this scale, this indicates a broad low energy phonon peak with a maximum at 17 meV and a shoulder near 8 meV. A small second peak is located at 24 meV. Such an $\alpha^2 F$ spectrum is very similar to that measured for Nb_3Sn by Shen,[9] with no clear evidence for any higher frequency modes. Unfortunately, it is

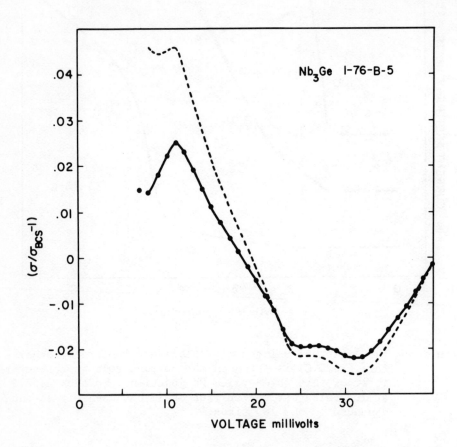

Figure 4: The density of states for Nb_3Ge-oxide-Pb junction # 1-76-B-5, normalized to the BCS density of states for a superconductor with the same gap (Δ_o = 3.7 meV), minus one. The solid points represent data taken directly from the derivatives of the I-V characteristic with a magnetic field applied to the junction to quench superconductivity in the Pb. The dashed line shows the effect of subtracting 20% of the normal conductance at each voltage, as an estimate of the fraction of low gap phase, from both superconducting and normal conductance traces.

clear from Fig. 2 that not all the tunneling conductance is into the high-gap phase, which is the phase responsible for these phonon effects. The deviations from a BCS density of states in the low gap material would be weaker by roughly $(\Delta_L/\Delta_H)^2$ and should not be observed on this scale. We have therefore attempted to regard the junction as representing two parallel conductance paths into the two phases and from Fig. 2 we estimate that only 80% of the conductance is into the Δ_H phase. Subtraction of 20% of the normal conductance from both the superconducting and normal conductance traces of the original data leads to a density of states for the high-gap phase as shown by the dashed line in Fig. 4. It can be seen that elimination of the conductance channel into the low gap material leads to a strong enhancement of the strength of the phonon effects, but they are still too weak for quantitative analysis and we do not wish to report further on the determination of $\alpha^2 F$ before further measurements have been made.

B. STRUCTURAL ANALYSIS

In view of the tunneling results discussed above, and other puzzling features of Nb_3Ge (e.g. the apparent need to stabilize the lattice with oxygen,[4,5] the metastability of the high T_c phase against annealing,[10] or simply the fact that high T_c material cannot usually be formed in bulk[11]) structural characterization of our films appeared as a necessity. The existence of structural features, such as defect arrays, has been proposed by numerous authors[12-17] to explain anomalous features of the softening of Nb_3Sn and V_3Si, which begins to occur near room temperature. If these two relatively stable materials exhibit peculiar structural distortions, it is reasonable to expect that Nb_3Ge will be an even more unusual case. We therefore report here the results of early work on structural analysis of getter-sputtered thin films of Nb_3Ge using Read camera configuration x-ray diffraction photography, x-ray diffractometer analysis and transmission electron diffraction microscopy.

Read camera x-ray photographs were taken with Cu radiation for a series of the getter-sputtered Nb_3Ge thin films that exhibited onset superconducting transition temperatures ranging from 3 to 23°K. Three basic types of patterns were observed, possibly the most interesting being the single broad diffraction ring observed in Fig. 5 for film 76-4, which was made with a 200°C table temperature. This can be interpreted as either very fine grained (< 50 A grain size) or amorphous material. From $Nb_{80}Ge_{20}$ films, an A15 pattern with sharp diffraction rings could be seen, for example in Fig. 6. By making a very Ge rich film, mixed phase Nb_5Ge_3 could be observed, as in Fig. 7 (this concentration was not included in the tunneling studies above).

Figure 5: Read camera photograph taken with Cu radiation showing a single diffuse diffraction line from a sample of Nb_3Ge deposited at $T_D = 473$ K. The spot pattern is characteristic of the single crystal Al_2O_3 substrate (Sample 4-76).

Figure 6: Read camera photograph, taken as above, of a film deposited from a $Nb_{80}Ge_{20}$ target onto an Al_2O_3 substrate (Sample 13-76).

Figure 7: Read camera photograph, taken as above, of a mixed phase Nb_5Ge_3 sample as deposited on an Al_2O_3 substrate (Sample 10-76).

Most of the high T_c Nb_3Ge films showed evidence of preferred orientation. The crystallite size was not uniform (300-8000 A), as reflected by broadened diffraction lines merging into regions of spotty and dashed seqments. Proper line intensity relationships were obtained from normal exposure photographs (1.5 hr), while 4-6 hr exposures were needed to record weak structural details and second phase patterns. It was never convincingly clear that the ring pattern of a high T_c film had an underlying broad diffuse ring (e.g. a superimposition of A15 lines and Fig. 5), as we expected from the tunneling results. However, unexplained diffraction lines were observed on the photographs of many high T_c films, for example, as in Fig. 8. It was not clear that these extra lines could be always ascribed to Nb_5Ge_3.

In an attempt to obtain better resolution of these extra features we also recorded diffractometer traces, as in Fig. 9 where the trace shown is characteristic of a 23°K onset T_c thin film of Nb_3Ge. Only two extra lines (marked A) are observed in addition to all symmetry permitted A15 phase lines.

We could not explain or identify the anomalous diffraction lines by line correlations with any of the known phases in the Nb-Ge phase diagram or any of the system oxides. In fact, these anomalous lines can be indexed by the relationship $nd_{210}/\sqrt{3}$, where n equals 1 or 2. Identification of foreign phases from one or two lines is, at best, one of many possible fits unless single crystal data is additionally available. Beneath the A15 lines of Fig. 9, particularly the {400}, {321}, and {320} lines, is a broad diffuse scattering background. It is tempting to assign this feature, noted also by Cox[18] in these conference proceedings, to the disordered or amorphous component of the films which gives rise to the low gap Δ_L.

It is by now well known, for example in the case of the layered compounds,[19] that transmission electron diffraction is a much more powerful tool than x-ray scattering for the study of structural abnormalities, and it was decided to utilize it in the case of the A15 superconductors.

Transmission electron diffraction photographs displaying single crystal patterns were successfully obtained from selected sample areas from 300 to 900°K. These areas were generally at the broken edges of free standing specimens, 5000 A thickness and mounted on fold-over copper mesh grids. Approximately 75 photographic exposures were made on six grid mounted samples. To date, all these samples have been taken from a single Nb_3Ge film (76-7). Each series of 15 photographs was preceeded and followed by a gold standard calibration plate such as that shown in Fig. 10.

Figure 8: Read camera photograph, taken as above, of A15 phase Nb_3Ge deposited at T_D = 1298 K under near ideal conditions of 600 um Ar on Al_2O_3 (Sample 52-75).

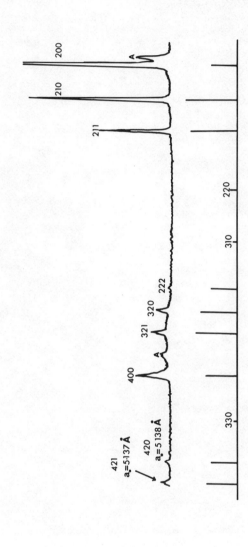

Figure 9: Diffractometer trace taken with CuKα radiation of diffraction lines of A15 phase Nb$_3$Ge, $T_c \sim 23$ K, showing two anomalous lines (A). Using the {420} and {421} peaks, the calculated cell size is $a_o = 5.137 \pm .002$ A (Sample 52-75).

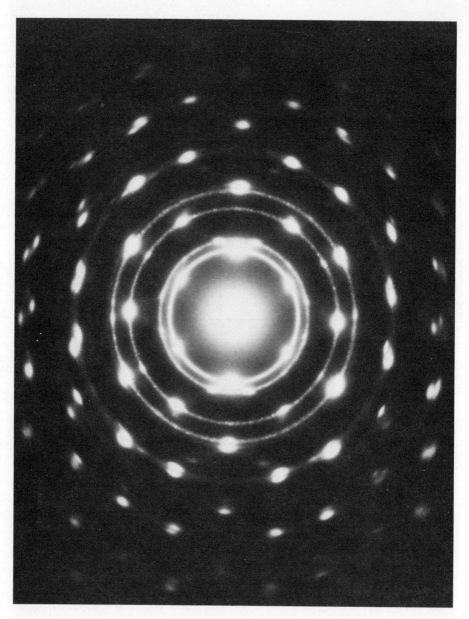

Figure 10: TEM pattern taken at 300 K of Au calibration plate. Displayed are diffraction rings from the polycrystalline part of the sample as well as a superimposed single crystal (111) spot pattern.

The single crystal pattern shown on plate #377 (Fig. 11) is believed to be the first of its kind on a high T_c (~20°K) thin film superconducting material. The pattern was taken with the specimen at 300°K.

Film analysis gives good agreement with measured diffractometer values of a_0 of 5.137 ± 0.002 A for A15 Nb_3Ge using d spacings as measured from the {110}, {220}, {330}, {440}, and {422} spots on the photographic plate. As can be seen from our assignment of spots in Fig. 12, the crystallite photographed is oriented with the <111> normal to the plane of the crystallite. The in-plane directions are as shown in Fig. 12.

The pattern is actually quite complex and is composed of symmetry allowed diffraction spots, nonallowed commensurate diffraction spots, localized diffuse scattering, and weak noncommensurate diffraction spots.

We interpret the pattern to reflect structural defects, for example superstructure formation, and speculate that weaker features might indicate electronic modulation (charge density waves).

The superstructure pattern observed at room temperature, that is the nonallowed commensurate diffraction spots, appears as a three-fold modulation of {242} in [21$\bar{1}$] and [11$\bar{2}$] directions, a six-fold modulation of {242} in the [1$\bar{2}$1] directions and a six-fold modulation of {220} in <211> directions. Defects on Nb chains in the form of antisite disorder, vacancies or gas impurity substitutions may be associated with this kind of lattice modulation. Weak noncommensurate diffraction spots are observed partially surrounding {220} spots in Figs. 11 and 12. From other diffraction plates, we sometimes see a complete hexagon shaped pattern around {220}. This is reminiscent of the way charge density waves are reflected in the TEM patterns of layered compounds.[17]

Localized diffuse scattering, observed at room temperature as the network of underlying lines in Fig. 11, may be an indication that some included Ar gas is present in the form of "sheets" on {110} planes.

A striking transition in the TEM patterns has been observed at 486°K. The transition is reflected by a simplification of the observed superlattice pattern to a simple three-fold modulation of {242} in <211> directions. Either pattern can be observed within a 1°K temperature variation about 486°K. The transition is completely reversible and can be observed repetitively. The noncommensurate diffraction spots also disappear above this temperature. Further details of this transition will be discussed elsewhere.

Figure 11: TEM pattern, plate #377, of a Nb$_3$Ge crystallite taken at 300 K. The crystallite orientation is ~ (111) (Sample 7-76).

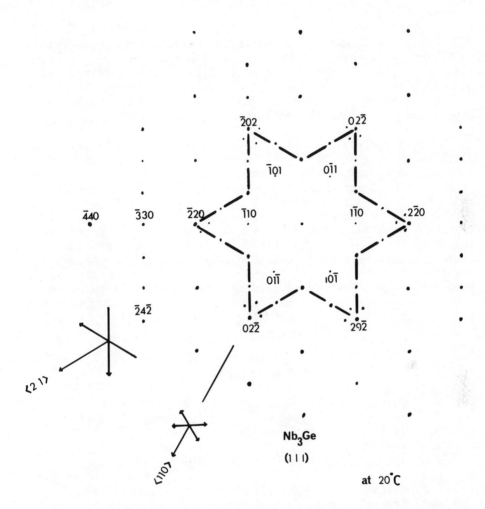

Figure 12: Identification of reciprocal lattice points and symmetry directions on plate #377. Solid and dashed lines indicate directions and location of regions of diffuse scattering. Note commensurate superstructure points as described in the text and the noncommensurate partial hexagon points around [220].

Although we have investigated only films prepared in Ar atmospheres, it is possible that partial O_2 atmospheres, known to stabilize A15 Nb_3Ge, may result in films showing similar or related structural effects. Rutherford backscattering studies, carried out for us by R. C. Dynes, indicate that any oxygen in our films must be less than 5 at.%.

It is possible that superlattice formation may be desirable (or essential?) for attaining high T_c's and may reflect, in part, the difficulties observed in high T_c film reproducibility. It should be stressed that lattice modulation and defect structures may vary from sample to sample and be influenced by changes in deposition conditions and film stoichiometry.

From these early results, we conclude that transmission electron diffraction analysis might well prove to be an ideal probe to measure independently the electronic and structural crystallographic parameters in superconducting materials, as well as related phase changes. It is not at present clear that any of the features reported above in the electron diffraction studies are directly related to the low energy gap that we consistently observed by tunneling into these Nb_3Ge films. It is possible that the electron microscope has been used to select only the high gap regions of the film, and that the disordered phase exists over other macroscopic regions. The more fascinating possibility is that the structural defects discussed above (or what these defects develop into at low temperatures) actually give rise to single crystal material exhibiting two energy gaps.

REFERENCES

1. J.R. Gavaler, Appl. Phys. Lett. <u>23</u>, 480 (1973); L.R. Testardi, J.H. Wernick and W.A. Royer, Sol. St. Comm. <u>15</u>, 1 (1974); J.R. Gavaler, M.A. Janocko and C.K. Jones, J. Appl. Phys. <u>45</u>, 3009 (1974).
2. R.H. Hammond, IEEE Trans. on Mag. <u>MAG-11</u>, 201 (1975).
3. L.R. Newkirk, F.A. Valencia, A.L. Giorgi, E.G. Szklarz and T.C. Wallace, IEEE Trans. on Mag. <u>MAG-11</u>, 221 (1975); A.I. Braginski and G.W. Roland, Appl. Phys. Lett.
4. J.R. Gavaler, J.W. Miller and B.R. Appleton, Appl. Phys. Lett. <u>28</u>, 237 (1976); J.R. Gavaler, Proceedings of this conference.
5. A.B. Hallak, R.H. Hammond and T.H. Geballe, to be published.
6. J.M. Rowell and P.H. Schmidt, to be published.
7. J.M. Poate, L.R. Testardi, A.R. Storm and W.M. Augustyniak, Phys. Rev. Lett. <u>35</u>, 1290 (1975).
8. J.M. Poate, R.C. Dynes, L.R. Testardi and R.H. Hammond, Proceedings of this conference.

9. L.Y.L. Shen, Phys. Rev. Lett. $\underline{29}$, 1082 (1972).
10. A.R. Sweedler, D.E. Cox, S. Moehlecke, R.H. Jones, L.R. Newkirk and F.A. Valencia, to appear in J. Low Temp. Phys. 1976.
11. Bulk Nb_3Ge with a T_c onset of > 22 K has been reported by N.N. Mikhailov, I.V. Voronova, O.A. Lavrová, E.V. Melnikov and M.N. Smirnova, Zh. ETF Pis. Red. $\underline{19}$, 510 (1974) [translation JETP Lett. $\underline{19}$, 271 (1974)].
12. C.M. Varma, J.C. Phillips, and S.-T. Chui, Phys. Rev. Lett. $\underline{33}$, 1223 (1974).
13. J. Noolandi and C.M. Varma, Phys. Rev. B $\underline{11}$, 4743 (1975).
14. J.C. Phillips, Sol. St. Comm. $\underline{18}$, 831 (1976).
15. R.N. Bhatt and W.L. McMillan, to be published.
16. G. Bilbro and W.L. McMillan, to be published.
17. L.R. Testardi, Bull. Am. Phys. Soc. $\underline{21}$, 220 (1976).
18. D.E. Cox, Proceedings of this conference. See also Ref. 10.
19. J.A. Wilson, F.J. DiSalvo and S. Mahajan, Adv. in Phys. $\underline{24}$, 117 (1975).

QUESTIONS AND COMMENTS

Discussion of this paper is presented at the end of the paper by Chu, Huang, Schmidt and Sugawara.

HIGH TEMPERATURE RESISTIVE PHASE TRANSITION IN A15 HIGH TEMPERATURE SUPERCONDUCTORS

C.W. Chu*

Department of Physics, Cleveland State University
Cleveland, Ohio 44115

C.Y. Huang[†]

Los Alamos Scientific Laboratory
Los Alamos, New Mexico 87544

P.H. Schmidt

Bell Laboratories
Murray Hill, New Jersey 07974

K. Sugawara[+]

Department of Physics, Case Western Reserve University
Cleveland, Ohio 44106

ABSTRACT

Resistive measurements were made on A15 high temperature superconductors. Anomalies indicative of a phase transition were observed at 433 K in a single crystal Nb_3Sn and at 485 K in an unbacked Nb_3Ge sputtered thin film. The results are compared with the high temperature transmission electron diffraction studies of Nb_3Ge films by Schmidt et al. A discussion of a possible instability in the electron energy spectrum is made.

---

* Research supported in part by the National Science Foundation under Grant No. DMR-73-02660 and Research Corporation.
[†] Work performed under the auspices of the U.S. ERDA.
[+] Research supported in part by the National Science Foundation under Grant No. DMR-74-0803.

It has been shown that a metal will exhibit a superlattice instability provided that the electron response function diverges at a certain wave vector $\bar{q}_0$, which need not be commensurate with the reciprocal lattice spacing of the metal-atoms. This can occur if the Fermi surface of the metal contains parallel pieces or nesting sections [1,2], or if two dimensional saddle points [3] in the band structure exist near the Fermi level. A charge density wave (CDW) will result when the electron-phonon interaction is strong enough to balance off the Coulomb repulsion, and a spin density wave when the exchange interaction dominates [2]. The quasi-one-dimensional arrangement of the transition metal atoms and the strong electron-phonon interaction in high temperature A15 superconductors suggest that this class of material may be susceptible to the aforementioned superlattice instability leading to the formation of CDW's. In fact, the low temperature martensitic transition in the A15 compounds, Nb_3Sn and V_3Si, has been speculated to be associated with the formation of the CDW's because of its similar phenomenology to the low temperature CDW-transition in some of the transition metal dichalcogenide layer compounds [3,4,5]. However, the high sensitivity of this martensitic transition on sample conditions is reminiscent of the incommensurate-commensurate (I-C) transition of the CDW states in layer compounds, instead of the normal-incommensurate (N-I) transition corresponding to the formation of CDW's. Recently, studies [6,7] on the CDW states in layer compounds have demonstrated that while the I-C transition is extremely sensitive to sample conditions, e.g. the strain and impurity level, the N-I transition is not. On the basis of these facts, the high temperature A15 superconductors would thus be expected to undergo another phase transition at a higher temperature due to the formation of CDW's.

By the transmission electron diffraction technique, Schmidt et al. [8] observed, above 300 K, three different kinds of superlattice in Nb_3Ge sputtered films: one incommensurate with the crystal lattice and the other two commensurate with periodicities of 3x and 6x the crystal lattice spacing. The incommensurate and the 6-fold commensurate superlattices disappear above 488 K, whereas the 3-fold one persists. They attributed the commensurate superlattices to the possible condensation of defects [9] and the incommensurate superlattices to either the formation of CDW's or the ordering of defects. Encouraged by these observations, we have measured the resistivity of a Nb_3Sn single crystal sample and an unbacked Nb_3Ge sputtered thin film. Anomalies in the curves of resistivity versus temperature have been detected.

The Nb_3Sn sample studied is an RCA single crystal [10] which undergoes a martensitic transformation at ~ 43 K and becomes superconducting at 17.78 K. It was spark-cut to the size ~ 1.3 mm x 1 mm x 5 mm. The resistance ratio between 300 K and 20 K was 7.

The a.c. resistivity was determined at 100 Hz by a standard four-probe technique between 300 K and 600 K, using a PAR 124 lock-in amplifier. Pressure contacts wetted with small amounts of indium were employed. The temperature dependence of the resistivity was continuously recorded. The relative resolution was estimated to be better than 0.2%. A small but distinct resistive anomaly was observed at 433 K as shown in Fig. 1. To simulate the sample conditions in the experiment carried out by Schmidt et al. [8], we measured the resistivity of an unbacked Nb_3Ge sputtered film. A Nb_3Ge film of ~ 6000 A thick was sputtered in an argon atmosphere on a copper film which was first sputtered onto a sapphire substrate. The thin film composite had a superconducting transition at 20 K and was immersed in a dilute nitric acid solution. A sample of unbacked Nb_3Ge film was thus obtained after the copper film dissolved away. To overcome the lead-contact problem, we then put the sample in a microwave cavity operating at 10 GHz. The relative resistivity of the sample as a function of temperature was determined by continuously monitoring the resonant frequency of the cavity [11]. The relative resolution was about 0.1%. As shown in Figure 2, a small anomaly of resistivity was recorded at 483 K. This resistive anomaly temperature coincides approximately with the temperature below which superlattice formulation occurs as detected by transmission electron diffraction in Nb_3Ge films. It seems therefore that the resistive anomalies in Nb_3Sn and Nb_3Ge can be taken as evidence for the superlattice formation. The minute size of the anomalies is not surprising, because of the large background resistance of these samples due to electron-phonon scattering at this high temperature.

In layer compounds where the existence of CDW's has been established, the N-I transition always occurs [12] at a higher temperature than the I-C transition if the latter exists. In view of the appearance of the incommensurateness of one of the superlattices below the transition temperature, it is tempting to propose that the presently observed phase transitions in high temperature A15 superconductors may correspond to the N-I transition in layer compounds and the low temperature martensitic transition to the I-C transition. The simultaneous occurence of the commensurate superlattice with a six fold modulation and the incommensurate superlattice in Nb_3Ge films below 488 K suggests that the existence of defects may help to stabilize the CDW states. Recent study [7] on the superconducting and the CDW states in layer compounds revealed that, in an increasing order, the pressure coefficients of the superconducting, the N-I and the I-C transition temperatures fall into three clearly different regions, with about an order of magnitude difference between each. Possible similar correlation is being sought in A15 compounds. Also in progress is the investigation of the occurence of this high temperature transition in A15 compounds prepared under different conditions and with different T_c's.

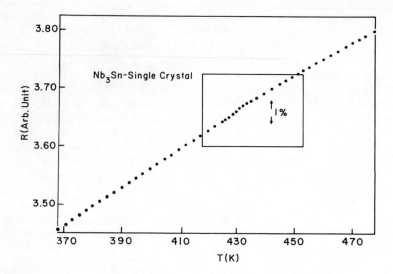

Figure 1: A.c. resistivity as a function of temperature of a single crystal Nb$_3$Sn sample.

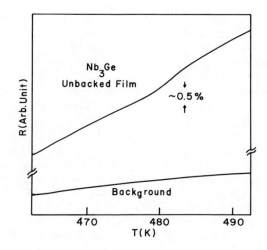

Figure 2: Microwave resistivity as a function of temperature for an unbacked Nb$_3$Ge sputtered thin film. Also shown is the background signal of the microwave cavity without the sample.

REFERENCES

1. W.M. Lomer, Proc. Phys. Soc. 80, 489 (1962); and A.W. Overhauser, Phys. Rev. 128, 1437 (1962).
2. S.K. Chan and V. Heine, J. Phys. F3, 795 (1973).
3. T.M. Rice and G.K. Scott, Phys. Rev. Lett. 35, 120 (1975).
4. L.P. Gorkov and O.N. Dorokhov, J. Low Temp. Phys. 22, 1 (1976) and references therein.
5. W.L. McMillan, Bull. APS 21, 362 (1976).
6. F.J. Di Salvo, J.A. Wilson, B.G. Bagley, and J.V. Wasczak, Phys. Rev. B12, 2220 (1975).
7. C.W. Chu, V. Diatschenko, C.Y. Huang and F.J. Di Salvo, preprint, and references therein.
8. P.H. Schmidt, E.G. Spencer, D. Joy and J. Rowell, preceding paper.
9. J.C. Phillips, Solid State Comm. 18. 831 (1976).
10. C.W. Chu, Phys. Rev. Lett. 33, 1283 (1974).
11. F.J. Rachford, thesis, Case Western Reserve University (1975).
12. W.L. McMillan, Phys. Rev. B12, 1187 (1975), and references therein.

QUESTIONS AND COMMENTS OF THE PAPERS BY SCHMIDT ET AL. AND CHU ET AL.

S. Sinha: [To Schmidt] What is the periodicity of the new phase seen here?

P. Schmidt: At room temperature?

S. Sinha: Yes.

P. Schmidt: The periodicity is as follows. We have two commensurate periodicities at room temperature, a three-fold and a six-fold directional defect pattern. At high temperatures we have only the three-fold modulation which means that at every third A-15 building block one anticipates either a defect in the form of an Argon substitution, an anti-site reversal or conceivably just a vacancy.

C. Varma: What about the period of the hexagon?

P. Schmidt: The period of the small hexagon?

C. Varma: Yes.

P. Schmidt: The small hexagon does not have a period since it is not modulated. The hexagon pattern of spots is centered about [220] principal A-15 spots. The measured d value obtained from the hexagon is ~ 18.5 Å. A commensurate hexagon centered about the (000) point in reciprocal space, assuming the smallest hkℓ, say (110), would yield a lattice value of $a_o \simeq 26$ Å. You can therefore see that it is a very small feature on the photographic TEM plate.

J. Phillips: How do you know the spot pattern for the small hexagon is not commensurate?

P. Schmidt: It is not totally positioned on the principal zones that are permitted for the symmetry of that pattern, for a single crystal of any orientation.

J. Phillips: It falls along high index directions? Literally non-commensurate means you have to distinguish between rational and irrational numbers.

P. Schmidt: The spots fall into fractional, really high index positions around [220] points.

P. Allen: Could you amplify on the arguments that favor an interpretation in terms of defect ordering? Now to me that sounds mysterious, especially if you had an ordering temperature that is reproducible from film to film where presumably the impurity, the defect concentration, is varying?

P. Schmidt: What I anticipate is that as temperature decreases, approaches let's say 220°C from say 223°C, I fully expect to see quite a different ordering to the vacancies or defects in a particular direction...
This is in fact what the experimental data shows. The local defect concentration can of course vary with temperature, be reproducible and the defects lock into specific geometries at a unique temperature.

J. Phillips: You have not done a variety of samples, yet. You do not know whether it is 213°C in 25 different samples!

P. Allen: Wasn't Paul Chu's sample different from your sample?

P. Schmidt: Right. For all grid mounted samples the high temperature transition was the same. Chu's sample also had the same transition temperature.

P. Allen: Do you anticipate that if you look at more films you will see that the disordering temperature varies for the films?

P. Schmidt: Perhaps, but I don't know how to predict that. I would not be surprised to see different phase change temperatures for different T_c's.

V. Diadiuk: [To Rowell] About the Nb_3Sn junction plot, what was the resistivity ratio? I have seen values of $\Delta \sim$ 3.5 meV and $2\Delta/kT_c \sim 4.5$ with even larger excess currents, so what was the resistivity ratio?

J. Rowell: Of the base film?

V. Diadiuk: No, of the junction---the ratio of the junction resistivity below and above the gap edge.

J. Rowell: It can vary enormously, from values like 2 up to 10^2 or 10^3. The resistivity ratio of the films can be as high as 7, which is getting up to the bulk values. [Note added in proof: This work on Nb_3Sn, some of which was discussed in the talk given by J.M. Rowell, will be published in Solid State Communications by Moore, Beasley, and Rowell.]

STRUCTURAL STUDIES OF ORDER, DISORDER, AND STOICHIOMETRY IN SOME

HIGH T_c Nb-BASE A-15 SUPERCONDUCTORS*

D.E. Cox, S. Moehlecke[†] and A.R. Sweedler

Brookhaven National Laboratory

Upton, New York 11973

and

L.R. Newkirk and F.A. Valencia

Los Alamos Scientific Laboratory

Los Alamos, New Mexico 87544

ABSTRACT

X-ray intensity measurements have been made on several Nb-base A-15 type superconductors. Order parameters have been determined with reasonable accuracy even in the presence of substantial amounts of impurity phases. Chemically vapor deposited Nb_3Ge of near-stoichiometric composition is characterized by a significant amount of disorder, corresponding to occupation of about 4% of the Nb sites by Ge. Annealing at 1350°C converts this material to a Nb-rich A-15 phase and Nb_5Ge_3. The former has essentially no Ge on Nb sites. From the lattice parameter trends in these compounds and the A-15 region of the Nb-Pt system, a B site radius of 1.43 Å for Nb is proposed. This is significantly different from the A site radius of 1.51 Å and closer to the normal metallic twelve-coordinated value.

---

*This work was performed under the auspices of the U.S. Energy Research and Development Administration

[†]On leave from Universidade Estadual de Campinas, Brazil, supported by Conselho Nacional de Pesquisas

INTRODUCTION

Recent studies of radiation effects on high T_c superconductors of A-15 type have revealed several very interesting features. Irradiation by fast neutrons (E > 1 MeV) with fluences in the region of 10^{19} neutrons/cm^2 produces a drastic decrease in the T_c's of Nb_3Sn, Nb_3Ga, Nb_3Al, and Nb_3Ge [1-5]. In the case of Nb_3Al, a neutron diffraction study [4] has shown that this decrease is accompanied by an increase in the amount of site-exchange disorder. In all cases, the T_c's can be restored to essentially their original values by annealing in the temperature range 700-850°C.

In chemically vapor deposited Nb_3Ge, a large decrease in T_c is also produced by annealing in the range 1100-1350°C [5]. X-ray phase analysis has shown that this heat treatment converts near-stoichiometric material to a more Nb-rich A-15 phase and tetragonal Nb_5Ge_3. In this case, it is the disorder resulting from the incorporation of excess Nb atoms on Ge sites which appears to be responsible for the decrease in T_c.

Large depressions in T_c have also been observed by Poate et al. [6] following bombardment of Nb-Ge thin films with 2 MeV helium particles. A pronounced correlation between T_c, the resistance ratio, and the lattice parameter, a_o, was noted in this work and in an earlier paper by Testardi et al. [7] The relationship between T_c and a_o was further discussed by Noolandi and Testardi [8] as a more general property of A-15 superconductors. The view point taken in these papers is that the attainment of high T_c's is not so much dependent on the elimination of site-exchange disorder or deviations from stoichiometry as of some other kind of "universal" defect, which is suggested to produce a loss of lattice symmetry as a result of microscopic strains on the scale of the unit cell.

Further papers dealing with the nature of the defects in Nb_3Ge and the effect of impurities on the stability of high T_c material are to be found in these Conference Proceedings [9-13].

The present paper describes some of the work which is currently in progress in an attempt to determine the relative importance of site-exchange disorder, stoichiometry, and other defects in the attainment of high T_c's. The x-ray diffraction measurements and data analysis used to determine the degree of order are described in some detail, and some observations and discussion concerning the lattice parameters and atomic radii in off-stoichiometric Nb compounds of this type are presented.

EXPERIMENTAL DETAILS

Nb_3Ge* samples were prepared by the chemical vapor deposition process described previously [14]. Appropriate mixtures of metal chorides were reduced by hydrogen, deposition taking place on the inside of copper tubes held at 900°C. After dissolution of the copper in HNO_3, the Nb_3Ge was left in the form of thin, brittle flakes.

The other samples mentioned were prepared by arc-melting, followed by high and low temperature anneals at appropriate temperatures. The compositions were assumed to be those of the starting materials, the precision being about 0.2 at.%. Debye-Scherrer x-ray patterns were taken with Cu K-α radiation, and the lattice parameters determined by a Nelson-Riley extrapolation. Guinier photographs were also obtained in some cases, and these proved to be particularly useful in identifying the various phases present in some of the Nb_3Ge samples. Superconducting transition temperatures were determined inductively as described previously [5].

The preparation of powder diffraction specimens for quantitative intensity measurements was influenced by the constraints likely to be imposed in the handling of irradiated samples in subsequent experiments which are planned. Small amounts of material (20-100 mg) were mixed with silicon powder and ground under cyclohexane to -400 mesh. The powder was transferred to an aluminum sample holder into which a shallow rectangular trough 2.5 x 1.5 x 0.025 cm had been machined, compacted and leveled with a glass slide. The specimen was mounted on a General Electric XRD 5 diffractometer with a fine-focus Cu target equipped with a Canberra semiautomatic 6511 step-scanning system for digital data collection. A graphite monochromator mounted on the detector arm was used to eliminate Cu K-β radiation and improve the signal-to-noise ratio. Scans over the angular region 20-120 deg. normally lasted about a day, estra time being devoted to weak peaks if necessary. With the procedure, peaks having about 0.1% the intensity of the strongest peak are easily detected.

In the case of the arc-melted samples, a slight shift in the position of the sample holder invariably resulted in quite significant changes (ca 5-20%) in the integrated intensities due to particle size effects. Additional scans were therefore made over all the peaks with the sample holder in four or five different positions,

---

*As in previous papers, the formula "Nb_3X" is used to denote the A-15 phase in the Nb-X system regardless of the actual chemical composition. The term "stoichiometric Nb_3X" is used to denot the ideal 3:1 composition.

and an average integrated intensity and standard deviation obtained for each peak. The latter was usually considerably larger than the errors due to counting statistics, uncertainties in background, and target stability, and was therefore taken as the error in the intensity. If this were not the case, an error of between 5-10% of the measured intensity was assumed.

The intensities measured for the vapor deposited materials showed very little variation of this type, presumably because of the much smaller particle size, estimated to be around 5 μm from photomicrographs. In these cases, the major source of error in the strong peaks was generally contained within the choice of background, while the accuracy of the weak peaks was governed by the counting statistics. The errors were assessed accordingly.

X-RAY DATA ANALYSIS

Least-squares refinement of the integrated intensity data was carried out with a computer program which permits each peak or composite peak to be specified in terms of the reflections from any phase which can possibly contribute. Appropriate structure factor expressions were used for the A-15 phase, Si, tetragonal Nb_5Ge_3 and other phases which were present. The expression for the A-15 structure, which has the space group Pm3n (O_h^3) with the A atoms in 6(c) sites and the B atoms in 2(a) sites is as follows

$$F = \{\overline{f}_B[1 + \cos 2\pi\left(\frac{h+k+1}{4}\right)] + 2\overline{f}_A[$$
$$\cos 2\pi\left(\frac{h}{4} + \frac{1}{2}\right) + \cos 2\pi\left(\frac{h}{2} + \frac{k}{4}\right) + \cos 2\pi\left(\frac{k}{2} + \frac{1}{4}\right)\}$$
$$\exp\left(-\frac{B \sin^2 \theta}{\lambda}\right) \quad \ldots\ldots\ldots\ldots\ldots\ldots\ldots \quad (1)$$

As before, a generalized composition $(A_{3-x}B_x)$ $[B_{1-x-y}A_{x+y}]$ can be written [4], where x is a measure of the order parameter and y represents the deviation from stoichiometry. The conventional order parameters for A and B sites are then

$$S_A = 1 - \frac{4x}{3(1-y)} \quad ; \quad S_B = 1 - \frac{4(x+y)}{3+y}$$

$\overline{f}_A$ and $\overline{f}_B$ represent average scattering factors for A and B sites

$$\bar{f}_A = \left(\frac{3-x}{3}\right) f_A + \left(\frac{x}{3}\right) f_B$$

$$\bar{f}_B = (1 - x - y) f_B + (x + y) f_A$$

The other symbols in Eq. 1 have their usual crystallographic significance.

Calculated values of the scattering factors were used appropriately corrected for dispersion effects [15]. Neutral atoms were assumed except in the case of NbO and NbO_2, where the values for Nb^{2+}, Nb^{4+}, and O^{2-} [16] were taken. In most cases, an overall Debye-Waller factor, B, was assumed, since attempts to refine individual values for the A and B sites did not usually produce any significant improvement.

It should be noted that in principle it is possible to refine with both x and y as variable parameters, since f_A and f_B have slightly different angular dependences. In practice, however, this turns out to be impossible, since f_A and f_B scale so closely over the experimental angular range that the refinement rapidly diverges with correlation coefficients of essentially unity between x, y, and the scale factor.

RESULTS

<u>Nb_3Al</u> -- This material was selected in order to provide a direct comparison with the results of the earlier neutron diffraction study [4]. An arc-melted sample with the composition $Nb_{0.77}Al_{0.23}$ was prepared and some of the x-ray data are shown in Fig. 1. Of particular note are the low background and the absence of any detectable amounts of impurities. Refinement of the data was carried out with scale and Debye-Waller factors for the A-15 phase and Si, together with an occupation factor k, defined as the percentage of Nb sites occupied by Al atoms (i.e., 100x/3). The final parameters and standard errors* are listed in Table 1, together with observed and calculated intensities.

Table 1 shows that the occupation factor is zero within a standard error of 0.7%. In other words, the excess Nb atoms occupy Al sites and order is as complete as possible in this material. The value of B agrees satisfactorily with the value obtained in the neutron

---

*Here and elsewhere these are given in parentheses and refer to the error in the least significant digit(s).

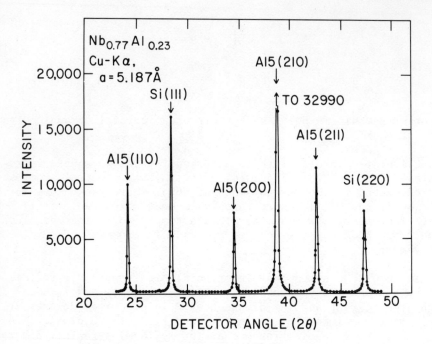

Figure 1: X-ray data for $Nb_{0.77}Al_{0.23}$.

diffraction study for a material with roughly the same composition [4]. It can also be seen from the comparison of observed and calculated intensities that there are no serious systematic discrepancies and that the crystallographic R factors are quite small*.

Refinement of the data was also carried out with scattering factors appropriate to the combinations Nb^{2+} and neutral Al, and Al^{3+} and neutral Nb. The changes in the final values of the parameters were well within the standard errors.

---

*These are defined in the usual manner as follows:
R factors = $\Sigma|I_{obs} - I_{calc}|/\Sigma I_{obs}$
Weighted R factor = $\{\Sigma w(I_{obs} - I_{calc})^2/\Sigma w(I_{obs})^2\}^{1/2}$, $w = 1/\sigma^2$

TABLE 1

Least-squares parameters and intensities for $Nb_{0.77}Al_{0.23}$ (a = 5.187Å, T_c = 18.6 °K). k is the percentage of Nb sites occupied by Al, B the Debye-Waller factor.

k (%)   −0.2 (7)

B (Å^2)   0.4 (2)

Phase	hkℓ	I_{calc}	I_{obs}	(σ)
A15	110	34.5	36.7	(3.5)
Si	111	65.0	66.1	(3.5)
A15	200	28.9	27.7	(3.5)
A15	210	185.0	173.5	(25.0)
A15	211	63.3	66.3	(4.5)
Si	220	42.2	39.5	(4.5)
A15	220	4.8	5.0	(1.0)
A15 / Si	310 / 311	6.6 } 24.3	30.1	(2.0)
A15	222	26.3	29.5	(4.5)
A15	320	39.0	35.8	(3.5)
A15	321	31.1	32.7	(4.0)
Si	400	6.3	6.8	(1.0)
A15	400	19.2	14.2	(2.0)
Si	331	9.3	9.5	(1.5)
A15 / A15	411 / 330	2.5 } 1.2	4.1	(1.0)
A15	420	8.7	9.4	(1.0)
A15	421	36.0	40.3	(5.0)

TABLE 1 (cont'd)

Phase	hkℓ	I_{calc}	I_{obs}	(σ)
A15	332	7.6 ⎫	23.8	(5.0)
Si	422	12.6 ⎭		
A15	422	1.7	1.6	(1.0)
Si	333	1.8 ⎫	7.1	(1.0)
Si	511	5.3 ⎭		
A15	510	1.6 ⎫	4.8	(1.0)
A15	431	3.2 ⎭		
A15	432	26.6 ⎫		
A15	520	13.3 ⎬	42.1	(3.5)
Si	440	4.5 ⎭		
A15	521	11.6	13.5	(1.5)

R factor 0.069   Weighted R factor 0.084

Nb_3Ge

Several samples have been examined and the results for two of these have been briefly summarized in a previous publication [5]. The first was a sample from a relatively large quantity of material designated $Nb_3Ge(II)$ with a T_c of 19.7°K and an a_0 of 5.142 Å. The results of annealing studies [5] showed that annealing of the as-deposited material at progressively higher temperatures resulted in a decrease in T_c, and an increase in a_0 and the amount of tetragonal Nb_5Ge_3 present. At 1350°C, this process was essentially complete, and resulted in an A-15 phase with a T_c of 5.8°K and an a_0 of 5.179 Å, quite similar to the bulk phase reported previously [17].

Parts of the diffractometer scans for $Nb_3Ge(II)$, both as-deposited and annealed at 1350°C, are shown in Fig. 2. In addition to the A-15 phase, there are substantial quantities of impurity phases in the as-deposited material (Fig. 2a). The chief of these is tetragonal Nb_5Ge_3 (a = 10.17$_3$ Å, c = 5.14$_3$ Å), but there are also significant quantities of hexagonal Nb_5Ge_3 (a = 7.72 Å, c = 5.34$_5$ Å), NbO (a = 4.210 Å), and NbO_2. Analysis by the carbon fusion technique revealed the presence of about 1.57 wt. % of oxygen, which originated from the starting materials, and gravimetric analysis for Nb give a figure of 75.05 wt. %.

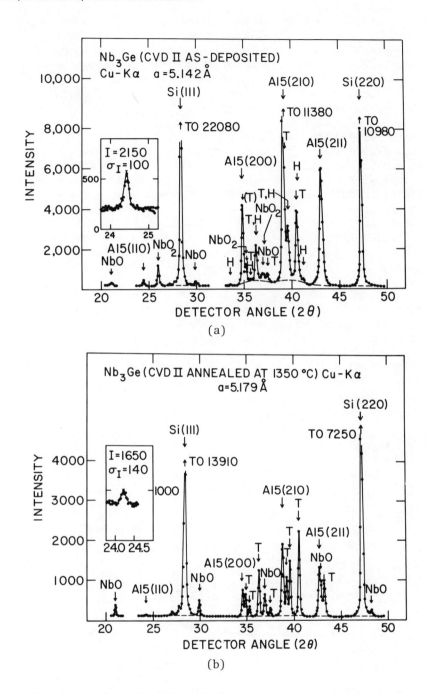

Figure 2: X-ray data for $Nb_3Ge(II)$. (a) As-deposited. (b) Annealed at 1350°C for 48 hours. The inset to each pattern shows the (110) peak of the A-15 phase in greater detail.

One obvious feature of both patterns is the number of overlapping peaks. However, it is emphasized that every single peak can be accounted for, and that the (110) peak of the A-15 phase is completely free from any overlap of this type. Although quite weak, it can be measured quite accurately, as shown in the inset to each figure. As pointed out earlier [4], this peak is a sensitive measure of the detree of order.

Another feature of the pattern for the as-deposited material previously mentioned [5] is the presence of significant line broadening and diffuse scattering. The latter is most pronounced in the angular region between 35-40 deg., as indicated by the broken line in Fig. 2a. Although this choice may appear to be rather arbitrary, there are good reasons to believe that is a reasonable one. From the least-squares fit of the 1350°C data shown in Fig. 2b, it follows from the intensities of the NbO and tetragonal Nb_5Ge_3 peaks that the narrow region at 38 deg in Fig. 2a is indeed background. Similar reasoning applies to the small hexagonal Nb_5Ge_3 peak at 35.6 deg. Furthermore, from the shape of the wings of some of the other A-15 peaks which are well-resolved, such as (211), it appears that the diffuse intensity is associated as much or more with the strong tetragonal Nb_5Ge_3 peaks. Data taken subsequently on a vapor deposited sample containing much less Nb_5Ge_3 show correspondingly less diffuse scattering of this type (Fig. 3). Further discussion on this point is deferred until the final section.

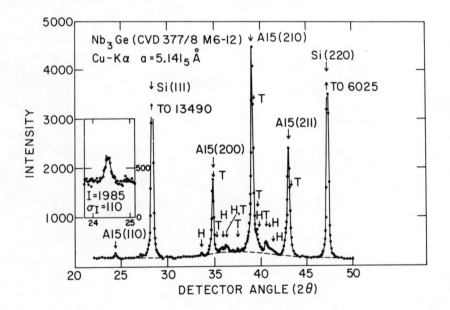

Figure 3: X-ray data for Nb_3Ge (377/8 M6-12). The inset shows the (110) peak of the A-15 phase in greater detail.

In spite of the number of impurity phases present, least-squares refinement of the intensity data presented no problems. The data from the sample annealed at 1350°C were first processed with variable scale and temperature factors for the four phases, an occupation factor k for the A-15 phase, and the three variable positional parameters for tetragonal Nb_5Ge_3. The positional parameters of NbO have been given previously by Bowman et al.[18]. The impurity phases were assumed to have their ideal stoichiometry.

The compositional parameter (y in $Nb_{3+y}Ge_y$) was assigned several values. Virtually identical fits were obtained in each case, since the scale factor and occupation factor adjust themselves accordingly, while the other parameters attain the same set of final values. From the respective values of the scale factors, the volume fractions of each phase were calculated from standard formulae, and hence the weight percentages of niobium and oxygen for each assumed value of y. By comparison with the chemical analysis figure, the Nb content of the A-15 phase was established as 83 $\pm$ 1 at. %. The oxygen content (1.75 wt. %) agrees quite well with the fusion analysis result of 1.57 wt. %.

Final values of the parameters for this material are given in Table 2, together with observed and calculated intensities. As indicated by the low R factor of 0.04, the overall fit is very good and each of the three phases is well-characterized in spite of the large number of overlapping peaks. The positional parameters found for Nb_5Ge_3 are in excellent agreement with those reported in a recent single crystal study by Jagner and Rasmussen[19], and the occupation of the A-15 phase is once again zero within error limits of 0.8%.

A similar procedure was followed for the as-deposited material. In this case, it is necessary to make allowance for NbO_2 and hexagonal Nb_5Ge_3 as well. The two positional parameters of the latter were permitted to vary, while those of NbO_2 were fixed at the values determined in a recent neutron diffraction study[20]. Both phases were assumed to be stoichiometric. The positional parameters of tetragonal Nb_5Ge_3, and the Debye-Waller factors of hexagonal and tetragonal Nb_5Ge_3, were fixed at the value given for the latter in Table 2. A similar constraint was placed on the Debye-Waller factors of NbO_2 and NbO.

This fitting procedure together with the chemical analysis figure for Nb yielded a self-consistent result of 76 $\pm$ 1 at . % for the Nb content of the A-15 phase. The oxygen content works out to be 1.97 wt. %, rather higher than the analysis figure. This is probably because there is no way to allow for the diffuse part of the scattering in the x-ray phase analysis, which would lead to an underestimate of the amounts of the three Nb-Ge phases. Allowance was made for this by scaling the fractions of NbO_2 and NbO so as to agree with the oxygen analysis. However, the relative proportions

TABLE 2

Least-squares parameters and intensities for Nb_3Ge (II) annealed at 1350°C. k is the percentage of Nb sites occupied by Ge, B the Debye-Waller factor. T denotes tetragonal Nb_5Ge_3.

$Nb_{0.83}Ge_{0.17}$ ($a_0 = 5.179$ Å, $T_c = 5.8$°K)

k (%)	0.2 (8)
B (Å^2)	0.8 (3)

Tetragonal Nb_5Ge_3 (a = 10.165 Å, c = 5.135 Å)

Nb(2):	k	0.0780 (5)
Nb(2):	y	0.2217 (5)
Ge(2):	x	0.1671 (6)
B(Å^2)		0.9 (2)

NbO (a = 4.210 Å)

B (Å^2)	0.7 (3)

Phase	hkℓ	I_{calc}	I_{obs}	(σ)
NbO	100	15.6	15.2	(0.7)
A15	110	2.6	2.7	(0.2$_5$)
T	220	0.1	0.3	(0.3)
T	211	1.5	1.5	(0.4)
T	310	2.7	2.2	(1.0)
Si	111	1073.7	1062.7	(50.0)
NbO	110	29.7	28.8	(1.5)
A15 T	200 002	63.0 } 41.6 }	105.1	(1.5)
T.	400	17.8	17.9	(0.7)
T	321	92.6	94.2	(2.0)

TABLE 2 (cont'd)

Phase	hkℓ	I_{calc}	I_{obs}	(σ)
NbO	111	47.1 ⎱	50.8	(1.2)
T	112	2.4 ⎰		
T	330	18.5	16.5	(0.8)
A15	210	213.6	195.4	(10.0)
T	202	90.3	90.3	(7.0)
T	420	110.6	115.5	(7.0)
T	411	193.8	192.7	(5.0)
A15	211	135.7 ⎱		
NbO	200	46.4 ⎬	262.7	(10.0)
T	222	74.2 ⎰		
T	312	0.2	0.4	(0.7)
T	510	1.1	0.5	(0.7)
Si	220	666.9 ⎱		
T	431	0.8 ⎬	703.0	(30.0)
NbO	210	7.4 ⎰		
A15	220	0.3	0.5	(0.7)
T	402	11.9 ⎱	13.1	(1.0)
T	440	1.8 ⎰		
T	521	21.3	19.9	(1.5)
T	332	7.6 ⎱	15.3	(1.5)
T	530	7.3 ⎰		
NbO	211	10.3	9.4	(0.7)
T	422	0.2	0.0	(0.5)
T	600	0.2	0.0	(0.5)
A15	310	0.5 ⎱	369.7	(18.5)
Si	311	373.9 ⎰		

TABLE 2 (cont'd)

Phase	hkℓ	I_{calc}	I_{obs}	(σ)
T	620	13.3 ⎫		
T	213	0.1 ⎭	11.8	(2.0)
T	611	1.0	0.7	(0.5)
T	512	4.2	4.3	(0.5)
T	541	3.0	3.0	(0.5)
A15	222	19.1 ⎫		
NbO	220	27.3 ⎭	42.7	(3.0)
T	442	11.9 ⎫		
T	323	15.9 ⎭	28.3	(3.0)
T	631	15.1	19.0	(2.0)
A15	320	42.2 ⎫		
T	532	13.1 ⎪		
T	710	11.5 ⎬	71.6	(4.0)
T	550	7.7 ⎭		
NbO	300	0.7 ⎫		
NbO	221	2.8 ⎪		
T	602	15.8 ⎬	60.0	(4.5)
T	640	0.0 ⎪		
T	413	39.5 ⎭		
A15	321	62.1	61.2	(4.6)
Si	400	93.1 ⎫		
T	622	4.2 ⎭	86.4	(6.5)
T	721	5.7	9.2	(2.0)
NbO	310	4.0 ⎫		
T	730	2.0 ⎭	7.2	(1.0)
T	433	0.2	0.6	(0.6)
A15	400	26.6	29.0	(2.9)
T	004	14.5	15.2	(2.0)

ORDER, DISORDER, AND STOICHIOMETRY

TABLE 2 (cont'd)

Phase	hkℓ	I_{calc}	I_{obs}	(σ)
NbO	311	17.8 ⎫		
T	800	0.0 ⎪	27.0	(2.7)
T	523	6.2 ⎬		
T	114	0.0 ⎭		
T	651	1.5	1.4	(0.7)
Si	331	133.7 ⎫		
T	712	9.0 ⎪	154.6	(11.6)
T	552	6.5 ⎬		
T	204	0.1 ⎭		
T	642	25.0 ⎫	45.0	(5.0)
T	820	12.6 ⎭		
A15	330	0.1 ⎫		
A15	411	0.2 ⎪	4.8	(1.5)
T	811	4.5 ⎬		
T	741	0.1 ⎭		
NbO	222	8.6 ⎫	8.3	(1.5)
T	224	0.0 ⎭		
T	660	11.3 ⎫		
T	613	0.4 ⎬	11.0	(0.7)
T	314	0.3 ⎭		
T	732	5.2 ⎫	5.2	(0.6)
T	705	0.4 ⎭		
NbO	320	1.5	1.2	(0.5)
A15	420	16.4 ⎫		
T	543	1.2 ⎬	34.3	(3.4)
T	831	15.6 ⎭		
T	404	2.3	3.1	(1.0)
A15	421	36.5 ⎫		
NbO	321	4.4 ⎪		
T	840	0.6 ⎪	63.9	(6.4)
T	633	6.2 ⎬		
T	334	2.7 ⎪		
T	802	9.7 ⎭		

TABLE 2 (cont'd)

Phase	hkℓ	I_{calc}	I_{obs}	(σ)
T	910	3.1		
T	424	18.4	30.4	(7.6)
T	822	0.0		
A15	332	14.1	206.1	(20.6)
Si	422	172.9		

R factor 0.039   Weighted R factor 0.049

found for the Nb-Ge phases should not be very much affected by the diffuse scattering. The final parameters are given in Table 3. Intensities have not been listed since there are over 150 reflections, but the overall agreement is quite satisfactory.

There is clearly a significant amount of disorder in the as-deposited material amounting to occupation of about 3% of the Nb sites by Ge ($S_A = 0.87$), and 14% of the Ge sites by Nb ($S_B = 0.82$). During the annealing process, it therefore appears as if Ge diffuses out of the Nb sites, resulting in the formation of Nb-rich material and excess Nb_5Ge_3.

The first of these was material with an unusually high a_o of 5.202 Å and low T_c of 3.8°K. The diffraction pattern showed that this was contaminated with about 3 wt. % of tetragonal Nb_5Ge_3. One small additional peak was observed with a d spacing of 2.40 Å and an intensity 2-3% of that of the strongest A-15 peak. This could

TABLE 3

Least-squares parameters for $Nb_3Ge(II)$ as-deposited. Composition of A15 phase $Nb_{0.76}Ge_{0.24}$ ($a_o = 5.142$ Å, $T_c = 19.7°K$). k is the percentage of Nb sites occupied by Ge, B the Debye-Waller factor

k (%)	3.2 (8)
B (Å^2)	0.8 (3)

R factor = 0.066   Weighted R factor = 0.089

conceivably arise from a sub-oxide of Nb with approximate composition Nb_6O [22] although no other peak characteristic of this phase could be seen. Refinement of the A-15 intensities corrected for tetragonal Nb_5Ge_3 yielded a set of occupation factors as a function of composition which are listed in Table 4. Although there was not enough sample for chemical analysis, the high a_o and low T_c strongly suggest that the Nb content is significantly higher than 83 at. %, and the results summarized in Table 4 are therefore indicative of little or no occupancy of Nb sites by Ge.

The second sample was material with an a_o of 5.141$_5$ Å and a T_c of 20.9°K. About 2 g. was prepared in order to carry out order measurements in neutron irradiated samples which are currently in progress. Part of the diffractometer scan is shown in Fig. 3, and refinement showed that the material contained about 10 wt. % of tetragonal and 8 wt. % of hexagonal Nb_5Ge_3 respectively, and a trace of NbO (ca 0.4 wt %). In view of the close similarity of the values of a_o and T_c with those of $Nb_3Ge(II)$, a Nb content of 76 at % was assumed. Except for the two positional parameters of hexagonal Nb_5Ge_3, the other parameters and Debye-Waller factors of the impurity phase were fixed at the values listed in Table 2. Final parameter values are listed in Table 5, together with observed and calculated intensities. A Nb site occupation factor of about 4% (S_A = 0.84) is obtained this time (this would be about 5% if the stoichiometric composition were assumed), quite consistent with the vaue obtained for $Nb_3Ge(II)$.

TABLE 4

Least-squares parameters for Nb_3Ge (a_o = 5.202 Å, T_c = 3.8°K). k is the percentage of Nb sites occupied by Ge, B the Debye-Waller factor.

	Assumed composition (at % Nb)		
	84	86	88
k (%)	4.3 (4)	2.2 (4)	0.2 (4)
B ($Å^2$)	0.6 (1)	0.6 (1)	0.6 (1)

R factor = 0.019   Weighted R factor = 0.036

Table 5

Least-squares parameters and intensities for Nb_3Ge. k is the percentage of Nb sites occupied by Ge, B the Debye-Waller factor. T and H denote tetragonal and hexagonal Nb_5Ge_3 respectively.

$Nb_{0.76}Ge_{0.24}$ ($a_0 = 5.141_5$ Å, $T_c = 20.9°K$)	
k (%)	4.0 (6)
B (Å^2)	0.7 (2)

Hexagonal Nb_5Ge_3 (a = 7.65 Å, c = 5.33 Å)	
Nb (2): x	0.256 (8)
Ge: x	0.619 (7)

Phase	hkℓ	I_{calc}	I_{obs}	(σ)
A15	110	4.9	4.9_3	(0.3_5)
H	200	1.5	1.4	(0.5)
Si	111	481.8	498.6	(25.0)
NbO	110	1.7	1.9	(0.6)
H	002	2.4	2.8	(1.0)
A15	200	79.1 ⎫		
T	002	3.6 ⎬	85.0	(5.0)
T	400	1.6 ⎭		
H	210	4.3	6.6	(2.0)
H	102	2.7 ⎫	13.3	(3.0)
T	321	8.1 ⎭		
NbO	111	2.7 ⎫	2.8	(1.0)
T	112	0.2 ⎭		
T	330	1.6	2.9	(1.0)

ORDER, DISORDER, AND STOICHIOMETRY

Table 5 (cont'd)

Phase	hkℓ	I_{calc}	I_{obs}	(σ)
A15	210	278.6		
T	202	7.9	323.2	(25.0)
T	420	9.7		
H	211	12.7		
H	300	9.5		
T	411	16.9	41.2	(7.0)
H	112	12.9		
A15	211	170.8		
NbO	200	2.7	177.9	(9.0)
T	222	6.5		
H	202	0.5		
Si	220	304.1	289.0	(15.0)
T	510	0.1		
A15	220	0.6	<0.8	
H	221	0.5		
T	440	0.2	0.9	(0.5)
T	402	1.0		
T	521	1.9	1.5	(0.5)
T	332	0.7	1.4	(0.5)
H	311	1.3	1.4	(0.5)
T	530	0.6		
Si	311	172.2	174.8	(8.0)
A15	310	0.9		
T	620	1.2	1.4	(0.7)
A15	222	25.8		
NbO	220	1.6	26.4	(2.6)
T	442	1.0		
T	323	1.4	3.9	(1.5)
H	321	2.0		

TABLE 5 (cont'd)

Phase	hkℓ	I_{calc}	I_{obs}	(σ)
T15	320	55.5		
H	410	0.8		
H	213	2.7		
H	402	1.7		
T	631	1.3		
T	532	1.1	70.6	(7.1)
T	710	1.0		
T	550	0.7		
T	602	1.4		
T	640	0.0		
T	413	3.4		
A15	321	78.8	75.4	(7.5)
Si	400	43.6		
T	622	0.4	43.5	(6.5)
T	721	0.5		
A15	400	34.5		
T	004	1.3	36.6	(3.7)
H	313	0.4		
Si	331	63.1		
T	712	0.8		
T	552	0.6		
T	204	0.0	68.7	(5.0)
T	642	2.2		
T	820	1.1		
H	331	0.5		
H	420	0.9		
A15	411	0.4		
A15	330	0.2	0.4	(0.6)
T	224	0.0		
A15	420	21.0		
H	511	1.3		
H	323	0.8		
H	332	0.3	26.6	(4.0)
H	304	1.8		
T	831	1.4		
T	404	0.2		
T	802	0.8		

TABLE 5 (cont'd)

Phase	hkℓ	I_{calc}	I_{obs}	(σ)
A15	421	48.9		
T	840	0.1		
T	633	0.3	58.6	(7.5)
T	334	0.2		
T	910	0.3		
T	424	1.6		
Si	422	83.0	90.8	(9.1)
T	822	0.0		
A15	332	18.1	18.5	(3.0)
A15	422	0.3		
Si	511	34.4		
Si	333	11.5	46.1	(4.6)
H	521	0.7		
T	842	0.7		

R factor 0.043   Weighted R factor 0.068

Since there has been considerable interest in and speculation about other kinds of defects and their role in stabilizing high T_c A-15 phases, the intensity data for this material were also refined on the basis of a vacancy model of the type $Nb_{3-x}\square_x Ge_{1-x-y} Nb_{x+y}$ in which the Nb site defects were assumed to be vacancies rather than Ge atoms. Since it is essentially the ratio of the average scattering factors in the two sites which is determined as discussed earlier, a virtually identical fit should be possible. This indeed is the case and the corresponding occupation factor for vacancies at the composition $Nb_{0.76}Ge_{0.24}$ is $3.7 \pm 0.4\%$.

LATTICE PARAMETERS OF Nb-RICH A-15 COMPOUNDS

In view of the success of the simple model proposed by Geller [23] for predicting A-15 lattice parameters, some consideration has been given to extending this scheme to non-stoichiometric phases of this type. An attractive feature of this is the possibility of drawing some conclusions about the composition and degree of order from the value of a_0 alone, since this is a very easy quantity to measure.

A brief discussion of the application of this model to the lattice parameters in irradiated Nb_3Al has been given previously [4]. An effective or average radius of type $\bar{r}_a = N_A^A r_A + N_B^A r_B$ can be defined, where N_A^A and N_B^A are the fractions of A and B atoms on A sites, and r_A and r_B are respective Geller radii. An analogous expression holds for the B sites. With the revised radii derived by Johnson and Douglass [24], this simple model gave a calculated increase in a_0 about 30% larger than that observed.

An analogous calculation for $Nb_3Ge(II)$ leads to values of 5.161 Å and 5.263 Å for the as-deposited and annealed samples respectively, compared to the observed values of 5.142 Å and 5.197 Å. While the former value might be regarded as acceptable in view of uncertainties in the composition, the latter is certainly not. However, if a B site radius of 1.43 Å is assumed for Nb, values of 5.142 Å and 5.194 Å are obtained, which is clearly a much more reasonable result.

Additional evidence for this idea is provided by a study of the Nb-Pt system currently in progress [25]. In this case, the solubility range extends from 19-29 at. % Pt [25,26], and a small linear increase is observed on the Nb-rich side, consistent once again with a B site radius of 1.43 Å for Nb. Assuming this value, the composition of the high a_0, low T_c, sample described in the previous Section would correspond to about 88 at. % Nb (see Table 4).

DISCUSSION

X-ray intensity measurements and appropriate refinement techniques are clearly capable of yielding reasonable accurate values of the order parameter in A-15 systems even in samples containing large amounts of impurity phases. It also appears possible to determine Debye-Waller factors with sufficient accuracy to detect any gross random static displacements of the atoms from their ideal positions. This kind of effect might be anticipated in irradiated samples with high concentrations of vacancies or interstitials, for example.

Of particular interest is the fact that in the two CVD samples of high T_c Nb_3Ge which have been studied, a significant amount of disorder has been found. It is not possible to distinguish between a simple site-exchange model and an A site vacancy model on the basis of the x-ray data; however, while the presence of some disorder is a well-established feature in many of these compounds [27], the few density determinations which have been made have not revealed any appreciable concentrations of vacancies [28-32], certainly not at the level of a few percent. A tendency towards disorder would not be unexpected in a low temperature method of preparation such as chemical vapor deposition, and may very well be an essential factor in the formation of near-stoichiometric Nb_3Ge.

Also noteworthy is the observation that the B site radius of Nb appears to be appreciably smaller than the Geller radius. This has been demonstrated in a rather different way in a plot by Vieland and Wicklund [33], who extrapolated lattice parameter data for Nb-Ga, Nb-Ge, and Nb-Sn, and derived a hypothetical a_o for "Nb_3Nb" of 5.24 ± 0.02 Å. They pointed out that the atomic volume of Nb in this hypothetical material would be identical to that of Nb in the b.c.c. metal. This value of a_o is also consistent with a B site radius of 1.43 Å, which is closer to the normal twelve-coordination radius of Nb [34] than the Geller radius of 1.51 Å, and probably reflects the more regular coordination of the B site in the A-15 structure. It also accounts for the small lattice contraction observed in Nb-rich Nb_3Sn [35,36]. In the case of the V-Ga system, however, the small linear decrease which is observed on the V-rich side corresponds to a B site radius for V of 1.37 Å [25], which is larger than the A site radius of 1.31 Å [23,24].

Although this simple substitutional picture seems to account very well for lattice parameter variations on either side of the stoichiometric composition, it does not explain the observed increases in a_o caused by irradiation. Recalculation of the changes in Nb_3Al with a B site radius of 1.43 Å for Nb gives essentially zero change compared to the observed increase of 0.23% [4]. The situation is even worse in Nb_3Sn, where a decrease would be predicted, but a similar increase is observed. Clarification of this discrepancy is being sought in irradiation studies of Nb_3Pt which are in progress [25].

Finally, we would like to speculate briefly about the origin of the diffuse x-ray scattering observed in $Nb_3Ge(II)$. Fig. 4 shows a projection of the structures of Nb_3Ge and tetragonal Nb_5Ge_3 which emphasizes a striking underlying structural similarity. Nb_5Ge_3 can be envisaged as being built up from unit cell "blocks" of the A-15 structure centered about Nb atoms at heights = 0.25 a_o. Immediately adjacent blocks are rotated 90° with respect to each other, and the "extra" Ge atoms are accommodated equidistant from the corners of four such blocks. Since the deposition of near-stoichiometric material is invariably accompanied by deposition of some tetragonal phase, it is likely that small fluctuations in the deposition parameters could result in a local excess of Ge atoms and the consequent nucleation and growth of small regions of tetragonal Nb_5Ge_3 having a close orinetational relationship with the A-15 regions. These could consist of as little as one unit cell and would accordingly contribute to diffuse scattering which would be most pronounced in regions of the pattern containing strong A-15 and tetragonal peaks.

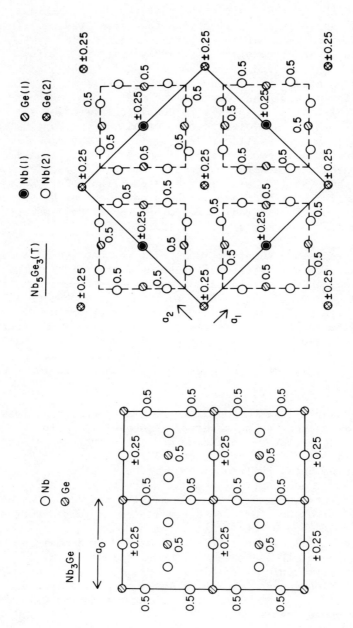

Fig. 4. Projection on (001) of the structures of (a) Nb_3Ge (b) Tetragonal Nb_5Ge_3. The small numerals denote the heights of the atoms in fractions of the vertical cell edge. The respective unit cells are shown by the solid lines, and the broken lines show the A15 unit cell "block" surrounding the Nb(1) atoms in Nb_5Ge_3.

ACKNOWLEDGEMENTS

We would like to thank R.H. Jones for expert technical assistance, and M. Strongin, G.J. Dienes, D.O. Welch, and R.M. Waterstrat for helpful discussions.

REFERENCES

1. A.R. Sweedler, D.G. Schweitzer, and G.W. Webb, Phys. Rev. Lett. 33, 168 (1974).
2. A.R. Sweedler, D. Cox, D.G. Schweitzer, and G.W. Webb, IEEE Trans. Magn. MAG 11, 163 (1974).
3. D.M. Parkin and A.R. Sweedler, IEEE Trans. Magn. MAG 11, 166 (1974).
4. A.R. Sweedler and D.E. Cox, Phys. Rev. B12, 147 (1975).
5. A.R. Sweedler, D.E. Cox, S. Moehlecke, R.H. Jones, L.R. Newkirk, and R.A. Valencia, J. Low. Temp. Phys. 24 (in press).
6. J.M. Poate, L.R. Testardi, A.R. Storm, and W.M. Augustyniak, Phys. Rev. Lett. 35, 1290 (1975).
7. L.R. Testardi, R.L. Meek, J.M. Poate, W.A. Roger, A.R. Storm, and J.H. Wernick, Phys. Rev. B11, 4304 (1975).
8. J. Noolandi, and L.R. Testardi, private communication.
9. J.C. Phillips, Second Rochester Conference on Superconductivity (1976).
10. J. Gavaler, Second Rochester Conference on Superconductivity (1976).
11. P. Schmidt, E.G. Spencer, D. Joy, and J. Rowell, Second Rochester Conference on Superconductivity (1976).
12. J. Poate, R. Dynes, and L. Testardi, Second Rochester Conference on Superconductivity (1976).
13. H. Lutz, H. Weismann, and M. Strongin, Second Rochester Conference on Superconductivity (1976).
14. L.R. Newkirk, F.A. Valencia, A.L. Giorgi, E.G. Szklarz, and T.C. Wallace, IEEE Trans. Magn. MAG 11, 221 (1974).
15. International Tables for X-ray Crystallography, Vol. 4, The Kynoch Press, Birmingham (1974).
16. M. Tokonami, Acta Cryst. 19, 486 (1965).
17. B.T. Matthias, T.H. Geballe, and V.B. Compton, Rev. Mod. Phys. 35, 1 (1963).
18. A.L. Bowman, T.C. Wallace, J.L. Yarnell, and R.G. Wenzel, Acta Cryst. 21, 843 (1966).
19. S. Jagner and S.E. Rasmussen, Acta Cryst. B31, 2881 (1975).
20. R. Pynn, J.D. Axe, and R. Thomas, Phys. Rev. B13, 2965 (1976).
21. L.R. Newkirk, F.A. Valencia, and T.C. Wallace, J. Electrochem. Soc. 123, 425 (1976).
22. N. Norman, J. Less Common Metals 4, 52 (1962); G. Brauer, H. Muller, and G. Kuhner, ibid, 533.
23. S. Geller, Acta Cryst. 9, 885 (1956), ibid 10, 380 (1957).

24. G.R. Johnson and D.H. Douglass, J. Low Temp. Phys. 14, 565 (1974).
25. S. Moehlecke, D.E. Cox, and A.R. Sweedler, (to be published).
26. R.M. Waterstrat and R.C. Manuszewski, "Noble Metal Constitution Diagrams", Part 2, NBS Report NBSIR 73-415 (1975).
27. See, for example, E.C. van Reuth and R.M. Waterstrat, Acta. Cryst. B24, 186 (1968); R. Flukiger, Thesis, University of Geneva (1972).
28. C.E. Lundin and A.S. Yamamoto, Trans. AIME 236, 864 (1966).
29. R.M. Waterstrat and E.C. van Reuth, "Ordered Alloys - Structural Application and Physical Metallurgy", Proceedings of the 3rd Bolton Landing Conference (1969).
30. A. Muller, Z. Naturforsch 25a, 1659 (1970).
31. S.M. Kuznetsova and G.S. Zhdanov, Sov. Phys. Crystallogr. 16, 1077 (1972). English transl. of Kristallografiya 16, 1230 (1971).
32. R.A. Meussner, private communication.
33. L.J. Vieland, and A.W. Wicklund, Phys. Lett. 49A, 407 (1974).
34. L. Pauling, "The Nature of the Chemical Band" 3rd Edition, Cornell University Press (1960), p. 403.
35. H. Pfister, Zeit. Naturforsch. 20a, 1059 (1965).
36. L.J. Vieland, RCA Review 25, 366 (1964).

QUESTIONS AND COMMENTS

B. Matthias: Am I right that sometimes if you take the niobium on the A site and the B site, or the vanadium on the A site and the B site, that the radius ratio is larger that one in the first case and smaller than one in the second?

D. Cox: In these niobium systems, the ratio of the A site radius to the B site radius is greater than one.

J. Phillips: And in vanadium it is less than one?

D. Cox: It is less than one. Correct.

B. Matthias: That doesn't make any sense.

J. Phillips: No way.

B. Matthias: I say it is impossible because according to the coordination, the larger the coordination, the larger the radius obviously.

D. Cox: If you calculate the lattice parameters in the excess vanadium part of the vanadium-gallium system with the 1.31 Å Geller radius then you do not get anything like the observed lattice dependence. You get a much steeper one.

J. Phillips: Sounds like the model is maybe a little incorrect.

D. Cox: Well, that is possible. It is a very crude model of course. It simply involves the sum of the averaged Geller radii, but this appears to be generally valid in A-15 systems.

J. Phillips: Maybe what this suggests is that in the vanadium system it makes sense, and in the niobium system it does not make sense. Maybe there is two phase behavior of one kind in one system and there is not the same kind of two phase behavior going on in the other system.

D. Cox: Well, maybe. But why can the metallic radius for niobium be anything from 1.43 to 1.46 Å ?

B. Matthias: You do not have to worry about that. The question is: is Zachariasen's rule about the coordination and the radii valid or not?

J. Phillips: That is very old. That has been used by crystallographers for 50 years.

B. Matthias: I have never seen it violated until today.

DEFECT PRODUCTION AND STOICHIOMETRY IN A-15 SUPERCONDUCTORS

J.M. Poate, R.C. Dynes and L.R. Testardi

Bell Laboratories

Murray Hill, New Jersey   07974

and

R.H. Hammond*

W.W. Hansen Laboratories, Stanford University

Stanford, California   94305

ABSTRACT

The dependence of T_c upon stoichiometry and 2 MeV ^{4}He induced damage has been investigated in Nb_3Ge, Nb_3Sn, V_3Si and V_3Ge superconducting thin films. Optimum values of T_c are obtained for Nb:Ge and Nb:Sn ratios throughout the range of 2.6:1 - 3:1 in 95% single phase A-15 films. This is inconsistent with anti-structure (anti-site) defects leading to great degradations in T_c. Similar T_c vs ^{4}He dose plots are observed for all materials; at high doses T_c saturates at values of 3.5 K (Nb_3Ge), 2.95 K (Nb_3Sn), 2.2 K (V_3Si) and 1.0 K (V_3Ge). Defects introduced by the ^{4}He bombardment produce T_c-resistance ratio correlations similar to that for as-grown films.

INTRODUCTION

Superconductivity in the A-15 materials is a subject of continuing interest with the crucial question being what determines or limits the superconducting transition temperature T_C. Clearly this interest is greatly stimulated by the fact that such materials have the highest recorded T_C's (Nb_3Ge ~ 23.2^{0}K) [1]. It has been

―――――――――――
*Work at Stanford sponsored by the Air Force, grant AFOSR73-2435.

conventionally assumed [2] that the dominant parameter in such materials in the order or stoichiometry of the material with the highest T_c material having the A_3B (A-15) configuration. T_c, in this interpretation, should therefore depend strongly upon the integrity of the A chains.

In previous work [3] on Nb-Ge films, however, we suggested that although stoichiometry is important, it is not the crucial parameter. T_c, resistance ratio and lattice parameter measurements before and after damaging the films with 2 MeV ^{4}He particles accentuated the role that defects play in determining the superconducting properties [4]. It did not appear though that simple antistructure defects (i.e., interchange of A and B sites) were solely the culpable defects with regard to T_c degradation as stoichiometries could be varied by some 40% while maintaining $T_c \sim 20$ K. Our interest in defect production had been stimulated by the comprehensive neutron damage studies of the Brookhaven group [5], who were the first to show a universal behavior of the T_c vs damage curves for the A-15 materials. While the general features of the ^{4}He and neutron damage curves agree well, there are significant differences. Firstly, ^{4}He damage in Nb_3Ge produced a saturated region in T_c after high doses at $T_c \sim 3.5$ K and this did not degrade with increasing dose; the neutrons appeared to depress T_c below 1.5 K with no saturation region. Secondly, the transition widths in the intermediate region, where T_c was rapidly degrading with dose, were found to increase significantly for ^{4}He damage but not for neutron damage. Important differences in interpretation also arose from the original Brookhaven observation that the degree of disorder obtained from neutron diffraction measurements correlated well with the assumption of simple antistructure defects being the dominant class of defect in degrading T_c.

In this work we have extended our measurements into the following A-15 materials: Nb_3Sn, V_3Si and V_3Ge. The motivation was to determine trends and to establish whether the Nb-Ge results were pathological or indeed representative of universal behavior in the A-15 materials. Ion beam techniques were used to analyze the films and induce defects. Analysis of thin films by high energy ion scattering has been reviewed in several recent papers [6]. Quantitative compositional profiles of thin films can be obtained simply and directly without recourse to sputter sectioning techniques. By use of these techniques, thin films, in many cases, can be better characterized with regard to composition than bulk materials.

EXPERIMENTAL TECHNIQUES

a. <u>Film Preparation</u>

The Nb_3Ge, V_3Si and V_3Ge films were prepared in the dc getter sputtering system described previously [3]. The V_3Si and V_3Ge

films were prepared with composition ratios of 3:1. The composition of the Nb-Ge films were varied by changing the composition of the sputtering targets. The majority of the Nb_3Sn films were produced at Stanford using the dual electron beam codeposition technique [7]. Films of varying composition were produced by this technique. Some Nb_3Sn films were also prepared at Bell by co-sputtering or sputtering and diffusion. All films, whether sputter or e beam produced, were deposited on clean sapphire substrates held at temperatures typically in the range 700-800°C. Film thicknesses were in the range 2000-3000 Å.

b. Resistance Measurements

Standard four-probe resistance measurements were made using pressure, silver paste or soldered contacts. Temperatures were measured to an accuaracy of ± 0.2 K with calibrated carbon and germanium resistors and/or a silicon diode. T_c's below 4 K were determined by pumping on the ^{4}He in the dewar (the width of the superconducting transition is taken as 95-5% of the normal state resistance and T_c is the mid-point).

c. Rutherford Scattering

Composition and impurity contents of the films were carried out using ^{4}He Rutherford scattering and nuclear reaction techniques discussed previously [3]. Beams were provided by the Bell 2 MeV Van de Graaff accelerator.

d. X-Ray Measurements

X-ray diffraction data were obtained on the films before and after the damage irradiations using either a Read or Seemann-Bohlin camera.

e. ^{4}He Irradiations

The samples were irradiated with 2 MeV ^{4}He particles at a typical particle density of 3×10^{15} cm^{-2} sec^{-1} (1 μA ^{4}He$^+$ on 2 mm beam spot size). It should be noted that no ^{4}He lodges in the films but penetrates ~ 5 μ into the sapphire substrates. The beam was swept both vertically and horizonally over a defining collimator of 5.5 x 5.5 mm². It is believed that the integrated dose is constant to within 5% over this area. Sample temperatures rose to no greater that 50°C during irradiations. Irradiations were carried out at pressures of 10^{-7} Torr.

RESULTS

We will concentrate principally on the aspects of the experiment relating to damage production and stoichiometric dependencies in the films: fuller details of the experiment and results will be given elsewhere.

Figure 1 shows two superimposed backscattering spectra for 1.9 MeV ^{4}He from Nb-Ge films with radically different T_c's of 21.4 and 12.9 K. Film thicknesses are approximately 2500 Å. The composition of the films within a depth interval defined by the depth resolution, ~ 200 Å, can be evaluated simply using the ratios Y_{Nb}/Y_{Ge} and suitable cross section and kinematic parameters as discussed previously [3]. The compositions of all films used in this work have been determined by this technique and the films have composition profiles that are uniform with depth, within a relative error ~ 5%.

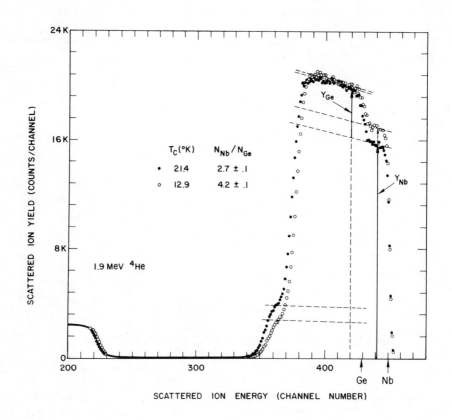

Figure 1: Superimposed backscattering spectra for two Nb-Ge films with T_c's of 21.4 and 12.9 K and respective stoichiometries of 2.7 and 4.2. After Poate [6].

Figures 2 and 3 show plots of T_c as measured by resistive transitions versus composition for Nb-Ge and Nb-Sn films where the composition has been deliberately changed. The error bars represent the transition widths (from 5-95%). In the compositional range Nb/Ge ~ 2.6 to 5.5 x-ray analysis indicated the films were predominantly single phase A-15 material. The very weak second-phase lines have been estimated to correspond to less than 5% second phase. Recent investigations [8] of similarly grown Nb_3Ge films have indicated in addition to the A-15 phase a disordered phase with T_c corresponding to the "saturated damage" phase obtained in the work (see Figure 4). Whether this disordered phase only pertains to the surface, or throughout the films, has not been established. It is likely that our films also contain this phase which is difficult to detect by x-ray analysis due to the diffuse lines. The Rutherford determination of average Nb/Ge composition is however believed to be essentially correct in describing the A-15 phase composition over the range 2.6 to 5.5. The Nb-Ge films prepared under non-optimum conditions were usually obtained at a substrate temperature ~ 50°C lower than optimum. The Nb-Sn films (Figure 3) were all prepared under optimum conditions. Similarly to Nb-Ge, the Nb-Sn films in the Nb/Sn compositional range > 2.5 were found to be predominantly single phase with any second phase being less than 5%.

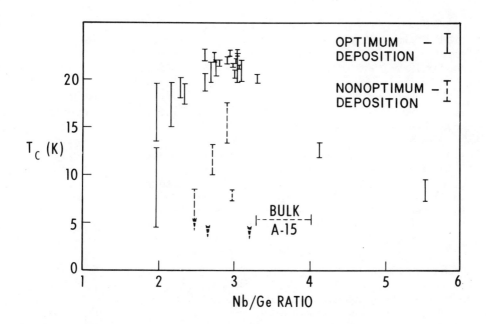

Figure 2: T_c vs composition for Nb-Ge. Single phase A-15 for Nb/Ge > 2.5. After Testardi et al. [3].

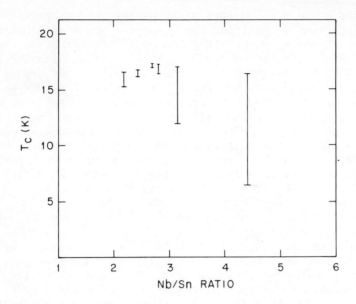

Figure 3: T_C vs composition for Nb-Sn single phase A-15 for Nb/Sn > 2.5.

The stoichiometries of the V_3Si and V_3Ge were not deliberately varied. Compositions were determined to be within 10% of 3:1 by Rutherford scattering and x-ray diffraction showed the films to be predominantly A-15 single phase.

The dependence of T_C upon ^{4}He damage for the four sets of A-15 films is shown in Figure 4. Nb_3Ge and Nb_3Sn films with compositions ~ 3:1 were only used for this study. All films show similar behavior. There is an initial plateau region where T_C is insensitive to low ^{4}He doses, T_C then decreases rapidly for doses in the region 10^{16} - 10^{17} cm^{-2}. Beyond this a saturation region is attained where T_C does not degrade with dose. The 5-95% values of the resistive transitions are plotted except where they are of the order of the size of the data points (mid points of the transitions). The initial, undamaged T_C values of the films and final saturated values of T_C are shown in Table 1. The x-ray diffraction data showed that on damaging the high angle lines for all materials tended to become less intense and that the lattices expanded. Table 2 shows the lattice parameter, a_0, of Nb_3Ge, Nb_3Sn and V_3Si before bombardment and the amount of lattice expansion, $\Delta a/a_0$, when the saturation region has been attained. The relative error in $\Delta a/a_0$ is ~ 30%. The lattice expansion data was not obtained in the V_3Ge case because in the process of etching off contacts the heavily damaged sample was destroyed.

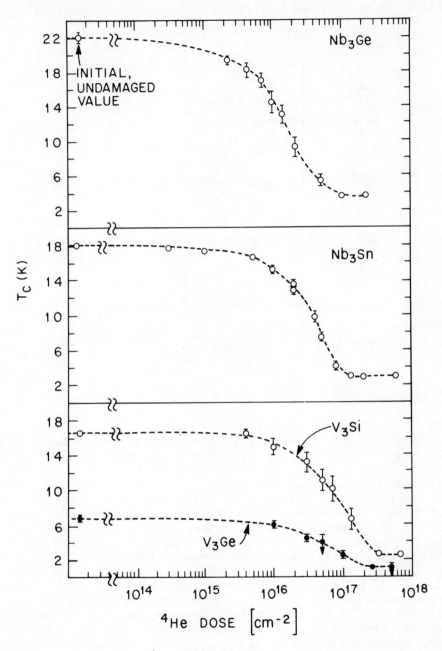

Figure 4: T_c vs 2 MeV ^{4}He bombardment for Nb_3Ge, Nb_3Sn, V_3Si and V_3Ge. Dashed lines are only meant to guide the eye.

Table 1: T_c values for initial and final states. Transition widths less than or the order 1 K.

	T_c (Initial)	T_c (Saturated)
Nb_3Ge	22.0 K	3.5 K
Nb_3Sn	18.1	2.95
V_3Si	16.8	2.2
V_3Ge	6.5	1.0

Table 2: Lattice parameter expansion following damage to saturation.

	a_0 (A)	$\Delta a/a_0$
Nb_3Ge	5.138	10×10^{-3}
Nb_3Sn	5.295	6×10^{-3}
V_3Si	4.727	1.7×10^{-3}

In the original Nb-Ge study we reported a T_c-resistance-ratio correlation for the as-grown films and found that ^{4}He damage of the films produced a correlation very similar to that for the as-grown films. We have produced resistance ratio plots for all the A-15 films of this study. Figure 5 shows the resistance ratio correlation for the NbSn films. We have not differentiated between the data and have plotted values from damaged and undamaged films, films with varying stoichiometry, films deposited under non-optimum deposition conditions and films damaged and then annealed at 700°C. The four encircled points come from a high T_c film (18 K) which has had 3 successive irradiations to depress T_c and indicate a typical experimental sequence. The cluster of points at a resistance ratio of unity and T_c of 3 K represent data, for example, from highly damaged films and also films deposited onto substrates held at room temperature. The resistance ratio vs T_c curves show similar features for all materials studied; the highest resistance ratios corresponding to optimum T_c. Maximum resistance ratios are $R(Nb_3Ge) \sim 3$, $R(Nb_3Sn) \sim 5$, $R(V_3Si) \sim 20$ and $R(V_3Ge) \sim 25$. These resistance ratio vs T_c curves and also resistivity vs T for the normal state will be discussed in detail elsewhere [9].

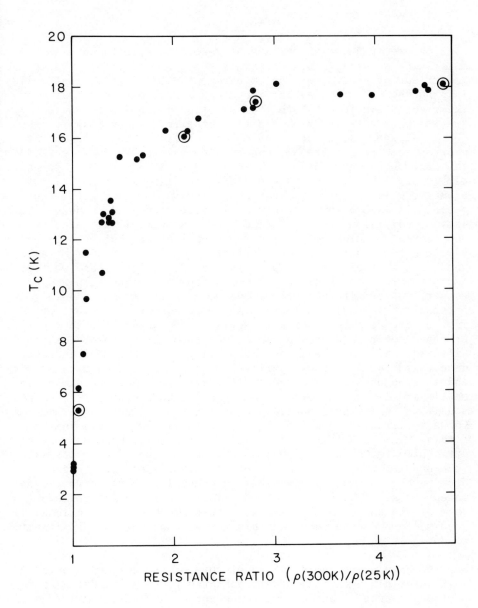

Figure 5: T_c vs resistance-ratio for Nb_3Sn.

DISCUSSION

a. Defect Production

Before speculating on the nature of the defects in A-15 materials, we will firstly consider the processes that produce the defect during neutron or ion bombardment. Disorder in most solids is produced by recoiling lattice atoms. For neutron bombardment the recoils can result both from elastic and inelastic nuclear scattering events and from elastic Rutherford scattering for 2MeV ^{4}He. The spectra of recoiling particles from either ^{4}He of fast neutron bombardment are not expected to be grossly different [4]. The maximum energy that can be imparted to Nb by a 2 MeV ^{4}He particle is 320 keV and the range of this recoiling Nb in Nb_3Sn, for example, will be approximately 500 Å [10]. Such energetic recoiling nuclei displace a large number of lattice atoms and produce a highly excited collision cascade. (The recoiling nuclei will also produce considerable inelastic or electronic excitation, but we have no evidence that such electronic effects produce damage in the A-15's.) The question therefore is how this cascade collapses and what defects remain. It is clearly a very complicated question and we have as yet, no definite picture of the defect structures in A-15 materials. The point defects, interstitials or vacancies, produced in the cascade may annihilate, agglomerate, interact with impurities or other defects such as dislocation; these are some of the possibilities. We can take a very broad view of bombardment induced defects in the A-15's by a brief comparison with damage in metals and semiconductors. The overriding difference between damage in metals and semiconductors is that the semiconductors are much more sensitive to damage. Semiconductors, for example, can be made amorphous by irradiation while this is virtually impossible in the metals. This sensitivity to damage in the semiconductors may be attributed to the directionality of the bonding. In metals without such bonding the damage may anneal relatively easily. Each lattice atom may be displaced many times in a metal with the crystalline integrity still being maintained. There is reason to believe that some of the bonds in the A-15 structure are highly directional [11], tending toward covalency and this may account for the sensitivity to damage.

We have as yet no detailed picture of the defect behavior in A-15 materials except for the Brookhaven group's [5] demonstration of the presence of antistructure defects. However, it is probably safe to assume that a whole gamut of defects remains following bombardment. We do know that the superconducting properties of the A-15's are very sensitive to damage. For example, 2MeV ^{4}He bombardment of Nb_3Ge at a dose of 10^{17} cm^{-2} degrades of T_c from 22 K to the saturated value of 3.5 K. An equivalent bombardment of Nb only reduces T_c from 9.3 to 9.1 K. The sensitivity of the T_c's to bombardment is not necessarily an indication of relative defect densities. Preliminary channeling experiments [12] indicate, however, that V_3Si resembles Si with regard to the density of defects resulting from bombardment. The defects observed in the channeling measurements are not antistructure defects.

The variation in the superconducting transition widths as a function of ^{4}He dose (Figure 4) may also contain significant information regarding defect production in the A-15's. The transition widths are very narrow in the undamaged and saturated states. In the intermediate region, however, where T_c is decreasing rapidly with dose, the widths increase quite markedly. This does not appear to be due to experimental artefacts, such as non-uniformity of dose, but rather a reproducible physical effect. The widths indicate inhomogeneities in the films on the scale of the coherence length ($\sim$ 50-100 Å) but these inhomogeneities cannot be due to gross changes in composition on that scale (^{4}He damage produces no compositional changes other than minor microscopic rearrangements due to the collision cascade, for example). Instead it would seem that there are inhomogeneities in the damage densities throughout the films. This can be pictured in terms of the individual collision cascade regions. When the density of cascades is low, there will be an uneven spatial distribution throughout the film due to the statistical nature of the process. At high cascade densities, the statistics improve and damage will be more evenly distributed.

Although the T_c vs ^{4}He dose curves have similar shapes there are obvious differences in the sensitivity to damage. For example the doses required to reduce T_c to 50% of the initial undamaged values are 2×10^{16} cm^{-2} (Nb$_3$Ge), 5×10^{16} (Nb$_3$Sn) and 10^{17} (V$_3$Si). This sensitivity to damage probably reflects the intrinsic structural stability of the materials with Nb$_3$Ge being the most unstable.

b. Saturated State

At high ^{4}He doses [13] T_c saturates as a function of dose for all four compounds with the values given in Table 1. This does not necessarily imply that the damage is also saturating. This assumption is based on our x-ray measurements of highly damaged, but initially high T_c Nb$_3$Ge and Nb$_3$Sn films and films that have been deposited on room temperature substrates. The highly damaged films while displaying considerable disorder retain some crystallinity whereas the films deposited at room temperature appear much more disordered. However the T_c's of the highly damaged or room temperature deposited films are identical (3.5 K for Nb$_3$Ge and 2.9 K for

Nb$_3$Sn) indicating a similar state. It would appear in fact that once a limiting value of damage is attained, T_c will not degrade any further.

Rowell and Schmidt [8] have now been able to perform tunneling measurements successfully on Nb$_3$Ge. Even in the highest T_C samples, two energy gaps are identified. The larger of these can be associated with well-ordered A-15 material. The smaller gap correlates extremely well with our highly damaged state (the gap value, $2\Delta_L$ = 1.0 meV, giving T_C ~ 3.3 K). Surprisingly enough even these tunneling measurements for the highest T_c films indicate the presence of considerable amounts of the "unwanted" phase at the surface regions of the films. The presence of this phase in the tunneling measurements was confirmed by sputter depositing a Nb$_3$Ge film on a 200°C substrate and observing the same Δ_L.

Similar highly disordered states for transition metals and transition metal alloy films were produced by Collver and Hammond [14] by evaporation onto substrates cooled to liquid helium temperatures. Meyer [15] has also been able to demonstrate the existence of such states by high dose implantations into transition metals.

c. T_C and Stoichiometry

In our original Nb-Ge study [3] we produced the T_C vs composition plot of Figure 2 and, as discussed previously, x-ray diffraction data showed the films to be essentially single phase for Nb/Ge ratios > 2.5. The implications of this data were very surprising. Either considerable densities of antistructure defects were possible or the structure could accommodate vacancy or interstitial concentrations of 20 atomic % and greater. This last possibility appears physically implausible. If we accept then that such a high density of antistructure defects is possible, we are faced with equally intriguing possibilities. For example, T_C's ~ 20 K are possible for Nb/Ge ~ 2.5. This ratio implies 5% of the sites on the Nb chains are occupied by Ge atoms. T_c therefore does not appear to be crucially dependent on the integrity of the Nb chains although some dependence clearly may exist.

We have extended this work to the Nb-Sn system as illustrated in Figure 3. The T_c vs stoichiometry plot is very similar to that for the Nb-Ge systems. Moreover x-ray diffraction data shows single phase behavior for Nb/Sn > 2.5. Figure 3 shows that the optimum composition for high T_c material would be at a Nb/Sn ratio of ≈ 2.7/1; a similar trend exists for the Nb-Ge data.

The evidence then from both the Nb-Ge and Nb-Sn experiments indicates that although stoichiometry is important, it is not crucial for attaining high T_c. It can be argued that the off-stoichiometry films contain filaments of the "good" 3:1 material which are not detectable by Rutherford backscattering. Critical current measurements [16] do not support this contention.

d. Resistance Ratio Correlations and Defect Universality

The T_c-resistance ratio plots of all four A-15 materials show the same trends (Figure 5 for Nb_3Sn) irrespective of whether T_c degradation is produced by the growth process or damage. This is persuasive evidence that the defects produced during ^4He bombardment are similar in their effect to those produced during the film growth process.

There are considerable differences between the maximum resistance ratios, R, for these materials; for example $R(Nb_3Ge) \sim 3$, $R(Nb_3Sn) \sim 5$ and $R(V_3Si) \sim 20$ corresponding to resistivities at 25 K of $\rho(Nb_3Ge) \sim 30$ $\mu\Omega$-cm, $\rho(Nb_3Sn) \sim 15$ $\mu\Omega$-cm and $\rho(V_3Si) \sim 5$ $\mu\Omega$-cm. Defect scattering therefore appears to be much more important in Nb_3Ge that in Nb_3Sn and V_3Si. No calculations of defect scattering exist for the A-15's. It will be intriguing to see what classes of defects can give rise to such large resistivities.

Recently Lutz et al. [17] have measured ρ between 4 and 575 K for Nb-Ge films with T_c's between 21 and 15 K and correlate certain features of the normal state resistance with T_c. They obtain values for ρ at 25 K of 40-45 $\mu\Omega$-cm for the highest T_c films.

e. What Role do Defects Play in A-15 Materials?

The results we have presented such as T_c vs dose and resistance ratio correlation indicate a certain universality of behavior. Clearly T_c can be lowered by damage and indeed the A-15 structure appears to be very sensitive to damage compared to Nb. Can any defect be solely culpable for T_c degradation? It now seems unlikely that antistructure (antisite) defects alone are responsible for such degradation. Probably any type of defect, in sufficient density, will affect T_c.

In earlier work by Bachner and Gatos [18] and Varma et al. [19] the concept was developed of an ordered array of point defects stabilizing crystals with the A-15 structure. More recently Phillips [20] has proposed that the long range order of defects develops by condensation from a defect cloud into a periodic structure. It is the strain fields produced by the defects that are important and not the individual characteristics of the defects. Our experimental results give no direct evidence on this model.

Gavaler [21] has suggested that the Nb$_3$Ge lattice might be stabilized by oxygen incorporation but we have no evidence to support this contention. Our Rutherford backscattering and nuclear reaction analyzes show that impurities such as C, N and O are at low levels (< 2 at %) in the high T_c films.

An interesting correlation, however, has emerged in our T_c vs stoichiometry plots for Nb$_3$Ge and Nb$_3$Sn. T_c appears to be optimized at composition ratios of 2.7:1 instead of the canonical 3:1. Lattice stabilization might therefore be effected by the simple mechanism of replacement of 3% of the Nb atoms by Ge or Sn atoms i.e., antistructure defects may be one of the ways of stabilizing the lattice.

ACKNOWLEDGEMENTS

We are indebted to A.R. Storm for the x-ray data and W.M. Augustyniak for expert technical assistance on the Van de Graaff.

REFERENCES

1. L.R. Testardi, J.H. Wernick, and W.A. Royer, Solid State Comm. 15, 1 (1974).
2. See, for example, A.R. Sweedler and D. Cox, Phys. Rev. B12, 147 (1975).
3. L.R. Testardi, R.L. Meek, J.M. Poate, W.A. Royer, A.R. Storm, and J.H. Wernick, Phys. Rev. B11, 4304 (1975).
4. J.M. Poate, L.R. Testardi, A.R. Storm and W.M. Augustyniak, Phys. Rev. Letters 35, 1290 (1975); also Institute of Physics Conf. Series No. 28, 176 (1976), Institute of Physics, London.
5. A.R. Sweedler, D.G. Schweitzer and G.W. Webb, Phys. Rev. Lett. 33, 168 (1974); A.R. Sweedler, D. Cox, and D.G. Schwietzer, IEEE Trans. Magn. 11, 163 (1975) and Ref. 2. Also, Ref. 22.
6. J.M. Poate, Proceedings of Karlsruhe Conf., "Ion Beam Surface Layer Analysis", Edited by O. Meyer, Plenum Press, N.Y. 1976.
7. R.H. Hammond, IEEE Trans. Magn. 11, 201 (1975).
8. J.M. Rowell and P.H. Schmidt, to be published.
9. L.R. Testardi, J.M. Poate and H.J. Levinstein, to be published.
10. Range calculated from LSS theory.
11. P.B. Allen and R.C. Dynes, Phys. Rev. B12, 905 (1975).
12. L.R. Testardi, J.M. Poate and J.H. Barrett, unpublished.
13. A dose of 10^{17} cm^{-2} corresponds to 0.5 displacements per atom (d.p.a.). That is, on the average, 1 out of 2 lattice atoms will be displaced during bombardment.
14. M.M. Collver and R.H. Hammond, Phys. Rev. Lett. 30, 92 (1973).
15. O. Meyer, Institute of Physics Conf. Series No. 28, 168 (1976).
16. L.R. Testardi, Solid State Commun. 17, 871 (1975).

17. H. Lutz, H. Weismann, O.F. Kammerer and Myron Strongin, to be published.
18. J.F. Bachner and H.C. Gatos, Trans. Metall. Soc. AIME 236, 1261 (1966).
19. C.M. Varma, J.C. Phillips and S.T. Chui, Phys. Rev. Lett. 33, 1223 (1974).
20. J.C. Phillips, Solid State Commun. 18, 831 (1976) and this meeting.
21. J.R. Gavaler, this meeting.
22. A.R. Sweedler, D.E. Cox, S. Moehlecke, R.H. Jones, L.R. Newkirk and F.A. Valencia, to be published.

QUESTIONS AND COMMENTS

P. Allen: There is indirect evidence in the Rowell and Schmidt work that the thermally damaged Nb-Ge is in fact amorphous. Do you have any direct evidence for that?

R. Dynes: No, not amorphous from the x-ray pictures. One sees several lines---not nearly as many as you do in the crystalline phase.

P. Allen: But this is a reproducible terminal damaged T_c of 3°K.

R. Dynes: Absolutely.

P. Allen: Has that occurred in other materials besides Nb-Ge?

R. Dynes: Yes, the four I just showed you. Niobium-germanium, niobium-tin, vanadium-silicon, and vanadium-germanium. This is not an unfamiliar result. Certainly, in amorphous transition metals one sees a similar kind of thing where you can get a smooth curve of T_c as a function of electrons per atom.

R. Hein: I am curious. In all cases where you have the transition temperature dropping with the radiation damage, do you always have a lattice deformation?

R. Dynes: Yes.

R. Hein: You also get a lattice deformation with composition, right? When you vary the composition so that it is--- do you know for sure whether you are getting A,B interchanges or are you putting interstitials into it? Let me put it the other way. If I were to change the lattice parameter by composition or radiation damage and give it to you to measure, could you tell me whether it was produced by radiation damage or by composition?

R. Dynes: We are irradiating with 2 MeV alpha particles, and that is pretty violent. I think they are doing all of that and more.

R. Hein: More what?

R. Dynes: Whatever damage you can imagine. I think there would be a correlation of T_c with whatever your pet damage is.

R. Hein: All right, suppose I took a damaged one and compressed it, what would happen? Would the disorder stay the same? Suppose I put it under pressure and now I shrink the order parameter, the lattice parameter, would the order go up? I know the transition temperature would go up.

R. Dynes: I do not know the answer to that.

S. Nam: You just mentioned you used 2 MeV alpha particles. I wonder if you studied some kind of correlation between the energy of the beam and the transition temperature?

R. Dynes: No.

S. Nam: Have you ever studied your sample using other beams such as Li, B, etc.?

R. Dynes: A lot of this work, to look at the rather general nature of the phenomenon, was because we were worried about the earlier neutron damage measurements from Brookhaven where they did not seem to show this saturated T_c. That worried me a lot because I have this preconceived prejudice that the T_c has to saturate with high damage. After all, amorphous transition metals do show this very smooth behavior and so we did these other materials to see if we did get this saturation at high damage but we only looked with alpha particles.

C. Koch: Have you done any transmission electron microscopy to look at the damage?

R. Dynes: I have one TEM picture of some as-grown films of Nb_3Ge; the same ones we did the Rutherford backscattering on. Here is a 12.9 K critical temperature film and here is a 21.4 K film [slides shown to audience]. You see two things. Firstly, you see smaller grain size in the lower T_c material. The second interesting point is that there are no new lines in these two materials. Here are spots and here are rings but there is a one-to-one correlation between the two photographs.

EMPIRICAL RELATIONS IN TRANSITION METAL

SUPERCONDUCTIVITY

C. M. Varma and R. C. Dynes

Bell Laboratories

Murray Hill, New Jersey 07974

ABSTRACT

The empirical relations among the parameters determining the superconducting transition temperature in transition metals and alloys and for some A-15 compounds are collected. These parameters are the density of states near the Fermi surface, the average phonon frequency, and an average electron-phonon matrix element. The empirical relations suggest a theory for transition metals that is based on the nature of the atomic orbitals, for example the tight binding scheme for the electronic structure. A simple single orbital non-orthogonal tight binding model is used to calculate parameters of importance in superconductivity. The empirical relations follow from such a model. It is argued that a realistic non-orthogonal tight binding model will yield similar results. One of the conclusions of the work is that T_c is higher for materials satisfying the conflicting requirements of large bandwidth, large density of states near the Fermi surface and with the Fermi level lying in a non-bonding part of the band. An application of theory to T_c degradation due to defects or radiation damage further substantiates the conclusions of this work.

1. INTRODUCTION

The BCS theory of superconductivity and its strong-coupling extensions tell us that given:

A. the (bare) electronic density of states near the Fermi surface, $N(0)$,

B. an average, $\langle\omega\rangle$, and a mean square average $\langle\omega^2\rangle$, of the phonon frequencies,

C. an average of the square of the electron-phonon matrix element over the Fermi surface, $\langle I^2\rangle$, and

D. an appropriately renormalized electron-electron Coulomb interaction, μ^*,

the superconducting transition temperature, T_c, can be accurately calculated for most metals.

It is actually found[1,2] sufficient to express T_c in terms of three parameters, $\langle\omega\rangle$, μ^* and λ, where

$$\lambda = \frac{N(0)\langle I^2\rangle}{M\langle\omega^2\rangle}, \qquad (1.1)$$

where M is the mass of the ion, assuming the metal under consideration to be monatomic. For simple (pseudopotential) metals, the quantities (A) to (D) can be calculated reasonably accurately. More important, the pseudopotential theory for electrons and phonons leads one to understand the functional relationships among the parameters (A) to (C) and the trends in the super-conducting and normal properties of the simple metals.

Similar understanding of transition metals (TM) and compounds is largely absent. A lot of effort[3,4] is currently being directed at obtaining the gross parameters, (A)-(C), that determine T_c through first calculating the microscopic properties of TM. This seems to us at this stage to be a very ambitious goal. The basic <u>starting</u> physical point of these theories is the same as that for simple metals. This is in contrast to the fact that while simple metals are nearly free-electron like, the d-electrons in TM are tightly bound. In fact, the "alchemists" rules for superconductivity in TM are quite different from those of simple metals.

The properties of transition metals have been experimentally investigated for a long time, and empirical relations among the three important quantities, (A)-(C) and other macroscopic properties like the cohesive energy, E_c can be drawn. These empirical relations are very suggestive in directing one to the starting point for an economical theory for TM. The simple "chemical" trends for superconductivity in TM suggest a theory that has a chemical concept as a starting point. This was recognized, for example, by Hopfield[5]. But while Hopfield took the atomic potential (or its derivative) as the basic starting point, we suggest that the relevant chemical entity is the bond between pairs of atoms. This in turn means that we find that the bandstructure, in so far as it reflects the symmetries of the bond (or crystal) are quite relevant to T_c, while Hopfield concluded they were not.

It also seems worth emphasizing that to understand the variation of T_c in TM, all the inter-relationships, (A) to (C) must be considered and not just of a pair of them. The latter, we believe, has engendered a lot of confusion in this field.

2. EMPIRICAL RELATIONS OF NORMAL STATE PROPERTIES

2.1 Constancy of $N(0) \langle I^2 \rangle$

McMillan[1] was the first to note that $N(0) \langle I^2 \rangle$ is approximately a constant for the pseudopotential metals and even more approximately for the transition metals. He proffered no explanation for the transition metals, but proved that such a relationship exists for the nearly free electron metals simply because the bare ion frequencies are much larger than the actual phonon frequencies. Actually for simple metals what McMillan really proved was not the constancy of $N(0) \langle I^2 \rangle$ but of $N(0) \langle I^2 \rangle / \Omega_p^2$ where Ω_p^2 is the bare ion plasma frequency. The latter is fairly well satisfied by the pseudopotential metals[6].

For transition metals two proofs of a qualitative nature for the constancy of $N(0) \langle I^2 \rangle$ have been given, by Hopfield[5] and by Barisic, Labbe and Friedel[7]. Hopfield's proof is based on an APW approach to electronic band-structure together with a deformation potential approach to the electron-phonon interaction. Detailed calculations on the basis of this approach have been performed by Gyorffy and his collaborators[4,8]. Barisic, Labbe and Friedel start with a orthogonal tight binding approach to the electrons together with a rigid-ion model for the electron-phonon interaction. This seems to us to be the correct starting point for transition metals although we find it important to include the non-orthogonality effects. In a model with a non-degenerate orbital and nearest neighbor electron transfer, they find

$$N(0) \langle I^2 \rangle = E_c / r_o^2$$

where E_c is the contribution to the cohesive energy of the d-orbital and r_o characterizes the exponential decay of the orbital. (We shall present a generalization of this approach to non-orthogonal orbitals in Sec. 3). It would be a bit hasty to conclude from this that $\lambda \sim E_c$, since to a first approximation $M\langle\omega^2\rangle$ is also proportional to E_c. Also the approximate constancy of $N(0)\langle I^2\rangle$ should not lead one to think $N(0)$ separately is unimportant to λ, since $\langle\omega^2\rangle$ depends strongly on $N(0)$, (see 2.3).

2.2 Connection between λ and $N(0)$

Contrary to the assertion sometimes made, there are very interesting and revealing correlations between λ and $N(0)$ in transition metals, alloys and compounds. Surprisingly, these have not received much attention. Consider the bcc transition metals and alloys. From the measured T_c, $\langle\omega\rangle$ or (θ_D), and the electronic contribution to the specific heat, γT, (and with a reasonable assumption for μ^* to which the results are not very sensitive) one can deduce λ and $N(0)$. We put λ against $N(0)$ in Fig. (1). This is the plot already shown by McMillan[1] and by Hulm and Blaugher[9]. We note that the points lie approximately on a set of straight lines, two each for the 4d and 5d metals and alloys one for the 3d metals and alloys. The dominant feature of the density of states[10] of transition metals is two peaks separated by a deep valley. On examination[11] of the position of the Fermi level for various transition metal alloys in a rigid band picture, one finds that $\lambda/N_B(0)$ has one value, C_{1s}, if the Fermi level lies in the lower of the peaks in the series s and another value, $C_{2s}>C_{1s}$, if it lies in the higher energy peak. The metals and alloys of the 3d series whose Fermi level lies in the higher energy peak are magnetic, and therefore there is only one line in Fig. 1 for the 3d series.

At this point we note[12] that in bcc transition metals and alloy, the lower peak in the density of states is determined by the bonding orbitals and the upper peak by the non-bonding orbitals.

One immediate conclusion from Fig. (1) is that transition metals and alloys fall into categories within which $\langle I^2\rangle/M\langle\omega^2\rangle$ is approximately a constant. The categories themselves are determined by (a) the series, 3d, 4d or 5d, and (b) the nature of the orbitals near the Fermi surface. From a practical point of view, it would seem therefore that the simplest way to understand electron-phonon interactions in transition metals and compounds is to start with a basis set for the electronic structure, which has the nature of the band orbitals built in.

How general is the constancy of $\langle I^2\rangle/M\langle\omega^2\rangle$? In Table 1, we give the values of λ and $N(0)$ for the V_3Si - V_3Ga - V_3Sn alloys. $N(0)$ in this case has been obtained from Fradin and Williamson's[14] NMR measurements. $N(0)$ determined from specific heat measurements are consistent with these. Fradin and Zamir, and Fradin and

TABLE I

Illustrating the constancy of $\lambda/N(0)$ in V_3Si-V_3Ga-V_3Sn alloys. Data taken from Ref. (14).

Alloy	λ	$N(0)$ (Ry V-spin)$^{-1}$	$\frac{\lambda}{N(0)}$ (× 10^{-1})
V_3Ga	0.86	38.8	0.22
V_3Si	0.76	35.0	0.22
$V_3Ga_{0.9}Sn_{0.1}$	0.73	33.9	0.22
$V_3Ga_{0.8}Si_{0.2}$	0.70	32.4	0.22
$V_3Ga_{0.2}Si_{0.8}$	0.67	30.6	0.22
$V_3Ga_{0.6}Si_{0.4}$	0.65	30.0	0.22
$V_3Ga_{0.4}Si_{0.6}$	0.64	29.3	0.22
$V_3Ga_{0.7}Sn_{0.3}$	0.63	28.8	0.22
$V_3Ga_{0.5}Sn_{0.5}$	0.60	26.4	0.23
$V_3Ga_{0.1}Sn_{0.9}$	0.53	21.3	0.25
V_3Sn	0.48	16.8	0.29

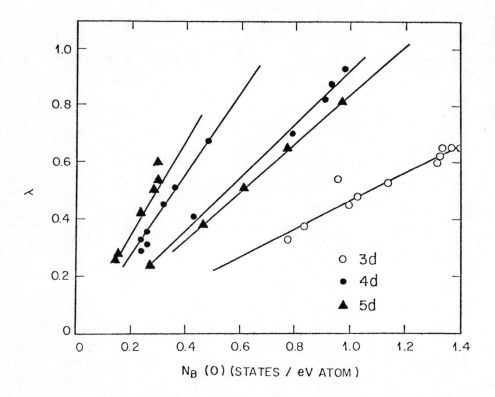

Figure 1: N(0) vs. λ for 3d, 4d and 5d alloys. Similar plots have been shown by McMillan, (Ref. 1) and by Hulm and Blaugher (Ref. 9).

Williamson have also determined that there is one orbital that is dominant near the Fermi surface for the entire V_3Si, V_3Ga and V_3Sn compounds and their alloys. For Nb_3Sn and the off-stoichiometric A15 compound, Nb_4Sn,[15] $<I^2>/M<\omega^2>$ has the same value. If however, we compare Nb_3Sn, Nb_3Al and Nb_3Sb, $<I^2>/M<\omega^2>$ is quite different. Fradin[16] has pointed out that the dominant orbital is not the same for Nb_3Sn, Nb_3Al and Nb_3Sb. Thus it would seem that $<I^2>/M<\omega^2>$ is a constant only among those compounds which have a dominant orbital of the same nature near the Fermi surface.

2.3 Behavior of the Average Stiffness $M<\omega^2>$

If $N(0) <I^2>$ is approximately a constant and $<I^2>/M<\omega^2>$ is also approximately a constant for a given set of materials, $M<\omega^2>$ must be approximately related to $N(0)^{-1}$. Such a behavior is of course found. We must also note that $M<\omega^2>$ is also related to the cohesive energy and that for transition metals and alloys a large part of the cohesive energy is due to the d-band. We illustrate these relations in Fig. 2. A related figure has been shown earlier by Hammond and Collver[17].

A great deal of theoretical effort has been expended[18] in explaining anomalous dips[19] in the dispersion of phonons, $\omega(q)$. These dips are observed to be large when the density of states near the Fermi surface is large. Within a given class of compounds $<I^2>/M<\omega^2>$ is approximately independent of $N(0)$; (Sec. 2.2), therefore λ is approximately proportional to $N(0)$. Hence within a given class of compounds dips in the phonon spectrum (and indeed on overall softening of phonons) and larger value of T_c occur for the same materials [i.e. those that have higher $N(0)$].

Having established that the three quantities, $N(0)$, $M<\omega^2>$ and $<I^2>$ that determine λ and thus T_c are intimately inter-related, it might seem a moot point as to which controls T_c, soft phonons or large $<I^2>$ or large $N(0)$. From a logical point of view, we think it is not simply a chicken and egg question. This is so, because of the adiabatic principle. The adiabatic principle states that we first determine an electronic structure with the atoms in their equilibrium position and use the variation of the electronic energy with lattice distortion as contributing to the stiffness of the lattice. Thus we prefer to regard the electronic structure as the independent variable and $<I^2>$ and $M<\omega^2>$ as being derived from it.

Finally, it is worth emphasizing that the empirical relations are all quite approximate and over a limited region of parameter space and under restricted conditions; they are meant as a guide to understanding and not as sacred rules.

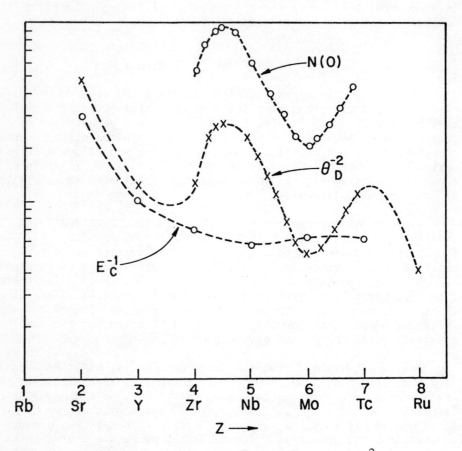

Figure 2: Illustrating that for transition metals $\langle\omega^2\rangle$ varies approximately as $E_c/N(0)$.

3. ELECTRON-PHONON MATRIX ELEMENTS IN TIGHT BINDING

The d-electrons in transition metals are best described in the tight binding scheme.[12] The orbitals at neighboring sites are in general non-orthogonal. Indeed, the non-orthogonality (i.e. overlap repulsion) together with the core-core repulsion lead to the strong short-range interatomic repulsion. This added to the "bonding" energy due to the band-structure has a minimum as a function of inter-atomic separation. In fact, the non-orthogonality parameters can be used to parametrize the effects of inter-atomic exchange and correlation to some extent.

It is worth emphasizing that non-orthogonality of orbitals is an essential physical ingredient of any tight binding description of electronic structure.[20-22] Without it, the tight-binding parameters are no more than a fitting scheme, and may often be quite unphysical.

It is complicated to proceed very far analytically with the realistic case of a non-orthogonal basis set of several orbitals. Therefore in this paper, we will do calculations with a single non-degenerate, nearest neighbor, but non-orthogonal orbital in a cubic lattice. We will see that several important physical points can be illustrated even with such a simple starting point.

An expression for $<I^2>$ in orthogonal tight binding has already been given by Barisic, Labbe and Friedel (BLF)[7]. The results of this section are thus just a generalization to non-orthogonal orbitals.

The Hamiltonian for the non-degenerate band of tight binding orbitals is

$$H_o = \sum_{i<j,\sigma} t_{ij} a^+_{i\sigma} a_{j\sigma} , \qquad (3.1)$$

where $a^+_{i\sigma}$ creates a particle with spin σ in the orbital $\phi(r-R_i)$ about the site i, and the restriction over the summation is that j be one of the nearest neighbors to i. $t_{ij} \equiv t$ is the transfer integral between the nearest neighbors, and we will denote the overlap integral of the orbitals about neighboring sites by S. H_o can be diagonalized to yield

$$H_o = \sum_{k\sigma} E_k c^+_{k\sigma} c_{k\sigma} ,\qquad(3.2)$$

where

$$E_k = \frac{\varepsilon_k}{(1+S_k)} ,\qquad(3.3)$$

$$\varepsilon_k = t \cos \underline{k} \cdot \underline{a}\qquad(3.4)$$

and

$$S_k = S \cos \underline{k} \cdot \underline{a}\qquad(3.5)$$

where $\underline{a}$ are the lattice vectors.

For tight binding to be valid, $S \ll 1$ (it can be as high as 0.25), so that

$$E_k \simeq \varepsilon_k (1-S_k) .\qquad(3.6)$$

The transfer integral t and the overlap integral will decay exponentially with the distance R, between atoms, and we will describe the decay of both by the same parameter, r_0. The electron phonon Hamiltonian can now be derived in the rigid ion approximation. We take that

$$t_{ij}(\underline{R}) \simeq t_{ij}(\underline{a}) + \left.\frac{dt_{ij}}{d\underline{R}}\right|_{\underline{a}} \cdot (\underline{u}_j - \underline{u}_i) ,\qquad(3.7)$$

where u_i is the displacement of the atom at site i from equilibrium, and

$$\frac{dt}{d\underline{R}} = \frac{t}{r_0} \frac{\underline{a}}{|\underline{a}|} ,\qquad(3.8)$$

just as in BLF. Then it is straight-forward to derive that the electron-phonon Hamiltonian is

$$H_{el \cdot ph} = \sum_{\underline{k},\underline{q},\nu} g^\nu_{\underline{k}\underline{q}} c^+_{\underline{k}} c_{\underline{k}+\underline{q}} (b_{\underline{q}\nu} + b^+_{-\underline{q}\nu}) ,\qquad(3.9)$$

where

$$g_{kq} = \sqrt{\frac{\hbar}{2M\omega_{q\nu}}} \, I_{kq}^{\nu} \, , \qquad (3.10)$$

$$I_{kq}^{\nu} = i \sum \frac{a_\alpha \cdot \varepsilon_q^\nu}{|a|} \frac{t}{r_o} \left(\sin(k_\alpha a_\alpha)(1-2S_{k\alpha}) \right.$$
$$\left. - \sin((k+q)_\alpha a_\alpha)(1-2S_{(k+q)_\alpha}) \right), \qquad (3.11)$$

$$= i \sum \frac{a_\alpha \cdot \varepsilon_q^\nu}{|a|} (V_{k\alpha} - V_{(k+q)\alpha}) \qquad (3.12)$$

In (3.9) - (3.12), α are the three cartesian directions, ν is the polarization of phonons of wave-vector q and the velocity of the electrons V_k is given by

$$V_{k\alpha} = \hbar^{-1} \frac{dE_k}{dk_\alpha} \simeq v_{k\alpha}(1-2S_{k\alpha}), \quad v_{k\alpha} = \hbar^{-1} d\varepsilon_k/dk_\alpha \, . \qquad (3.13)$$

Equation (3.12) is formally similar to that of BLF. The difference is that because of the non-orthogonality corrections, the band structure is skewed; the velocities of electrons for the less than half-filled band are decreased and those of the more than half-filled band are increased.

From (3.12) it is easy to see that for k and $k+q$ on the Fermi surface

$$\sum_\nu |I_{kq}^\nu|^2 = \sum_\alpha (r_o a)^{-2} \left[V_{k\alpha}^2 + V_{(k+q)\alpha}^2 \right] \, . \qquad (3.14)$$

Now the parameter $<I^2>$ that occurs in λ is given by

$$\langle I^2 \rangle = \frac{\int d\sigma_k \int d\sigma_{k+q} \sum_\nu |I^\nu_{k,k+q}|^2 \, v_k^{-1} \, v_{k+q}^{-1}}{\int d\sigma_k \int d\sigma_{k+q} \, v_k^{-1} \, v_{k+q}^{-1}} \tag{3.15}$$

where the integrals are over the Fermi surface. Also the density of states at the Fermi surface is

$$N(0) = \int d\sigma_k \, v_k^{-1} \tag{3.16}$$

Using (3.14) and (3.16) in (3.15), one gets

$$N(0) \langle I^2 \rangle \simeq (ar_o)^{-2} \int d\sigma_k \, \nabla E_k \tag{3.17}$$

It is easy to derive the results of the rest of this paper with (3.17), but they are rendered much more transparent if we approximate

$$E_k \simeq \epsilon_k (1-S) \;,$$

for lower half of the band

$$\simeq \epsilon_k (1+S) \;, \tag{3.18}$$

for the upper half of the band.

This is a reasonable approximation except near the middle of the band. On using Gauss' theorem in (3.17), one gets

$$N(0) \langle I^2 \rangle = \frac{E_o}{r_o^2} (1 \mp S) \;, \tag{3.19}$$

where

$$E_o = \sum_{k<k_F} \epsilon_k \;.$$

From now on, in any factor of the form $(1 \mp S)$, the upper sign will refer to Fermi level in the lower half of the band, and the lower sign to the Fermi level in the upper half of the band.

4. PHONONS IN THE TIGHT-BINDING SCHEME

The customary way of calculating phonons in metals is to calculate the bare ion frequencies and then screen them by the electrons.[23] For transition metals, where the electrons are tightly bound, this requires calculation of screening to extremely high precision. The natural zeroeth order phonon frequencies for this situation are those determined from a dynamical matrix which contains the "adiabatic" variation of the electronic energy levels with the deformation. By adiabatic, we mean that an electronic energy level $\underset{\sim}{k}$ with energy ε_k remains under deformation in a state $\underset{\sim}{k}$ but with an altered energy $\bar{\varepsilon}_k$. Next, such phonon frequencies must be corrected due to scattering of states from $\underset{\sim}{k}$ to $\underset{\sim}{k}$, due to deformation. In real space, the first corresponds to calculating the phonon frequencies for rigid ion motion and the second to corrections due to the deformability of the ion. Formally the first corresponds to first order perturbation of energy in second order displacements, while the second to second order perturbation of energy in first order displacements.

4.1 Zero-order Phonon Frequencies

We shall be concerned here with only the mean square average phonon frequencies, $<\omega^2>$. If the total energy of the transition metal is $E(R)$, the zero-order (rigid-ion) phonon frequencies are given by

$$M<\Omega^2> = \sum_\alpha \left. \frac{d^2E}{dR_\alpha^2} \right|_{R=a} \qquad (4.1)$$

Contributions to $E(R)$ arise from
(1) The sum of the one-electron (band-structure) energy.
(2) Core-core repulsion <u>minus</u> the electron-electron repulsion energy.

The sum of the one-electron energies calculated in the Hartrec-Fork approximation includes the electron-electron repulsion twice. Hence it is subtracted in (2) from the core-core repulsion terms. Now, for any long wave-length property, (2) sums up to zero. We believe that for the derivatives of the energy (2) can be neglected compared to (1), because the variation of (1) is on the scale of the variation of the overlap integral. In fact chemists[24] calculating the structure and the vibration frequencies of molecules have found that best results are obtained by ignoring (2) altogether. For transition metals, we may note the further justification that about 60-80% of the cohesive energy is provided by the d-electrons above. Even the crystal structure of transition metals[25] can be correctly predicted on the basis of (1) alone.

This brings us to the contribution (1) to $E(R)$. This will come from both s electrons and d electrons. The contribution of s electrons to the average phonon frequencies of a transition metal (and of the contribution (2)) may be estimated to be the same as the average phonon frequency of the corresponding alkali metal of the series. These have phonon frequencies an order of magnitude less than the transition metals.

In (4.1), the second derivative is to be calculated at the value $R=a$, at which the first derivative

$$\frac{dE(R)}{dR}\bigg|_a = 0 \text{ , at equilibrium} \qquad (4.2)$$

where

$$E = \sum_{k<k_F} \varepsilon_k (1-S_k) \text{ .} \qquad (4.3)$$

We then have

$$S \sum_{k<k_F} \cos^2(\underline{k}\cdot\underline{a}) = \frac{1}{2} \sum_{k<k_F} \cos(\underline{k}\cdot\underline{a}) \text{ .} \qquad (4.4)$$

We may take (4.4) to be determining S at the known lattice constant. Taking the second derivative of $E(R)$ and using (4.4), we get

$$M\langle\Omega^2\rangle = -r_o^{-2} \sum_{k<k_F} \varepsilon_k \equiv |E_o| r_o^{-2} \text{ ,} \qquad (4.5)$$

$$= 2 S |t| r_o^{-2} \sum_{\underline{k}} \cos^2\underline{k}\cdot\underline{a} \text{ .} \qquad (4.6)$$

These simple equations say that the average stiffness in the lattice is proportional to the leading contribution (i.e. the orthogonal terms) to the cohesive energy. This is in turn proportional to the overlap between the orbitals. This makes physical sense - the more the overlap, larger the bonding energy and larger the lattice stiffness.

4.2 Phonon Frequencies

Now we consider the renormalization of the average stiffness due to particle-hole creation by lattice deformation. For a phonon of wave-vector q, the change from Ω_q^2 is given by

$$\omega_q^2 \simeq \Omega_q^2 + 2\Omega_q \, \Pi(q,0) \tag{4.7}$$

where $\Pi(q,0)$ is the self energy contribution:

$$\Pi(q,0) = \sum_k \left| g_{k,k+q} \right|^2 \chi(k,k+q) \tag{4.8}$$

where $\chi(k,k+q)$ is the electronic polarizability at zero frequency. If $U_{k,k+q}$ is the residual interaction between the tightly bound d orbitals,

$$\chi_{k,k+q} = \chi_q^o / (1 + U_{k,k+q} \, \chi_q^o) , \tag{4.9}$$

where χ_q^o is the zero-order electronic polarizability which is given in terms of the electronic Green's functions by

$$\chi_q^o = \sum_k \int d\varepsilon \, G_o(k,\varepsilon) G_o(k+q,\varepsilon) \tag{4.10}$$

For the moment, let $U_{k,k+q} = 0$. Now employing the Gomersall and Gyorffy[27] trick,

$$\sum_q 2M\Omega_q \, \Pi_o(q,0) \equiv 2 \sum_{k,q} M\Omega_q \left| g_{k,k+q} \right|^2 \chi_{k,k+q}^o \tag{4.11}$$

$$\simeq - N^2(0) \, W \, \langle I^2 \rangle$$

where W is a cut off energy for the electronic excitations, which is of the order of the bandwidth. Since, by (3.19) and (4.5) respectively, $N(0)\langle I^2\rangle \simeq \dfrac{E_o}{r_o^2}$, $M\langle\Omega^2\rangle \simeq E_o/r_o^2$, and $N(0)W$ is larger than 1 for most superconducting metals of interest, it is necessary to consider the reduction in the polarizability due to $U_{k,k+q}$.

First we note that by virtue of (3.19) and (4.5), we can rewrite (4.11) as

$$2 \sum_q \Omega_q \sum_k |g_{k,k+q}|^2 \chi^o_{k,k+q} \simeq -N(0)W(1\mp S) \sum_q \Omega_q^2 . \qquad (4.12)$$

so that we estimate

$$2/\Omega_q \sum_k |g_{k,k+q}|^2 \chi^o_{k,k+q} \simeq -N(0)W(1\mp S) \qquad (4.13)$$

Next, we estimate $U_{k,k+q}$. $U_{k,k+q}$ is the net interaction between two tightly bound orbitals considering the electron-electron, the ion-ion and the electron-ion interaction. Thus it is a weak and quite short-range interaction. We guess that

$$U_{k,k+q} = \frac{2|g_{k,k+q}|^2}{\Omega_q} \qquad (4.14)$$

For jellium, a relation of this form exists between the electron-bare phonon interaction g_q the bare ion frequency, Ω_o and the bare electron-electron (Coulomb) interaction V_q. The physics[28] in that case is that $-2g_q^2/\Omega_o$, the effective electron-electron attraction due to lattice motion at zero frequency, plus the electron-electron repulsion must add to zero. A similar physical argument does not exactly work for tightly bound electrons, though it indicates that (4.11) must be approximately true. With (4.14), (4.13) and (4.9) we then get

$$2M \sum_q \Omega_q \Pi(q,o) = -\frac{N(0)W\langle\Omega^2\rangle(1\mp S)}{1+N(0)W(1\mp S)} \qquad (4.15)$$

(4.15) shows that (4.14) is equivalent to estimating $U \simeq W$, and estimate arrived at by Kanamori[29] by other arguments. Equation (4.15) used in (4.7) yields

$$M\langle\omega^2\rangle = \frac{M\langle\Omega^2\rangle}{1+N(0)W(1\mp S)} \qquad (4.16)$$

Using (3.19), and (4.16), we have the relation between $M\langle\omega^2\rangle$ and $\langle I^2\rangle$,

$$\frac{<I^2>}{M<\omega^2>} \simeq \frac{(1\mp S)[1+N(0)W(1\mp S)]}{N(0)}, \qquad (4.17)$$

$$\simeq W(1\mp 2S), \qquad (4.18)$$

for $S \ll 1$ and $1 \ll N(0)W$, and we remind that the upper sign is for Fermi level in the lower half of the band and the lower sign for Fermi level in the upper half of the band. If we had not made the approximations (3.18) to the band-structure, we would find the result (4.18), for the Fermi-level far from the middle of the band, with the two lines joining smoothly for Fermi-level in the middle of the band.

5. RELATION OF THEORY TO EMPIRICAL RELATIONS

The expression for $<I^2>$ and $M<\omega^2>$ derived in Sec. 3 and 4 are on an oversimplified model. They can hardly bear <u>quantitative</u> comparison with the corresponding quantities in real systems. However, we believe that the qualitative features of our results will continue to hold in realistic models for the transition metals. This belief is based on the fact that we have expressed our results in terms of certain general parameters, which have the same meaning in a realistic situation as they have in our simplified model. These parameters are: $N(0)$, the density of states near the Fermi energy; W, the characteristic cut-off energy for electronic excitations; E_c, the cohesive energy; and the reduction or increment of the velocity of electrons in the less than or more than half filled band respectively. This latter fact we have expressed through the factor $(1 \mp S)$. Our results of course are relatively more convincing in real situations where a single orbital dominates near the Fermi surface, such as occurs in many transition metals and compounds.

With these points in mind, we can try to relate the general features of our results to the three empirical relations discussed in Sec. 2: our derivation of $N(0) <I^2>$ deviates from that of BLF by factors $(1 \mp S)$, which is quantitatively a small correction. Thus the remarks made by BLF about the slowness of the variation of the product $N(0) <I^2>$ compared to either $N(0)$ or $<I^2>$ remain.

We have derived the approximate relationship, (4.16)-(4.5) between $M<\omega^2>$, E_o and $N(0)$. Since E_o is apart from non-orthogonality corrections approximately equal to the cohesive energy E_c, Fig. (2) shows the qualitative correctness of (4.16)-(4.5).

The relationship discussed in Sec. 2.2 and illustrated in Fig. (1), the constancy of $\langle I^2 \rangle/M\langle\omega^2\rangle$ in a class of materials, is the most intriguing and the richest of any of the empirical relations. On the basis of Eq. (4.18) for $\langle I^2 \rangle/M\langle\omega^2\rangle$ we can easily understand the empirical relations of Fig. (1). First we note that the rigid band scheme applies in transition metals and their alloys. For our purposes it means that the parameters W and S have approximately one value for all 3d metals and alloys, another for 4d's and yet another for 5d's. Since the bandwidth increases in going from 3d to 4d to 5d metals and alloys, we expect the slopes in the λ vs $N(0)$ plot of Fig. (1), (i.e. the value of $\langle I^2 \rangle/M\langle\omega^2\rangle$ to increase in going from 3d to 4d to 5d. The further division of the metals and alloys of the 4d and 5d series into two lines each comes about due to the factor $(1\mp S)$. As discussed in Sec. 2, the peak below the big dip in the density of states corresponds to a bonding band and the peak above to a non-bonding or an antibonding band. The former has the electronic velocity reduced due to non-orthogonality and the later has it increased. We can understand the order in which the lines of 4d and 5d fall in Fig. (1) only if

$$S_4/S_5 > W_5/W_4 \quad ,$$

where S_4/S_5 and W_4/W_5 are the ratio of the overlap and bandwidth parameters respectively for the 4d and 5d series. This now begins to be a fine point and we will have to wait till calculations on realistic models are available before commenting on it.

We can also understand the results of Table 1, on the V_3Si-Ga-Sn alloys on the basis of our model since a single orbital is dominant near the Fermi surface for them and the rigid band model applies.[3,14]

6. T_c DEGRADATION

We have presented plausible evidence that for two different superconductors in the same series, $\langle I^2 \rangle/M\langle\omega^2\rangle$ is the same if the same orbital is dominant near the Fermi surface. Some heuristic justification for this behavior has also been given. The same arguments also indicate that for superconductors with the same relative contribution of different orbitals near the Fermi surface, $\langle I^2 \rangle/M\langle\omega^2\rangle$ will also be nearly the same. With this point of view, we can try to understand the degradation in superconducting transition temperature due to defects or damage.[30-32]

It is hard to imagine that neutron or α-particle irradiation damage can shift bands relative to each other. The same is true for small deviations from stoichiometry or other defects. Therefore

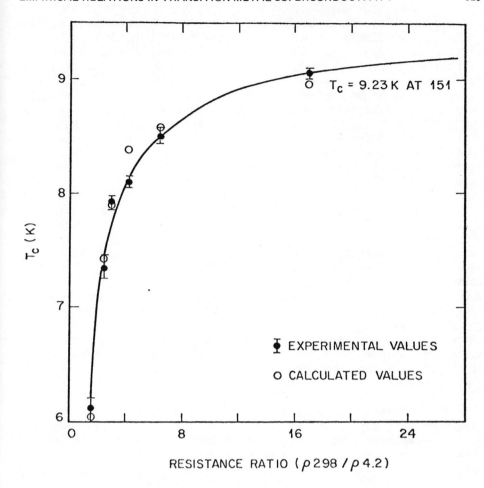

Figure 3: Variation of T_c with the residual resistance ratio in Nb-O system. Closed circles are experimental points, open circles are calculated.

according to our hypothesis $<I^2>/M<\omega^2>$ will not change on damaging a superconductor. The change in T_c can come about only through a variation in $N(0)$. It is possible to test this idea if an independent estimate of $N(0)$ is available. For Nb doped with varying concentration of oxygen, both T_c and specific heat measurements have been made.[33] Assuming that $<I^2>/M<\omega^2>$ is a constant, we have tried to predict T_c from the specific heat measurements. In Fig.(3), we compare the prediction with the experimental result and conclude that our assumption that $<I^2>/M<\omega^2>$ is a constant is verified.

In Fig. (3), we have plotted, T_c vs. Resistivity ratio to emphasize the similarity of the behavior of Nb with defects to that of the A-15 compounds with defects, which are often plotted in this fashion.[31,32] We believe that in the A-15 compounds also, the depression of T_c occurs mostly due to a change in $N(0)$. The saturation at higher defect concentration occurs when all the fine features in the electronic band-structure are washed out, i.e. when the electronic mean free path is a few lattice spacings. We also note that the depression of T_c and the behavior of the phonon contribution to the resistivity are mutually consistent.[34] The latter varies at high temperatures as $\lambda_{tr}T$, where λ_{tr} is to a good approximation proportional to λ.

An interesting question is the identification of the defects that arise on irradiating the A-15 compounds, and whether there is a cooperative defect producing mechanism; i.e. whether the rate of production of defects is non-linear with radiation fluence for high fluence. As regards the latter, plots of the impurity contribution to the resistivity (i.e. resistivity in the limit $T \to o$) vs. fluence, show that the rate of production of defects does not change significantly with fluence. Another interesting plot is that of T_c vs. impurity resistance,[35] Fig. (4). In this region $T_c \sim \lambda$, and agreement with our earlier hypothesis, Fig. (4) indicates that $N(0)$ decreases (almost linearly) with increasing defect concentration. At very high defect concentration, the amorphous limit, we expect $N(0)$ to saturate.

The radiation incident on the superconductors is at high enough energy that in principle all possible defects will be produced - vacancies, anti-structural defects - interstitials - stacking faults, etc. The barrier energy for each of these, which will determine their relative concentration is a difficult question. One clue is that damage increases the average lattice spacing - this can happen both due to anti-structural defects and vacancies. An interesting possibility is that the transition metal atoms increase their relative distance by forming a zig-zag arrangement, (we could call these interstitial vacancy pairs). This would be favored since the transition metal atoms in the A-15 compounds are closer to each other than the sum of their metallic radii, and thus under enormous compression.

7. CONCLUDING REMARKS

The basic premise of this paper is that the d-electrons in transition metals and compounds interact with ions strongly enough that the tight-binding approximation for the electronic structure is valid. For the vibrational frequencies, we must then consider the motion of the ions plus their tightly-bound electrons

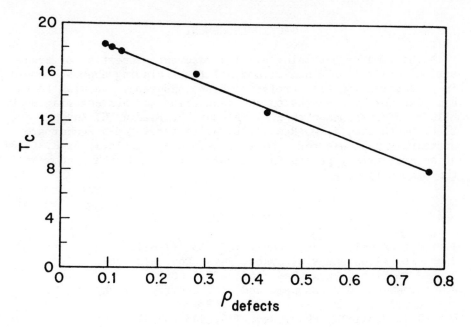

Figure 4: Plot of T_c vs. the defect contribution to resistivity for Nb_3Ge with different amounts of damage. Data is taken from Ref. (31), (32) and (34).

together. The major contribution to the stiffness of the lattice comes from the variation of electronic energy and the redistribution of electronic density due to strain. The electron-phonon interaction is in turn closely related to the lattice stiffness.

All our answers are expressed in terms of general features of the electronic band-structure: 1) the overall width, 2) the total energy, 3) the density of states (average inverse velocity) at the Fermi surface, and 4) the average velocity at the Fermi surface. We have been able to qualitatively understand the empirical relations in superconductivity in terms of these, and the phenomena of T_c degradation. Our conclusion is that for the highest λ, we need materials with the mutually contradictory requirements of large overall width of the band, large density of states near the Fermi level and with the Fermi level lying in an "anti-bonding" band. In this paper we have not touched the issue of how these same parameters determine questions of crystal stability.

ACKNOWLEDGEMENTS

The idea of obtaining all the relevant properties of transition metals from a non-orthogonal tight binding approach arose in a conversation with Professor P. W. Anderson. We wish to thank him also for his remarks on the manuscript and his encouragement. Helpful discussions with E. I. Blount, F. Fradin, G. Knapp, L. F. Matthiess, J. M. Rowell, L. R. Testardi and W. Weber are gratefully acknowledged. We were helped particularly by E. I. Blount at several tricky points in the calculation and by his comments on the manuscript.

REFERENCES

1. W.L. McMillan, Phys. Rev. 167, 331 (1968).
2. P.B. Allen and R.C. Dynes, Phys. Rev. B12, 905.
3. S.K. Sinha, Phys. Rev. 169, 477 (1968).
4. R. Evans, G.D. Gaspari and B.L. Gyorffy, J. Phy. F, 3, 39 (1973).
5. J.J. Hopfield, Phys. Rev. 186, 443 (1969).
6. R.C. Dynes & J.M. Rowell, Phys. Rev. B11, 1884 (1975).
7. S. Barisic, J. Labbe, and J. Friedel, Phys. Rev. Lett. 25, 919 (1970).
8. S.K. Sinha in Second Rochester Conference on d and f-band Superconductivity, to be published.
9. J.K. Hulm and R.D. Blaugher, AIP Conference Proceedings #4 Edited by D.H. Douglass (1972).
10. L.F. Matthiess, Phys. Rev. 139, A1893 (1965).
11. R.C. Dynes and C.M. Varma, J. Phys. F 6, L1 (1976).
12. J. Friedel, in The Physics of Metals, Edited by J.M. Ziman, Cambridge University Press Cambridge (1969).
13. F.Y. Fradin and D. Zamir, Phys. Rev. B7, 4861 (1973).
14. F.Y. Fradin and T.D. Williamson, Phys. Rev. 10, 2803 (1974).
15. L.J. Vieland and A.W. Wicklund, Phys. Rev. 166, 424 (1968).
16. F.Y. Fradin, private communication.
17. R.H. Hammond and M.M. Collver, in Low Temperature Physics LT 13, Vol. 3, 532, edited by K.D. Timmerhaus, W.J. O'Sullivan and E.F. Hammel (1972).
18. W. Weber, Phys. Rev. B8, 5082 (1973): W. Weber, H. Bilz and U. Schroder, Phys. Rev. Lett. 28, 600 (1972); S.K. Sinha and B.N. Harmon, Phys. Rev. Lett. 35, 1515 (1975).
19. H.G. Smith and W. Glaser, Phys. Rev. Lett. 25, 1611 (1970).
20. W.H. Adams, J. Chem. Phys. 34, 89 (1961).
21. T.L. Gilbert, in Molecular Orbitals in Chemistry, Physics and Biology, edited by P. O. Lowdin and B. Pullman (Academic, New York 1964).
22. P.W. Anderson, Phys. Rev. 181, 25 (1969); Phys. Rev. Lett. 21, 13 (1968).

23. L.J. Sham in Dynamical Properties of Solids, edited by G.K. Horton and A.A. Mardudin (North Holland Amsterdam, 1974) Vol. 1, Chap. 5.
24. See for example, R. Hoffman, J. Chem. Phys. 39, 1397 (1963).
25. D.G. Pettifor, J. Phys. C, 5, 97 (1972), F. Cyrot-Lackmann, J. Phys. Chem. - Solids 29, 1253 (1968).
26. J. Friedel, Ref. (12).
27. R. Gomersall and B.L. Gyorffy, Phys. Rev. Lett. 33, 1286 (1974).
28. See for example J.R. Schrieffer in Superconductivity, W.A. Benjamin, New York (1964).
29. J. Kanamori, J. Appl. Phys. 36, 929 (1965).
30. A.R. Sweedler, D.G. Schweitzer and C.W. Webb, Phys. Rev. Lett. 33, 168 (1974).
31. L.R. Testardi, R.L. Meek, J.L. Poate, W.A. Royer, A.R. Stern and J.H. Wernick, Phys. Rev. B11, 4304 (1975).
32. J.M. Poate, L.R. Testardi, A.R. Storm and W.M. Augustyniak, Phys. Rev. Lett. 35, 1290 (1975).
33. C.C. Koch, J.O. Scarbrough and D.M. Kroeger, Phys. Rev. B9. 888 (1974).
34. L.R. Testardi, private communication. M. Strongin (preprint).
35. Data taken from Ref. (31), (32) and (34).

QUESTIONS AND COMMENTS

M. Cohen: You could use the jellium picture which gives the same results that you got in the beginning. But with jellium you reach different conclusions, i.e., jellium is a good model for the s-p simple metals.

C. Varma: No, that is, I guess, the basic point of my presentation. One should not do transition metals as if it was just a harder version of jellium. In fact, the empirical relations that exist show that the transition metals do not behave at all as s-p metals; that is point one. Point two: this kind of method implicitly depends on s being a small parameter. That will not hold for pseudopotential metals. For pseudopotential metals jellium is just fine.

F. Mueller: Unaccustomed as I am to publicly agreeing with Jim Phillips on anything; I must, in this instance, say that in fact I think the physics here is closer to what he was presenting than what you gave. Let me just remind us that there are a couple of papers in the literature which address themselves to this question and that in fact Harold Myron and myself were interested in precisely the problem you pointed out here and was discussed by John Hulm at the last Rochester conference. On the left hand side of your slide [a relation between λ and $N(0)$. Fig. 1 of this paper. See also Fig. 3 of Hulm and Blaugher of the first Rochester Conference.] one sees λ which gives roughly 0.4 of the units plotted and that λ divided by $N(0)$ is a straight line for a series of alloys. On the right hand side one has 0.86 in the same units. So something different is happening on the two sides of the dip in the Nb-Mo-Re density of states structure. What I believe is happening, as is alluded to in a paper with "approximate d-band quantum numbers" by Phillips and Mueller in 1967 in Physical Review, and also has been reported in a Physical Review Letter by Dimitri Papaconstantopoulos. They did factorizations of the density of states into different kinds of orbital symmetries. Theories of the electron-phonon interaction which we at Nijmegen have been working on suggest the utility of such factorization in terms of Wannier functions---which of course have no overlap whatsoever. This suggests that the physical picture is really very simple---the picture I alluded to yesterday. Namely that on the left hand side one has T_{2g} orbitals---as in chemists' bonding picture whereas on the right one has E_g... These two differences are just bonding and anti-bonding. The differences are that for one case the lobes are pointing out, and in the other case the lobes are along the nearest neighbor distance---it is this delicacy and the fact that I think the interaction is important in the interstitial regions---that is what is producing the difference between the two effective bonding constants. We hope to present results on this soon. We are working actively on the problem, using computers of course.

C. Varma: It is very nice that there are other people working with similar ideas. Let me answer the points Fred has raised: first, I do not understand your remarks about preferring Jim Phillips' work to mine. He was talking about effect of defects, while I am discussing pure materials. Secondly, I disagree with your statement that the T_{2g} and E_g orbitals are bonding and anti-bonding orbitals. In fact tight binding fits show that the ratio of T_{2g} to E_g orbitals varies smoothly across the bonding-anti-bonding peaks.

F. Mueller: Let me just comment briefly on that to explain what I have in mind. There is a difference in pictures here. One is generated by means of atomic-like functions and individual nuclear sites...that is the picture you are speaking about. There is another which is in terms of the Wannier function representations and is a factorization by symmetry. It is the factorization by symmetry that I wish to stress, and that is where I think the real physics is different. In fact, these kinds of symmetry arguments in r-space which have not been applied much---I mean, after all, symmetry exists in r-space as well as in k-space. We are used to talking about symmetry of various kinds of functions in Bloch space. But the Wannier functions can be factorized in the same way and it is that precise difference that I wish to address myself to.

B. Klein: I would like to make a small criticism of your theory. Since you found that λ is proportional to the band width, the theory would predict that T_c would always go up with the pressure for transition metals, and I do not think that is the case.

C. Varma: In my theory λ is proportional to both the band-width W and the density of states $N(0)$ near the Fermi surface. The product $N(0)W$ does not necessarily go up with pressure.

D. Papaconstantopoulos: Well, I just wanted to say what Gyorffy said: that other theories also correlate with the Matthias rules.

B. Gyorffy: If you want a simple chemical picture of the relation between T_c and e/a there is such a theory already by Pauling. It gets the two peaked structure as a consequence of bonding and anti-bonding orbitals being separated in energy.

C. Varma: If I can do a tenth as well as Linus Pauling often did, I would be very happy.

B. Gyorffy: You did it six years later.

C. Varma: My theory for transition metals is completely different from Pauling's.

J. Phillips: There is a basic problem and that is the parameters W and S which are so convenient---as somebody said about the pressure. You ask 'How do these parameters vary under pressure?' The answer is: you have to fiddle with them because they vary in different ways and the smiple picture you get breaks down as soon as you try to interpret these W and S parameters as they depend on pressure. When somebody asks you how it depends on non-stoichiometry you run into the same question---will it predict, for instance, these two-phase superlattice behaviors and so on.

C. Varma: Well, Jim since I am your co-author on the paper on defects I would have to say the theory that I presented does not really address itself to the question of what defects would do to the lattice structure, the question that you have addressed earlier in your inimitable style.

J. Phillips: And you in yours.

C. Varma: However, Jim has eluded to the question of what would stoichiometry do to this thing. This is the point at which Bob Dynes and I started working on this problem. You do not expect, in damage, for the d-bands to shift relative to each other. So the orbital that is dominant in the good material is going to remain dominant in the damaged material and our hypothesis would then say that I^2/ω^2 should remain the same in the damaged material and only the density of states should vary. We can in fact fit the data for "damaged" niobium with the hypothesis that variation in T_c in Nb is due purely to variation in $N(0)$. About the A-15 compounds, Frank Fradin told me that for the V_3Si, V_3Ga alloys, he can do NQR measurements. He finds a resonance width (which is a measure of the disorder in the system) which scales inversely with the density of states just as T_c goes on dropping. So we have a beautiful correlation: that as you go off stoichiometry the density of states goes down, disorder increases. Nothing else happens in the system as far as T_c is concerned; I^2/ω^2 remains the same.

D. Papaconstantopoulos: The bandwidth always increases with pressure. I does not vary in different ways. Another comment is that I assume you do not have any numbers yet from this theory and it is probably too early to claim success.

C. Varma: My earlier comment on the effect of pressure on T_c still stands. I do not have numbers but I have physical understanding.

D. Papaconstantopoulos: Yes, and a third comment is: I would like to ask you what is the meaning of your statement, 'let's forget about the muffin-tin'?

C. Varma: Because, believe it or not, there are not muffin-tins inside the solid, there are atoms. I merely meant that the interesting correlations in transition metal superconductivity can be derived in a transparent fashion from a simple tight binding scheme. Very sophisticated schemes of electronic structure can tend to obscure the simplicity of the results in a mass of details.

STUDIES OF THE TRANSITION TEMPERATURE AND NORMAL STATE RESISTIVITY

OF Nb_3Ge AND Nb FILMS*

H. Lutz, H. Weismann, M. Gurvitch, A. Goland, O.F. Kammerer and M. Strongin

Brookhaven National Laboratory

Upton, New York 11973

ABSTRACT

Correlations between T_c and specific features of the normal state resistance vs temperature curves are discussed for both Nb_3Ge and ion damaged Nb films. Of particular interest is the correlation between T_c and ρ_o in Nb_3Ge films.

INTRODUCTION

In this paper we discuss correlations between the superconducting transition temperature, T_c, and the normal state resistivity of films of Nb_3Ge and Nb. In recent work Testardi [1] and Poate et al. [2] have already indicated the correlation between T_c and the resistance ratio in as grown films of Nb_3Ge and also in films where T_c is lowered by He^4 damage. The present work extends these observations, and emphasizes that T_c appears to correlate

---

* Work performed under the auspices of the U.S. Energy Research and Development Administration.

with the residual resistivity at 25 K, ρ_o, in Nb_3Ge films. Initial data on the T_c depressions and normal state resistivity in Nb films that have been severely damaged by oxygen ions are also discussed.

EXPERIMENTAL

The Nb_3Ge samples were prepared by electron beam co-deposition in pressures in the mid 10^{-6} torr range, onto polished sapphire substrates held at $1000^\circ C$. The Nb film was prepared at 10^{-7} torr at an evaporation rate of a few thousand Å/min onto an ambient temperature slide. In the case of the Nb_3Ge films, composition and temperature gradients along the substrates provide a series of samples, the transition temperatures of which change from 21 K to 17 K in one set and from 19 K to 15 K in the second set. X-ray diffraction analysis indicates almost all A-15 phase in the first set of samples. The thicknesses of the Nb_3Ge films were about 1.4 μm and the thickness of the Nb films was about 1200 Å. In these samples the correlation of T_c with resistance ratio is as reported by Testardi et al. and this shown in the insert in Fig. 1. This correlation does not hold in our films that are strongly two phase.

DISCUSSION OF ρ vs T DATA

The resistance vs temperature curves for Nb_3Ge and other strongly coupled superconductors differ from the normal behavior described by the approximate equation [3]

$$\rho(T) - \rho_o = \rho_{e-p}(T) = \left(\frac{m}{n}\right)_{eff} \left(\frac{2\pi k_B}{e^2 \hbar}\right) \lambda_{tr} T \qquad (1)$$

for $T \ll \theta_D$ where λ_{tr} is related to λ_{sup} in the McMillan theory of the superconducting transition. In the case of Nb the extrapolation of the room temperature data, where the curves are fairly linear, to 0 K does essentially give ρ_o. However the data in Fig. 1 shows that this is not the case for Nb_3Ge. The anomalous shape of the normal state resistivity curve in high T_c superconductors has already been pointed out by Lawson and Fisk [4], and the general features have been recently discussed [5].

Some of the interesting features of the data in Fig. 1 are the following. First, there appears to be a correlation between T_c and ρ_o, i.e. the high T_c films have lower values of ρ_o. Second, there is a correlation between the slope of the ρ vs T curves at

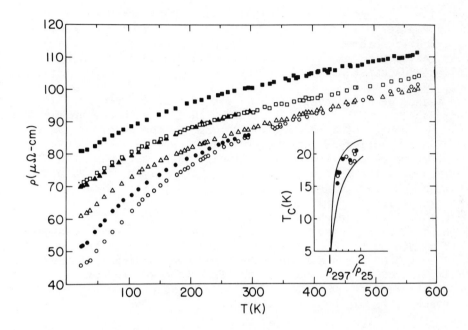

Figure 1: Set 1: o - T_c=20.5, △ - T_c=19.4, □ - T_c=17.4
Set 2: ● - T_c=19, ▲ - T_c=17, ■ - T_c=15.4
Transition widths from 5% to 90% are about 0.5 K and T_c's are ½ way points. T_c's in Set 1 ranged from 21 K to 17 K with a concentration gradient of less than 10% as determined from microprobe data. Only the highest T_c, the lowest T_c and one T_c in the middle of the slide are shown. Other points along the slide exhibit similar behavior and are not shown for clarity in the main part of the figure. The range of T_c's on Set 2 was from 19 K to 15 K; again only three curves are shown for clarity. The insert shows the range of T_c vs resistance ratio given in Ref. 1. The solid circles are the data from the main part of the figure, and the open circles show other T_c measurements along the slide.

high temperatures and the T_c, i.e. samples with lower T_c's have smaller slopes. Finally, the data also shows that the large temperature dependent part of the resistivity due to phonon scattering decreases as T_c decreases.

The slope at high temperatures can be discussed using Eq. (1) which implies that $d\rho/dT$ at high temperatures is proportional to λ_{tr}, and therefore is related to λ_{sup}. Of course, this assumes that the factor $(m/n)_{eff}$ in Eq. (1) which is equal to $[N(0)<V_F^2>/3]^{-1}$ [6] remains constant.

Hence we argue that the observation that T_c decreases as ρ_o increases, further implies a smaller electron-phonon coupling with increasing ρ_o, which in turn would lead to, the observed, smaller $d\rho/dT$ and a series of generally merging curves as shown in Fig. 1. These features are also consistent with the observation that T_c correlates with resistance ratio [1]. It is important to mention that the changes in $d\rho/dT$ taken from the data for films of different transition temperatures are larger than the values of λ calculated from the changes in T_c. This is true in the case of the Nb_3Ge samples and also in the case of the Nb film which was damaged with oxygen ions. The data in the table illustrates this result. Some possible causes for this are: the non-constancy of the $N<V_F^2>$ factor in equation (1), the fact that we are not yet in the linear regime, or possible effects due to saturation as the mfp ℓ approaches its minimum value [7].

CHANGES OF T_c AND NORMAL STATE PARAMETERS

In Fig. 2 we show how reduced transition temperature varies as a function of the residual resistivity, plotted on a reduced scale. The main point here is to qualitatively indicate that the depressions in T_c for Nb_3Ge are not too dissimilar to those for pure Nb. The Nb data shown here represents the effect of damage due to 200 keV O^- ions whose range was about twice the Nb film thickness. The total dose was about 10^{17} atoms over the film area of about 1 cm^2. We estimate that only a small fraction of the oxygen ions remained in the film. This is being checked by back scattering measurements. The depressions in T_c, as a function of ρ_o, are not the same for all Nb samples and the T_c's of bulk Nb samples with oxygen additions [8] fall off at a much slower rate with $\rho_o/\rho_o(initial)$ than the work presented here. Cryogenically deposited films of Nb also show a smaller depression in T_c with $\rho_o/\rho_o(initial)$ [9]. Apparently the various scattering mechanisms which contribute to both ρ_o and $\rho_o(initial)$ are different in the Nb samples prepared in different ways and there is no particular relationship between ρ_o and the transition temperature. This is, in fact, what one would expect since many factors could influence

Table 1

Values of the slope of the resistivity at 600 K for four different T_c Nb$_3$Ge samples normalized to $(d\rho/dT)_o$, which is the slope at 600 K for the 20.5 K sample. Values of λ were calculated from the McMillan theory [7] and are normalized to the 20.5 K value. The table is shown to illustrate that $d\rho/dT$ changes more than λ. In the case of Nb the slope was measured at 300 K and normalized to the 8 K sample.

Nb$_3$Ge	$\dfrac{d\rho/dT}{(d\rho/dT)_o}$ at 600 K	$(\lambda/\lambda_o)_{calc.}$
20.5	1	1
19.7	.77	.96
17	.69	.89
15.5	.65	.84

Nb	$\dfrac{d\rho/dT}{(d\rho/dT)_o}$ at 300 K	$(\lambda/\lambda_o)_{calc.}$
8	1	1
6	.762	.89
6	.723	.89
3.5	.516	.76

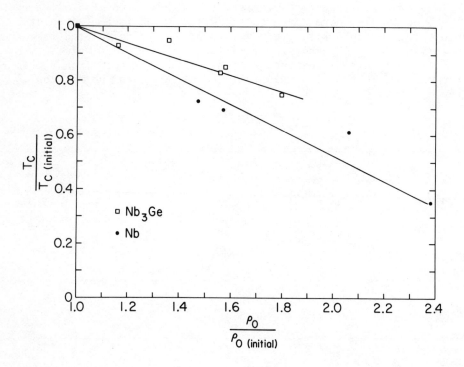

Figure 2: T_c/T_c(initial) vs ρ_0/ρ_0(initial) for Nb_3Ge and Nb films. For Nb_3Ge, T_c(initial) ~ 20.5 K and ρ_0(initial) ~ 45 μΩ-cm. For Nb, T_c(initial) ~ 8.0 and ρ_0(initial) ~ 17 μΩ-cm.

ρ_0 which one would not necessarily expect to influence T_c. However, in the A-15 Nb$_3$Ge films it appears that there is actually a strong correlation between T_c and ρ_0.

The correlation between T_c and ρ_0 can, perhaps, be understood in A-15 materials by the assumption that $N(o)$, the density of states at the Fermi level, decreases as ρ_0 increases. This behavior in $N(o)$ is actually evident from the behavior of H_{c2} in materials where T_c is decreased due to the effects of neutron irradiation. This can be seen in the following way. If we write down the expression for the condensation energy of a superconductor at 0 K we have

$$N\Delta^2/2 = H_c^2/8\pi . \qquad (2)$$

H_c can be expressed as $H_{c2}/\sqrt{2}\kappa$ and $\Delta \propto kT_c$. Then we have

$$N \propto H_{c2}^2/T_c^2\kappa^2 . \qquad (3)$$

In A-15 Nb$_3$Al it has been shown [10] that $H_{c2}(o)$ decreases as T_c decreases, due to the effects of neutron irradiation, such that $H_{c2}(o) \propto T_c$, and similar results have been found for Nb$_3$Sn [11]. Hence if we assume Nb$_3$Ge is a similar case and that $H_{c2}(o) \propto T_c$. Then

$$N \propto 1/\kappa^2 . \qquad (4)$$

In the dirty limit $\kappa = \kappa_0 + \kappa_\ell$, where $\kappa_\ell \propto N^{\frac{1}{2}}\rho_0$. It is immediately seen that if κ_ℓ is comparable to κ_0 then N will decrease as ρ_0 increases. In Nb$_3$Sn, κ_0 and κ_ℓ are comparable [12], but in Nb$_3$Ge or Nb$_3$Al where ρ_0 is larger it is probably reasonable to assume that $\kappa \propto N^{\frac{1}{2}}\rho_0$. Hence to a zeroth order approximation

$$N \propto 1/\rho_0 . \qquad (5)$$

A further factor which could lead to some inaccuracy in this expression is the change in the constant of proportionality between Δ and T_c as the coupling changes. The approximate result given by equation (5) might especially be expected to hold in the limit of strongly damaged samples when $T_c/T_{co} \sim 0.5$. This behavior of the density of states on ρ_0 can generally explain why T_c decreases as ρ_0 increases. The dependence of T_c on the density of states has been previously noted in various systems [9,13] and the problem has been most recently discussed by Dynes and Varma [14]. It is interesting that in the case of niobium $H_{c2}(o)$ increases in the bulk as oxygen is added and T_c decreases.

Although in Nb_3Al, Nb_3Sn, and presumably other A-15's, H_{c2} decreases as T_c decreases, studies with oxygen impurities in Nb [8] have indicated that as T_c drops due to oxygen impurities, H_{c2} still rises. In the Nb case, in the dirty limit, κ_o is small compared to κ_ℓ and to a good approximation $\kappa \propto N^2 \rho_o$, which leads to $N^2 \propto H_{c2}^2/T_c^2\rho_o^2$. However, in contrast to the case of the A-15's, $(H_{c2}/T_c)^2$ rises as ρ_o rises and hence N does not decrease as much. This is verified in the specific heat measurements of Koch et al. [8]. Thus the critical field data indicates that the density of states apparently falls off faster in the A-15's, as compared to Nb, for increases in ρ_o. Specific heat measurements would unambiguously determine this.

In concluding it should be mentioned that ρ_o appears to approximately correlate with the degree of residual disorder in various A-15 compounds. For well ordered compounds like V_3Si and Nb_3Sn, ρ_o is about 2-10 $\mu\Omega$-cm [15] and about 10 $\mu\Omega$-cm respectively. On the other hand in Nb_3Al and Nb_3Ge, compounds which both neutron and x-ray diffraction [16] indicate have a sizable degree of residual disorder, $\rho_o \sim 40$ $\mu\Omega$-cm. Hence, although an exact correlation is not expected it appears that ρ_o in the A-15 compounds is a sensitive function of the residual defect structure.

ACKNOWLEDGEMENTS

We thank Arthur Paskin, P.B. Allen, D.O. Welch, G.J. Dienes and D.E. Cox for many discussions.

REFERENCES

1. L.R. Testardi, IEEE Trans. on Magnetics MAG-11, 197 (1975).
2. J.M. Poate, L.R. Testardi, A.R. Storm, W.M. Augustyniak, Phys. Rev. Lett. 35, 1290 (1975).
3. J.J. Hopfield, Superconductivity in d- and f-Band Metals, Ed. by D.H. Douglass, 358, AIP, New York (1972); P.B. Allen, Phys. Rev. B3, 905 (1971).
4. Z. Fisk and A.C. Lawson, Solid State Comm. 13, 277 (1973).
5. R.W. Cohen, G.D. Cody, J.H. Halloran, Phys. Rev. Lett. 19, 840 (1967); F.Y. Fradin, Phys. Rev. Lett. 33, 158 (1974); P.F. Allen, J.C.K. Hui, W.E. Pickett, C.M. Varma and Z. Fisk, Solid State Comm. To be published.
6. Theory of the Properties of Metals and Alloys, N.F. Mott and H. Jones, 97, Dover Pub., New York.
7. Z. Fisk and G.W. Webb, Phys. Rev. Lett. 36, 1084 (1976).

8. C.C. Koch, J.O. Scarborough, and D.M. Kroeger, Phys. Rev. $\underline{9}$, 888 (1974).
9. J.E. Crow, M. Strongin, R.S. Thompson and O.F. Kammerer, Phys. Lett. $\underline{30A}$, 161 (1969).
10. A.R. Sweedler, D.E. Cox, S. Foner and E.J. McNiff, Bulletin of APS $\underline{20}$, 461 (1975).
11. L. Snead, private communication.
12. See for example: G. Cody, Proc. of 1968 Summer Study on Superconducting Devices and Accelerators, BNL-50155(C-55), Brookhaven National Lab. (1968).
13. See for example: G. Gladstone, M.A. Jensen and J.R. Schrieffer, Superconductivity, Ed. by R.D. Parks, Vol. $\underline{2}$, 742, Marcel Dekker (1969).
14. R.C. Dynes and C.M. Varma, Superconductivity and Electronic Density of States, Preprint.
15. H. Taub and S.J. Williamson, Solid State Comm. $\underline{15}$, 181 (1974).
16. A.R. Sweedler and D.E. Cox, Phys. Rev. $\underline{B12}$, 147 (1975).
    D.E. Cox, private communication.

QUESTIONS AND COMMENTS

Discussion of this paper appears after that of Bader and Fradin.

SATURATION OF THE HIGH TEMPERATURE NORMAL STATE

ELECTRICAL RESISTIVITY OF SUPERCONDUCTORS

Z. Fisk* and G.W. Webb[†]

Institute for Pure and Applied Physical Sciences
University of California, San Diego

La Jolla, California 92093

The high temperature, normal state resistivity of strong coupled superconducting transition metal compounds and other materials with high resistivities saturates at a value corresponding to an electron mean free path of order the interatomic spacing. This accounts for the nonlinear temperature dependence of the resistivity observed in these compounds at high temperature.

We know that the room temperature electrical resistivity (ρ) of transition metal superconductors is often large - typically 50-70 µohm-cm consistent with strong electron-phonon coupling (λ). In addition, the temperature dependence of ρ for these superconductors in the normal state is quite unlike that predicted by Bloch-Grüneisen theory or Wilson's modification thereof [1-4]. $d^2\rho/dT^2$ is negative for temperatures above an inflection point quite generally of order $100°K$, giving a high temperature resistivity rising less rapidly than a linear T-dependence. We ask whether this curvature is related to the occurrence of superconductivity.

---

*Research supported by U.S. Air Force Office of Scientific Research, Air Force Systems Command, USAF, under AFOSR Contract No. AFOSR/F44620-C-0017.
[†]Reseach supported by the National Science Foundation under Grant No. DMR75-04019.

Two plausible explanations have been proposed for this curvature. One is the Cohen, Cody and Halloran model [3]: a rapid temperature dependence of the density of states at the Fermi level resulting from a nearly empty or full high density of states d-band overlying a low density of states s-band leads to resistance curves of the kind observed. This model successfully correlates the resistance with a number of other temperature dependent properties of Nb_3Sn, V_3Si and V_3Ga. The other explanation is that of Allen et al. [5]: hardening of lattice vibrational modes as T increases results in a resistivity which rises with less than a linear T-dependence. On the basis of neutron data for V_3Si, they can account for about half the curvature observed in the resistivity of V_3Si from T_c to room temperature.

While the above treatments certainly have their domains of applicability, we believe that a different effect dominates at high temperature. The figure shows data on the electrical resistivity of A-15 structure Nb_3Sn ($T_c = 18°K$) and single crystal Nb_3Sb ($T_c = 0.2°K$). Data for Nb_3Sn were read from a graph in ref. 1. Knapp et al. have measured the heat capacity of both these compounds between $2°K$ and $400°K$ [6]. The electronic specific heat γ for Nb_3Sb is 1.1 $mJ/°K^2$ gr-atom, more than a factor ten less than that of the iso-structural Nb_3Sn. In addition, analysis of the anharmonic contribution to the heat capacity shows that the Nb_3Sb lattice softens as T increases, opposite to Nb_3Sn. It is unlikely, therefore, that the previous two explanations apply to Nb_3Sb, yet its resistivity is very similar to that of Nb_3Sn in both shape and magnitude above room temperature. The resistivity for both compounds appear to be saturating at $\rho_{sat} \sim 150$ μohm-cm [7]. One may interpret such a limit as a lower limit to the electron mean free path in the solid. Ioffe and Regel [8] and Mott [9] point out that this lower limit would be of order the interatomic spacing of the material. There is no theory at present predicting a functional form for the approach to saturation [10].

The value we find for ρ_{sat} in these compounds is not unreasonable in light of the detailed calculations for Nb by Yamashita and Asano [11]. They find that a resistivity which they calculate at $290°K$ to be about 20 μohm-cm corresponds roughly to a mean free path of 30 Å. This implies a ρ_{sat} for Nb of some 200 μohm-cm. We expect ρ_{sat} for A-15's to be of the same order of magnitude. A free electron estimate using the Drude formula yields a similar value.

The large room temperature resistivity of the poor superconductor Nb_3Sb is somewhat surprising and is something for which we do not have a straightforward explanation [12]. At low temperature the phonon resistivity of Nb_3Sn is much larger than that of Nb_3Sb. But it is difficult to make a comparison based on λ's which differ

HIGH TEMPERATURE NORMAL STATE RESISTIVITY

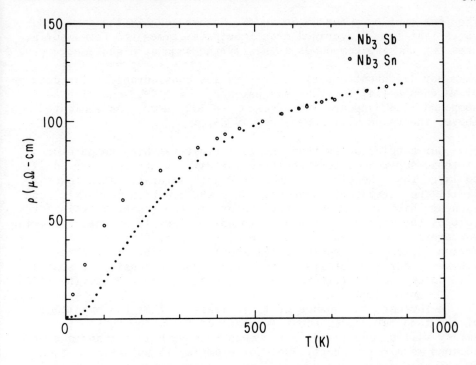

Figure 1: The resistivities of Nb_3Sb and Nb_3Sn. The data for Nb_3Sn were taken from reference 1. Not all points are plotted. The T_c of Nb_3Sn is $18°K$ while that of Nb_3Sb is $0.2°K$. Note that the two curves are nearly identical above $500°K$. At lower temperatures, e.g. $50°K$, different residual resistivities can account for less than half the difference between the two resistivities; the remainder of the difference we ascribe to the stronger electron phonon interactions in Nb_3Sn.

by a factor of 3.3 [6] since the functional form of the resistivity for the two compounds is completely different, being T^2 for Nb_3Sn and $T^{3.6}$ for Nb_3Sb [13].

For a cubic metal we can write

$$\rho = (12\pi^3\hbar/e^2)/(\int \tau v \, dS).$$

The relaxation time τ and the magnitude of the electron velocity v are functions of the wavevector. The integral is over the Fermi surface. At saturation the mean free path, τv, will be a function of wavevector determined by the interatomic distances in the crystal.

The fact that ρ_{sat} for Nb_3Sn and Nb_3Sb are roughly the same indicates that the averaged mean free path times the Fermi surface area for the two compounds is nearly the same at high temperature.

For different materials ρ_{sat} varies considerably. The data of Marchenko [14] for V_3Si suggest a $\rho_{sat} \sim 170$ $\mu\Omega$-cm. The A-15 compound Ti_3Sb appears to have $\rho_{sat} \sim 400$ $\mu\Omega$-cm. As indicated above, the value will depend on $\int \bar{v} dS$.

Our conclusion is that the resistivity at high temperatures of those superconductors whose resistivity is strongly curved is so large that the electron mean free path saturates at a lower limit, the interatomic spacing. When the electron mean free path is near this limit, negative curvature in the resistivity is a result of the mean free path not depending linearly on the scattering perturbation. At lower temperature, the effects of Fermi level motion [3] and anharmonicity [5] should be visible in the resistivity. Near T_c, details of the actual phonon density of states will determine the temperature dependence of the resistivity [13].

A further consequence of these ideas is that the magnitude of the temperature dependent (phonon) contribution to the resistivity will be decreased if the mean free path in the absence of phonons is already short, due, for example, to spin-fluctuations or the atomic disorder produced by alloying or defects. This is observed. In the case of atomic disorder, Mooij has noted that $d\rho/dT$ for concentrated transition metal alloys decreases to zero in the vicinity of 150 μohm-cm and concludes that the mean free path is then of order the interatomic spacing [15]. The approach to saturation also seems to be evident in many rare earth and actinide intermetallic compounds where magnetic effects are important. While such compounds are not superconductors, they do all have high room temperature resistivities.

We wish to thank P.B. Allen, A.C. Lawson, B.T. Matthias and F.M. Mueller for numerous useful discussions, and W.A. Fertig for measuring the T_c of Nb_3S.

REFERENCES

1. D.W. Woodward and G.D. Cody, RCA Review 25, 405 (1964).
2. D.W. Woodward and G.D. Cody, Phys. Rev. 136, 166A (1964).
3. R.W. Cohen, G.D. Cody and J.J. Halloran, Phys. Rev. Lett. 19, 840 (1967).
4. Z. Fisk and A.C. Lawson, Solid State Comm. 13, 277 (1973).
5. P.B. Allen, J.C.K. Hui, W.E. Pickett, C.M. Varma and Z. Fisk, Solid State Comm. (in press).
6. G.S. Knapp, S.D. Bader and Z. Fisk, Proceedings of International

Conference on Low Lying Lattice Vibrational Modes and Their Relationship to Superconductivity and Ferroelectricity (San Juan, 1975); to be published.
7. Z. Fisk and G.W. Webb, Phys. Rev. Lett. 36, 1084 (1976).
8. A.F. Ioffe and A.R. Regel, Progr. in Semiconductors 4, 237 (1960).
9. N.F. Mott, Metal-Insulator Transitions (Taylor and Francis, London, 1974), p. 23ff.
10. Empirically we find that over the range $500^{\circ}K$ to $1000^{\circ}K$ the resistivity of both compounds appears to be approaching saturation as T^{-1}. Clearly this might be only the leading term in a series expansion.
11. J. Yamashita and S. Asano, Prog. Theor. Phys. 51, 317 (1974).
12. Part of the answer for the high room temperature resistivity of Nb_3Sb is probably to be found in the fact that its Fermi surface is quite complicated, having at least four sheets (A. Arko and Z. Fisk, to be published).
13. G.W. Webb, Z. Fisk, J.J. Engelhardt and S. Bader, to be published.
14. A. Marchenko, Sov. Phys. Solid State 15, 1261 (1973).
15. J.H. Mooij, Phys. Stat. Sol. (a) 17, 521 (1973).

QUESTIONS AND COMMENTS

Discussion of this paper is presented at the end of the paper by Bader and Fradin.

EVIDENCE FOR SELECTIVE ELECTRON-PHONON SCATTERING IN THE A-15

SUPERCONDUCTORS Nb_3Sn, Nb_3Sb, AND V_3Si

S.J. Williamson and M. Milewits

Department of Physics, New York University

New York, New York 10003

ABSTRACT

Analysis of existing data for the electrical resistivity of single crystals and polycrystals of Nb_3Sn, Nb_3Sb, and V_3Si shows that over a wide temperature range the data can be fit by an empirical formula $\rho(T) = \rho_0 + b_1 T + d_1 \exp(-T_0/T)$, where ρ_0, b_1, and T_0 are adjustable parameters. The value of b_1 is sufficiently small for Nb_3Sn that for $T_c \leq T \leq 100$ K the temperature dependence can be described by an exponential alone. The characteristic energies T_0 are 90 K, 210 K, and 175 K for Nb_3Sn, Nb_3Sb, and V_3Si, respectively. An explanation advanced by Milewits, Williamson, and Taub for an exponential term at low temperature based on selective electron-phonon scattering is reviewed. An analysis of the resistivity of Nb_3Sb measured by Fisk and Webb suggests that a crucial test for this explanation would be provided by neutron scattering experiments on this compound.

I. INTRODUCTION

Binary and pseudobinary transition metal compounds with the chemical formula A_3B and the A-15 crystal structure exhibit anomalous normal state characteristics which to an extent are correlated with the type of transition metal atom A and the value of T_c. The anomalies include unusual variations with temperature for the magnetic susceptibility, Knight shift, specific heat, and phonon mode frequencies, as well as evidence for strong lattice anharmonicity. Crystals of V_3Si and Nb_3Sn having a high residual resistivity ratio RRR = $\rho(295\ \text{K})/\rho(18\ \text{K})$ experience a lattice instability on cooling

and transform from the high temperature cubic phase (A-15) to a
low temperature tetragonal phase at $T_M \approx$ 21 K and 43 K, respectively.
Microscopic explanations for these anomalies and the lattice instability have been based on the nature of the d-band electronic density of states $N(E)$, which is taken to be large and to vary appreciably within an energy range of $\sim$ 100 K near the Fermi energy E_F.
The approaches include the linear chain model of Weger [1] and
Labbé and Friedel [2]; the step-function model of Cohen, Cody, and
Halloran [3]; the X-point model of Gor'kov [4]; and the coupled
chain model of Gor'kov [5], as elaborated by Gor'kov and Dorokhov
[6]. Features of the lattice dynamical properties such as the
anomalous softening of [110] shear mode phonons and strong harmonic
generation in ultrasonic propagation have also been explained on
the basis of lattice defect models, as proposed by Varma et al.[7]
and Phillips [8].

Recently, interest has developed in the unusual transport
properties of A-15 materials. Anomalous behavior in the electrical
resistivity $\rho(T)$ is evident in the earliest measurements on a number
of polycrystalline samples by Sarachik et al.[9] They found that
$\rho(T)$ does not have positive curvature ($d^2\rho/dT^2$) at all temperatures,
approaching a linear dependence at high temperature, as expected
from the Bloch-Grüneisen theory [10]. Instead, an inflection is
observed at an intermediate temperature, and above that the curvature is negative. This feature is not confined to high-T_c materials, but is observed in V_3Ge (T_c = 6 K) as well as in V_3Si (T_c =
17 K) and Nb_3Sn (T_c = 18 K). The low temperature behavior of $\rho(T)$
is also anomalous, since it does not follow a T^3 or T^5 law as
predicted by simple models for electron-phonon scattering. Instead,
if a power law fit is sought, the best results are obtained with a
T^2 dependence for both V_3Si [11] and Nb_3Sn [12]. Taub and Williamson [13] have demonstrated through measurements on different V_3Si
samples that the anomalous behavior at low temperatures cannot be
ascribed to pronounced phonon mode softening, so its origin must
lie in other electron or phonon characteristics.

One approach in developing an understanding of the physics
which determines $\rho(T)$ is to determine whether the essential features
of the temperature dependence can be described by an empirical formula. This was first done successfully by Woodard and Cody [14]
(WC) who analyzed their data on ploycrystalline films of Nb_3Sn and
found agreement to within 1% over the range $T_c \leq T \leq$ 850 K for the
expression

$$\rho(T) = \rho_0 + b_1 T + d_1 e^{-T_0/T} . \qquad (1)$$

The exponential term describes the sharp rise in $\rho(T)$ at low T and
the quasi-saturation with negative curvature at high T. The non-exponential term was needed in part to describe the increase in $\rho(T)$

at the highest temperatures, which the exponential alone cannot do. Milewits, Williamson, and Taub [15] (MWT) obtained similar success in applying Eq. (1) for their data on a number of single crystals of V_3Si of various quality. The emphasis of MWT was on testing the precision with which Eq. (1) could be applied, and they found that agreement to better than ± 0.15% could be obtained for the range $T_c \leq T \leq 50$ K, where $\rho(T)$ has positive curvature.

We report here an extension of these fits to data on single crystals of Nb_3Sn provided by Gunsalus [16], data on a single crystal of Nb_3Sb provided by Fisk and Webb [17], and high temperature data on ploycrystalline V_3Si reported by Marchenko [11]. We find that Eq. (1) gives an accurate characterization for all these samples within the wide temperature ranges of the measurements, provided that the parameters are appropriately chosen. The implications of this success will then be discussed in light of a proposal by MWT that the exponential term of the Fermi surface permits phonons in only certain modes to be particularly effective. The characteristic energy kT_0 is interpreted as the energy of these phonons.

II. ANALYSIS FOR Nb_3Sn

Fig. 1 shows $\rho(T)$ data for two single crystals of Nb_3Sn kindly provided by G. Gunsalus [16]. The crystals are from adjacent portions of a parent crystal, and one was subsequently annealed at 1000^0 C for 2 hours in a high vacuum. The annealed sample undergoes a cubic-to-tetragonal transformation at $T_M = 46$ K, where the slope of $\rho(T)$ displays a discontinuity. We have determined the best fit parameters of Eq. (1) for the data through use of a nonlinear least squares fit routine. In all cases the results are found to be insensitive to the choice of initial trial parameters in the routine.

Data for the unannealed sample can be fit to ± 1% over the illustrated range of $T_c \leq T \leq 100$ K by a simplified form of Eq. (1) in which $b_1 = 0$. Thus the regions of both positive and negative curvature are described simultaneously by a two parameter fit, with the residual resistivity ρ_0 serving as a third parameter. The values of the best fit parameters are given in Fig. 1 and in Table I. The characteristic temperature is $T_0 \approx 90$ K. To see whether this fit is a significant compromise between somewhat different dependences at low and high T, we have fit the data separately in the ranges $T_c \leq T \leq 40$ K and 40 K $\leq T \leq 100$ K using Eq. (1) with $b_1 = 0$. The results are virtually identical to those of the overall fit. Table I lists the best parameters for each range.

Data for the annealed sample cannot be fit over the full range be-

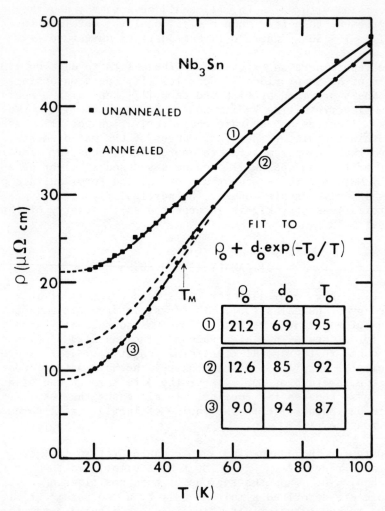

Figure 1: Observed $\rho(T)$ for annealed and unannealed samples of Nb_3Sn for $18 \leq T \leq 100$ K as provided by Gunsalus (Ref. 16). Solid lines 1, 2 and 3 are fits with parameters displayed in the inset. Units for both ρ_o and d_o are $\mu\Omega$ cm and that for T_o is K.

TABLE I

Characterization of Nb$_3$Sn

Fits for unannealed single crystals of Nb$_3$Sn; data of Gunsalus [16].

$$\rho(T) = \rho_o + d_o\, e^{-T_o/T}$$

	$18 \leq T \leq 100K$	$18 \leq T \leq 40K$	$40 \leq T \leq 100K$
$\rho_o(\mu\Omega\ cm)$	21.2	21.2	20.9
$d_o(\mu\Omega\ cm)$	68.6	69.3	70.2
$T_o(K)$	95.3	95.3	95.9

$\rho(T) = \rho_o + b_1 T + d_1 e^{-T_o/T}$ where $b_1 = 46.6 \times 10^{-3}\ \mu\Omega\ cm\ K^{-1}$

	$18 \leq T \leq 100K$
$\rho_o(\mu\Omega\ cm)$	20.3
$d_1(\mu\Omega\ cm)$	60.4
$T_o(K)$	97.0

Fit for annealed sample of Nb$_3$Sn; data of Gunsalus (Ref. 16)

	$18 \leq T \leq 40K$	$50 \leq T \leq 100K$
$\rho_o(\mu\Omega\ cm)$	9.0	12.6
$d_o(\mu\Omega\ cm)$	94	85
$T_o(K)$	87	92

Fit for a polycrystalline sample of Nb$_3$Sn; analysis of Woodward and Cody (Ref. 14)

$$\rho(T) = \rho_o + b_1 T + d_1 e^{-T_o/T}$$

	$18 \leq T \leq 850K$
$\rho_o(\mu\Omega\ cm)$	10.8
$b_1(10^{-3}\ \mu\Omega\ cm\ K^{-1})$	46.6
$d_1(\mu\Omega\ cm)$	74.7
$T_o(K)$	85

cause of the anomaly associated with the lattice distortion below T_M. However, the agreement between low T and high T fits for the unannealed sample implies that a representative characterization of the cubic phase can be obtained from data in the range 50 K $\leq$ T $\leq$ 100 K alone. There are only slight differences between the results for the two samples, as indicated by the tabulation in Table I.

Below T_M the annealed sample is tetragonal, and it is not known whether the sample is in a single domain or what direction the measuring current is along. A further complication in interpreting an empirical analysis is the marked temperature dependence of lattice parameters, especially just below T_M, with accompanying changes in band structure and phonon mode energies. When Eq. (1) is fit to data for $T_c \leq T \leq 40$ K, reasonable good agreement can be obtained, but with substantially different parameters than for the higher temperature range. The fit fails within ~ 5 K of T_M, and presumably this failure is associated with marked changes in band structure or phonon characteristics near T_M. The overall effect of these changes is substantial, since the tetragonal phase has a residual resistivity ρ_0 which is markedly lower than ρ_0 for the cubic phase.

The parameters obtained from analysis of these single crystal data do no differ significantly from the parameters reported by WC for polycrystalline film data. The main difference is that in order to describe high temperature behavior above ~ 100 K the linear term in Eq. (1) must be included. The WC parameters are shown in Table I. Whether the linear term is present at low temperatures cannot be determined from the available data, because of the dominance of the exponential term. To determine the effect of including a linear term in the low T analysis, we have forced a fit of Eq. (1) using $b_1 = 46.6 \times 10^{-3}$ $\mu\Omega$ cm K^{-1} to the data for the unannealed sample, and find that the resulting parameters are $\rho_0 = 20.3$ $\mu\Omega$ cm, $d_1 = 60.4$ $\mu\Omega$ cm and $T_0 = 97$K. These do not differ appreciably from the values listed in Table 1 for the case $b_1 = 0$, and so whether or not the linear term is included is irrelevant at low T.

III. ANALYSIS FOR Nb_3Sb

Fisk and Webb [17] recently have reported measurements of $\rho(T)$ for a single crystal of Nb_3Sb, which has a low superconducting critical temperature ($T_c = 0.2$ K). Qualitatively, the temperature dependence of $\rho(T)$ is similar to that of Nb_3Sn except that the rapid increase is found at higher temperatures. We have fit Eq. (1) to data that they kindly provided for the range $4.2 \leq T \leq 890$ K and obtain the parameters listed in Table II. The fit is accurate to within 2% over most of the range.

For Nb_3Sb the value of b_1 is smaller and d_1 larger than for

TABLE II

Characterization of Nb$_3$Sb

Fit for single crystal of Nb$_3$Sb; data of Fisk and Webb (Ref. 17)

$$\rho(T) = \rho_o + b_1 T + d_1 e^{-T_o/T}$$

$18 \leq T \leq 890K$

RRR	44
T_c (K)	0.2
ρ_o ($\mu\Omega$ cm)	1.3
b_1 (10^{-3} $\mu\Omega$cm K^{-1})	22
d_1 ($\mu\Omega$ cm)	127
T_o (K)	210

Nb$_3$Sn. The differences counterbalance to the extent that $\rho(T)$ at high temperatures is quantitatively similar for the two systems. An outstanding difference are the numerical values for T_o. For Nb$_3$Sb we find $T_o \approx 210$ K, which is more than twice as large as for Nb$_3$Sn. This larger value describes the sharp increase in $\rho(T)$ occurring at higher temperatures for Nb$_3$Sb than for Nb$_3$Sn.

IV. ANALYSIS FOR V$_3$Si

MWT have reported extensive and detailed analyses of $\rho(T)$ for a number of V$_3$Si single crystals of differing quality as indicated by the RRR. The value of T_o is relatively insensitive to RRR, ranging from T_o = 160 K for RRR = 5 to 180 K for RRR = 39. The value of b_1 is identical in all specimens, but there is a dramatic increase in d_1 with increase in RRR. Table III taken from MWT illustrates this trend.

Because T_o is much higher in V$_3$Si than Nb$_3$Sn, and therefore the exponential term is much smaller at low T for V$_3$Si, it was possible for MWT to identify unambiguously a non-exponential term in $\rho(T)$ in the low T range. Such an identification has not been possible for Nb$_3$Sn using either the WC data or Gunsalus data. MWT considered various forms that the non-exponential term might have in V$_3$Si through determining fits to Eq. (1) with the linear term replaced by $b_n T^n$, with n = 1, 3/2, or 2. Subsequently, Sarachik et al. [18] were able to measure $\rho(T)$ of the normal state in one sample (VS14) down to 11.5 K by applying a high magnetic field

TABLE III

Characterization of V_3Si

Fit for various single crystals of V_3Si in range $T_c \leq T \leq 50$ K (or $T_n \leq T \leq 50$ K); analysis of Milewits et al. (Ref. 15).

$$\rho(T) = \rho_0 + b_1 T + d_1 e^{-T_0/T}$$

	VS5	VS8	VS14	VS39
RRR	5	8	14	39
T_c (K)	15.5	16.1	16.8	16.8
ρ_0 ($\mu\Omega$ cm)	15.5	8.47	4.28	0.854
b_1 ($10^{-3} \mu\Omega$cm K^{-1})	57.2	58.0	57.4	57.2
d_1 ($\mu\Omega$ cm)	72.0	87.2	87	96.8
T_0 (K)	163	172	170	180

Fit for polycrystalline V_3Si with RRR = 26; data of Marchenko [11].

	$40 \leq T \leq 1200$K	$T_c \leq T \leq 28$ & $40 \leq T \leq 1200$K
ρ_0 ($\mu\Omega$ cm)	1.51	2.76
b_1 ($10^{-3} \mu\Omega$cm K^{-1})	25.2	21.5
d_1 ($\mu\Omega$ cm)	115	120
T_0 (K)	175	187

to suppress the superconducting transition. By fitting the observed magnetoresistance to a quadratic field dependence, the corresponding resistivity at zero field was deduced. The data were found to lie between the predictions for the cases of n = 1 and n = 3/2 extended below T_c = 16.8 K. The data agreed with both predictions to better than the ± 0.2% experimental uncertainties. However the data were not consistent with the case n = 2. As we shall show, the case n = 1 gives also a reasonably good fit for $\rho(T)$ at high temperatures, whereas n = 3/2 does not. Thus in Table III we include only the n = 1 results because they are accurate over a wider range.

For completeness we note that $\rho(T)$ alternatively can be fit by only a quadratic (and no exponential) in the range 11.5K $\leq$ T $\leq$ 25K with a precision of ± 0.2%. Above that range, $\rho(T)$ increases much more rapidly than would be predicted by an extension of the quadratic dependence. Thus a pure T^2 dependence can be obtained, but only over a limited range.

Now we turn to consider the high T behavior in V_3Si, using data reported by Marchenko (Fig. 1 of Ref. 11) for a polycrystalline sample. This sample has RRR = 26, which is normally sufficiently large that a single crystal specimen would exhibit a cubic-to-tetragonal lattice transformation at $T_M \approx$ 21 K. Marchenko, however, apparently did not observe an anomaly in $\rho(T)$ near this temperature, because he reported that $\rho(T)$ could be described to better than 1% by a T^2 dependence for $T_c \leq$ T $\leq$ 29 K.

We have fit Eq. (1) to the data of Marchenko using two different procedures to obtain separately a "high T fit" and a "full range fit". For the former we took data in the range 40K $<$ T $<$ 1200 K, within most of which the curvature is negative, and obtained the parameters given in Table III. For the latter we added to this data values at degree intervals in the range $T_c \leq$ T $\leq$ 28 K as predicted by Marchenko's quadratic fit (where accurate values could not be read from Fig. 1 of his paper). This full range of data requires agreement both where the curvature is negative and where it is positive (see Fig. 2). The best fit parameters are also shown in Table III, and it can be seen that they are in good agreement with the low T parameters reported by MWT. The overall fit for $T_c \leq$ T $\leq$ 1200K is within 4% for all data points.

V. DISCUSSION OF THE FITS

Before proceeding to possible explanations for an exponential term in $\rho(T)$ it is worthwhile offering a few remarks about trends in the empirical parameters. For all three A-15 compounds, regardless of the temperature range in which a fit has been made, Eq. (1) is a remarkably successful empirical description. Parameters such as T_o and d_1 differ very little when obtained from just low T data or data from a wide temperature range. Examination of the Tables will reveal however a small systematic trend in that T_o is generally smaller and d_1 larger when high T data is used. This we believe may arise from the fact that the non-exponential term is actually not a linear function of T but increases somewhat less rapidly at high T than low T. Evidence for this occurring in V_3Si is the fact that b_1 is ~ 30% smaller for high T fits than for low T fits (Table III). If this is the case, then we can understand

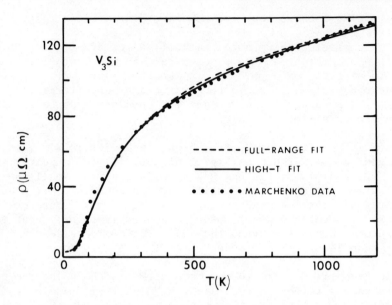

Figure 2: Observed $\rho(T)$ for V_3Si (RRR = 26) for $40 \leq T \leq 1200K$ as obtained from Marchenko(Ref. 11). Solid line is the "high T fit" and the dashed line is the "full range fit" as discussed in the text. Parameters of fits are given in Table III.

why T_o is smaller and d_1 larger at high T, because the exponential term in Eq. (1) is adjusted to account for this variation in the actual non-exponential term when in the fits we force the latter to be described by a linear function of temperature.

A second point to emphasize is the fact that the numerical values for the parameters b_1 and d_1 are very nearly the same (within a factor of two) for all three A-15 compounds. This quantitative similarity exists despite wide variations in ρ_o, T_c, and T_o for the samples.

Finally, we close this section with a word about the accuracy with which the empirical parameters can be determined. We have evaluated the sensitivity of their values to the accuracy of the data in the following way. For the full-range data on Nb_3Sn we carried out a fit with the additional stipulation that T_o be fixed

30% below or above what we had previously determined to be the least squares fit value. This resulted in an increase in the error of the fit from 1% to 4%. A similar procedure was carried out by fixing d_1 30% too low or high, and an error of 6% was obtained. Thus we deduce that if the input data are accurate to 1% we could expect the best fit parameters d_1 and T_0 to be determined with an accuracy of 5 and 10% respectively. The fitting uncertainties in determining parameters by our non-linear least squares routine are always much smaller than 5%.

VI. INTERPRETATION

The quality of the empirical formula in Eq. (1) has been evaluated in two ways: (1) MWT have demonstrated that it can yield high precision results to better than ± 0.15% for single cyrstals of V_3Si below 50 K; and (2) WC and the present work demonstrates that it has a precision of several percent for applications over the full temperature range in V_3Si, Nb_3Sn, and Nb_3Sb. Thus there is motivation to assume that the exponential term has physical significance and is characteristic of a particular physical mechanism that dominates the temperature dependence of the electron mobility.

A term of the form $\exp(-T_0/T)$ in $\rho(T)$ suggests a thermally activated process involving an energy gap kT_0 that reduces mobility. [This is not to be confused with a different situation in other materials such as semiconductors where thermal activation of electrons into conduction band states sees the conductivity vary as $\exp(-T_0/T)$.] An exponential term in $\rho(T)$ can arise in a conductor if a low mobility band edge lies kT_0 above the Fermi level of a high mobility band. An increase in T will see an effective trapping of more carriers in the low mobility band as they are excited. In this view, the non-exponential term b_1T could be an approximation for the effect of electron-phonon scattering. At this time, however, there is insufficient knowledge of the details of the A-15 band structure to provide independent support for this explanation of the exponential term, and therefore we will not pursue it further.

MWT have proposed two different explanations based on electron-phonon scattering that would account for an exponential term, at least at low temperatures. They were motivated by the close correspondence between the respective values of kT_0 for Nb_3Sn and V_3Si and a characteristic phonon energy in each system. For both cases, kT_0 is very close to the energy for [100] TA phonon modes at the Brillouin zone boundary. Table IV summarizes the measured phonon energies and representative values for T_0. That these phonons with the same wavevector $q \approx \pi/a$ could be particularly effective in

TABLE IV

Phonon energies for [100] TA modes at the Brillouin zone boundary as determined by inelastic neutron scattering studies, compared with characteristic energies T_o obtained from resistivity measurements.

Samples	T_o Values	TA PHONON ENERGIES AT B Z			Sources
		46K	80K	295K	
V_3Si	160-180K	-	182K	190K	A
Nb_3Sn	85-95K	79.5±9K	-	88.8±4K	B
Nb_3Sb	210K	-	-	-	

A G. Shirane, J.D. Axe and R.J. Birgeneau, Solid State Commun. **9**, 397 (1971).

B J.D. Axe and G. Shirane, Phys. Rev. B1 **8**, 1965 (1973).

scattering electrons can be understood if there is a momentum selection rule which is imposed by the topology of the Fermi surface. One possibility is Umklapp scattering between opposite faces of a large carrier pocket with $2k_F \approx \pi/a$. The exponential is the Boltzmann factor for the number of phonons with energy kT_o that just satisfy this condition [19]. The band structure calculations of Mattheiss indicate that such a pocket could exist at Γ in the center of the cubic Brillouin zone for a number of A-15 compounds, but the dimensions of the Fermi surface cannot be predicted with sufficient accuracy to establish whether $\pi/\underline{a}$ is a proper caliber.

The second proposal by MWT is that the exponential term is produced by interband scattering of electrons between two small pockets of the Fermi surface. If the pockets are separated by $q \approx \pi/a$ only phonons with essentially this wavevector can participate, and again an exponential term in $\rho(T)$ describes the temperature dependence for the number of phonons that are available for scattering. This process can be particularly important if one band has a high mobility and the other a low mobility with a high density of states. This explanation based on interband scattering is consistent with band structure calculations of Mattheiss [20] in that one feature of the A-15 structure is the occurrence of carrier pockets at the zone corners (R) and midpoints of the edges (M). These locations are separated by π/a. They are not the only possibilities, and a more complete discussion has been given by MWT. The precision of the current band structure calculations is not sufficient to warrant more detailed consideration of particular bands at this time. There is some evidence from de Haas-van Alphen studies [21] on Nb$_3$Sb that small pockets are indeed located at M and R, but the nature of the bands has not been established. We note that the coupled chain model of Gor'kov predicts a logarithmic singularity for N(E) associated with 6-fold degenerate bands at R, so there is some theoretical basis for the interband scattering explanation which requires a high density of states for one of the carrier pockets.

The similar behavior of $\rho(T)$ for Nb$_3$Sn where $N(E_F)$ is large and for Nb$_3$Sb where $N(E_F)$ is small might seem at variance with the interband scattering explanation. But we draw attention to the fact that the relevant density of states which determines the scattering rate is not $N(E_F)$ but $N(E_F + kT_0)$, since a phonon of energy kT_0 is annihilated in the process. The difference can be important, because band models which have been proposed to explain the anomalous normal state properties of V$_3$Si and Nb$_3$Sn contain a strong energy dependence for N(E), with appreciable variations in magnitude over an energy range of ~ 100 K. Thus it is possible to understand why Nb$_3$Sb has a low value of $N(E_F)$ as determined from low temperature experiments but still exhibits a large exponential term in $\rho(T)$ if $N(E_F + kT_0) \gg N(E_F)$.

The interband scattering mechanism fails to explain why $\rho(T)$ is not linear at high T. The Bose distribution for the phonons with $q \approx \pi/a$ is approximated by an exponential (Boltzmann) term only at low T and converts to a linear term for $T \gg T_0$. Presumably the energy dependence of N(E) will affect to some extent the temperature dependence of $\rho(T)$ at high T, as Fradin and colleagues have shown for Pt and Pd [22] and will make $\rho(T)$ less sensitive to T. But as suggested by Fisk and Webb [17] there may be limiting mean free path effects as well that establish an upper limit for the magnitude of $\rho(T)$. Allen et al. [23] have proposed that the quasi-saturation in Nb$_3$Sn could be due to the temperature dependence of phonon mode energies. The observed increase in energy with temperature for virtually all modes implies that the number of phonons at high T will increase less repidly than linearily in T. A test of this explanation would be provided by neutron scattering measurements of phonon energies in Nb$_3$Sb to see whether there is a corresponding hardening with increase in T.

Neutron scattering measurements on Nb$_3$Sb would also provide a test of the explanation for the exponential term based on selective electron-phonon interband scattering. If phonons with the same q are responsible for scattering in Nb$_3$Sb as for Nb$_3$Sn and V$_3$Si, Table IV predicts that the [100] TA phonons near the zone boundary in Nb$_3$Sb will be found to have an energy of ~ 210 K.

A substantially smaller value would contradict the explanation. This predicted energy is more than twice as large as the observed energy in Nb_3Sn, so its measurement should provide a clear cut test.

VII. ACKNOWLEDGEMENTS

We are grateful to G. Gunsalus for providing his unpublished data on the resistivity of Nb_3Sn and to Z. Fisk and G.W. Webb for their data on Nb_3Sb. We have benefitted from numerous discussions with M. Mourino, H. Taub, and G.W. Webb.

REFERENCES

1. M. Weger, Rev. Mod. Phys. 36, 175 (1964).
2. J. Labbé and J. Friedel, J. Phys. (Paris) 27, 153, 303, 708 (1966).
3. R.H. Cohen, G.D. Cody and J.J. Halloran, Phys. Rev. Lett. 19, 840 (1967).
4. L.P. Gor'kov, Zh. Eksp. Teor. Fiz. 65, 1658 (1973) [Sov. Phys. - JETP 38, 830 (1974)].
5. L.P. Gor'kov, Zh. Eksp. Teor. Pisma 20, 571 (1974).
6. L.P. Gor'kov and O.N. Dorokhov, J. Low Temp. Phys. 22, 1 (1976).
7. C.M. Varma, J.C. Phillips and S.T. Chui, Phys. Rev. Lett. 33, 1223 (1974).
8. J.C. Phillips, Solid State Commun. 18, 831 (1976).
9. M.P. Sarachik, G.E. Smith and J.H. Wernick, Can. J. Phys. 41, 1542 (1963).
10. F. Bloch, Zeits. f. Phys. 52, 555 (1928).
11. V.A. Marchenko, Fiz. Tverd. Tela. 15, 1893 (1973) [Sov. Phys. Solid State 15, 1261 (1973)].
12. G.W. Webb, Z. Fisk, J.J. Engelhardt and S.D. Bader, Rochester Conference and these proceedings, to be published.
13. H. Taub and S.J. Williamson, Solid State Commun. 15, 181 (1974).
14. D.W. Woodward and G.D. Cody, RCA Rev. 25, 392 (1964).
15. M. Milewits, S.J. Williamson and H. Taub, Phys. Rev. B15, XXX XX (1976).
16. G. Gunsalus, private communication of unpublished data.
17. Z. Fisk and G.W. Webb, Phys. Rev. Lett. 36, 1084 (1976).
18. M.P. Sarachik, M. Milewits and S.J. Williamson, Bull. Am. Phys. Soc. 21, 230 (1976).
19. A.H. Wilson, Proc. Roy. Soc. (London) 167, 580 (1938).
20. L.F. Mattheiss, Phys. Rev. B15 12, 2161 (1975).
21. A. Arko and Z. Fisk, Bull. Am. Phys. Soc. 21, 391 (1976).
22. F.Y. Fradin, Phys. Rev. Lett. 33, 158 (1974).
23. P.B. Allen, J.C.K. Hui, W.E. Pickett, C.M. Varma and Z. Fisk, to be published.

QUESTIONS AND COMMENTS

Discussion of this paper appears after that of Bader and Fradin.

INTERBAND SCATTERING CONTRIBUTIONS TO THE RESISTIVITY

OF A-15 METALS*

S.D. Bader and F.Y. Fradin

Argonne National Laboratory

Argonne, Illinois 60439

ABSTRACT

We calculate the temperature dependence of the resistivity of Nb_3Sn and V_3Si below 50 K in the interband-scattering model using realistic phonon densities of states and find agreement with experiment. For V_3Si we also find that phonon-mode shifting is in part responsible for, though not the dominant cause of, the observed negative deviation from linearity of the high-temperature resistivity. We present calculations to suggest that this negative deviation from linearity of the high-temperature resistivity is predominantly due to sharp structure in the electronic density of states $N(\varepsilon)$ near the Fermi energy ε_F. We further argue that the unusually large magnitude of the high-temperature resistivity of Nb_3Sb, a low-T_c (~ 0.2 K), low density-of-states A-15 superconductor, is due to thermally-accessible structure in $N(\varepsilon)$ within a characteristic phonon energy from ε_F.

The temperature dependence of the electrical resistivity $\rho(T)$ of strong-coupled A-15 structure superconductors, such as Nb_3Sn [1] and V_3Si [2], does not obey the simple predictions of the Wilson interband-scattering model [3], which is usually appropriate for other transition metals [4]. Is this due to shortcomings of the

---

*Work supported by the U.S. Energy Research and Development Administration.

interband-scattering model itself, or to the assumptions made when actually testing the model? We address ourselves to this question to gain insight into the electron-phonon interaction, which enters both the scattering matrix elements of $\rho(T)$ and the electron-pairing interaction responsible for superconductivity. We present the results of simple model calculations using the Wilson theory to assess (1) the role of non-Debye-like structure in the phonon density of states in producing deviations from a $\rho \propto T^3$ law at low temperatures ($T \gtrsim T_c$) for Nb_3Sn and V_3Si, (2) the role of phonon-mode shifting in producing the negative deviations from the $\rho \propto T$ law expected above room temperature (i.e., V_3Si), (3) the role of structure in the electronic density of states $N(\varepsilon)$ near the Fermi energy ε_F in contributing to the high-temperature negative deviation of ρ from linearity for A-15 structure superconductors, and (4) the role of structure in $N(\varepsilon)$ within a characteristic phonon energy from ε_F in producing the very large value of the resistivity of Nb_3Sb, a low T_c, low $N(\varepsilon_F)$ A-15 compound.

In the Wilson model for phonon-assisted interband scattering [3], if we keep the electronic density of states $N(\varepsilon)$ near the Fermi energy ε_F constant,

$$\rho(T) \propto \int \frac{X}{\sinh^2 X} F(\omega) \, d\omega \qquad (1)$$

where $X = \hbar\omega/2k_BT$, and $F(\omega)$ is the phonon density of states. The proportionality constant contains an average electron-phonon coupling strength, denoted α_{tr}^2, where the subscript indicates that this is the α^2 appropriate for transport phenomena [5]. The close relationship between α_{tr}^2 and α_{sc}^2 used in defining the superconductivity parameter $\lambda = 2\int d\omega \, \alpha_{sc}^2 F(\omega)/\omega$, is discussed by Allen [5]. Eq. (1) is sufficient to explore the temperature dependence of the interband scattering contribution to ρ at low temperatures. Physically, the scattering process under consideration involves a light-mass "conduction electron" (i.e., an s-electron) scattering by a phonon into a heavy-mass state or sink that does not conduct (i.e., a d- of f-state). Hence, this type of scattering, which predominates in transition metals, is often referred to as s-d scattering and denoted ρ_{sd}. In Wilson's idealization of the process there is a minimum phonon wave vector q_{min} necessary to connect s and d parts of the Fermi surface. This minimum-wavevector requirement would give rise to an initial exponential rise in ρ_{sd}, and, at the lowest temperatures, other scattering processes, such as electron-electron scattering ($\rho_{ee} \propto T^2$, as $T \to 0$), or the usual Block-Grüneisen scattering ($\rho_{BG} \propto T^5$, as $T \to 0$) and/or impurity scattering might be expected to dominate ρ. In reality ρ_{sd} is usually found to initially rise as T^3, indicating the q_{min} is very small, and that, as transition metal band-structures have shown, pure s-like and d-like

INTERBAND SCATTERING CONTRIBUTIONS

parts of the Fermi surface tend to be mixed together.

The properties of Eq. (1) are shown in Figure 1 using for $F(\omega)$ a Debye phonon spectrum $F_D(\omega) \propto \omega^2$ for $\omega \leq \omega_D$, otherwise $F(\omega) = 0$, where ω_D is the Debye cut-off frequency. The T^3 law [shown by the lower-temperature straight line in the bilogarithmic plot of Fig. (1)] holds below ~ 12% of Θ_D, the Debye temperature, where $\Theta_D = \hbar\omega_D/k_B$. The high-temperature linear region of $\rho(T)$ occurs for $T \gtrsim \Theta_D$ [shown by the higher-temperature straight lines in Fig. (1)]. In general, if we go beyond a Debye model for $F(\omega)$ we find that the linear region in $\rho(T)$ at high temperature is a property of Eq. (1), as written, for any harmonic $F(\omega)$. The T^3 law is also always expected to hold as $T \to 0$, because at low temperatures only the lowest-energy phonon states are thermally populated, and $F(\omega)$ will always be proportional to ω^2 as $\omega \to 0$ for any three-dimensional crystal, i.e., at long wavelengths the vibrations of real crystals simulate those of the Debye elastic-continuum model. However, the T^3 law may now only be expected to appear below roughly 2% of the appropriate characteristic phonon temperature Θ. This property of real, as opposed to Debye, phonon spectral density functions is also responsible for deviations from a T^3 law in the lattice heat capacity at reduced temperatures T/Θ as much as an order-of-magnitude lower than expected from the Debye model. In the first part of this paper we exploit these well-known observations to show that the T^3 law in ρ_{sd} is not expected to persist to $T > T_c$. The resistivity has not been measured below ~ T_c because of the large upper critical fields of the materials of interest.

First we consider Nb_3Sn. Webb and co-workers [6] have recently shown that between T_c and 40 K (1) the resistivity of Nb_3Sn can be represented by a constant residual resistivity term ρ_0 plus a T^2 term, and (2) the lattice heat capacity is not proportional to T^3. In order to use Eq. (1) we need a realistic model $F(\omega)$. We use Shen's superconductive tunneling [7] determination of $\alpha_{sc}^2 F(\omega)$ and renormalize it to put more spectral weight at higher frequencies, in order to recover calorimetrically-determined moments of $F(\omega)$. Representing

$$F(\omega) = \frac{[\alpha_{sc}^2 F(\omega)]_{Shen}}{1 - A\omega},$$

we calculate $\hbar/k_B \exp[\int (\ln\omega) F(\omega) \, d\omega]$, the logarithmic moment of $F(\omega)$ at $T = 0$, which has been extrapolated from an analysis of the measured high-temperature lattice entropy [8]. The extrapolated moment value of 201 K [8] yields $A = 0.033$ meV^{-1}. The model $F(\omega)$ appears as insert (a) of Fig. 2. It is important to note that not only does this phonon density of states describe the shape and magnitude of the lattice heat capacity over a decade in temperature [see insert (b) of Fig. 2 and discussion in Ref. 6], but the peaks

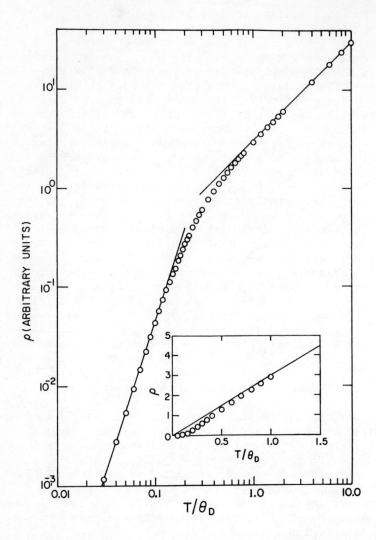

Figure 1: Bilogarithmic plot of the interband scattering contribution to the resistivity calculated using Eq. (1) and a Debye-model phonon density of states (circles). The lower-temperature straight line indicates that $\rho \propto T^3$ for $T \lesssim 0.12\,\Theta_D$, while the higher-temperature straight line indicates that $\rho \propto T$ for $T \gtrsim \Theta_D$. The insert shows the calculated ρ versus T/Θ_D (circles) and its high-temperature slope extrapolated back to $T = 0$. Note that even at rather low reduced temperatures ($T/\Theta_D \gtrsim 0.2$) ρ gives the misleading appearance of being a linear function of temperature.

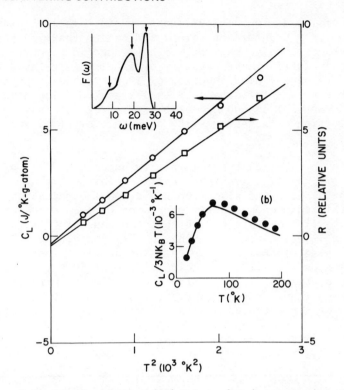

Figure 2: Calculated values of the interband scattering contribution to the resistivity R (squares) and the lattice heat capacity C_L (circles) plotted at 5 K intervals between 20 and 50 K for Nb_3Sn. Note that the straight lines indicate that over this temperature interval both quantities are apparent linear functions of T^2, as found experimentally [6,8]. Note also that the negative $T^2 = 0$ intercepts indicate that there is no physically meaningful T^2 law in effect here. Insert (a) shows the $F(\omega)$ used in the calculations. (See text for details.) Insert (b) shows a comparison of the measured [8] (solid curve) and calculated (filled circles) C_L/T between 20 and 200 K. (See text and Ref. 6 for further discussion.)

in the model $F(\omega)$ appear at the same frequencies [see arrows in insert (a) of Fig. 2] as the peaks in the weighted or generalized phonon density of states $G(\omega)$ determined using inelastic neutron scattering [9] from polycrystalline Nb_3Sn. Hence, the model, though not unique, is certainly realistic. Using this $F(\omega)$ and Eq. (1), ρ_{sd} is plotted versus T^2 in Figure 2 (squares), and the straight line indicates that below 40 K we find that, indeed, ρ is a linear function of T^2, as found experimentally [6]. The calculation yields a negative $T^2 = 0$ intercept for ρ, indicating that the apparent T^2 dependence of ρ has no physical signigicance in terms of a "T^2 law". The harmonic lattice heat capacity C_L is also calculated (circles) and found to be a linear function of T^2 over this temperature range, as is the experimental data [6,8], This calculation demonstrates that the unusual T dependence of ρ (and of C_L) at low temperatures is a reflection of non-Debye-like structure of $F(\omega)$, and nothing more.

We now present another example to demonstrate the effect of unusual $F(\omega)$ on $\rho(T)$. V_3Si has a $\rho(T)$ below 50 K that differs from that of both Nb_3Sn and also from Debye-model expectations. Using the generalized phonon density of states $G(\omega)$ measured at 4.2 K by Schweiss [10], we again find that the unusual temperature dependence of ρ is accounted for using Eq. (1), (see Figure 3). Note that in Fig. 3, to compare the calculation with experimental data of Milewits et al. [11], the calculated ρ at 50 K is normalized to the experimental $\rho(50) - \rho_0$. The use of $G(\omega)$ in performing these calculations requires some comment. The atom that has more scattering power (as measured by its thermal neutron scattering cross-

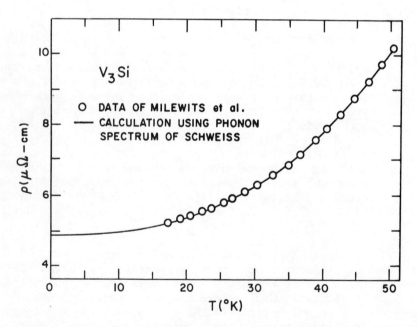

Figure 3: The resistivity of V_3Si. Solid line is the interband scattering contribution, calculated using Eq. (1). (See text for details.)

INTERBAND SCATTERING CONTRIBUTIONS

section to mass ratio σ/M) has its displacements more heavily weighted in $G(\omega)$ than in $F(\omega)$. Fortunately in V_3Si $(\sigma/M)_V \simeq (\sigma/M)_{Si}$, and it is not necessary to introduce any adjustable parameters to rescale $G(\omega)$ into a realistic $F(\omega)$. Also based on the heat-capacity criterion used in the Nb_3Sn example, above, Schweiss's $G(\omega)$ yields an adequate representation of $F(\omega)$ for our purposes. The logarithmic moment of $F(\omega)$ is within 7% of the value extracted from calorimetric measurements [6] for $F(\omega)$. Given the combined experimental uncertainties in these experiments [and the additional eigenvector and Debye-Waller factor weighing of $G(\omega)$] adjustments to improve agreement are unwarranted. Hence, we see that it is unnecessary to invoke unusual mechanisms for $\rho(T)$ below 50 K in order to understand the physical significance of its temperature dependence.

Now that it is clear that non-Debye-like structure in the phonon density of states is responsible for unusual ρ-versus-T curves below ~ 50 K for the high-T_c A-15 superconductors, what can we say about the high-temperature region? In particular these same materials also exhibit negative deviations [1,2] from the expected linearity of ρ versus T. By 1000 K these deviations are strong enough that ρ appears to be approaching a constant value, for Nb_3Sn, for instance [12]. Allen and co-workers [13] have recently suggested that this bending away from linearity of ρ is due to the quartic anharmonicity that causes the phonons of these high-T_c materials to stiffen on the average as temperature is raised from 4.2 K to 300 K. (This feature of the high-T_c A-15 superconductors is quite striking when one keeps in mind that anharmonicity usually causes phonon softening as temperature is increased, until the material eventually melts.) Since at high temperatures [13] $\rho \propto T/\Theta^2$, the stiffening of Θ qualitatively accounts for the negative deviations from linearity. To quantitatively assess the effect, we calculate $\rho(T)$ using Eq. (1) and Schweiss's $G(\omega)$ spectra for V_3Si [10] at 412, 77, and 297 K. The family of curves is displayed in Figure 4. As temperature is increased the measured ρ_{sd} might be expected to cut across from the upper through the lower curve of Figure 4 in some manner, touching the 77 K curve at 77 K and the 297 K curve at 297 K. If we calculate the ratio ρ_{297}/ρ_{77} using $G(\omega)$ for 77 K only, we obtain the value 8. Using the $G(\omega)$ appropriate to each temperature to calculate the ratio ρ_{297}/ρ_{77} we find a value of 7. Hence, the mode-shifting effect is clearly seen. However, when compared to the experimental ratio $(\rho_{295}-\rho_0)/(\rho_{77}-\rho_0)$, where it is, of course, necessary to correct for the residual resistivity ρ_0 term, we find, using data of Milewits, et al. [11], for two different samples, (see Table I) values of ~ 5. Hence, phonon-mode shifting accounts for only 30% of the observed negative deviation from linearity of the resistivity of V_3Si. This result leads us to look elsewhere for the predominant cause of the high-temperature curvature in ρ for the A-15 superconductors. We can safely arrive at this conclusion because there is general agreement that on the

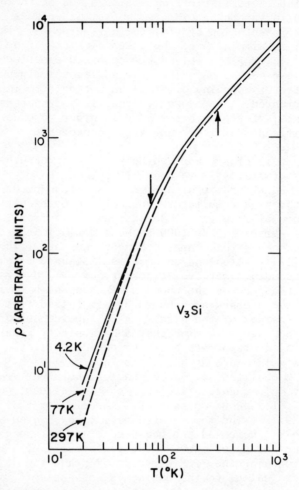

Figure 4: The interband scattering contribution to the resistivity [c.f. Eq. (1)] of V_3Si on a bilogarithmic plot calculated using the 4.2, 77, and 297 K spectra of Schweiss. The down arrow locates the value of ρ at 77 K using the 77 K phonon spectrum, and the up arrow locates the value of ρ at 297 K using the 297 K phonon spectrum. The ratio of these values appears in Table I.

Table I

Effect of Phonon-mode Shifting on the High-Temperature Resistivity of V_3Si

V_3Si[a]	$T_c(^0K)$	$\rho_o(\mu\Omega\text{-cm})$	$\dfrac{\rho_{295}-\rho_o}{\rho_{77}-\rho_o}$	$\dfrac{\rho_{295}}{\rho_{77}}$ calc.
#14	16.8	4.8	4.9	7.1[b]
#39	16.8	1.4	4.8	

[a]Reference 11.

[b]Using spectra of Reference 10.

average V_3Si is more anharmonic than the other A-15 superconductors of interest [6,9,10,13]. Hence, if we can not attribute the bulk of the unusual ρ versus T curve at high temperatures to mode shifting for V_3Si, we also cannot for the other A-15 superconductors.

We now further generalize the Wilson model by relaxing the constant-$N(\varepsilon)$-near-ε_F assumption used in Eq. (1). This also was done almost a decade ago by Cohen, Cody, and Halloran [14], who proposed what is now referred to as the "RCA model" [15] to describe the unusual temperature dependence of the resistivity of A-15 superconductors. They introduced sharp structure in the electronic density of states near ε_F, i.e., a step-function $N(\varepsilon)$ with the step ~ 100 K from ε_F/k_B. The strong energy dependence of $N(\varepsilon)$ has the effect of making the integration over electronic states, which contributes to the proportionality constant α_{tr}^2 of Eq. (1), non-factorable and temperature dependent. The step-function $N(\varepsilon)$ has the advantage that it yields a simple analytical expression for the resistivity, but its disadvantages are that it has too few degrees of freedom to simultaneously fit the temperature dependence of other electronic properties such as the spin susceptibility, and, of course, the actual $N(\varepsilon)$ is expected to be more complicated. For transition-metals such as Pd and Pt for which realistic $N(\varepsilon)$ information is available the effect of structure in $N(\varepsilon)$ near ε_F, based on *ab initio* RAPW band-structure calculations, has actually been shown [16,17] to be responsible for the measured negative deviations of $\rho(T)$ from linearity. [The peak in the susceptibility of Pd near 80 K is also understood as due to structure in $N(\varepsilon)$, and the temperature dependence of the nuclear spin-lattice relaxation rate [18] of platinum was, likewise, accurately predicted.]

To definitively treat the A-15 superconductors of interest here we need high-accuracy high-resolution densities of states. Unfortunately this information is not available, so we will use various model densities of states for the remainder of our calculations. The generalization of Eq. (1) is given [16] for phonon absorption by

$$\rho_{s-d} = \frac{\rho_o}{k_B T} \int_o^{\omega_{max}} d\omega F(\omega) n(\omega) \int_o^\infty d\varepsilon \frac{N_d(\varepsilon+\omega)}{N(0)} f(\varepsilon)[1-f(\varepsilon+\omega)] \qquad (2)$$

where ρ_o is a constant proportional [13] to $\lambda_{tr} = 2\int d\omega\, \alpha_{tr}^2 F(\omega)/\omega$, $n(\omega)$ is the Bose function, $f(\varepsilon)$ is the Fermi function, and $N_d(\varepsilon)$ is the d-fraction of the total density of states $N(0)$. The contribution to ρ_{s-d} from phonon emission is given by a similar expression with $n(\omega)$ replaced by $n(\omega)+1$ and $\varepsilon+\omega$ replaced by $\varepsilon-\omega$. [We note that due to the additional integral over the phonon energy, ρ_{s-d} is not nearly as sensitive as the temperature dependence of the spin susceptibility to details of $N(\varepsilon)$.] For our purposes, we can now use Debye phonon spectra to evaluate Eq. (2) for $T \gtrsim \Theta_D/2$, since the details of $F(\omega)$ do not play an important role at high temperature. Since the Fermi functions $f(\varepsilon)$ in Eq. (2) depend on the chemical potential $\mu(T)$, we iteratively calculate μ at each temperature using the constraint that the number of occupied states is constant. An interesting feature of Eq. (2) is that as $N_d(\varepsilon)$ approaches a step-function in shape with edge at ε_o, the high-temperature limiting form is given, for $|\varepsilon_o-\mu|$ and $|\varepsilon_o-\omega_{max}-\mu| \ll T$, and $|\varepsilon_o - \mu| \approx \omega_{max}$, by a 1/T approach to a constant value $\rho_\infty \approx (3/8)\rho_o N(\varepsilon_o)/N(0)$.

In Figure 5, we illustrate the effect on ρ_{s-d} of successively sharpening an edge in $N(\varepsilon)$. Here we have used a Gaussian-edged density of states given by

$$N(\varepsilon) = \begin{cases} N_o\{\frac{1}{3} + \frac{2}{3}\exp[-(\varepsilon-\varepsilon_o)^2/\Delta^2] & \text{for } \varepsilon \geq \varepsilon_o \\ N_o & \varepsilon \leq \varepsilon_o \end{cases}$$

with $\varepsilon_F \equiv \mu(0) = \varepsilon_o$ and Δ ranging from 20,000 to 200 in degree-Kelvin units. As the edge sharpens the negative deviations of ρ_{s-d} from linearity at high temperature become more pronounced. To attribute the remainder of the curvature in the previously-discussed V_3Si ratio $(\rho_{295}-\rho_o)/(\rho_{77}-\rho_o)$ to structure in $N(\varepsilon)$ would require a Δ-value between 200-600 °K using the $N(\varepsilon)$ model of Figure 5. It is not meaningful to quote characteristic $N(\varepsilon)$ parameters, such as Δ, to better than about a factor of three, since, as mentioned, the resistivity cannot be used to uniquely determine the form of $N(\varepsilon)$ versus ε [15]. Suffice it to mention that using this same model

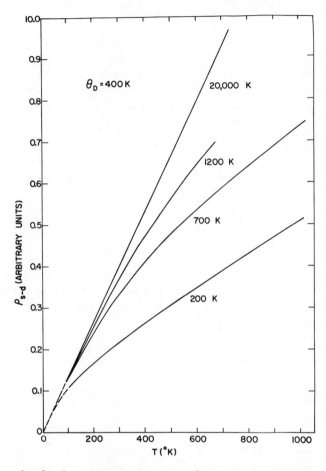

Figure 5: The high-temperature interband resistivity ρ_{s-d} calculated using Eq. (2) with the Gaussian-edge electronic density of states model discussed in the text. The labels indicate the widths Δ (in $^\circ$K) of the Gaussian-edge in $N(\varepsilon)$. Note that over the temperature range plotted the curve with the most gradual structure in $N(\varepsilon)$ near ε_F is linear in T, while for the more sharply structured $N(\varepsilon)$ models, the resistivity negatively deviates from linearity increasing dramatically at the higher temperatures. The position of $\mu(0) \equiv \varepsilon_F$ is fixed at ε_0.

$N(\varepsilon)$ with $\Delta \sim 200$ K yields a reasonable description of the temperature dependences of other physical properties, such as the magnetic susceptibility and the ^{51}V nuclear spin-lattice relaxation time [19].

Fisk and Webb recently attributed the flattening of the ρ-versus-T curve for A-15 superconductors to a saturation effect [12]. The resistivity at high temperature (300 < T < 1000 K) is so large that they suggest the mean free path of an electron is on the order of an interatomic spacing. Since it is not meaningful to think of scattering lengths shorter than this, saturation occurs. The important physical point is the mechanism by which the resistivity approaches saturation. Consider Nb_3Sb, which has a low T_c (~ 0.2 K), an $N(\varepsilon_F)$ (~ 0.4 states/eV-atom for both spin directions), that is a factor of ~ 5 lower than Nb_3Sn, and normal phonon behavior, i.e., phonon-mode softening as the melting temperature is approached (again, unlike its high-T_c homologues) [8]. Although at low temperatures $\rho(T)$ is much smaller for Nb_3Sb than for Nb_3Sn, at high temperature $\rho(T)$ is approaching saturation in a similar manner in both materials [12]. We propose that for Nb_3Sb, $\rho(T)$ becomes large enough to eventually saturate because structure in $N(\varepsilon)$ within about ω_{max} of ε_F makes a high density of d-state sinks available as temperature is raised, (i.e., for Nb_3Sb, ε_F is in a low-density of states region, adjacent to a thermally accessible high-$N_d(\varepsilon)$ region). To illustrate this type of effect we have calculated ρ_{s-d} using Eq. (2) with a parabolic-model density of states given by

$$N(\varepsilon) = \begin{cases} 5.5 \, N_0[1 - (\varepsilon-\varepsilon_0)^2/\Delta^2] & \text{for } |\varepsilon-\varepsilon_0| \leq \Delta \\ N_0 & \text{for } |\varepsilon-\varepsilon_0| \geq \Delta \end{cases}$$

where $\Delta = 300$ K and $\mu(0) \equiv \varepsilon_F$ takes on values from ε_0 to $\varepsilon_0 + 450$ K. The results are shown in Figure 6. We note that for $\varepsilon_F \geq 300$ K the values of $N(\varepsilon_F)$ are small and the resistivity starts to increase slowly with T, while for $0 < \varepsilon_F < 300$ K the $N(\varepsilon_F)$ is large and ρ_{sd} initially increases very rapidly with T. As shown in Figure 6, the approach to saturation of the resistivity is a natural consequence of this type of structured $N(\varepsilon)$ model, and saturation is approached more rapidly as ε_F approaches ε_0. The magnitude of the high-temperature resistivity for $|\varepsilon_0-\mu| \lesssim \omega_{max}$ is proportional to $\rho_0[N(\varepsilon_0)/N(\varepsilon_F)] \sim \lambda_{tr}[N(\varepsilon_0)/N(\varepsilon_F)]$. Thus for similar average phonon properties we expect the high-temperature resistivities to scale as $\lambda_{tr}/N(\varepsilon_F)$. For example, for Nb_3Sn, $[\lambda_{tr}/N(\varepsilon_F)] \approx [\lambda/N(\varepsilon_F)] = 0.05$, while for Nb_3Sb, $[\lambda_{tr}/N(\varepsilon_F)] \approx [\lambda/N(\varepsilon_F)] = 0.07$. The ratio of parabolic peak height to background level in the $N(\varepsilon)$ model was chosen to roughly simulate the values of $N(\varepsilon_F)$ for Nb_3Sn and Nb_3Sb, respectively. We note that the curve in Figure 6 for $\varepsilon_F - \varepsilon_0 = 150$ K approximates the temperature dependence of ρ for Nb_3Sn, whereas $\rho(T)$ for Nb_3Sb is similar to the curve in Figure 6

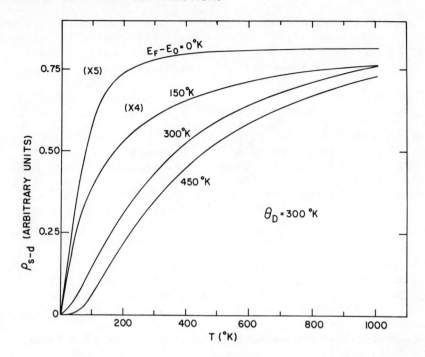

Figure 6: The high-temperature interband resistivity calculated using Eq. (2) with constant ρ_0. The electronic density of states model used is based on a parabolic peak of width $\Delta = 300$ K discussed in the text. The effect of different placement of ε_F relative to the peak centroid ε_0 is shown. Note that the values of ρ_{s-d} for $\varepsilon_F - \varepsilon_0 = 150$ K have been multiplied by a factor of 4, and, for $\varepsilon_F - \varepsilon_0 = 0\,^0K$, by a factor of 5 for clarity.

labeled $\varepsilon_F - \varepsilon_0 = 450$ K. If these curves are multiplied by the values of λ_{tr} for Nb_3Sn and Nb_3Sb, respectively, we obtain qualitative agreement in the measured ratio of the high-temperature resistivities [12,6].

In summary, the s-d or interband scattering model for the resistivity is capable of explaining the unusual resistivity of the A-15 compounds both at low temperatures and at high temperatures. At temperatures between T_c and a few times T_c, non-Debye-like structure in the phonon density of states clearly causes the unusual ρ versus T behavior. At intermediate temperatures, it is necessary to consider the role of anharmonicity and even more importantly, of fine structure in $N(\varepsilon)$ near ε_F in producing deviations from $\rho \propto T$. Finally, at very high temperatures compared to the characteristic phonon temperature Θ, $\rho(T)$ approaches saturation. The saturation effect is a natural consequence of fine structure in $N(\varepsilon)$ near ε_F (see Fig. 6), but might also arise because the mean free path approaches an interatomic spacing.

ACKNOWLEDGEMENTS

We are indebted to Dr. P. Schweiss for kindly providing us with his V_3Si spectra. One of us (S.D.B.) would like to thank Dr. George Webb for initially kindling his interest in this problem, and for numerous discussions. We also thank Professor S.J. Williamson for a preprint of Reference 11.

REFERENCES

1. D.W. Woodward and G.D. Cody, Phys. Rev. 136, 166A (1964).
2. V.A. Marshenko, Sov. Phys. Solid State 15, 1261 (1973).
3. A.H. Wilson, Proc. Roy. Soc. London, Ser. A 167, 580 (1938); also, see A.H. Wilson, *Theory of Metals* (Cambridge University, Cambridge, 1954).
4. G.T. Meaden, *Electrical Resistance of Metals* (Plenum Press, New York, 1965).
5. P.B. Allen, Phys. Rev. B3, 305 (1971). See appendix therein.
6. G.W. Webb, Z. Fisk, J.J. Engelhardt, and S.D. Bader, to be published; also see G.W. Webb and Z. Fisk in the Rochester proceedings.
7. L.Y.L. Shen, Phys. Rev. Lett. 29, 1082 (1972).
8. G.S. Knapp, S.D. Bader, and Z. Fisk, Phys. Rev. B (in press).
9. W. Reichardt, Progress Report of the Teilinstitut Nukleare Festkorperphysik, KFK2054, September 1974, Kernforshungszentrum, Karlsruhe, Karlsruhe, Germany; also see W. Reichardt in another paper of the Rochester proceedings.
10. P. Schweiss, Prog. Report of the Teilinstitut Nukleare Festkorperphysik, KFK2054, September 1974, Kernforshungszentrum Karsruhe, Karlsruhe, Germany.
11. M. Milewits, S.J. Williamson, and H. Taub, Phys. Rev. B (in press).
12. Z. Fisk and G.W. Webb, Phys. Rev. Lett. 36, 1084 (1976).
13. P.B. Allen, J.C.K. Hui, W.E. Pickett, C.M. Varma, and Z. Fisk, Solid State Commun. (in press).
14. R.W. Cohen, G.D. Cody and J.J. Halloran, Phys. Rev. Lett. 19, 840 (1967).
15. M. Weger and I.B. Goldberg, in *Solid State Physics, Vol. 28*, edited by H. Ehrenreich, F. Seitz and D. Turnbull (Academic Press, New York, 1973).
16. F.Y. Fradin, Phys. Rev. Lett. 33, 158 (1974).
17. F.Y. Fradin, D.D. Koelling, A.J. Freeman, and T.J. Watson-Yang,

Phys. Rev. B12, 5570 (1975).
18. D. Zamir, to be published.
19. G.S. Knapp, S.D. Bader, H.V. Culbert, F.Y. Fradin and T.E. Klippert, Phys. Rev. B11, 4331 (1975).

QUESTIONS AND COMMENTS

J. Phillips: Could we have that slide? [Fig. 1 of the Fisk and Webb paper]. This is an idea which I tried to publish a year ago. I do not know physics anymore. It is hard for a theorist to predict new things if everytime you predict...For those of you who happen to be interested in these anomalies you may notice that these two things which are chemically rather similar are basically the same at high temperature but around 400° or 450° they break apart and you get the bulge which is supposed to have many mysterious explanations and all of them have failed so far. Now the interesting thing about this is that the two break apart just at the point where the new two phase behavior which has been discovered by Paul Schmidt was discussed this morning. So I would like to suggest to all of you thinking in terms of single phase models that you may not be dealing with a single phase system. This may be a two phase system. What may be happening there is the tetragonal microdomains which are forming around 450° may be responsible for all the extra scattering here. So that is the first point. The second point is if you are forming these two phase systems with these microdomains then what will happen? Well, what will happen is that at the interface between the microdomains and the bulk material you will get additional scattering. What kind of additional scattering will you get? You will get scattering from the surface modes at the interface. What will be the characteristic frequency of these surface modes, which are basically shear modes vibrating perpendicular, parallel to the surface? What will the frequencies of those be? They will be the short wavelength, transverse acoustic frequencies of the zone boundaries which are exactly the frequencies that Sam Williamson identified in his talk. So it is simply a macroscopic effect as I emphasized earlier this morning. You have a two phase system. You produce these microdomains and the scattering from the interface between the microdomains and the bulk material. The extra scattering at that interface, which is obviously going to be very strong, is responsible for this entire anomaly. In my opinion.

F. Fradin: You know, I would like to point out that this bulging, this curvature in the resistivity, is a very general phenomena. It happens in transition metals that have nothing to do with superconductivity (in nearly magnetic systems) in the paramagnetic regime and also the effect has been calcualted form first principles band structure calculations for palladium metal [F. Y. Fradin Phys. Rev. Lett. 33, 158 (1974)] and platinum metal [F. Y. Fradin et al., Phys. Rev. B 12, 5570 (1975)]. So I do not think we have to look for mysteries when it is a very general phenomena that occurs in many different kinds of pure metals, e.g., Pt, Pd, U, Np, and Pu.

J. Phillips: Why is it present in Nb_3Sn and not in Nb_3Sb?

F. Fradin: It is present in both of them.

J. Phillips: Why is it bigger in Nb_3Sn that in Nb_3Sb?

G. Knapp: Nb_3Sn has 10 times the density of states of Nb_3Sb.

J. Phillips: Why does it grow and vary under pressure? You saw that in Fisk's talk. The correlation with superconductivity is very high.

F. Fradin: Do we have microdomains in palladium metal?

J. Phillips: We have microdomains or precursors in all soft lattices. A good discussion of them is given by A. Nagasawa, J. Phys. Soc. Japan 40, 93 (1976). They are present in Nb_3Sn, but not Nb_3Sb.

M. Cohen: Now you know the referee.

S. Bader: [Added in proof] The real question of fundamental significance here is why Nb_3Sb, a low-T_c, low density of states compound, has as large a resistivity at high temperatures as its high-T_c, high density of states homologue Nb_3Sn. This question is answered in the text of our paper.

SUPERCONDUCTIVITY IN METALLIC HYDROGEN*

R. P. Gupta

Northwestern University

Evanston, Illinois 60201

and

S.K. Sinha

Argonne National Laboratory

Argonne, Illinois 60439

ABSTRACT

The occurrence of superconductivity in the metallic phase of hydrogen is re-examined. While the band structure is quite free electron-like, the wavefunctions look like atomic wavefunctions near the protons. A consideration of the electron-phonon interaction reveals that the motion of the bound part of the electron wavefunctions considerably decreases the effective electron-phonon coupling constant over previous estimates. The alternative derivation of the electron-phonon coupling constant in terms of phase shifts is also discussed and the effects of screening are considered. Again the estimate of T_c are found to be considerably reduced by screening effects.

---

*Supported by the National Science Foundation, Air Force Office of Scientific Research, and the Energy Research and Development Administration.

The metallic phase of hydrogen has attracted considerable attention over the last few years.[1-5] It has been proposed that this phase, which is believed to be stable at high densities, should be a high temperature superconductor with estimates of T_c ranging from 55 K to 226 K.[3-6] Such high values of T_c are supposed to arise from the strong electron-proton coupling in this solid. In this note, we shall not concern ourselves with the question of the stability of the metallic phase, but rather address ourselves to the question of whether it might be a good superconductor assuming it is stable.

It might be thought that the simple e^2/r potential between electrons and protons renders this problem rather simple, but such is not the case. The reason is that the strength of this potential distorts the electronic wavefunctions in the unit cell considerably from plane waves. This complicates considerably the treatment of the electron-phonon interaction, how this interaction is screened, and the calculation of the phonon frequencies, all of which are necessary ingredients in a microscopic calculation of T_c. A definitive answer as to what T_c might be is not yet attainable. However, in this paper, we discuss considerations which indicate that optimistic predictions of a high T_c in this material obtained by previous workers should be treated with extreme caution. In the first part of this paper, we point out that calculations based on linear response theory applied to plane wave electron states are invalid, and show that if one assumes instead one augmented plane wave to represent the electron wavefunction (a much more reasonable assumption) then the electron-phonon interaction is renormalized to the extent that T_c is very low. In the second part of the paper, we discuss predictions based on the Gaspari-Gyorffy formalism for calculating T_c, which does not suffer from the defects of the perturbative approaches, but ignores important aspects of electronic screening.

Previous calculations of the phonon frequencies and the electron-phonon interaction in this system have been based on the assumption that the electron-phonon matrix element is adequately represented by the matrix element of the gradient of a screened Coulomb potential between plane wave states. The reasoning has been that the absence of cores states precludes the use of a pseudopotential, but that nevertheless the screened Coulomb potential of the protons is such as not to appreciably perturb the plane wave character of the electronic wavefunctions. It has become increasingly evident, however, from results of recent more sophisticated band calculations[1] and from the results of nonlinear screening theory,[7-8] that the electron wavefunctions in the vicinity of the protons are not at all free electron like, but rather show the characteristic cusp more appropriate to atomic hydrogenic wavefunctions. We have carried out an augmented plane-wave (APW) energy band calculation for metallic hydrogen in both fcc and bcc phases for two different densities ($r_s=1.29$ and $r_s=1.6$) which are in the vicinity of that expected[1-5] for the metallic phase. The difference between the band structure approach and the earlier perturbative approaches is that the former does not rely on the smallness of the crystal potential Fourier coefficients. The APW calcuations was done with an assumed Wigner-Seitz crystal potential, and also with the standard muffin tin crystal potential constructed from superimposed atomic hydrogenic charge densities with exchange being taken into account in the $\rho^{1/3}$ approximation. While more accurate crystal potentials have been used[1] and are probably essential in determining quantities such as the total ground state energy of metallic hydrogen, they do not appear to be necessary for discussing the properties we wish to concentrate on. Properties such as the phonon spectrum and the electron-phonon interaction depend only on the relative energies of the electron states and on the electronic wavefunctions. As far as these quantities are concerned, our results are in good agreement with the more accurate band calculation of Harris et al.[1] in that we get an energy band structure for which the lowest band and the Fermi surface is highly free electron-like in all cases, except near the zone boundaries. However, the wavefunctions are not free-electron-like over the whole unit cell, as also found by Harris et al.[1] We shall restrict our explicit discussion in this paper to the case of an assumed fcc lattice at a density corresponding to r_s = 1.64 a.u. Fig. 1 shows the spherically averaged electron density inside a Wigner-Seitz sphere around a proton site for three values of $\vec{k}$, corresponding to wave vectors on the Fermi surface along the three principal symmetry axes. Also shown is the free atom charge density. It may be seen that the wavefunctions behave more like atomic orbitals in the vicinity of the protons, but behave like smooth linear combinations of plane waves in the interstitial regions. Let us assume for the moment that only one APW is important in the eigenfunctions. This is in the spirit of the previous calculations[3-5] of the phonon frequencies and T_c based on plane waves, but we are going to take into account the effect of the "core-like" part of the wavefunctions in a more rigorous fashion.

When we consider the screened electron-phonon matrix element (EPME) it is intuitively obvious that the main effect of the large potential change in the region of the proton due to the lattice displacements will be to move the "atomic" part of the charge density shown in Fig. 1 rigidly with the protons, and will thus lead to a considerable decrease in the effective EPME. This idea was formulated rigorously in an earlier paper,[9] where it was shown that the self-consistent EPME between states $\vec{k}$ and $\vec{k} + \vec{Q}$ could be written as

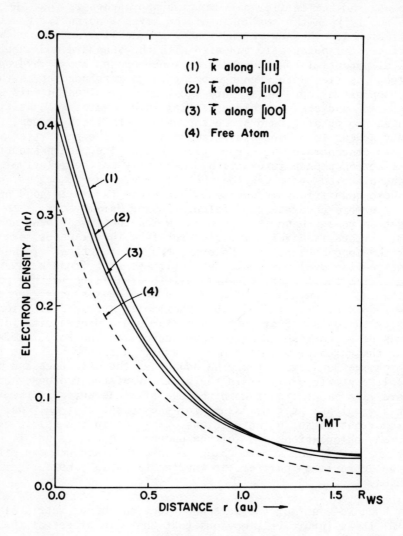

Figure 1: Spherically averaged electron density as a function of distance from the proton for three different wavevectors, on the Fermi surface along the [100], [110] and [111] axes. Also shown is the free atom electron density.

SUPERCONDUCTIVITY IN METALLIC HYDROGEN

$$M_{\vec{k}+\vec{Q},\vec{k}} = i \sum_\alpha A_\alpha [I_\alpha(\vec{k}+\vec{Q},\vec{k}) - Q_\alpha \frac{W(\vec{Q})}{\epsilon(\vec{Q})} + \frac{v(\vec{Q})}{\epsilon(\vec{Q})} \sum_\kappa \tag{1}$$

$$\frac{n_{\vec{\kappa}} - n_{\vec{\kappa}+\vec{Q}}}{E_{\vec{\kappa}} - E_{\vec{\kappa}+\vec{Q}}} I_\alpha(\vec{\kappa}+\vec{Q},\vec{\kappa})] - (E_{\vec{k}} - E_{\vec{k}+\vec{Q}}) \sum_\alpha A_\alpha \int_{V_o} \psi^*_{\vec{k}+\vec{Q}} \nabla_\alpha \psi_{\vec{k}} \, d\vec{r}$$

where we are working in the extended zone scheme, $\vec{A}$ is the phonon amplitude vector, $\epsilon(\vec{Q})$ the dielectric function, $v(\vec{Q})$ the Fourier transform of the electron-electron interaction, $W(\vec{Q})$ the Fourier transform of a potential which is constant inside the muffin-tin sphere V_o and outside is equal to the electrostatic potential due to the total charge inside V_o. $I_\alpha(\vec{k}',\vec{k})$ is given by a surface integral involving the wave functions over the muffin-tin shpere,

$$I_\alpha(\vec{k}+\vec{Q},\vec{k}) = \frac{1}{\Omega} \frac{h^2}{2m} \int_{S_o} ds \{ \nabla_n \psi^*_{\vec{k}+\vec{Q}} \nabla_\alpha \psi_{\vec{k}} - \psi^*_{\vec{k}+\vec{Q}} \nabla_n \nabla_\alpha \psi_{\vec{k}} \} \tag{2}$$

where Ω is the unit cell volume and ∇_n signifies a derivative normal to the surface. In Eq. (1), we have neglected the off-diagonal components of the dielectric screening function which give rise to the local field corrections. Again, this is in the spirit of previous calculations [3-5] which have assumed free-electron screening. We shall discuss local field effects later. For now, we are concerned with the effect of the renormalized electron-phonon matrix elements. (Note, however, that neglecting the effect of the atomic parts of the $\psi_{\vec{k}}$ in the electron-phonon matrix element itself would lead to a large error in this quantity because of the large value of V in these regions.) The formalism in Ref. (9) was set up so that the $\epsilon(\vec{Q})$ in Eq. (1) screens not the bare proton potential but the much weaker potential $W(\vec{Q})$ so that the objections to using linear screening theory do not arise in this case. This is because Schrödinger's Equation has been used to obtain the principal effects of ∇_V on the wavefunctions, rather than linear response. Physically, the effects of ∇V are rigid displacements of the charge density with the protons, while $W(\vec{Q})$ causes further deformations.

We now use the plane wave representation of the ψ_k at the muffin tin radius, and Eq. (2) simplifies to

$$I_\alpha(\vec{k}+\vec{Q},\vec{k}) = k_\alpha(E_{\vec{k}} - E_{\vec{k}+\vec{Q}}) \frac{V_o}{\Omega} G(Qr_{MT})(1+\alpha) \tag{3}$$

where $G(x) = 3(\sin x - x\cos x)/x^3$, r_{MT} is the muffin-tin radius, and $(1+\alpha)$ is a renormalization factor for the plane wave part ψ_k due to the "sucking in" of charge near the protons (see Fig. 1). From our calculations, $\alpha \simeq -0.025$ for the fcc phase with $r_s = 1.64$ a.u.

Substituting this into Eq. (1), it may be shown to reduce to

$$M_{\vec{k}+\vec{Q},\vec{k}} = -i \sum_\alpha A_\alpha Q_\alpha \frac{W_{eff}(\vec{Q})}{\varepsilon(\vec{Q})} - (E_{\vec{k}} - E_{\vec{k}+\vec{Q}})$$

$$\sum_\alpha A_\alpha [\int_{V_o} \psi_{\vec{k}+\vec{Q}} \nabla_\alpha \psi_{\vec{k}} \, d\vec{r} - ik_\alpha \frac{V_o}{\Omega} G(Qr_{MT})(1+\alpha)] \quad (4)$$

where

$$W_{eff}(Q) = W(Q) - \frac{4\pi e^2}{Q^2} Z^* \frac{V_o}{\Omega} G(Qr_{MT}) \quad (5)$$

$$W(\vec{Q}) = -\frac{4\pi e^2}{Q^2} Z^* (1 - \frac{V_o}{\Omega}) \frac{\sin Qr_{MT}}{Qr_{MT}} \quad (6)$$

and Z^* is the effective proton charge, $= Z(1+\alpha)$. The second term in Eq. (4) does not contribute to the electron-phonon coupling constant λ since only transitions across the Fermi surface are involved. Thus $W_{eff}(\vec{Q})/\varepsilon(\vec{Q})$ acts as an effective pseudopotential form factor between plane wave states from which λ may be calculated. In Fig. 2 we plot this form factor as a function of $\vec{Q}$ and compare it with the Fourier transform of the screened Coulomb potential. Note that for larger values of $\vec{Q}$, the magnitude of the former is considerably smaller than that of the latter. This will considerably decrease the value of λ over that of previous estimates, since the larger Q values of the pseudopotential dominate in the expression for λ, which is given, for a spherical Fermi surface by

$$\lambda = \frac{m}{16\pi^3 Mnk_F} \int_{Q<2k_F} \frac{d\vec{Q}}{Q} \left(\frac{W_{eff}(\vec{Q})}{\varepsilon(\vec{Q})}\right)^2 \sum_j (\vec{\varepsilon}_{Qj} \cdot \vec{Q}/\hbar\omega_{Qj})^2 \quad (7)$$

where m is the electron mass, M the proton mass, n the electron density, k_F the Fermi wavevector, and $\vec{e}_{\vec{Q}j}$ and $\omega_{\vec{Q}j}$ are the phonon

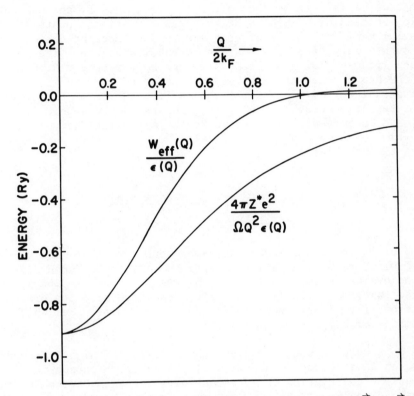

Figure 2: The screened effective pseudopotential $W_{eff}(\vec{Q})/\epsilon(\vec{Q})$ vs. $\vec{Q}$ shown as the full curve. The lower curve represents the screened Coulomb potential $-4\pi Z^*/(\Omega Q^2 \epsilon(Q))$.

eigenvector and frequency for the mode j at wavevector Q in the repeated zone scheme.

The phonon frequencies and eigenvectors may be calculated by the formalism developed in Ref. (9). With the approximations made in deriving Eq. (4), they may be shown to be obtained by the standard expressions[10] with $W_{eff}(Q)$ playing the role of the pseudopotential form factor and an effective charge Z_* on the ions. The dispersion curves showed no instability at $r_s = 1.64$ as found in Ref. 4, a result we believe to be due to the incorrect use of the Coulomb interaction instead of $W_{eff}(Q)$. Neither was any strong quantum crystal behavior manifest, the root mean square displacement coming out to be roughly only 9% of the nearest neighbor distance. The maximum phonon frequency ν_o at this density was approximately 81.4×10^{12} c/s and the Debye temperature θ_o was evaluated to be 2850°K. λ was evaluated from Eq. (7) and the phonon frequencies by performing the integration over a mesh of 8000 points in the sphere of radius $2k_F$, and convergence with regard to the mesh size was verified. The value of λ came out to be 0.261. The value of T_c was calculated using the McMillam formula[11]

$$T_c = 0.62 \frac{h\nu_o}{k} \exp[- \frac{1.04(1 + \lambda)}{\lambda - \mu^*(1 + 0.62\lambda)}] \quad (8)$$

where μ^* is the usual Morel-Anderson pseudopotential[12] describing the Coulomb repulsion between electrons. At the assumed density, μ^* was evaluated to be 0.116, and T_c was obtained as 0.08 K, in marked contrast to earlier estimates. We have also performed calculations for higher densities of metallic hydrogen for which the estimates of T_c go down ever further.

We thus conclude that previous estimates of high T_c's were based on an incorrect evaluation of the electron-phonon matrix element by calculating the gradient of a screened Coulomb potential between plane waves states. Not only is this based on an invalid use of linear screening theory for 1/r potential, but it ignores the effect of the "bound" portion of the hydrogenic wavefunctions near the protons. For instance, if one used the screened Coulomb form factor shown in Fig. 2 instead of $W_{eff}(\vec{Q})$ between plane waves, one obtains $\lambda = 0.92$.

The one APW approximation used in Eq. (3), however, is probably not a very accurate one, especially since second derivatives of the wavefunction at r_{MT} are involved. This is also borne out by the fact that the expression for $I_\alpha(\vec{k}',\vec{k})$ for $\vec{k}'$, $\vec{k}$ on the Fermi surface given by Eq. (3) vanishes identically, whereas the phase-shift method[6] yields sizeable contributions arising from the large s-wave phase shifts at E_F. We accordingly examine how screening effects may be approximately incorporated using the phase-shift approach and including local field corrections.

A rigorous formulation for λ in terms of angular momentum decomposition of the wavefunctions and including local field corrections is given by Sinha and Harmon[13] in these proceedings in Eq. (4.1) of that paper. The effect of screening is included in the two $(1 + VN)^{-1}$ factors appearing in that expression. It is of course difficult to estimate these factors without a rather elaborate calculation, but preliminary estimates indicate a reduction of a factor of at least three in λ as a result of screening. If we consider Papaconstantopoulos and Klein's[6] or Switendicks's[14] value of ~2.7 for λ, allow for screening and for a further decrease of λ due to an $<\omega^2>^{\frac{1}{2}}$ of 2000K as estimated from our phonon frequencies rather than the value of 1316K used by these authors,[15] we find an overall decrease of λ to roughly 0.45, and T_c to be reduced to about 20K. Thus we believe that, while the question of T_c in metallic hydrogen is still not resolved satisfactorily on a microscopic level, we should expect it to be at least revised downwards rather drastically from previous estimates.

We wish to acknowledge useful discussions with B. N. Harmon.

REFERENCES

1. D.E. Ramaker, L. Kumar, and F.E. Harris, Phys. Rev. Letters 34, 812 (1975); F.E. Harris, L. Kumar and H.J. Monkorst, Phys. Rev. 87, 2850 (1973).
2. E.G. Brovman, Yu Kagan and A. Kholas, Zh. Ekso. Teor. Fiz. 61, 2429 (1971); 62, 1492 (1972) [Sov. Phys. - JETP 34, 1300 (1972); 35, 783 (1972)].
3. T. Schneider, Helv. Phys. Acta 42, 957 (1969).
4. L.G. Caron, Phys. Rev. B9, 5025 (1974).
5. N.W. Ashcroft, Phys. Rev. Letters 21, 1748 (1968); J. Hammerberg and N.W. Ashcroft, Phys. Rev. B9, 409 (1974).
6. D.A. Papaconstantopoulos and B.M. Klein, Proceedings of Conference on Soft Modes, Superconductivity and Ferroelectricity, Univ. of Puerto Rico (1975).
7. P. Bhattacharya and K.S. Singwi, Phys. Rev. Letters 29, 22 (1972).
8. P. Vashishta, P. Bhattacharya and K.S. Singwi, Phys. Rev. B10, 5108 (1974).
9. S.K. Sinha, Phys. Rev. 169, 477 (1968).
10. L.J. Sham, Proc. Roy. Soc. (London) A283, 33 (1965); W.A. Harrison, Phys. Rev. 129, 2503 (1963); 129, 2512 (1963).
11. W.L. McMillan, Phys. Rev. 167, 331 (1968).
12. P. Morel and P.W. Anderson, Phys. Rev. 125, 1263 (1962).
13. S.K. Sinha and B.N. Harmon, Proceedings of this conference.

14. A.C. Switendick, Proceedings of this conference.
15. The authors of reference 6 use a Debye Θ obtained from the calculations of Caron[4]. However, as discussed previously Caron's treatment bases as it is on a free-electron model, overestimates the renormalization due to the electron-phonon interaction. Thus we believe that the true phonon frequencies are likely to be quite a bit higher.

QUESTIONS AND COMMENTS

Discussion of this paper appears after that of Switendick.

ELECTRON-PHONON INTERACTION AND SUPERCONDUCTIVITY IN

METALLIC HYDROGEN*

A.C. Switendick

Sandia Laboratories

Albuquerque, New Mexico 87115

The energy bands of face-centered-cubic hydrogen at a r_s value of 1.64 a.u. were calculated. This corresponds to a density of 3.65×10^{23} hydrogen atoms/cm^3 and would require pressures of the order of a megabar to achieve. We have calculated the electron-phonon matrix element using the formalism of Gaspari and Gyorffy and parameters derived from our calculations. Using various theoretical estimates of the phonon properties of metallic hydrogen we have calculated the electron-phonon coupling constant or mass enhancement factor, λ, and the superconducting transition temperature. Those calculations indicate that this system would be a superconductor with a transition temperature in excess of 200 K.

INTRODUCTION

Interest in the properties of hydrogen atoms at high densities goes back over 40 years [1]. This interest renews itself periodically [2] as the computational ability and experimental technique increase and at present seems to have a sizable positive derivative. Previous treatments [3,4] have largely been perturbative approaches based on a free-electron gas. Our approach is a nonperturbative one which has found useful application in the treatment of metal hydrides [5]. Although we cannot agree with the analogy [6] made

---

*Research prepared for the U.S. Energy Research and Development Admininstration under contract AT(29-1)-789.

of these materials to metallic hydrogen we do feel that our techniques are capable of correctly treating many aspects of hydrogen in lattices. For the purpose of this communication we shall leave aside questions of (meta) stability or achievability and concentrate on the detailed electronic character and its implications on superconductivity for this system. Somewhat out of context for this meeting, we shall see that this system is truly an s-p superconductor. Similar conclusions have been reached by Papaconstantopoulos and Klein [7] using a similar approach. We shall concentrate on the comparison with the free electron gas. An inclusion of "screening corrections" in this system leads to different physical conclusions [8] from those of this work. Their inclusion does not seem to be warranted by the current level of predictive capability of the theory of electron-phonon coupling [9] nor is the magnitude and manner of inclusion well established. We shall return to this point in our discussion.

ENERGY BAND CALCULATION

For a postulated face-centered-cubic structure with a cubic lattice constant of 0.419656 nm the energy bands of metallic hydrogen were calculated using standard augmented plane wave (APW) energy band programs [10]. This spacing corresponds to a Wigner-Seitz sphere radius, $R_{WS} = r_s$, of 1.64 atomic units and a density of 3.65×10^{23} hydrogen atoms/cm^3. The potential was derived from a superposition [11] of neutral hydrogen atom charge densities and gave 0.82 electrons inside the APW sphere radius, R_S, of 1.484 a.u. as compared with 0.57 electrons inside the free atom. Slater exchange [12] was used for the crystalline exchange potential. The muffin-tin zero potential corresponds to -1.354 Ry on the free atom scale and results in a discontinuity of 0.053 Ry at the APW sphere indicating slight non-muffin-tin corrections.

The energy bands are shown in Figure 1. Although they are very wide and appear quite free electron-like, there are several differences. The bottom of the energy bands at Γ is over one-half a Rydberg below the muffin-tin zero. The degenerate states along the directions W-X, L-W, W-K are similarly split by over half a Rydberg. It is as if the whole bottom band had been lowered by this amount. In fact, detailed comparison of our energy bands for metallic hydrogen shows that up to the Fermi energy our energy bands are given by:

$$E = k^2 - 0.59 \pm 0.04 . \tag{1}$$

While in the region around the free electron Fermi energy of

$$E_F^0 = k_F^2 = \left(\frac{9\pi}{4}\right)^{1/3} \frac{1}{R_{WS}} , \tag{2}$$

METALLIC HYDROGEN

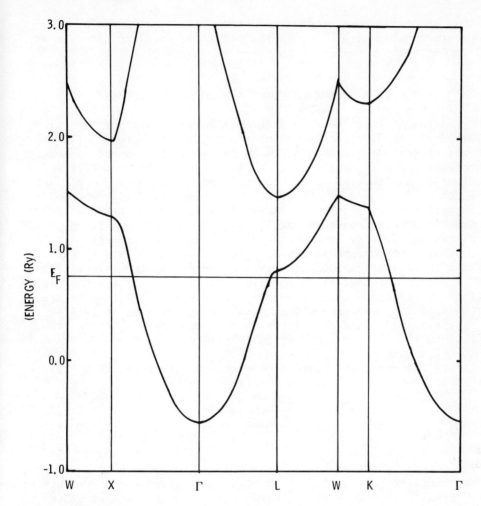

Figure 1: Energy bands for face-centered-cubic hydrogen corresponding to $R_{WS} = 1.64$.

the metallic hydrogen Fermi energy is given by

$$E_F = E_F^0 - 0.62 \pm 0.01 = 0.75 \pm 0.01 \text{ Ry}. \qquad (3)$$

We shall in fact rely on this similarity to determine the total density of states for our metallic hydrogen to be 0.55 spin states/ Ry cell. This approximation results in an error of less than 1% for this value. As the subsequent discussions and comparisons will show, we cannot however use (adjusted) free electron values for our other parameters. Our energy bands for metallic hydrogen are in substantial agreement with those of others [7,13].

ELECTRON-PHONON COUPLING AND SUPERCONDUCTIVITY

Although there have been several formulations for the calculation of the electron-phonon mass enhancement factor, λ [14-16]. Until recently only those based on free-electron, pseudo-potential approaches had found any applicability. The cumbersomeness and calculational *tour de force* required had prevented any application of formalisms based on APW or KKR techniques [14, 15], whose appropriateness for transition metal-based systems has however adequately been demonstrated [17,18]. It was not until a paper by Gaspari and Gyorffy [19] which provided a calculationally simple procedure that one could consider these systems. This approach found rapid application and relative success in the calculation of λ and with the McMillan formula [20] or variants thereof [21] of the estimation of the superconducting transition temperature, T_C. Application has been made to a variety of elemental [22] transition metal and transition metal compounds [23,24] and their pressure [25,26] and composition [27] dependence.

In the rigid ion, strong coupling theory of McMillan [20], the electron-phonon coupling constant is given be

$$\lambda = \frac{n(E_F) \langle I^2 \rangle}{M \langle \omega^2 \rangle} . \qquad (4)$$

The numerator subsequently denoted by η is given by the product of the density of states at the Fermi energy and the square of the electron-phonon matrix element averaged over the Fermi surface. The renormalized phonon frequency, $\langle \omega^2 \rangle^{1/2}$, contribution to the denominator has yet to be calculated in any realistic fashion for transition metals and compounds although steps [28] are being made in this direction within the same framework as that used to calculate the numerator. Experimental or model results are usually used to obtain these phonon averages.

The result given for η by Gaspari and Gyorffy is

$$\eta = \frac{2\varepsilon_F}{\pi^2} \frac{1}{n(\varepsilon_F)} \sum_e \frac{(\ell+1)\sin^2(\delta_{\ell+1}-\delta_\ell) n_\ell n_{\ell+1}}{n_\ell^1 n_{\ell+1}^1} \qquad (5)$$

in atomic units, where ε_F and $n(\varepsilon_F)$ are the Fermi energy and density of states at the Fermi, respectively. δ_ℓ is the phase shift of the potential, n_ℓ is the ℓ component of the density of states and n_ℓ^1 is the single site density of states; all determined at the Fermi energy. In contrast to more complicated formulas usually given [22, 29] n_ℓ^1 is quite simply given be [30]

$$n_\ell^1 = -\frac{\sqrt{E}}{\pi}(2\ell+1)R^2 J_\ell^2(R)\frac{\partial \gamma_\ell}{\partial E} \quad (6)$$

where γ_ℓ is the logarithmic derivative of the ℓth interior solution J_ℓ with matching exterior solution

$$J_\ell(R) = \cos\delta_\ell j_\ell(\sqrt{E}R) - \sin\delta_\ell n_\ell(\sqrt{E}R) \quad . \quad (7)$$

All these quantities are easily accessible from our potential and band structure calculation for metallic hydrogen and can be compared with a free electron calculation.

As is the usual in most band structure calculations, one obtains eigenvalues, and eigenfunctions on a small finite set of $\vec{k}$ points. To obtain statistically meaningful densities and component densities of states one has to resort to some sort of interpolation procedure. As stated above we feel the total density of states is given by the (shifted) free electron value. To obtain the component densities of states we followed a prescription based on smoothness arguments and analogy with the free electron results. In Figure 2 we show the free electron single site scattering densities of states, n_ℓ^1, determined from our APW programs and equation (6). Their sum is n_t^1 the total single site density of states which differs from the free electron density of states $\pi/2(a/2\pi)^3$ just by the ratio of the APW sphere volume to the Wigner-Seitz sphere volume since there are no band structure (multisite) effects. We also show in Figure 3 the single-site scattering densities of states for our metallic hydrogen potential. Since there are no bound states in the potential, these quantities are undefined below the muffin tin zero. The phase shifts are also straightforwardly obtained and are plotted in Figure 4. Note the extremely large ($\delta_0 \sim \pi/2$) s-wave scattering characteristic of the hydrogen potential and very little p-wave scattering. Recall that the phase shifts are all zero in the free electron case.

The only missing ingredients are the component densities of states, n_ℓ, for the case of metallic hydrogen, for which we need the interpolation procedure alluded to above which we now detail.

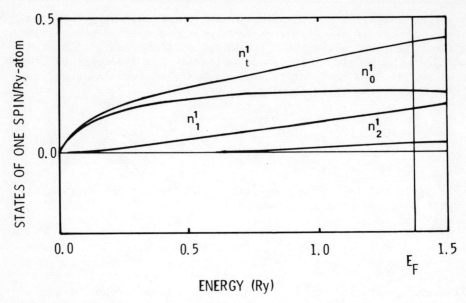

Figure 2: Single site component densities of states for free electrons.

In the case of free electrons

$$n_\ell^0 = n_t^0 \, f_\ell^0 \tag{8}$$

where f_ℓ^0 is the fraction ℓ character/state, which has both energy and R dependence. Following John [31] we shall choose $R = R_{APW}$. This leads to physically correct ratios n_ℓ/n_ℓ^1 but to a slight overall trouble in normalization due to $\Sigma \, n_\ell \neq n(E_F)$. This leads to the exact result for plane waves

$$n_\ell(R_{APW}) = f_\ell^0(R_{APW}) \, n_t \tag{9}$$

In Figure 5 we show $f_\ell^0(R_{APW})$ for plane waves which is just

$$\propto (2\ell + 1) \int_0^{R_{APW}} j_\ell^2(\sqrt{E}r) \, r^2 \, dr, \tag{10}$$

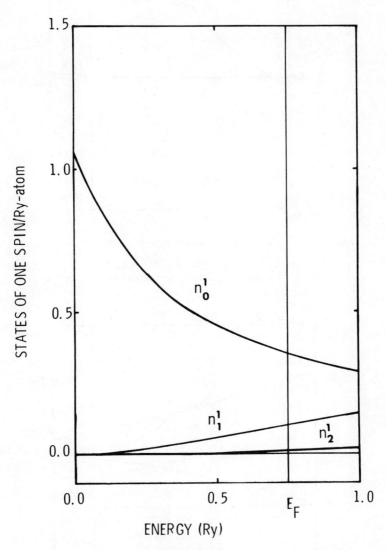

Figure 3: Single site component densities of states for metallic hydrogen.

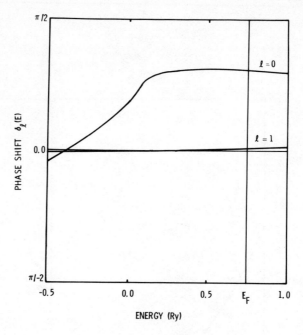

Figure 4: Phase shifts for metallic hydrogen.

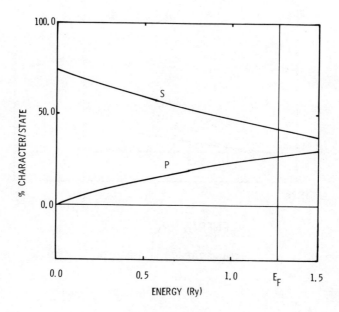

Figure 5: State component charge character for free electrons.

the amount of ℓ-like charge associated with the plane wave inside the APW sphere with the total normalized to one in the cell. It is a smooth function of the energy with comparable parts of s and p character for one electron/cell. This provides the motivation for our determination n_ℓ for the metallic hydrogen case. In Figure 6 we plot $f_\ell(R_{APW})$ for this case, as determined from our discrete $E(k)$ calculation. All of the results fall very near the smooth curve except at zone boundary points where there are symmetry restrictions. At the Fermi energy this interpolation should be very good to give

$$n_\ell = f_\ell n_t \simeq f_\ell n_t^0 \qquad (11)$$

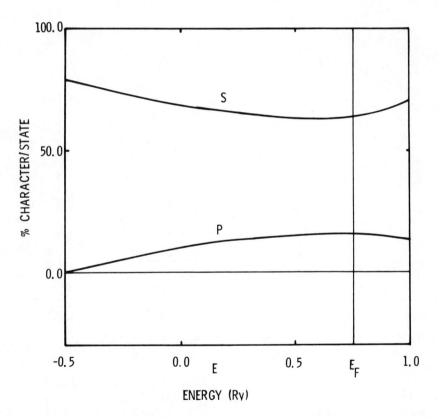

Figure 6: State component charge character for metallic hydrogen.

Figure 6 shows that the s-like character is 62% while the p-like character is 11% compared with the free electron values of 40% and 20%, respectively, at the Fermi energy.

The results are summarized in Table I. The value of η calculated is 7.56 eV/A^2 which is to be compared with values of 2-10 observed [22-25] for the transition metals and transition metal

TABLE I

r_s(a.u.)	1.64
E_F(Ry)	0.747
$n(E_F)$ states/atom Ry-spin	0.548
n_s	0.345
n_p	0.088
n_s/n_s^1	0.980
n_p/n_p^1	0.898
δ_s	0.956
δ_p	0.028
η	7.56 eV/A^2
Θ_D	1782
$<\omega^2>^{\frac{1}{2}}$	1260
λ	2.65
T_c	250

compounds. This value is probably correct to 10% within the rigid ion approximation.

The final quantity indicated in Eq. (4) is $<\omega^2>$ for which we have used the theoretical results of Carọn [4] for the Debye temperature Θ_D = 1782 K and taken $<\omega^2>$ = $\Theta_D^2/2$ to yield 1260 K. Since $<\omega^2>$ occurs in the denominator of λ and Θ_D as a factor in T_c, increasing Θ_D by 50% actually increases T_c by 40% while decreasing it by 50% decreases T_c by 25 %. The phonons are not crucial because of the low mass and our value of η.

DISCUSSION AND CONCLUSION

The value of η calculated within the rigid ion approximation of Gaspari and Gyorffy is numerically very reasonable. The s-like density of states is very comparable to that calculated for hydrogen in PdH by Papaconstantopoulos and Klein [32] as are other parameters relating to the hydrogen site. The major difference being that ~ 60% of the total density of states is associated with the hydrogen while for PdH it is only 10% of the total. This, coupled with a factor of 2 in the n_p/n_p^1 ratio for the metallic hydrogen, accounts for the difference in η's for the two systems.

This result coupled with the fact that λ_H in PdH for the same $\mu^* = 0.10$ as used in our hydrogen calculation gives a calculated T_c of 25 K in contrast to the experimental value of ~ 10 K. This indicates that λ_H might be over estimated by 50% due to inaccuracies in the results or due to corrections beyond the rigid ion and one-electron approximations. Corrections yielding factors of 5-10 do not seem warranted. The physical simplicity and reasonable success of the Gaspari-Gyorffy approach coupled with this type of uncertainty strongly indicates that metallic hydrogen of these densities and reasonable phonon properties would be a high-temperature superconductor as indicated in Table I. Entirely comparable values and conclusions for other densities and lattices have been reported by Papaconstantopoulos and Klein [7].

REFERENCES

1. E. Wigner and H.B. Huntington, J. Chem. Phys. 3, 764 (1935).
2. H. Beck and D. Strauss, Helvetica Physica Acta, 48, 655 (1975) for recent work and a fairly comprehensive reference list.
3. N.W. Ashcroft, Phys. Rev. Lett. 21, 1798 (1968).
4. L.G. Caron, Phys. Rev. B9, 5025 (1974).
5. A.C. Switendick, Int. J. Quantum Chemistry 5, 459 (1971).
6. B. Baranowski, T. Skoskiewicz, and A.W. Sxafranski, Fizika Nizkikh Temperatur 1, 616 (1975).
7. D. Papaconstantopoulos and B. Klein, Conf. Proceedings Low-Lying Lattice Vibrational Modes and their Relationship to Superconductivity and Ferroelectricity, to be published.
8. R. Gupta and S.K. Sinha, this conference.
9. I.R. Gomersall and B.L. Gyorffy, Phys. Rev. Lett. 33, 1286 (1974).
10. L.F. Mattheiss, J.H. Wood, and A.C. Switendick, Methods in Computational Physics (Academic Press, New York) Vol. 8.
11. L.F. Mattheiss, Phys. Rev. 133, A1399 (1964).

12. J.C. Slater, Phys. Rev. 81, 385 (1951).
13. F.E. Harris, L. Kumar, and H.J. Monkhorst, Phys. Rev. B7, 2550 (1973).
14. D.C. Golibersuch, Phys. Rev. 157, 532 (1967).
15. S.K. Sinha, Phys. Rev. 169, 477 (1968).
16. P.B. Allen and M.L. Cohen, Solid State Comm. 7, 677 (1969), J.M. Ziman, Phys. Rev. Lett. 8, 272 (1962).
17. T.L. Loucks, "Augmented Plane Wave Method", Benjamin, N.Y., 1967.
18. J.O. Dimmock, Solid State Phys. 26, 103 (1971).
19. G.D. Gaspari and B.L. Gyorffy, Phys. Rev. Lett. 28, 801 (1972).
20. W.L. McMillan, Phys. Rev. 167, 331 (1968).
21. R.C. Dynes and J.M. Rowell, Phys. Rev. B11, 1884 (1975). These authors indicated the McMillan formula may underestimate T_c. In view of the uncertainties in phonon averages this result should be adequate for our purposes.
22. R. Evans, G.D. Gaspari, and B.L. Gyorffy, J. Phys. F3, 39 (1973).
23. I.R. Gomersall and B.L. Gyorffy, J. Phys. F. 4, 1204 (1974).
24. B.M. Klein and D.A. Papaconstantopoulos, Phys. Rev. Lett. 32, 1193 (1974).
25. R. Evans, V.K. Ratti, and B.L. Gyorffy, J. Phys. F 3, L199 (1973).
26. V.K. Ratti, R. Evans, and B.L. Gyorffy, J. Phys. F 4, 371 (1974).
27. A.C. Switendick, Bull. Am. Phys. Soc. 20, 420 (1975).
28. W.E. Pickett, this conference.
29. K. Schwartz and P. Weinberger, J. Phys. C 8, L573 (1975).
30. P.W. Anderson and W.L. McMillan in "Theory of Magnetism in Transition Metals", Proceedings of the International School of Physics, "Enrico Fermi" edited by W. Marshall (Academic, New York, 1967) and Eqs. 2.16, 2.20.
31. W.L. John, J. Phys. F 3, L231 (1973).
32. D.A. Papaconstantopoulos and B.M. Klein, Phys. Rev. Lett. 35, 110 (1975).

QUESTIONS AND COMMENTS

P. Vashishta: [To Switendick] The value of Θ_D is 1780°, right?

A. Switendick: Yes.

P. Vashishta: And in the previous calculation, 2800°, right?

A. Switendick: I think that our r_s values might have been different, I am not sure.

R. Gupta: It is for the same value of r_s.

P. Allen: The same value of r_s. We ask for a comparison of the η's and the λ. What is your η?

R. Gupta: I didn't calculate η.

P. Allen: You must have.

R. Gupta: I mean, we calculated λ without separating it as $\eta/M\langle\omega^2\rangle$. I can make a point here. The value of Θ_D which we have used in this calculation was based on the phonon frequencies which came out of this model. Whereas in the other case, the Θ_D is based on a calculation using a screened e^2/r potential which Caron used, so that certainly makes a difference.

A. Switendick: I don't think the difference in Θ_D is going to make that much difference in the transition temperature.

B. Klein: Θ_D occurs both in the denominator of λ (as a square), and as a multiplier in the T_c equation; so that small changes in Θ_D tend to compensate at least as far as T_c is concerned. Therefore, T_c might oscillate between 150°K and 250°K, but not between one and 250°K due to the different Θ_D's. There is a discrepancy in the calculations of η well beyond the different Θ_D's.

P. Vashishta: What do you mean by high temperature?

B. Klein: Metallic hydrogen is a room temperature superconductor, but the room might either be in Alaska or the Sahara.

S. Sinha: I think the main purpose of most of the work Dr. Gupta presented was to show that if you use plane waves as people did to calculate λ for hydrogen, and you get a large T_c, such as a 100° superconductor, and that you cannot believe the result. If one uses plane waves, but taking into account the core-like part of the wavefunction, one gets a very low-temperature superconductor. Now obviously a better way to do it is to use Gaspari-Gyorffy methods but now you have to worry about what ω_D, and what $\langle\omega^2\rangle$ to put in and how much the screening decreases λ. If it is a factor of 2 or so, you do not decrease T_c too much, but if you come down to a λ of about 0.4 or 0.3, go down by more than a factor of 5, then your T_c rapidly approaches zero. So I think it is really an open question. I am not saying that the single-APW calculation is good. We just don't know, yet.

B. Matthias: I don't want to destroy an engaging picture but there is a recent measurement by Velashagin below 10° where the hydrogen does become metallic but not superconducting.

DIRECT MEASUREMENT OF $\alpha^2 F$ IN NORMAL METALS USING POINT-CONTACTS:

NOBLE METALS*

A.G.M. Jansen, F.M. Mueller and P. Wyder

Physics Laboratory and Research Institute for Materials
University of Nijmegen

Toernooiveld, Nijmegen, the Netherlands

ABSTRACT

A new technique of forming tiny point junctions, first discussed by Sharvin, is described. By measuring the second derivative of voltage with respect to current, using techniques similar to tunnel junction spectroscopy, structure is found which is consistent with bulk phonon densities of states derived from neutron scattering. The same results were reported at LT 14 by Yanson using a shorted film technique.

Firstly an explanation of the title change from "can theorists calculate T_c?" to the present one: in my view previous speakers through out this conference have adequately answered this question ---so I present my alternate talk---an experimental one.

---

* This work was performed as part of the research program of the "Stichting voor Fundamenteel Onderzoek der Materie" (FOM) with financial support from the "Nederlandse Organisatie voor Zuiver Wetenschappekijk Onderzoek" (ZWO).

Now the history of the relationship of phonon spectra to transport in metals is a long one - so I can only deal with high-lights. We have all read "Conductivity of Monovalent Metals" by J. Bardeen [1] and the earlier papers of Bloch [2] and particularly Peierls [3] where the importance of "Umklapprozesse" was first stressed. Recent attention has shifted to the measurement of phonon spectra, particularly the relationship between electronic structure, phonon spectra and superconductivity [4]. This has certainly been true of this conference. (I do not give references.) Neutron scattering, tunnel junction spectroscopy, and transport properties have all been mentioned. I was surprized to see no further work on the temperature dependence of electronic masses as measured, for example, by Goy [5] and co-workers and predicted by Phil Allen [6] and Grimvall [7] presented here.

At Nijmegen, Harold Myron and myself have recently [8] taken up again the question first posed by Fowler and Prange [9] of the magnetic field dependence of orbital masses in big magnetic fields. As many of you may know the National Magnet Laboratory is building some big magnets for Nijmegen. We hope that when these are complete, experimentalists may be stimulated to measure such effects again. In 1971 Palin [10] had already seen big hints of such effects at Cambridge in Hg at low fields.

At Jülich, Werner Schilling and co-workers have been using diffuse X-ray scattering to measure defect structures. C.B. Walker [11], in 1956, was able to turn these into one of the first complete presentations of the phonon spectra of Al. I have wondered aloud with Schilling whether now-a-days this can be done on systems more complicated than one-atom-per-unit-cell. As a theorist and a computational physicist, I am stimulated to try, especially in view of the rather complete data, spanning many Ewald spheres, Schilling has automatically available.

So there are already a number of methods.

What I wish to present this afternoon as briefly as possible are some experiments being carried out at Nijmegen by my two colleagues Louis Jansen and Peter Wyder using a new technique. I might just mention that Jansen is a new student performing, as far as I know, his first experiment - so he is young and energetic and able to work the whole night through (which turns out to be the best time). He is also from the province of Brabant - my European colleagues will understand the significance of that.

Let me begin with a personalized history of point-contacts. In 1965 Sharvin (a Soviet theorist) proposed [12] that by measuring the longitudinal magneto-resistivity of two such point contacts, information could be found on the Fermi surface of pure metals.

DIRECT MEASUREMENT OF $\alpha^2 F$

Fig. 1 illustrates the geometry. The idea is that selected electrons are focused between the contacts as a function of the external magnetic field. This gives information (as discussed by Sharvin) on the radius of curvature of the Fermi surface. Sharvin and Bogatina [13] have successfully carried out such experiments and illustrated the further dependence of the focusing effect on the mean free path, the temperature, and on the current density. The most important result for us here is Sharvin's formula for the resistance R_o of one point contact (slightly re-written):

$$R_o \approx p/e^2 D^2 N = \rho l/D^2 \qquad (1)$$

where ρ is the bulk resistivity, l is the mean free path and D is the diameter of the contact. The formula presumes that $l \gg D$ and so the situation resembles the effusion of a dilute gas through a small orifice - one is then in the extreme Knudsen regime.

Two consequences are important:
1. by making the contacts small, for even modest voltages one has raised current densities to a new and unusual height - 10^{10} to 10^{11} Ampères/cm^2;
2. one has done something normally impossible in "good" metals - made a large electric field. Now the first of these leads to the phenomena of electro-migration - of some interest as a production problem in the semi-conductor industry. But about this I know almost nothing. It is the second which is important for us here today.

In Karkov a Soviet experimentalist, I.K. Yanson, has been measuring [14] the resistance of shorted tunneling junction films - these certainly do not fulfill the Rowell criterion. In fact if they have not shorted spontaneously, he carefully shorts them with a burst of voltage, coupled through a high resistance. Peter Wyder, I believe, first became aware of Yanson's work when Yanson presented [15] it at LT 14, in July of 1975 as paper L.131 "Non-linear effects in electrical conductivity of point contacts and electron-phonon interaction in normal metals". Yanson has been working on this phenomena, as near as I can tell, off and on for about ten years.

What he has discovered [14] is illustrated in the second figure. As one goes from curve (a) to (d) the resistance of the point contact is systematically decreasing - so by the Sharvin formula the orifice diameter is increasing. Thus the phenomena is further and further from the Knudsen regime - and one sees the systematic build up of a "background effect". But let's focus for the rest of this talk only on the high resistive, "good" point contacts.

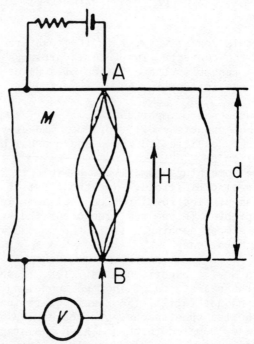

Figure 1: Experimental arrangement for observation of electron focusing in a metal. (From Sharvin and Bogatina, Ref. 13).

DIRECT MEASUREMENT OF $\alpha^2 F$

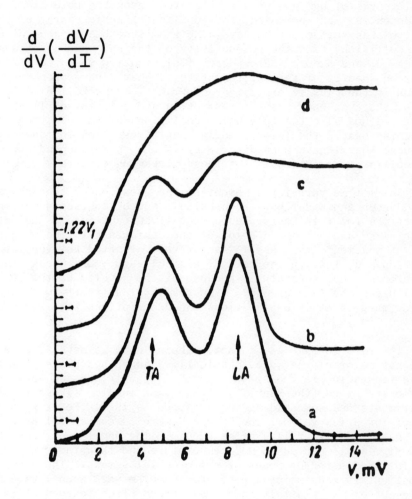

Figure 2: -As listed on figure - (From Yanson, Ref. 14)

The phenomena is clear - a systematic change from diffusive "resistance" to one of hydrodynamic streaming - brought about by the small orifice. But what about the structure? Such structure has been seen before in another experiment we all know - and is illustrated in Fig. 3. Here, slow electrons were introduced into a dilute gas of Hg, and were accelerated through a potential. When the potential was increased, the resultant kinetic energy of the electron was sufficient to have a Hg atom absorb a quantum of energy by a collision, then the conductivity of the gas chamber strongly decreased, and so on. Later Hertz [16] actually measured [17] the emission spectra of the excited states. What I have been describing to you is the classic experiment of Franck and Hertz [18]. It showed that not only was the absorption and emission of light quantized but also the (presumed) collisions of the electron gas streaming passed the (slowly moving) Hg atoms. Now in modern language what I have been describing is the energy dependence of the collision cross-section of those streaming electrons.

But let me turn for a moment to the question of the apparatus that Jansen and Wyder have been using. Now of course in our instructions for this conference our chairman cautioned us to "stick to physics" and not spend a lot of the conference's time on discussion of monuments to oneself---apparatus. I noted however, as I'm sure you did, that in the talk of his colleague, Dr. Ghosh, they chose to discuss apparatus - because of the novelty of the set up - and I plead the same excuse. I must also plead thatI am a theorist - so you must treat my explanations and descriptions with some caution.

In order to make clear what is being done we have adopted a somewhat colorful language. Basically a sharpened piece of metal whose tip radius is ~ ½µ (the "spear")is gently pressed into a stationary piece (the "anvil") of the same metal. (These first experiments use some old Ag and Au wires laying in Peter Wyder's drawer for a few years. The "copper" wire is commercial Brown and Sharp #10 wire rolled off a spool.) The anvil-spear ensemble (see Fig. 4) is cooled to He^4 temperature and then their separation carefully adjusted by means of differential screws. The final adjustment is made using a piezo-active element. When I gave [19] a pre-cursor of this talk to a small group of colleagues at La Jolla immediately after the December '75 San Juan conference, a student there, David C. Johnson, thought of the idea of a piezo-electric substage. So I wrote to Wyder about it. Meanwhile he and Jansen had also been working at Nijmegen and arrived at the same solution. He told me later when I got back to stick to thinking and collating my computer cards - and leave the experiments to them. So I now call the piezo-idea a collective discovery.

Anyway, technique has now improved to the point where we believe that we can control the spear-anvil separation to a few

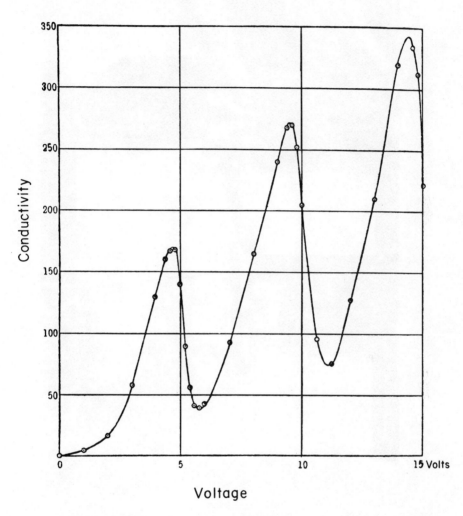

Figure 3: The differential conductivity as a function of voltage for electrons streaming in a dilute gas of mercury. (From Franck and Hertz, Ref. 18.)

Figure 4: The "spear" and "anvil" used to form point contacts. On the right is the mechanical separator - adjusted by the differential screws. On the left is the sub-assembly mounted on a piezo-electric substage. As discussed in the text, the final separation is controlled to about 10^{-6} cm by means of the small voltage adjustments on the piezo-electric element.

hundredths of a micron ($\sim 10^{-6}$ cm).

Now there is no doubt that a great many complicated things are happening inside those tiny junctions. However, by averaging the results over rather long times a consistent and reproducible picture emerges. (Working during the quiet part of the night helps, too.)

In Fig. 5a is given the voltage dependence of the resistance ($\equiv dV/dI$) and in Fig. 5b the current derivative. Because of signal averaging requirements, it takes about 2-3 hours to go from the left part of the tracing to the right. These data were taken on the night of April 23-24, 1976, and are photographs of direct experimental out-puts plotted on a x-y-recorder. The second derivative was derived by the usual second-harmonic and lock-in techniques-familiar from tunneling junction spectroscopy. I would like to make two points:

1. The resistance 5a is symmetric and has a slight positive curvature about the zero point. Based on the symmetry and curvature, we conclude that we are actually measuring a <u>resistive</u> phenomena and not, for example, tunneling [20] through an adhered oxide (which, we believe, would yield a shifted and inverted parabola). The same results were seen by Yanson [14, 15] in his experiments.
2. Some of the structure in Fig. 5b is reproducible and some is an artifact due to unknown causes. On a vertical scale of ca.10% (which is a measure of our accuracy in the second derivative), the large and anti-symmetric peaks near ± 17 meV and the smaller peaks near ± 27 meV are reproducible - and we believe significant. The smaller structure (at higher (lower) voltages and) near to the origin and the narrow dip at + 17.5 meV we interpret as noise. Again the same results were seen by Yanson.

We have, of course, no independent check at this time of the size of our little contacts, other than the Sharvin formula. Yanson however, has found that by using the superconductivity of his films, he "finds a result consistent with the estimate". In addition, the phenomena is complicated by the extreme conditions of the experiment. Now, Ap van Gelder and I <u>have</u> been discussing non-linear Boltzmann equations and all that. The difficulty is that for sure - if you can use the concepts - the electronic and phononic distribution functions are changing on a scale of the size of the orifice. And the Peltier effect must play a significant role. But that's a problem for theorists - at least for us - for the next year or so.

I would like to return to the simple picture of those streaming electrons with excess kinetic energy seeking a channel, a coupling in which to dump their excess kinetic energy. Electron-electron

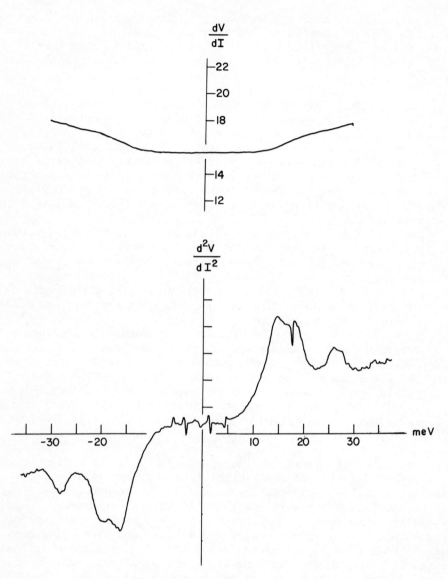

Figure 5: (a) The junction resistance as a function of voltage. Within experimental accuracy the resistance is independent of current flow direction. (b) The current derivative of the junction resistance. This was a Cu-Cu junction measured on April 23, 1976. Temperature = 1.2 K.

processes? Unlikely. Surface plasmons? Possibly. The most likely interaction - and the simplest - is coupling to the phonons through the electron-phonon interaction. This is the picture Yanson presents and we agree with him.

In Figs. 6,7, and 8 are given results of the second current derivative of the junction voltage (plotted as the solid lines) with respect to voltage (taken earlier than Fig. 4 and smoothed somewhat) for the noble metals - Cu, Ag, and Au. The dashed curve, in each case, is the phonon density of states derived from neutron scattering data fit to a Born-von Karmen model. Because its resistance was the lowest (5.7 Ω), the Cu result shows, we believe, the largest "background" effect. Subtracting these out mentally, one sees that in every case, the first peak coincides nicely with peaks in $F(\omega)$. The lower one which corresponds to "transverse modes" and the second to "longitudinal" modes (this second coincidence is worst for Ag). This shows, we believe, that bulk phonons are playing a significant role. The same results were seen by Yanson [14,15].

Harder to understand is the different "scaling" of the two peaks. In the work of Yanson [14] I presented on Pb (Fig. 2), the scaling was the other way around - transverse coupling was small or smaller than longitudinal coupling.

The difference has, I believe, a theoretical explanation or basis. At the last Rochester Conference (in 1971) I spoke [24] about the thesis work [25] of S.G. Das - on self-energy effects in the electronic structure of the noble metal copper due to phonons. At that time I emphasized that copper - as a noble metal - was more like a transition metal - Nb or Mo than like a "simple" metal - Al or Pb. The d-electrons of copper played a significant role in the anisotropy of renormalization effects. These were manifest in the variation of λ as one moved from belly to bulge to necks around the Fermi surface, and depended on where one began and ended one's phonon scattering or coupling. The important point was that umklapp transverse phonons coupled more strongly to d-electrons than longitudinal phonons coupled s-electrons - a result possibly of significance in understanding superconductivity in d- and f-band metals. This result was also implicit in the work on umklapps, discussed by Peierls [3] and Bardeen [1].

I believe that if one reads Das' thesis in the Physical Review with this in mind one can see the origin of the differences of the scaling of transverse and longitudinal peaks - as a strong, effective energy dependence in the α^2 part of $\alpha^2 F$. I also believe that other transition metals - Nb or Mo - or possibly transition metal compounds will show similar effects. We hope to perform at Nijmegen experiments on these similar to those I have described for the noble metals. But we will probably need the big magnets I spoke about

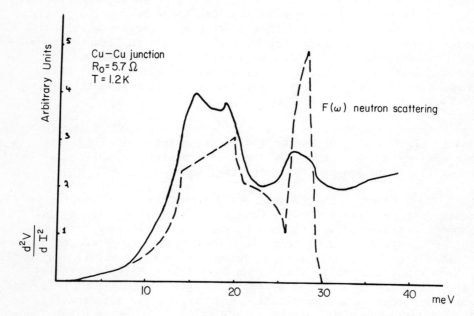

Figure 6: The current derivative of the resistance of a Cu-Cu junction of 5.7 Ω resistance: The dashed curve is the phonon density of states from Ref. 21.

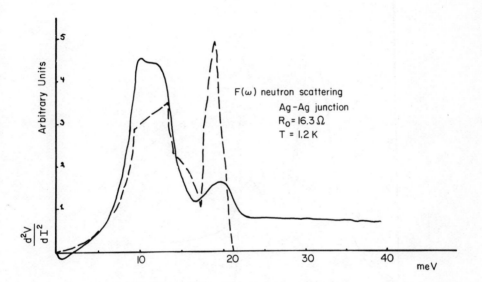

Figure 7: The current derivative of the resistance of a Ag-Ag junction of 16.3 Ω resistance: The dashed curve is the phonon density of states from Ref. 22.

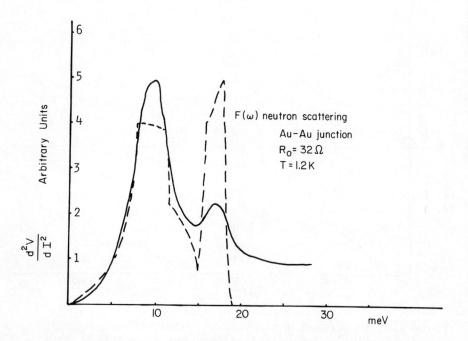

Figure 8: The current derivative of the resistance of a Au-Au junction of 32.0 Ω resistance: The dashed curve is the phonon density of states from Ref. 23.

before to help drive these and other superconducting materials normal. Perhaps at the next Rochester Conference - hopefully in less than five years - I - or more properly my colleagues - will report some results to you.

ACKNOWLEDGEMENTS

We would like to thank S. Doniach for helpful discussions and H.W.H.M. Jongbloets for help with setting up the experiment.

REFERENCES

1. J. Bardeen, Phys. Rev. 52, 688 (1937).
2. F. Bloch, Zeits. f. Physik 52, 555 (1928).
3. R. Peierls, Ann. d. Physik 12, 154 (1932).
4. J. Bardeen, L.N. Cooper, J.R. Schrieffer, Phys. Rev. 108, 1175 (1956).
5. B. Castaing and P. Goy, Jour. Phys. C 6, 2040 (1973).
6. P.B. Allen and M.L. Cohen, Phys. Rev. 81, 1329 (1970).
7. G. Grimvall, Phys. Cond. Mat. 9, 283 (1969).
8. F.M. Mueller and H.W. Myron, Communications on Physics (in press).
9. M. Fowler and R. Prange, Physics 1, 315 (1965).
10. C.J. Palin, Proc. Roy. Soc. (London) A329, 17 (1972).
11. C.B. Walker, Phys. Rev. 103, 547 (1956).
12. Yu.V. Sharvin, Zh. E.T.F. 48, 984 (1965).
13. Yu.V. Sharvin and N.I. Bogatina, Zh. E.T.F. 36, 772 (1969).
14. I.K. Yanson, Zh. E.T.F. 66, 1035 (1974).
15. I.K. Yanson, paper presented as L.131 at LT 14.
16. G. Hertz, Zeit. f. Phys. 22, 18 (1924) - quoted in plate VIII in M. Born, Moderne Physik (Springer, Berlin, 1933).
17. J. Franck and G. Hertz, Phys. Zeit. 20, 132 (1919).
18. J. Franck and G. Hertz, Verhand. Deut. Physik Ges. 16, 457 (1914); ibidem p. 512.
19. F.M. Mueller - "pinholes and peeking" - seminar organized for L.J. Sham.
20. See J.M. Rowell, W.L. McMillan and W.L. Feldman, Phys. Rev. 180, 658 (1969), for a particularly clear discussion of this point.
21. R.M. Nicklow, G. Gilat, H.G. Smith, L.J. Raubenheimer and M.K. Wilkinson, Phys. Rev. 164, 922 (1967).
22. W.A. Kamitakahara and B.M. Brockhouse, Phys. Lett. 29A, 639 (1969).
23. J.W. Lynn, H.G. Smith and R.M. Nicklow, Phys. Rev. B8, 3493 (1973).
24. F.M. Mueller, "Is copper a transition metal?" (unpublished).
25. S.G. Das, Phys. Rev. B7, 2238 (1973).

QUESTIONS AND COMMENTS

R. Dynes: I am a little worried about your discussion of dimensions, physical dimensions. It is known in lead that the mean free path for phonon emission when you get up to energies comparable to peak positions in the phonon spectrum is 150 Å or so. I can conceive that one can make a pinhole by sparking or something like that in a tunnel junction that is on the scale of 150 Å. But now you are talking about the dimensions that are of the order of microns.

F. Mueller: No, if I gave you that impression then I was misleading you in my discussion. Because, in fact, they are much smaller than that. You have to control them carefully and you have to average over (probable) motion of the spear and anvil relative to each other. That is certainly happening and is in the physics. Now the only way we have at the moment of telling the size of our pinholes is through the Shavin formula. Yanson has compared his results by looking at two things. Both the Shavin formula which gives you a representation for the size of the hole and also by superconducting experiments which are also hole-dependent and reported in one paper. At least he gets reasonable agreement between the two.
There are difficulties with this new technique. I would be lax if I said this was the final solution to all problems in life. There are problems with thermal broadening, problems with getting out the right derivative, there are problems with shaking. We are arranging to have things made quieter so please do not believe that we think this is the last word on doing everything. We think we do have some interesting [and reproducible] results.
Now to address the explicit question you were asking: Yanson's estimates of his point contact sizes for resistances which are similar to the ones we are presenting is of the order of 20 Å or so.

R. Dynes: There is another point here. The emission length is energy dependent.

F. Mueller: That is the whole point of what we are discussing!

R. Dynes: That is right. So your comparison of transverse and longitudinal modes in the discussion of coupling parameters in the transition metals depends in detail on the size of this emission length relative to your pinhole.

F. Mueller: The numbers---I do not have the numbers in my head. In fact, Dynes, I did not do the emission length calculation and what Yanson, for noble metals, quotes is something of the order of 1000 Å in his paper. So I actually have not looked at that consequence of what we are doing yet. The idea here is that in order to get this Knudsen regime---where you have the streaming behavior you want to have the scattering length large compared to the orifice size. I think that what you may be worried about is that in other materials we may have a length which is changing so rapidly that the assumptions of using the simple Sharvin formula are invalid. I do not think so, but I am not going to be so strong about that. Let me just say I am working with Van Gelder at Nijmegen looking at this problem in detail---it is a rather difficult problem theoretically. One has to solve non-linear Boltzman equations as functions of distances and so on. We hope to have something useful done on that soon. The essential physics is to look at the rate of entropy creation in this non-equilibrium regime and to try and solve that. The paper by Wexler is very helpful, but it is not the final solution either.

W. Butler: Do you get a value for λ?

F. Mueller: We are working on that based on the assumptions that I presented here. The paper has not probably left my desk yet so I do not think I can discuss it in detail. I don't know, to answer you explicitly. The problem here is---let me just explain. Clearly, one has an interesting vertical axis [in the derivatives]. As you noticed in this talk the vertical axis, in each case, was labeled with arbitrary units. Now we believe those units are not, in fact, arbitrary at all. We hope on the basis of such experiments to derive explicit λ's and α's and maybe compare with niobium, for example. George Webb has given us a piece and we will see if we can throw some light on this tunneling question and the problem of the inversions and so on. We did think of doing that.

CALORIMETRIC OBSERVATION OF A PHASE TRANSITION IN THE SUPERCONDUCTING STATE IN $Gd_{1.2}Mo_6Se_8$*

R.W. McCallum, D.C. Johnston, R.N. Shelton and M.B. Maple

Institute for Pure and Applied Physical Sciences
University of California, San Diego

La Jolla, California 92093

ABSTRACT

A lambda-type anomaly in the specific heat of the Chevrel phase compound $Gd_{1.2}Mo_6Se_8$ has been observed at 3.5K, below its superconducting transition temperature of 5.5 K. The anomaly is not reflected in the temperature- or magnetic field-dependence of the magnetization, and low temperature X-ray powder diffraction measurements indicate that the anomaly is not due to a crystallographic phase transformation. The anomaly appears to be due to a phase transition which is associated with the presence of the localized 4f electrons of the Gd^{3+} ions, but the exact nature of the transition remains to be established.

Following the discovery of superconductivity in the rare earth (RE) molybdenum sulfides ($RE_xMo_6S_8$) by Fischer and coworkers [1], Shelton, McCallum and Adrian [2] synthesized the series of rare earth molybdenum selenides ($RE_xMo_6Se_8$). The rare earth molybdenum selenide compounds were found to have superconducting transition temperatures (T_c's) which are substantially higher than the corresponding sulfide compounds. McCallum, Shelton, Maple and Adrian [3] concluded that the characteristic variation of T_c with substituted RE in both the $RE_xMo_6S_8$ and $RE_xMo_6Se_8$ series could be accounted for in terms of pair breaking effects due to the interaction of the

---

* Research supported by the U.S. Energy Research and Development Administration under Contract No. ERDA E(04-3)-34 PA227.

RE magnetic moments with the conduction electron spins via the usual exchange interaction

$$H = -2I(g_J - 1)\underline{J} \cdot \underline{s} .$$

Here I is the exchange interaction parameter, g_J and J are respectively the Landé g-factor and total angular momentum vector for the RE ion's Hund's rule ground state and $\underline{s}$ is the conduction electron spin density at the site of the RE ion. With the exception of Ce and Eu, it was found that each RE ion depressed T_c from a value estimated for the absence of magnetic interactions at a rate which scaled roughly with the deGennes factor [4] $(g_J - 1)^2 J(J + 1)$ in accordance with the predictions of the theory of Abrikosov and Gor'kov [5]. The inferred depressions of T_c for the four Ce and Eu compounds, which were not superconducting down to 50 mK [6], were anomalously large. The value of the exchange interaction parameter for all of the other rare earth molybdenum sulfide and selenide compounds was found to be an order of magnitude smaller than in most systems previously studied such as (LaRE)Al$_2$. [7]

The exchange interaction which is responsible for pair breaking in the superconducting RE$_x$Mo$_6$S$_8$ and RE$_x$Mo$_6$Se$_8$ systems should lead to magnetic ordering of the RE ions at sufficiently low temperatures via the indirect RKKY interaction [8]. Since the RE ions in these Chevrel phase compounds occupy fixed lattice sites rather than being distributed randomly throughout the lattice as a dilute impurity, the magnetic ordering of the RE ions should be of long range rather than diffuse "spin glass" type. Thus the RE$_x$Mo$_6$S$_8$ and RE$_x$Mo$_6$Se$_8$ compounds appeared to be promising candidates in which to study the coexistence of superconductivity and magnetic order. This prompted us to initiate an extensive investigation of this interesting class of materials.

Of the rare earth molybdenum sulfide and selenide compounds, Gd$_{1.2}$Mo$_6$Se$_8$ appeared to offer the greatest potential for exhibiting the simultaneous occurrence of superconductivity and long-range magnetic order. A selenide compound was chosen because of the significantly higher T_c's of the rare earth molybdenum selenides relative to the sulfides. The rare earth Gd was selected for two reasons. First, it is expected to exhibit the highest magnetic ordering temperature in the RE$_x$Mo$_6$Se$_8$ series since, like the depression of T_c, the magnetic ordering temperature should scale with the deGennes factor of the RE ion which is largest for Gd. Second, Gd is an S-state ion (L = 0, S = 7/2 where L is the orbital angular momentum and S is the spin angular momentum) so that crystal field splitting of the J = 7/2 Hund's rule ground state should be negligible. In this paper we report specific heat, magnetic susceptibility, and X-ray powder diffraction measurements on a Gd$_{1.2}$Mo$_6$Se$_8$

sample.

The $Gd_{1.2}Mo_6Se_8$ sample was prepared in the form of a sintered pressed pellet in a manner previously described [2]. The specific heat measurements were carried out with a semi-adiabatic He^3 calorimeter [9] using a standard heat pulse technique. Static magnetic susceptibility measurements were made with a Faraday magnetometer [10] in the temperature range from 1.3 to 295 K. X-ray diffraction measurements were made using CuK_α radiation at 2.1, 4.4 and 300 K with a Janis varitemp dewar fitted with beryllium X-ray windows. A standard ac inductance technique (20 Hz) was used to monitor the ac susceptibility of the sample in the small background field of the earth.

Shown in Fig. 1 are plots of the specific heat as a function of temperature for $Gd_{1.2}Mo_6Se_8$, $La_{1.0}Mo_6Se_8$ and $Lu_{1.2}Mo_6Se_8$. The lanthanum and lutetium molybdenum selenide compounds are nonmagnetic and exhibit specific heat jumps at their respective superconducting transition temperatures. The gadolinium molybdenum selenide sample shows a specific heat jump at 5.5 K which is associated with its transition into the superconducting state at 5.64-5.40 K as verified in a separate experiment by means of a standard ac inductive technique. A plot of the ac susceptibility vs temperature showing the transition to the superconducting state is given in Fig. 2. The most remarkable feature of the specific heat of $Gd_{1.2}Mo_6Se_8$ is the pronounced lambda-type anomaly which occurs in the superconducting state at about 3.5 K, which is followed by an upturn at lower temperatures. In addition to the zero field data shown in Fig. 1, specific heat measurements were performed in magnetic fields of 1.5, 3 and 6 k$\emptyset$. However, no significant change in the magnitude or shift in the temperature of the lambda anomaly could be detected in this magnetic field range.

It is tempting to ascribe the lambda anomaly in the specific heat of $Gd_{1.2}Mo_6Se_8$ below T_c to the sought-after long range magnetic ordering in the superconducting state. In order to explore this possibility further, the specific heat of the $Gd_{1.2}Mo_6Se_8$ compound was corrected for lattice and electronic contributions which we assumed to be the same as for the nonmagnetic $Lu_{1.2}Mo_6Se_8$ compound. With this assumption, the excess entropy between 1.2 and 10 K was deduced to be equal to 8.0 J/mole-K. This represents a substantial fraction (~ 0.4) of the entropy $S = 1.2\ R\ \ln 8 = 20.75$ J/mole-K which would be associated with complete order of the Ge moments.

In an attempt to further elucidate the nature of the specific heat anomaly below T_c in the $Gd_{1.2}Mo_6Se_8$ sample, magnetization measurements were made between 1.3 and 295 K. The results are displayed in Figures 3 and 4. In Fig. 3 the inverse low field susceptibility is plotted vs. temperature. The magnetic susceptibility

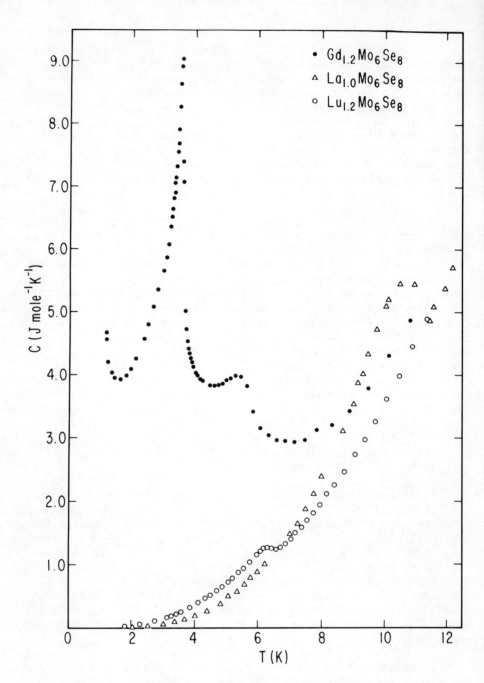

Figure 1: Specific heat vs temperature for $Gd_{1.2}Mo_6Se_8$, $La_{1.0}Mo_6Se_8$ and $Lu_{1.2}Mo_6Se_8$.

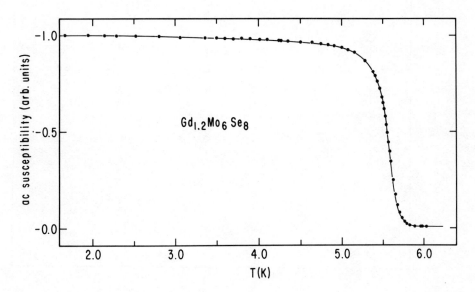

Figure 2: Zero field ac susceptibility vs temperature for $Gd_{1.2}Mo_6Se_8$.

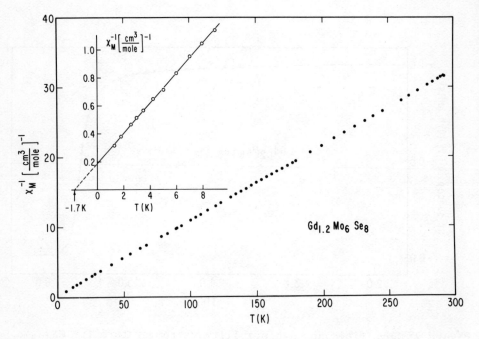

Figure 3: Inverse low field magnetic susceptibility vs temperature for $Gd_{1.2}Mo_6Se_8$. The low temperature data are shown in the inset.

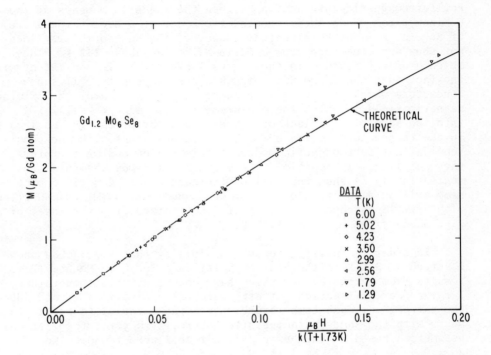

Figure 4: Magnetization plotted vs the function $x \equiv \mu_B H/k(T-\theta)$ for temperatures below 6 K. The theoretical curve was calculated from $M = (7\mu_B)B_J(7x)$, where $B_J(7x)$ is the Brillouin function for $J = 7/2$ and $g_J = 2$.

follows a Curie-Weiss law

$$\chi = \frac{N\mu_{eff}^2}{3k(T-\theta)}$$

with an effective moment μ_{eff} of 7.9 μ_B per Gd ion, in good agreement with the free ion value of 7.94 μ_B, and a Curie-Weiss temperature of -1.7 K. The negative Curie-Weiss temperature indicates an antiferromagnetic interaction between the magnetic moments of the Gd ions, from which antiferromagnetic order at a Néel temperature of several K might be expected. However, the susceptibility does not show any departure from a Curie-Weiss law above 1.3 K. The absence of any feature in the χ^{-1} vs T plot at T_c is consistent with the fact that the $RE_xMo_6Se_8$ compounds are extreme type II superconductors which permits almost complete penetration of the external magnetic field in which the measurements were made (> 1.47 k∅) into the sample. The lack of a break in slope of the χ^{-1} vs T curve in the temperature region below T_c is inconsistent with the occurrence of antiferromagnetic order in this region and is therefore very puzzling. In addition, the magnetization measurements presented in Fig. 4 show that all of the data can be described by the same Brillouin function on a plot of magnetization (M) vs $\mu_B H/k(T-\theta)$, with a field-indenpendent value of θ determined from the inset of Fig. 3 and given by θ = -1.73 K.

In order to investigate the possibility that the lambda anomaly observed at 3.5 K in the heat capacity data for $Gd_{1.2}Mo_6Se_8$ (Fig. 1) results from a crystallographic phase change at this temperature, X-ray powder diffraction patterns were taken at 2.1, 4.4, and 300 K. The patterns at each temperature showed the material to be single phase with the same characteristic Chevrel structure; the lattice parameters based on the hexagonal unit cell were found to be a_H = 9.49 Å, c_H = 11.86 Å at 300 K, and a_H = 9.40 Å, c_H = 11.86 Å at 2.1 K. The observed and calculated lattice d-spacings for the 300 K measurement are shown in Table I.

In conclusion, the specific heat measurements on $Gd_{1.2}Mo_6Se_8$ reveal a pronounced lambda anomaly in the superconducting state which is unaffected by magnetic fields up to 6 k∅. The entropy associated with the excess specific heat in the temperature interval 1.5 - 10 K is large and of the order of 0.4 of the value expected for complete ordering of the Gd spins. While the negative Curie-Weiss temperature obtained from the magnetic susceptibility measurements suggests that the Gd spins should order antiferromagnetically at several K, there is no evidence for an antiferromagnetic or any other type of magnetic transition in the temperature dependence of the low field magnetic susceptibility, in the magnetization vs applied magnetic field isotherms, or in the zero field ac suscepti-

Table I

Observed and calculated lattice d-spacings at room temperature for $Gd_{1.2}Mo_6Se_8$. The calculated values were computed from hexagonal lattice parameters $a_H = 9.49$ Å, $c_H = 11.86$ Å.

$hkl_{hexagonal}$	CuK$_\alpha$ Radiation	
	$d_{obs.}$ (Å)	$d_{calc.}$ (Å)
101	6.722	6.716
012	4.797	4.789
110	4.714	4.710
003	3.941	3.943
021	3.857	3.857
202	3.363	3.358
113	3.025	3.024
211	2.986	2.984
122	2.734	2.734
300	2.720	2.719
024	2.393	2.394
015	2.272	2.272
131	2.223	2.223
214	2.134	2.135
312	2.114	2.113
223	2.022	2.022
006	1.968	1.972
321	1.848	1.849
134	1.796	1.797
232	1.785	1.785

bility. Finally, X-ray powder diffraction measurements give no indication of a crystallographic phase transformation in the neighborhood of the lambda anomaly and the magnitude of the entropy associated with the anomaly precludes that it arises from any subtle lattice tranformation which would have escaped detection by this method. Thus it would seem that if the specific heat anomaly is due to magnetic ordering or the Gd spins in the superconducting state as suggested by the sharpness of the anomaly and the large amount of entropy associated with it, the type of magnetic order must be such that it does not result in a change in the temperature dependence of the magnetic susceptibility in the neighborhood of the transition. Another possibility is that the anomaly is due to some other type of phase transition of unknown origin. Irrespective of its source, however, the presence of the magnetic Gd ions in $Gd_{1.2}Mo_6Se_8$ is apparently required for the occurrence of the specific heat anomaly, since a lambda-type anomaly was not observed in the heat capacity data for the isostructural compounds $La_{1.0}Mo_6Se_8$ and $Lu_{1.2}Mo_6Se_8$. It is clear that additional experiments are required to clarify the nature of these striking physical properties of this interesting material.

It is a pleasure to thank Professor B.T. Matthias for stimulating and informative discussions regarding this work.

REFERENCES

1. Ø. Fischer, A. Treyvaud, C. Chevrel and M. Sergent, Solid State Commun. 17, 721 (1975).
2. R.N. Shelton, R.W. McCallum and H. Adrian, Phys. Letters 56A, 213 (1976).
3. R.W. McCallum, R.N. Shelton, M.B. Maple and H. Adrian, Bull. Am. Phys. Soc. 21, 383 (1976).
4. P.G. deGennes, J. Phys. Rad. 23, 510 (1962); H. Suhl and B.T. Matthias, Phys. Rev. 114, 977 (1959).
5. A.A. Abrikosov and L.P. Gor'kov, Zh. Eksp. Teor. Fiz. 39, 1781 (1960); Sov. Phys. JETP 12, 1243 (1961).
6. L.E. DeLong and W.A. Fertig, private communication.
7. M.B. Maple, Solid State Commun. 8, 1915 (1970).
8. M.A. Ruderman and C. Kittel, Phys. Rev. 96, 99 (1954); T. Kasuya, Progr. Theoret. Phys. (Kyoto) 16, 45 (1956); K. Yosida, Phys. Rev. 106, 893 (1957).
9. C.A. Luengo, Ph.D. Thesis, Universidad Nacional de Cuyo, Argentina (1972) (unpublished).
10. D. Wohlleben and M.B. Maple, Rev. Sci. Inst. 42, 1573 (1971).

SOME SURPRISES IN SUPERCONDUCTIVITY

Bernd T. Matthias

Institute for Pure and Applied Physical Sciences*
University of California, San Diego

La Jolla, California 92093

and

Bell Laboratories

Murray Hill, New Jersey 07974

Many times in the recent past--also here--I have pointed out how superconductivity and magnetism are very closely related. Bill McCallum, who works with Brian Maple in La Jolla, has now done some measurements which show for the first time both phenomena in the very same crystal; one clean cut crystal which is no longer a mixture of compounds. This is only one of the various surprises. You will hear about it for the next seven minutes, as I want Bill McCallum to speak. [See McCallum's paper--this Conference Proceedings].

This morning I wasn't asleep when Varma mentioned the e/a ratio. The reason why I didn't want to say anything was because there were so many comments anyhow, and I was just too depressed for the following reason. Once upon a time, the electron per atom ratio (e/a) as a criterion for transition temperatures, worked beautifully. Well, it doesn't work so well any longer. That is not to say

---

*Research in La Jolla sponsored by the Air Force Office of Scientific Research, Air Force Systems Command, USAF, under AFOSR contract #AFOSR/F-44620-72-C-0017.

that what we did in the past was wrong. What I am trying to say is
that what we are discovering right now, that is the transition temperatures of the molybdenum sulfides and selenides, just cannot be
described in a coherent way by the usual e/a ratio.* Mind you,
in general the number of electrons still works beautifully--the
melting point correlation--all of this is still true. For instance,
when you read the latest issue of the Proceedings of the National
Academy you will see something really stunning. Namely, the
measurements of the viscosity of liquid metals [1]. The convincing
fact is that the highest viscosity found is for Fe, Co and Ni.
The lowest viscosity they find for K, Na and the alkalies. In
between, you find the whole range of the superconducting elements.
So, not only the melting point, but even the properties above the
melting point show that, in general, the superconducting transition
temperatures are really determined by <u>all</u> the outer electrons.
But the simple picture which I once gave you, namely, e/a with the
two maxima, doesn't hold so exclusively any longer.

Actually, the first indication came a while ago when Giorgi
and Szklarz [2] found the $(Y,Th)_2C_3$ high temperature superconductors.
There was a maximum at 4 and I did not think that this was an
isolated fact. True enough, when other compounds were made, many
of them showed this same maximum at 4 provided the structure was
right--tetragonal or Pu_2C_3 structures [3]. Then we found other
compounds which had a maximum at 4, but with different transition
temperatures. At least in this and all other systems we could
always trace the highest transition temperatures with a variation
of the e/a ratio. But now, and I don't know whether you noticed
it, all of these sulfur and selenium compounds are a total violation of this rule according to which the transition temperature is
a function of e/a. The correct formula is one metal atom, 6 Mo's,
8 S and Se's, respectively. And, therefore, it can be seen immediately that whatever the compound is the e/a ratio is always exceedingly close to 6 from the Mg, Zn and Cd compounds and their
superconductivity at two degrees to temperatures close to 16^0's
for Sn and Pb compounds; yet, the e/a ratio cannot be juggled
to fit in a way as to be held accountable. The way out from this
is, at present, absolutely obscure. Except, I was very much intrigued when I listened to all of you who want to measure the phonon
spectrum of these MoS's. Strangely enough, something seems to
have escaped all of you that appears to be symptomatic for all
these molysulfides and molyselenides. In a strange way it comes
back to the intrinsic instabilities of high transition temperature
superconductors. Instabilities are, of course, an old story ever
since Fröhlich and Bardeen. And, empirically, it was realized
shortly thereafter. So, even though we have known this now for a
long time, we just didn't know really how far it went.

---

*This is what prompted the title of my talk.

SOME SURPRISES IN SUPERCONDUCTIVITY

One of the instabilities in this world is immiscibility. As you know, compounds of elements can be made which don't want to form properly, if rapid quench techniques are used. However, this case is then in itself an unstable situation. But, the immiscibilities in themselves are of a different kind of instability, which is actually in a comparative sense quite stable. This then is the crucial feature of all these superconducting molybdenum sulfides and selenides. Namely--only those elements which will not react at all with molybdenum alone form superconducting compounds with Mo_3S_4 and Mo_3Se_4, S or Se--the latter being in a way the ternary binding element--or glue. And, that is the criterion. This became obvious to me the moment I saw Fischer's beautiful data. Rare earths of course are totally immisicible with Mo. So, altogether only those elements which will not react with Mo, but in turn introduce an instability on a microscopic level is, in my opinion, at present one of the answers for the superconductivity of these compounds.

Let me quickly show you now how well this works. Usually when something combines with Cu and Ag, it certainly combines with Au. The Cu and Ag compounds are superconducting from 12^0K or so on down to 8^0K. The Au compounds doesn't even form. In "Hansen", under binary phase diagrams, Mo and Au are reported as immisicible. In reality, they are not. A long time ago, we made MoAu alloys and they were superconducting at $\sim 4^0K$ indicating that there are compounds or solid solutions and it is for this reason Au won't even combine with Mo_3S_4. What happens when we choose trivalent elements such as Al, Ga, In or Tl? All of them react with Mo-- Mo alloys with Ga and Al very easily. Because of this they will not form the superconducting Chevrel phases. Instead they form a new structure, the Spinel structure, which is ferrimagnetic. So, you see, not just any trivalent element will form superconducting phases only those that will not react with pure Mo and they are the rare earths. The same is also true for the four valent elements. For instance, Ti and V do form, but they are not superconducting because Ti and V react with Mo. Fe, Co and Ni do form the phases and as you know very well, they again all react with Mo, and thus again there is no superconductivity. But, Pb and Sn won't react with Mo by themselves and therefore they give rise to superconducting transition temperatures near 16^0K for the sulfides.

And, so I think the <u>immisicibility</u> with Mo is one of the crucial features of their superconducting properties. This then is-- that is, the molysulfides and molyselenides--the second example, and the third I will show you shortly. The first was $LiTi_2O_4$, which has the Spinel structure and is superconducting [4], again Li and Ti are immisicible.

The latest example of superconductivity due to immisicibility which we have now discovered came strictly from this same hypothesis. We always knew there were traces of superconductivity in the Sc-Cr system. Through specific heat measurements, we knew it was only a trace which we just could not sort out. Furthermore, again in Hansen: scandium and chromium are immisicible. Well, we decided, maybe these two elements would combine if there were traces of a third element. So, we tried just about everything under the sun, but only in little quantities. And believe it or not, we found that $Sc_{2+x}Cr_{1-x}B_\epsilon$ [5]--it took very little boron, for us to get out of this immisicibility range--became superconducting at 6^0K. Again, McCallum measured the specific heat and while it was not a complete bulk effect, the superconducting phase was now in the range of 20 to 30%.

So, I think the future of superconducting compounds has really been made a little bit brighter, but far more difficult by the fact that much which we couldn't do in binary systems, we can now do in ternary systems. In ternary systems we can combine two elements which from a binary point of view would be hopeless, but in a ternary phase have a finite chance. And, so in the future I think this will be the direction we are going to take. The one thing that really bothers us at present, aside from the immisicibility, is that we have no idea what really determines the transition temperature. As mentioned before in the sulfides and selenides the first column of the periodic system, Cu and Ag give transition temperatures in the range of 11^0K. The second column is between $2-3^0K$. The third between $9-12^0K$ and the fourth near 16^0K. So obviously, the e/a ratio will no longer work in true ternary compounds.

And that is where we are at present. Since the time is just about right I think, for all of you, I want to thank David for a marvelous conference. Many of us have come to the conclusion that if he had more conferences of this kind, superconductivity would be in an entirely different shape.

REFERENCES

1. J.H. Hildebrand and R.H. Lamoreaux, Proc. National Academy of Sciences 73, 988 (1976).
2. A.L. Giorgi, E.G. Szklarz, M.C. Krupka, T.C. Wallace and N.H. Krikorian, J. Less-Common Metals 14, 247 (1968).
3. A.S. Cooper, E. Corenzwit, L.D. Longinotti, B.T. Matthias and W.H. Zachariasen, Proceedings of the National Academy of Sciences 67, 313 (1970).
4. D.C. Johnston, H. Prakash, W.H. Zachariasen and R. Viswanathan, Mat. Res. Bull. 8, 777 (1973).
5. J.M. Vandenberg, B.T. Matthias, E. Corenzwit and H. Barz, Journal of Solid State Chemistry (in press).

SUPERCONDUCTIVITY IN d- AND f-BAND METALS:

SECOND ROCHESTER CONFERENCE

ATTENDEES

B.N.N. Ackar
Pennsylvania State University

P.B. Allen
SUNY - Stony Brook

M. Ashkin
Westinghouse Electric
R. & D. Center

S.D. Bader
Argonne National Laboratory

J. Bardeen
University of Illinois

G. Barsch
Pennsylvania State University

L.L. Boyer
Naval Research Laboratory

J. Bostock
Massachusetts Institute
of Technology

A.I. Braginski
Westinghouse Electric
R. & D. Center

E. Brown
Rensselaer Polytechnic Institute

J. Budnick
University of Connecticut

W.H. Butler
Oak Ridge National Laboratory

F.J. Cadieu
Queens College

T. Castner
University of Rochester

R. Caton
Brookhaven National Laboratory

B. Chakraborty
SUNY - Stony Brook

R. Chaudhari
SUC - Oswego

N. Cheung
EXXON

C.W. Chu
Cleveland State University

M.L. Cohen
University of California-Berkeley

E.W. Collings
Battelle Memorial Institute

D.E. Cox
Brookhaven National Laboratory

S.G. Das
Argonne National Laboratory

V. Diadiuk
Massachusetts Institute of
Technology

D.H. Douglass
University of Rochester

R.C. Dynes
Bell Laboratories

L. Elbaum
New York University

J. Engelhardt
University of California
San Diego

D. Farrell
Case Western Reserve

E. Fawcett
University of Toronto

Ø. F_scher
University of Geneva

Z. Fisk
University of California
San Diego

S. Foner
Massachusetts Institute of
Technology

F.Y. Fradin
Argonne National Laboratory

A. Freeman
Northwestern University

B.N. Ganguly
Xerox Corporation

J. Gavaler
Westinghouse Research Labs.

A.K. Ghosh
University of Rochester

D. Ginsberg
University of Illinois

A. Giorgi
Los Alamos Scientific Laboratory

A.M. Goldman
University of Minnesota

A.M. Gray
Watervliet Arsenal

D.M. Gray
Watervliet Arsenal

J. Gregory
Massachusetts Institute of
Technology

M. Gupta
Northwestern University

R.P. Gupta
Northwestern University

M. Gurvitch
Brookhaven National Laboratory

B. Gyorffy
H.H. Wills Physics Laboratory

B. Harmon
Iowa State University

R. Hein
University of California
San Diego

J.G. Huber
Tufts University

E. Jacobsen
University of Rochester

H. Jones
Indiana University

V. Keith
University of Waterloo

H.R. Kerchner
Oak Ridge National Laboratory

C.W. Kimball
Northern Illinois University

B. Klein
Naval Research Laboratory

ATTENDEES

G.S. Knapp
Argonne National Laboratory

R. Knox
University of Rochester

C.C. Koch
Oak Ridge National Laboratory

F. Kus
McMaster University

V. Landman
University of Rochester

A.C. Lawson
University of California
San Diego

T.K. Lee
City College of New York

J.D. Leslie
University of Waterloo

C.S. Lim
University of Waterloo

J.D. Livingston
General Electric R. & D.
Center

H. Lutz
Brookhaven National Laboratory

M.L.A. MacVicar
Massachusetts Institute of
Technology

C.K. Manheimer
University of Rochester

M.B. Maple
University of California
San Diego

B.T. Matthias
University of California
San Diego

R.W. McCallum
University of California
San Diego

E.J. McNiff
Massachusetts Institute of
Technology

M. Milewits
New York University

S. Moehlecke
Brookhaven National Laboratory

S. Mohabir
University of Waterloo

F. Mueller
Katholieke Universiteit
Netherlands

A.D. Nagi
University of Waterloo

S.B. Nam
Dayton, Ohio

K.L. Ngai
Naval Research Laboratory

J. Noolandi
Xerox Research Center of Canada

D.A. Papaconstantopoulos
Naval Research Laboratory

R.D. Parks
University of Rochester

B. Patton
University of California
San Diego

S. Perkowitz
Emory University

J.C. Phillips
Bell Laboratories

W.E. Pickett
H.H. Wills Physics Laboratory

W. Reichardt
Karlsruhe, Germany

B. Roberts
General Electric - Schenectady

B. Robinson
Stanford University

Carol Rosen
Queens College

C. Rossel
Temple University

J.M. Rowell
Bell Laboratories

P. Schmidt
Bell Laboratories

R. Shelton
University of California
San Diego

P. Vashishta
Argonne National Laboratory

R.C. Shukla
Brock University

S.K. Sinha
Argonne National Laboratory

H.G. Smith
Oak Ridge National Laboratory

E.G. Spencer
Bell Laboratories

M. Strongin
Brookhaven National Laboratory

A.C. Switendick
Sandia Laboratories

H. Teichler
Max-Planck Institute - Germany

C.M. Varma
Bell Laboratories

R. Viswanathan
Brookhaven National Laboratory

S.F. Wang
University of Waterloo

A. Webb
Naval Research Laboratory

G.W. Webb
University of California
San Diego

W. Weber
Bell Laboratories

H. Weisman
Brookhaven National Laboratory

S.J. Williamson
New York University

J.A. Woollam
NASA Lewis Research Center

Index

A

A-15 compounds
 anharmonicity of, 301,548
 defects in, 413,501
 electron density of states, 298,541
 high temperature resistivity saturation of, 546,563
 phonon spectra of, 191,298
 superlattice patterns in, 417
 temperature dependence of phonon density of states of, 197
 upper critical field, 165
Anisotropy
 of the energy gap, 73
 of H_{c2}, 126
 of H_{c2} for niobium, 130
Anomalous phonon dispersion, 223,269,513
Augmented plane wave, 584

B

B-1 compounds
 structural defect, 415
Band, bonding and antibonding, 524
Band structure
 electronic, 526,562
 for NbC and TaC, 318
 of hydrogen, 594
 of PdD, 344
 of refractory carbides, 345
 of V_3Si, 354
 self consistent, relativistic APW calculation, 340,509, 575
Band width, 521
Born-von Kármán force constant for $PdD_{0.63}$, 232

C

Charge density wave, 271,417,454
Charge fluctuation, 269,271,273, 286

Chemical potential, 576
Chevrel phases, 137,167,175,202
 electronic density of states, 305
 electronic parameters of, 310
Cohesive energy, 509,520
Conductivity, 561
 differential, 613
Coulomb coupling constant μ^*, 51,368
Coulomb interaction, 273,399
 screened, 275,584,588
Critical field
 effect of an exchange field, 181
 of anisotropic superconductors, 170
 of β-W compounds, 165
 of Li-Ti-O compounds, 170
 of rare earth molybdenum chalcogenides, 179
 of $Sn_{1-x}Eu_xMo_6S_8$ and $Pb_{1-x}Eu_xMo_6S_8$, 182
 of ternary molybdenum chalcogenides, 167,175
 of type II superconductors, 161
Crystal field splitting, 626

D

Debye phonon spectrum, 568,576
Debye temperature, effective, 302,305
Defect production in A-15, 498, 526
Deformation potential, 509
Density matrix expansion coefficients T_J^α, 99
Density of states
 angular momentum decomposition of, 43
 electron, 568
 model for a bcc transition metal, 262
 of HfC, 345
 of NbC and TaC, 320,345
 structure in, 568,575

temperature dependence of, 546
Density response function, 251, 255,282
Dielectric function
 Fermi Thomas, 3
 including local field effects, 12
 of semiconductors, 8
Dielectric screening, 315-316, 587
Dynamical matrix, 273,316,519

E

Electron beam depositon, 536
Electron density response matrix, 272
Electron-electron interaction, 254,317,33,522,587
Electron focusing, 609
Electron-ion interaction, 251
Electron plasma frequency, 4
Electron-phonon coupling, 596
 function $\alpha_2(\omega)$, 202,568
 constant λ, 34,48,91,283,371, 392,608,538,545,588,596
Electron-phonon interaction
 theory, 29-53,272,316,340,349, 515,552,568,584
Electron-phonon matrix elements, 255,272,316,402,515,584,596
 screened, 585
Electronic susceptibility, 227, 258,521
Eliashberg function $\alpha_2(\omega)F(\omega)$,
 of Nb_3Sn, 201
Eliashberg theory
 reanalysis to strong coupling, 18-25
Eta, η
 band structure effects on, 42
 calculation of, 35,49,97,395, 596
 estimates of, for Nb,V,Ta, 44,45
 phase shift analysis of, 41-42
Exchange interaction, 626
Excitonic superconductivity, 1,67
 mechanisms of, 1,3,4,7

F

Fermi surface
 of Nb, 322
 of NbC and TaC, 349
 topology of, 562
Fermi surface harmonics, 33,75
 construction of, 91
Force constants
 band structure contribution to, 255
 harmonic, 253
 "neutral object", 254
Frohlich instability, 415

G

Gap anisotropy
 for niobium, 112
 theory, 73
$Gd_{1.2}Mo_6Se_8$, 626
 magnetic susceptibility of, 630
 specific heat of, 628
 x-ray diffraction measurements of, 633
Generalized susceptibilities, 325,349
 of NbC, TaC, ZrC and HfC, 325-332,352
Germanium, 2,10,59
Gorkov, coupled chain model for A-15, 552,563
Green's function
 electron, 38,256,521
 phonon, 32,282,284

H

Heat capacity
 electronic, 298
 of Chevrel phases, 306
Hydrogen
 density of states, 599
 metallic, 584,594
 phase shifts, 600
 stability of, 584
 superconductivity in, 585,602

INDEX

I

Immiscibility, 637
Impurity contribution to resistivity, 526,538,568
Interband scattering, 562,567
   phonon assisted, 568
   s-d, 568,578
   temperature dependence of, 568
Interfacial phases, 69,417
Ion-ion interaction, 47,251
Ion-plasma frequency, 509,291

K

K.K.R. method, 37,596

L

Lattice distortion, 556
Lattice dynamics
   band structure effects in, 255
   density functional formalism of, 252
   theory of, 251
Lattice instability, 270,413, 551,636
   electronically driven, 276, 282
Lattice parameter
   expansion of A-15, 496
   of A-15 compounds, 481
Linear chain model, 552,426
Local field corrections, 12,270, 587

M

Magnetic order, 633
   and superconductivity, 626, 635
Martensitic structure, 416
Mass enhancement λ, 309
Matsubara frequency, 33
Mean free path
   of electron, 545,548,578
   saturation of, 548,563

Microdomain, 416,581
Mo_3S_4-Mo_3Se_4-Mo_3Te_4, 138
   effect of hydrostatic pressure, 140-145
Mössbauer effect experiments, 301
Muffin tin approximation, 30, 33,393,585
   rigid, 255,283

N

Nakagawa-Woods spectra, 376
$NbGe_2$-Ge
   superconductivity, 59,417
Nb_3Ge
   density of states, 438
   effect of impurities on T_c of, 425
   effect of O_2 on T_c of, 423, 431
   energy gap of, 433
   impurity stabilization of, 421
   normal state resistivity, 535
   radiation damage of, 538
   resistive anomaly, 455
   sputtered films of, 422
   structural analysis of, 439
   tunneling measurements, 432
Nb_3Sb
   de Hass-van Alphen studies of, 563
   electrical resistivity of, 546,556
   heat capacity of, 546
Nb_3Sn
   electrical resistivity of, 546,552,554,569
   heat capacity of, 546,571
   lattice instability of, 551
   polycrystalline films of, 552
   resistance ratio dependence on T_c, 497
   resistive anomaly, 455
   single crystal of, 554
Neutron scattering, inelastic, 190,211,225,563,571
Niobium
   $\alpha_2^2F(\omega)$ of, 373,388,397
   anisotropic mass renormalization, 112

energy gap, 368,385
normal state resistivity, 535
oxidation of, 384
reduced density of states
of, 371,386
tunneling studies on, 368
Niobium-base A-15 compounds
^{4}He bombardment of, 491
order-disorder of, 461
stochiometry of, 461,500
x-ray intensity measurements on, 463
Niobium-tin
tunneling measurements of, 430
Nuclear spin-lattice relaxation, 298

O

Orthogonal tight binding, 509

P

Pauli paramagnetic limiting, 163,179
PbTe, 2
Phonon anomalies, 223,269,274, 349
Phonon density of states, 35, 190,568
generalized, 190,211
temperature dependence of, 197
Phonon dispersion curves
of Nb with 2 a% oxygen, 236
of Nb-Mo system, 233
of NbC, 279
of NbSe$_2$, 232
of PdH (D), 229
of the Mo-Re system, 236
of the Ta-Mo and Nb-Zr systems, 236
of Tc, Zr and Y, 243
of transition metal carbides, 225
Phonon frequency 519,590
renormalized, 34,272,52
Phonon mode shifting, 568,573
Phonon self-energies, 271
Phonon softening, 47,270
of PbMo$_6$S$_8$, 205
of SnMo$_6$S$_8$, 216
Phonon spectra, 189
of A-15 compounds, 191
of Mo$_6$Se$_8$, 213
of ternary molybdenum chalcogenides, 202,213
Phonon spectrum, 585
calculation of λ using Schrödinger Equation, 289
Point contacts, 608
Preferential orientation of flux line lattice relative to the crystal lattice, 131
Pressure effects on T_c
of rare earth molybdenum selenides, 152
of rare earth molybdenum sulphides, 149
of ternary molybdenum selenides, 145
of ternary molybdenum sulphides, 145
Pseudopotential
metals, 509
Morel-Anderson, 590
screened effective, 589

Q

Quartic anharmonicity, 573

R

Radiation damage, 524,538
in A-15 compounds, 462, 495
Rare earth molybdenum chalcogenides, 149,152,179,626
Relaxation time, 547
Resistance ratio, 551
dependence on T_c, 497,526
Resistive phase transition in A-15 superconductors, 453
Resistivity, ρ
anomalous behavior, 552
Bloch-Grüneisen theory of, 545,552
high temperature normal state, 545
phonon, 546
residual, 538,553

saturation of, 546,578
temperature dependence of, 545, 552, 567
temperature dependent part, 538,548
Wilson's theory of, 545,568
Resonant modes, 238,249
Rigid band scheme, 524
Rigid ion model, 272,509,596
Rutherford scattering, 492

S

$Sc_4Ge_{.1}C_3$
  pressure effect on T_c of, 365
Scattering
  Bloch-Grüniesen, 568
  electron-electron, 568
  impurity, 568
  p-wave, 596
  rate of, 563
  s-wave, 597
  surface mode, 581
Scattering matrix elements, 568
Scattering path operator, 257
Scattering phase shift, 257
Screening effects
  of d-electrons, 399
Semiconductor-metal
  interface, 1,8
  mixture of Ge-NbGe$_2$, 59
Shell model
  double, 227,286,315
  parameters for NbC, 281
  screened, 225
  three body force, 315
Silicon, 2
Spectral weight function $\alpha^2 F$, 34,372,391,608
  transverse and longitudinal contribution to, 399
Spectroscopy
  Auger, 2
  LEED, 2
Spin density wave, 454
Spin orbit scattering, 163
Spin susceptibility, 298,576
Superconductivity
  p-wave, 51

Superlattice
  instability, 454
  structure, 447
Susceptibility
  Coulomb enhanced charge fluctuation, 273
  electron, 258
  generalized, 313,399

T

Ternary molybdenum chalcogenides, 137,167,175,202
  anharmonicity in, 218
  effects of high pressure on T_c of, 137-156
  electronic parameters for, 178
  intracluster distance in, 176
  molecular crystal model of, 205,210
  phonon spectra of, 202
  upper critical filed of, 167
Tetrahedron integration method, 326,394
Th_2C_3, $(Th_xY_{1-x})_2C_3$
  pressure effect on T_c of, 363
Tight-binding approximation, 519
Transition-temperature
  effect of gap anisotropy on, 74,80
  superconducting, 16,30,72,388 590
Transmission electron diffraction
  of Nb_3Ge, 443
Tunnel junction, I-V characteristics
  of Ag-Ag, 619
  of Al-Al oxide-Nb, 381
  of Au-Au, 620
  of Cu-Cu, 618
  of Nb-O-In, 369
  of Nb-O-Pb, 407
  of Nb-P-Au, 408
  of Nb_3Ge-oxide-Pb, 433
  of Ta-O-Pb, 405
  point contact, 615
Tunneling measurements, 368,381, 430
Two band model, 88,91

U

Umklapp
  process, 3,8,419
  scattering, 562,617

V

V_3Ga
  H_{c2} vs T of, 169
  phonon density of, 194
V_3Si
  Debye temperature of, 302
  density of states, 300
  electrical resistivity of, 548,552,557,572
  high temperature resistivity of, 575
  lattice parameter expansion of, 496
  martensitic transformation of, 416,551
  phonon density of states of, 193,198
  radiation damage of, 495
  single crystal of, 552
Valence bonds, 4
Vanadium compounds, 423,300,552
  Debye temperature of, 302
  λ and $N(0)$ values of, 511
  phonon density of states of, 193,198

Vibrational modes
  acoustic and torsional, 204, 219
Viscosity, 636

W

Wannier function representation, 333

X

X-ray
  diffuse scattering, 483
X-ray data
  of $Gd_{1.2}Mo_6Se_8$, 633
  of Nb_3Al, 466
  of Nb_3Ge, 439,469-481
X-ray data analysis, 464
X-ray measurements, 62,439,461, 491

Y

Y_2C_3
  pressure effect on T_c of, 361